Serviceology for Designing the Future

Takashi Maeno • Yuriko Sawatani •
Tatsunori Hara
Editors

Serviccology for Designing the Future

Selected and Edited Papers of the 2nd International Conference on Serviceology

Springer

Editors
Takashi Maeno
Graduate School of System
Design and Management
Keio University
Yokohama
Japan

Yuriko Sawatani
Graduate School of Entrepreneurship
Tokyo University of Technology
Tokyo
Japan

Tatsunori Hara
The University of Tokyo
Kashiwa
Japan

ISBN 978-4-431-55859-0 ISBN 978-4-431-55861-3 (eBook)
DOI 10.1007/978-4-431-55861-3

Library of Congress Control Number: 2016942046

Printed on acid-free paper

This Springer imprint is published by Springer Nature
The registered company is Springer Japan KK

Preface

Services are not merely key economic activities, but also major factors that improve our quality of life, make local communities prosperous, and then provide a foundation for solving emerging issues. In an increasingly globalized market, it is necessary to increase the economic value of products and services, as well as to enrich their value (life value) for every individual citizen using those services. In addition, in order to bring solutions to emerging social issues, such as an aging society and social security, and to global challenges, such as energy and environmental issues, it is necessary to design a system that facilitates co-creative consensus-building efforts among the stakeholders in the services sector.

Traditionally, service-related research has developed in individual fields such as management, marketing, information engineering, and design engineering. However, to provide better services to our society, it is critical that social sciences, human sciences, and engineering sciences work together as well as establish a strong partnership between industry and academia. There, we need to create an academic understanding of the activities that relate to social and economic services, which means it is necessary to establish an understanding of the comprehensive services that include not only the narrowly defined services industry but also the development of services by manufacturers. Moreover, it is necessary to develop a framework to cocreate high customer satisfaction in alliance with customers.

The Society for Serviceology (SfS) was launched in Japan in October 2012 and is expected to be developed globally. SfS aims to contribute to efforts concerning various industrial issues by organizing the vast knowledge of services and to establish "academics for society" relating to services.

The Second International Conference on Serviceology (ICServ2014) was held September 14–16, 2014, in Yokohama. It covered (1) fundamental research in serviceology, such as mechanism design for services, service innovation and design, service management and marketing, service theory, service economy and productivity, system design and management, and product service system (PSS), and (2) technological research into services such as data assimilation and human modeling, enhancing service analysis, and testing with VR/AR/MR. Some concrete applications and business implications related to tourism and hospitality, healthcare services, public and urban services, regional development, and policymaking were also discussed.

The conference was sponsored by the Graduate School of System Design and Management, Keio University. We would like to thank the members of the organizing committee, the program committee, and all conference participants for their contribution to the success of the conference.

General Chair
ICServ2014
Yokohama, Japan

Takashi Maeno

Contents

Part I
Service in General

A Survey of Business Models in Japanese Restaurant and Retail Industries

Kenju Akai, Keita Kodama, and Nariaki Nishino

Abstract This study explores the service models of Japanese restaurant and retail chains using publicly available data on companies. The sample includes 18 restaurants and 8 retail stores. The centralized kitchen in some of the restaurant chains reduces the time for cooking and preparing foods, which decreases the price of the foods and increases the quality of the services. Meanwhile, retail companies like AEON MALL Co., Ltd. and Fast Retailing Co., Ltd own the original supply chain used by their affiliated companies, which is then adapted to the companies' circumstances such as customer demands and economic cycles.

Keywords Service model • Restaurants • Retails • Survey

1 Introduction

The low productivity of services in Japan requires research into the services industries. However, since this area of research is a fairly new one for most researchers, and industries and sectors vary significantly from each other, systematizing or discerning the various service models can be difficult. Additionally, the service practice includes various business categories and human factors that are affected by economic cycles, which makes identifying present and future issues in service difficult. As a result, solutions to the complexity of service and the direction of this solution are still unknown. Japanese traditional service involves a high-quality, detailed response to customers' requirements. This approach is based on human experiences and tacit knowledge but is sometimes not rational and efficient in the practice. This approach, which relies heavily on Japanese human behaviors, hinders the overseas expansion of "Japanized" service. Thus, to facilitate this expansion, a clear process for service provision is needed.

For example, fast food chain McDonald's is highly successful in many parts of the world because its products, customer relations, commercials, and service quality are good, but more importantly, because its product and service delivery

K. Akai (✉) • K. Kodama • N. Nishino
School of Engineering, The University of Tokyo, Tokyo, Japan
e-mail: akai@css.t.u-tokyo.ac.jp

T. Maeno et al. (eds.), *Serviceology for Designing the Future*,
DOI 10.1007/978-4-431-55861-3_1

systems are defined by the individual stores and are well-known and appreciated by the customers. In Japan, however, no such service approaches have been found suitable for overseas markets. Further, its rapidly aging society poses the risk of insufficient human resources, on which the Japanese economy heavily relies. In addition, this insufficient labor force causes work-related health issues like stress, and worse, death. In fact, the OECD [1] reports that the labor productivity per hour in Japan for 2012 is US$41.6, less than that in the U.S. (US$60.2), France (US $57.7), and Germany (US$55.8). Thus, for today's economy, Japan needs a new systematic service approach that benefits employers, employees, and customers not just in Japan but in other parts of the world. Such an approach will increase Japan's competitiveness. This limitation in the services industry motivated us to explore the service models of restaurant and retail chains in Japan. These two industries provide products and services simultaneously. They are good case for considering the business model for service; how to provide products to the customers.

The rest of the paper is organized as follows. Sections 2 and 3 describe the service models of the restaurant and retail chains in the study, respectively, while Sect. 4 presents the conclusions of the study.

2 Related Literatures

Cook et al. [2] surveyed historical service definitions and service classifications from 1964 to 1996. Judd [3] wrote the first article about classifying services and classified services into three categories: rented goods services, owned goods services, and non-goods services. The advantage of this classification is it defines each category of services and distinguishes between products and services.

Lovelock [4] classified services according to the common characteristics, used in marketing strategy. He used five 2×2 or 3×2 matrixes for the classifications, based on the following criteria: (i) service act (tangible or intangible) and service receiver (people or things), (ii) delivery (continuous or discrete) and relationship between provider and receiver ("membership" or no formal relationship), (iii) extent of customization and judgment in meeting receiver needs, (iv) supply constraints and demand fluctuations, and (v) interaction between provider and receiver (receiver goes to provider, provider goes to receiver, or both transact) and availability of service outlets (single or multiple).

Shostack [5] viewed services as processes and focused on their complexity and divergence. He defined a service's complexity as the number of steps required completing a process and a service's divergence as the degree of freedom allowed or inherent in a process step or sequence. Classification by these two characteristics is useful for describing structural change.

Bowen's [6] study provided an empirically based classification using a cluster analysis of receiver perceptions. Ten services were analyzed and classified into four categories by seven characteristics: importance of employees, customization, ability to switch firms, employee/customer contact, services directed at people or

things, continuous benefits, and difference. The results offer several insights into strategic marketing, especially in to product or service receiver perceptions.

Several studies have also attempted to develop service classification schemes. Cook et al. [2] classified services from both a macro view and a micro view. In a macro sense, services are classified based on organizational ownership or intent (for-profit, private not-for-profit, and public). In a micro sense, organizations are interested in two aspects of services (product and process). The product and process packages are affected by marketing-oriented service dimensions and operations-oriented service dimensions, respectively.

Liu and Wang [7] clearly described services as a process by defining four dimensions of service and integrating them into an open system. The "provider" as an "input" has three attributes: people, equipment, and knowledge. The "process" as an "operation" has three attributes: customization, standardization, and contingency. The "patron" as an "output" has three attributes: humans, things, and information. The "place" as an "environment" has three attributes: front field, back field, and virtual space. With this system, a service provider can integrate these modules into its business strategy.

In summary, these works have taken characteristics of services into consideration: intangibility, perishability, simultaneity, and heterogeneity in developing classifications of services. The objective of such classifications is mainly "to facilitate the development of meaningful strategies and guidelines for service marketing and operations" (Cook et al. [2]).

3 Restaurant Service Models

3.1 Sample of Restaurant Chains

This study focuses on some firms that represent various types of restaurant chains (Table 1).

3.2 Service Models of the Restaurant Chains

The firm information provided in this subsection was obtained from the asset securities reports from EDINET, a corporate disclosure system developed by the Financial Services Agency [8]. The study focused on the two types of flows in the firms: product flow and operation flow. These flows were extracted from the annual security reports and modified with additional information from other resources.

Table 1 Sample firms that represent various types of restaurant chains

Enterprise name	Store name	Classification	Number of stores
Plenus Co., Ltd.	Yayoi	Eating places, except specialty restaurants	Japan: 257, Thailand:104, and Singapore: 4
Saizeriya Co., Ltd.	Saizeriya	Eating places, except specialty restaurants	Japan: 1010
Matsuya Foods Co., Ltd.	Matsuya	Japanese restaurants, fast food restaurants	Japan: 973; China:1
Zensho Co., Ltd.	Sukiya	Japanese restaurants, fast food restaurants	Japan:1962; China: 35, Thailand: 8, and Brazil: 1
Hiday Corp.	Chuka Soba Hidakaya	Chinese restaurants	Japan: 319
Ohsho Food Service Corp.	Gyoza no Ohsho	Chinese restaurants	Japan: 661 and China: 4
Kourakuen Corp.	Kourakeun	Ramen (Chinese noodle) restaurants	Japan:513 and Thailand: 3
Amiyaki Tei Co., Ltd.	Amiyaki Tei	Yakiniku (grilled meat) restaurants	Japan: 119
Ichibanya Co., Ltd.	CoCo Ichibanya	Miscellaneous specialty restaurants (curry)	Japan: 1269, China: 44, Taiwan: 21, Korea: 21, Thailand: 22, Singapore: 3, and Indonesia: 1
Sagami Chain Co., Ltd.	Sagami	Soba and udon (Japanese noodle) restaurants	Japan:128
Kappa Create Co., Ltd.	Kappa-Sushi	Sushi bars	Japan: 359
Watami Co., Ltd.	Watami	Drinking houses and beer halls	Japan: 191, China: 54, Taiwan: 18, Korea: 2, Malaysia: 3, Singapore: 8, and Philippines: 3
Starbucks Coffee Japan, Ltd.	Starbucks Coffee	Coffee shops	Japan: 985
Doutor Coffee Co., Ltd.	Doutor	Coffee shops	Japan: 1096
McDonald's Hold-ings Co. Japan Ltd.	McDonald's	Fast food restaurants	Japan: 3146
B-R 31 Ice Cream Co., Ltd.	Baskin-Rob-bins 31 Ice Cream	Eating and drinking places, n.e.c. (ice cream)	Japan: 1157
KFC Holdings Japan Ltd.	Kentucky Fried Chicken	Eating and drinking places, n.e.c. (fried chicken)	Japan: 1180
Fujiya Co., Ltd.	Fujiya	Eating and drinking places, n.e.c. (cakes)	Japan: 732

3.2.1 Plenus Co., Ltd. – Yayoi

Yayoi is a chain of Japanese combination meal restaurants. An associated company operates and sells ingredients and shipping supplies to the captive stores and affiliated stores, from which it earns royalties (Fig. 1).

3.2.2 Saizeriya Co., Ltd. – Saizeriya

Saizeriya is a chain of Italian-style restaurants and cafés Operated by Saizeriya Co., Ltd., the factory functions as a centralized kitchen and oversees production and logistics. The significant efficiency and systemization of this company enables it to charge low prices. The company has a private, 816.8 acre factory farm called "Commissary" and a cold chain system that can store and transport vegetables at 4 °C after harvesting [9]. This system makes it more efficient for the restaurants to serve foods [10] (Fig. 2).

3.2.3 Matsuya Foods Co., Ltd. – Matsuya

Matsuya is a restaurant chain that mainly serves beef-on-rice and set meals. Matsuya Foods Co., Ltd. has two consolidated subsidiaries and one non-consolidated subsidiary. The consolidated subsidiary M.T.T Co., Ltd. performs all the maintenance and sanitation in and sells kitchen equipment to all the Restaurants, while the consolidated subsidiary M.L.S Co., Ltd. launders and repairs staff uniforms and collects garbage for the restaurants. The non-consolidated subsidiary, Matsuya Farm Co., Ltd., farms ingredients used in the stores (Fig. 3).

3.2.4 Zensho Co., Ltd. – Sukiya

Sukiya is a restaurant chain that mainly serves beef-on-rice meals. The central office of Zensho Co., Ltd. directly operates and develops the meals for the restaurants. Zensho Co., Ltd. has four consolidated subsidiaries: Techno Support Co., Ltd., which performs the maintenance in all the restaurants; Global Foods Co., Ltd., which purchases the stock of ingredients and sells them to the restaurants; Global

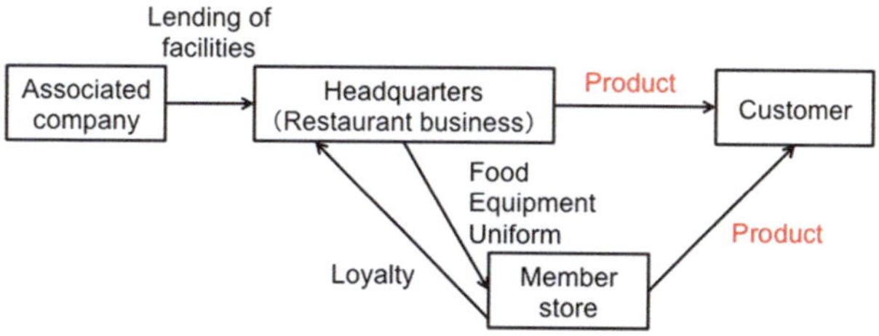

Fig. 1 Service model of Plenus Co., Ltd. – Yayoi

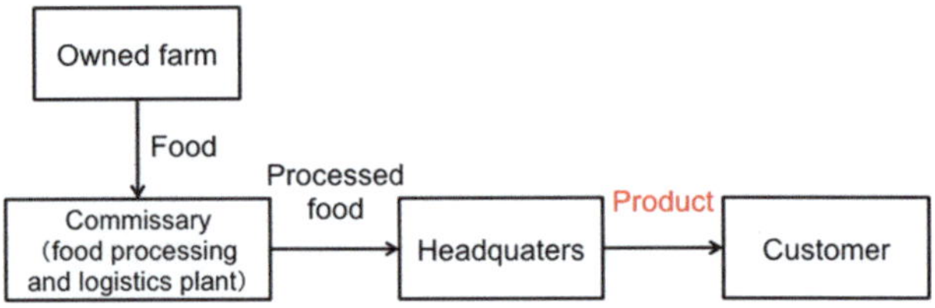

Fig. 2 Service model of Saizeriya Co., Ltd. – Saizeriya

Fig. 3 Service model of Matsuya foods Co., Ltd. – Matsuya

Fresh Supply Co., Ltd., which transports the stock of ingredients to the restaurants; and Global Table Supply Co., Ltd., which sells consumable products and kitchen equipment to all the restaurants. The company procures ingredients in bulk for the restaurants [11] (Fig. 4).

3.2.5 Hiday Co., Ltd. – Hidakaya

Hidakaya is a chain of Chinese-style restaurants. Hiday Co., Ltd. has a farm in Gyoda City, Saitama that functions as a centralized kitchen for all the restaurants. This farm oversees the production, ingredient orders, and logistics for all Hidakaya restaurants. The company has both captive stores and franchise stores (Fig. 5).

3.2.6 Ohsho Food Service Co. – Gyoza no Ohsho

Gyoza no Ohsho is a chain of Chinese-style restaurants. Ohsho Food Service Co. operates its own captive stores and sells ingredients to the franchise stores. The company provides each store discretion in developing its own menu, unlike in other companies [12] (Fig. 6).

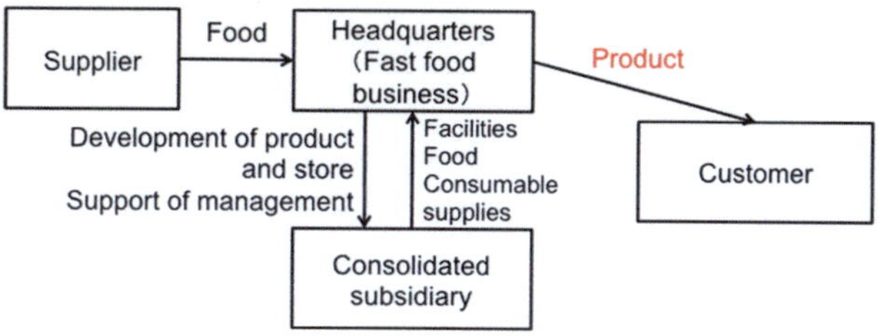

Fig. 4 Service model of Zensho Co., Ltd. – Sukiya

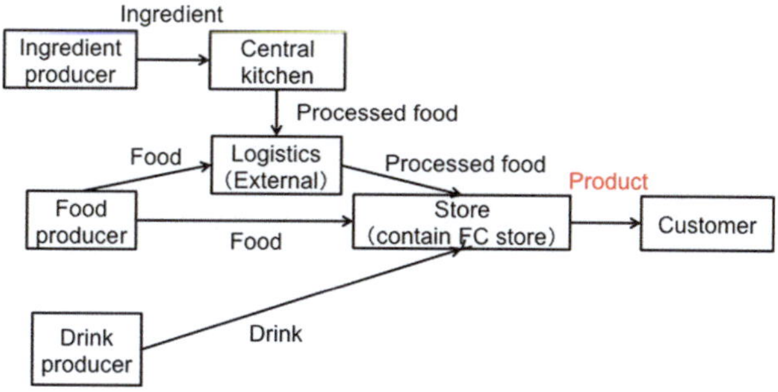

Fig. 5 Service model of Hiday Co., Ltd. – Hidakaya

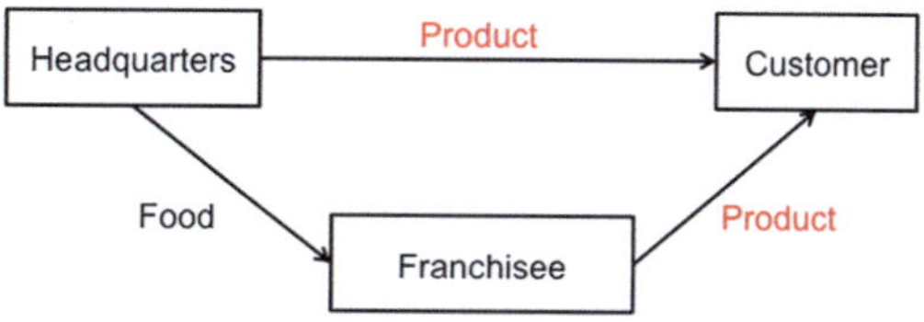

Fig. 6 Service model of Ohsho Food Service Co. – Gyoza no Ohsho

3.2.7 Kourakuen Co. – Kourakuen

Kourakuen is a restaurant chain that mainly serves ramen. Kourakuen Co. operates its own captive stores. It also sells ingredients such as noodles and soups, consumable products, and kitchen equipment to its franchise stores; provides administrative guidance; and oversees construction management and administration and store design for such stores. Its consolidated subsidiary, SCREEN Co., Ltd., creates print promotions, makes a TV commercial, and organizes events for all the restaurants. It is rare for a restaurant chain to have its own advertising company, and there is not much information available on this advertising company's promotion strategy for the restaurant chain (Fig. 7).

3.2.8 Amiyaki Tei Co., Ltd. – Amiyaki Tei

Amiyaki Tei is a chain of grilled meats restaurants. The central office and subsidiary Suehiro Restaurant System Co., Ltd. operates the restaurants. The central office

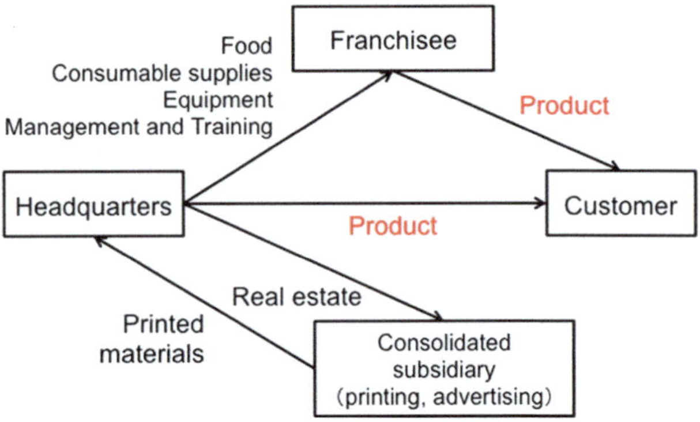

Fig. 7 Service model of Kourakuen Co. – Kourakuen

operates two kinds of restaurants. They procure ingredients and perform processes in bulk by sharing a centralized kitchen (Fig. 8).

3.2.9 Ichibanya Co., Ltd. – CoCo Ichibanya

CoCo Ichibanya is a chain of curry house restaurants. Ichibanya Co., Ltd. operates both captive stores and franchise stores. The company provides administrative guidance and kitchen equipment to its franchise stores in return for a fee. In addition, the company earns royalties by licensing its brand to the franchise stores.

This company uses its own original franchise system, called the "blue system." Generally, beginner restaurateurs can easily open a restaurant by simply signing a franchise contract. In contrast, the blue system requires interested people to take at least 2 years to acquire a qualification for opening a store In order to ensure that such people have sufficient management knowledge and skills for operating a store. Moreover, the blue system provides continuous support to the franchise restaurants but does not charge them royalties. In this way, the blue system is unique [13] and can more effectively increase the number of stores [14] (Fig. 9).

3.2.10 Sagami Chain Co., Ltd. – Sagami

Sagami is a chain of Japanese-style noodle restaurants. The central office operates both captive stores and franchise stores. It has two consolidated subsidiaries. One is A.S. Sagami Co., Ltd., which supplies ingredients to the stores. Because this subsidiary also provides administrative guidance to another restaurant chain, the central office assists in supplying ingredients to the restaurants. The other is Sagami Service Co., Ltd., which provides Sagami Chain Co., Ltd with insurance (Fig. 10).

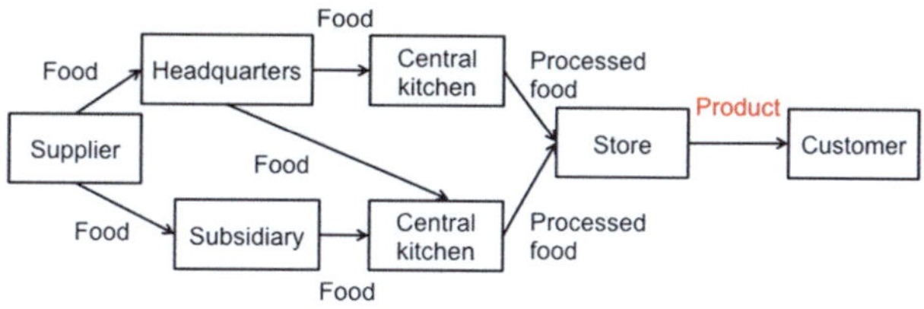

Fig. 8 Service model of Amiyaki Tei Co., Ltd. – Amiyaki Tei

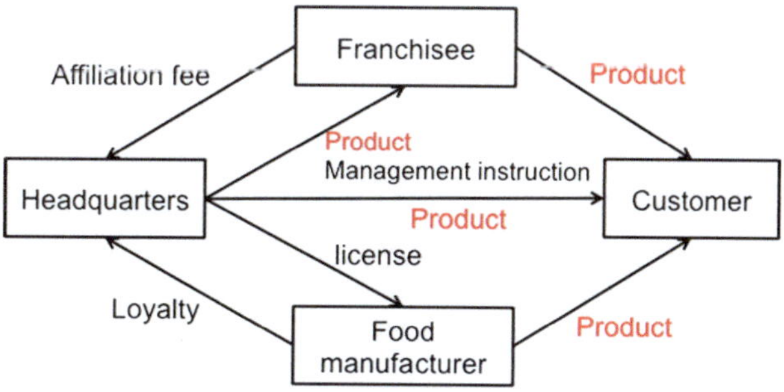

Fig. 9 Service model of Ichibanya Co., Ltd. – CoCo Ichibanya

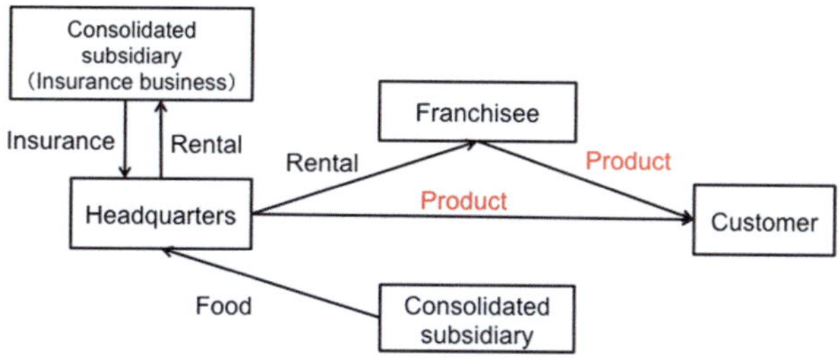

Fig. 10 Service model of Sagami Chain Co., Ltd. – Sagami

3.2.11 Kappa Create Co., Ltd. – Kappa-Sushi

Kappa-Sushi is a chain of restaurants that serve conveyor-belt-style sushi. Two consolidated subsidiaries jointly operate the restaurant chain: Kappa Create Co., Ltd. and Kappa Create Supply Co., Ltd., which purchase the stock of ingredients and supply them to the restaurants. There is not enough information available regarding the subsidiaries' roles (Fig. 11).

3.2.12 Watami Co., Ltd. – Watami

Watami is a chain of Japanese casual dining restaurants. The consolidated subsidiary Watami Food Service Co., Ltd. operates the restaurants, while the consolidated subsidiaries Watami Farm Co., Ltd. and Touma Green Life Ltd. oversee farm production of ingredients used in the restaurants. Watami Merchandising Co., Ltd. oversees the production and wholesale of ingredients. Watami Co., Ltd.

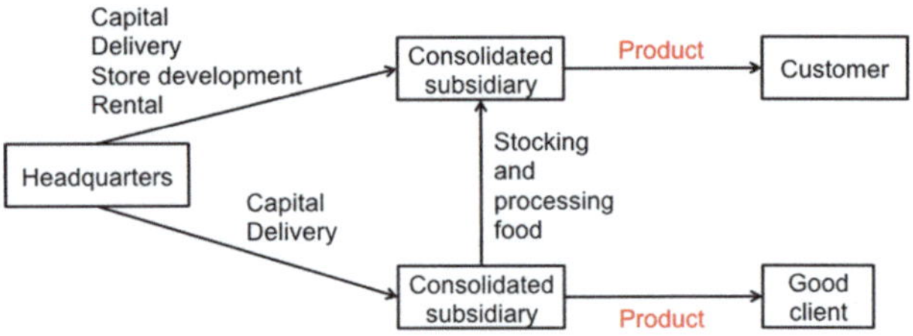

Fig. 11 Service model of Kappa Create Co., Ltd. – Kappa-Sushi

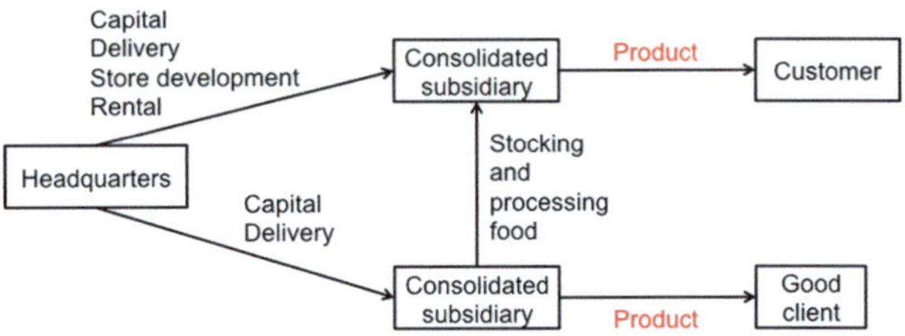

Fig. 12 Service model of Watami Co., Ltd. – Watami

operates both captive stores and franchise stores. A notable point is that Watami Co., Ltd. totally conducts from primary sector to third one (Fig. 12).

3.2.13 Starbucks Coffee Japan, Ltd. – Starbucks Coffee

Starbucks Coffee is a chain of coffee shops. Starbucks Coffee Japan, Ltd. has been granted the right to open and operate Starbucks coffee shops in Japan by Starbucks International, Inc. and the right to use the brand, design, labels, skills, and know-how by SBI Nevada, Inc. Starbucks Corporation in the United States supplies coffee beans to the company, while, SAZABY LEAGUE Ltd. Provides the company with store management skills and know-how. In addition to running the coffee shops, Starbucks Coffee Japan, Ltd. designs clothing and other non-food items sold at its shops.

Although most coffee beans are traded at the commodity futures market in New York City [15], Starbucks Corporation does not procure its beans through this market in order to protect itself from the risk of price fluctuations [16]. Since Starbucks Corporation operates only captive stores, it is better able to maintain the quality of the products and services at its stores [17] (Fig. 13).

3.2.14 Doutor Coffee Co., Ltd. – Doutor

Doutor is a chain of coffee shops. The consolidated subsidiary Doutor Coffee Co., Ltd. operates the coffee shops and wholesale for franchise stores and usual stores. In addition, Magna, Inc. and D&N Confectionery Co., Ltd. provide this company with kitchen equipment and cakes, respectively. Doutor Coffee Co., Ltd. earns royalties from its franchise stores (Fig. 14).

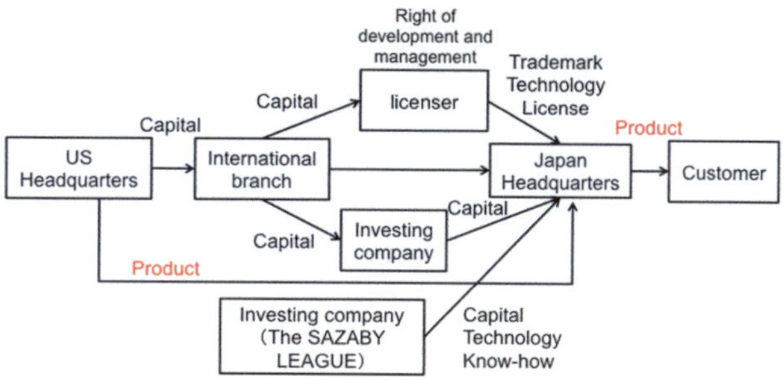

Fig. 13 Service model of Starbucks Coffee Japan, Ltd. – Starbucks Coffee

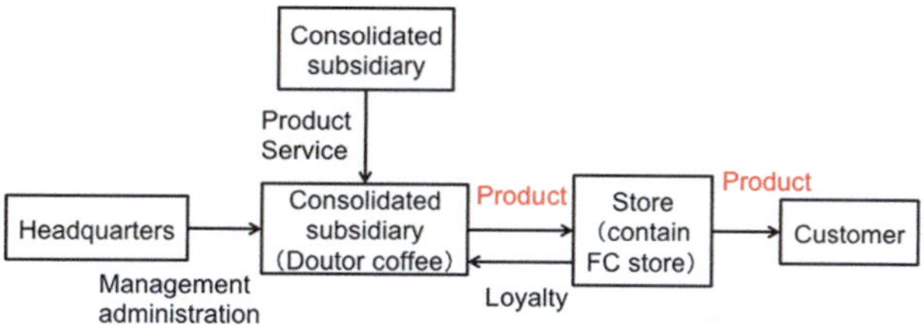

Fig. 14 Doutor Coffee Co., Ltd. – Doutor

3.2.15 B-R 31 Ice Cream Co., Ltd. – Baskin-Robbins 31 Ice Cream

Baskin-Robbins 31 Ice Cream is a chain of ice cream shops. B-R 31 Ice Cream Co., Ltd. has a contract with Baskin-Robbins Franchising LLC that enables it to produce and sell ice cream in Japan and receive technical assistance and franchise management know-how, including information on earning royalties. As shown in Fig. 15, Fujiya Co., Ltd. supplies the ice cream to the stores. The company lend the stores by franchise contract with Fujiya Co., Ltd. and other related companies.

3.2.16 McDonald's Holdings Company (Japan), Ltd. – McDonald's

McDonald's is a chain of fast food restaurants well known for its hamburgers. McDonald's Holdings Company (Japan), Ltd. has a contract with McDonald's Corporation in the United States, to which it pays royalties. The contract enables the company to operate captive stores and franchise stores and earn royalties from the franchise stores. The consolidated subsidiary EveryD Mc, Inc. provides support to the restaurants and customers, while the consolidated subsidiary JV Co., Ltd. manages promotional activities for the restaurants.

A notable point in the company's operations is the significant efficiency of its supply chain, from procuring stock to logistics, through the use of IT systems. This type of operations is unlike those of other restaurant companies [18]. For this reason, the company is No. 2 in Gartner, Inc.'s "Supply Chain Top 25 for 2013" [19] (Fig. 16).

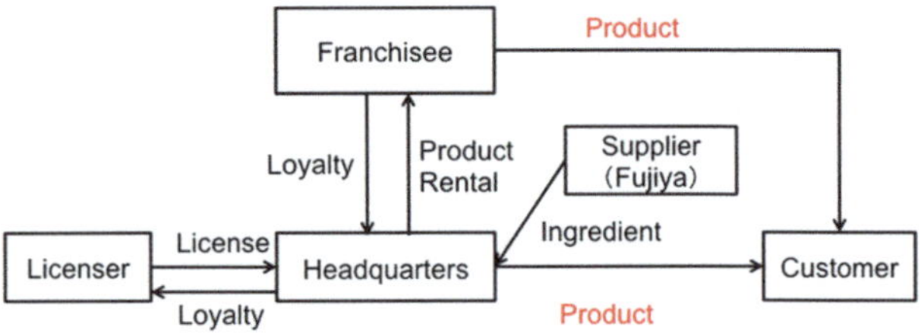

Fig. 15 Service model of B-R 31 Ice Cream Co., Ltd. – Baskin-Robbins 31 Ice Cream

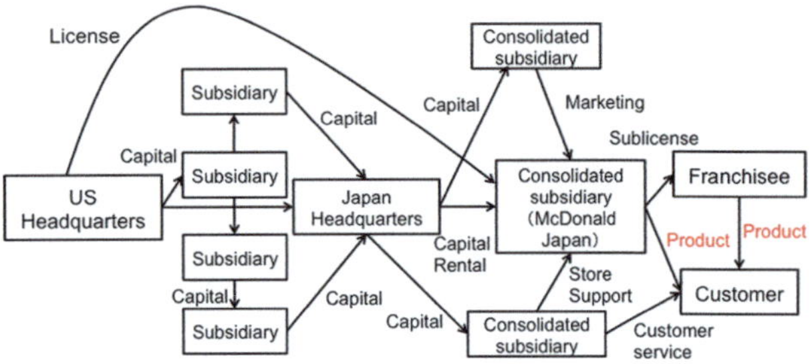

Fig. 16 Service model of McDonald's Holdings Company (Japan), Ltd. – McDonald's

3.2.17 KFC Holdings Japan Ltd. – Kentucky Fried Chicken

Kentucky Fried Chicken, also known as KFC, is a fast food chain that mainly serves fried chicken. Two consolidated subsidiaries, Kei Ado Co., Ltd. and Kei Foods, as well as KFC Holdings Japan Ltd. operate the restaurants. KFC Holdings Japan Ltd. has a master franchise and licensing contract with Yum! Restaurants Asia Pte. Ltd. in Singapore. These two entities do not form a capital alliance (Fig. 17).

3.2.18 Fujiya Co., Ltd. – Fujiya

Fujiya is a chain of cake shops. It is operated by several consolidated subsidiaries, namely Fujiya Food Service Co., Ltd., Dalloyau Japon Co., Ltd., and Fujiya-Hohoku., Ltd. Fujiya Co., Ltd. and the consolidated subsidiaries sell cakes to and manage the locations of the restaurants. Fujiya Co., Ltd. is in partnership with B-R 31 Ice Cream Co., Ltd. regarding the production and sales of ice cream and with Yamazaki Baking Co., Ltd., the parent company, regarding the production and sales of baked goods. It is rare for a restaurant company to be run by several subsidiaries (Fig. 18).

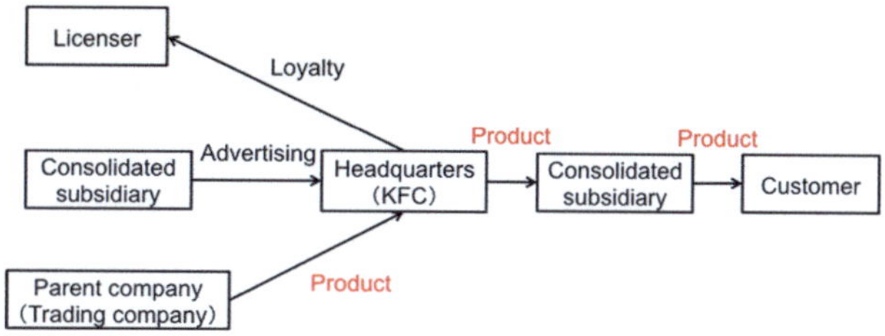

Fig. 17 Service model of KFC Holdings Japan Ltd. – Kentucky Fried chicken

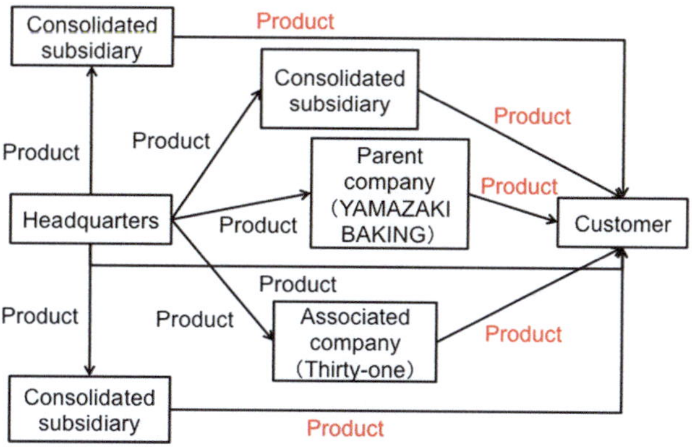

Fig. 18 Service model of Fujiya Co., Ltd. – Fujiya

3.3 Comparison of Restaurant Specifications

Restaurant chain Saizeriya is supported by a flat-out streamlined process and systematization. This chain has a more efficient centralized system than other restaurants chains. This system cooks most of the food served in the restaurants.

Chinese restaurant chain Zensho procures ingredients in bulk for all of its restaurants. On the other hand, Ohsho gives each restaurant significant discretion in developing its own menu. Kourakuen is unique because it is the only chain that has its own advertisement company.

In the world wide restaurants, Starbucks has the original market and imports it from that route, which protects it from the risk of price fluctuations. It also runs only captive stores and does not have franchise stores. On the other hand, McDonald's strength is in its streamlined supply chain, which uses IT systems, from procurement to commodity distribution, to ensure a more efficient service. Finally, Fujiya differs from these two businesses because its restaurants are run by several affiliated companies simultaneously.

4 Retail Service Models

4.1 Sample of Retail Chains

This study focuses on some firms that represent various types of retail companies (Table 2).

4.2 Service Models of the Retail Chains

This subsection describes the service model in each firm using data from the firms' annual securities reports.

4.2.1 Takashimaya Co., Ltd. – Takashimaya

Takashimaya is a large chain of department stores. Takashimaya Co., Ltd. provides retail spaces for the boutiques and its own brands. It procures products for its private brand from wholesale and suppliers from the fabricated clothing sector. It sells

Table 2 Samples firms that represent various types of retail companies

Enterprise name	Store name	Classification	Number of stores
Takashimaya Co., Ltd.	Takashimaya	Department stores	Japan: 20, China: 1, Taiwan: 1, and Singapore: 1
AEON MALL Co., Ltd.	AEON MALL	General merchandise stores	Japan: 61 and China: 3
Heiwado Co., Ltd.	Heiwado	General merchandise stores	Japan: 151 and China: 4
Don Quijote Co., Ltd.	Don Quijote	Discount stores	Japan: 240 and US: 3
Fast Retailing Co., Ltd.	UNIQLO	Clothing superstores	Japan: 849, China: 260, Hong Kong: 22, Taiwan: 43, Korea: 116, Singapore: 16, Malaysia: 14, Thailand: 14, Philippines: 11, Indonesia: 3, UK: 10, US: 17, France: 4, and Russia: 4
The Maruetsu, Inc.	Maruetsu	Supermarkets	Japan: 271 and China: 1
Yamada Denki Co., Ltd.	Yamada Denki	Machinery and equipment stores	Japan: 634 and China: 3
Shimachu Co., Ltd.	Shimachu	Home centers	Japan: 54

products to customers in three ways: through (i) an entire purchase (5–10 % on average among the department stores), in which the customers are not able to return the products; (ii) an entrustment (20–25 %), in which customers can return the products, and (iii) a digest (70 %), in which the products are sold to customers at the same time the next stock is purchased [20]. In addition, the company entrusts its stores' interior design to the architectural sector (Fig. 19).

4.2.2 AEON MALL Co., Ltd. – AEON MALL

Like Takashimaya, AEON MALL Co. Ltd. provides retail spaces for boutiques (tenants) and related companies and provides products to customers through the same three methods. Unlike Takashimaya, however, the company does not manage the space for the boutiques directly.

In addition, the company functions as a logistics center that procures products directly from manufacturers and supplies them to its related company [21]. This system differentiates it from other retail companies and improves its logistics information systems. It earns two kinds of rent income: fixed rent income and percentage [22]. The percentage rent income depends on the sales of a tenant. This percentage is currently set at half of sales (Fig. 20).

4.2.3 Heiwado Co., Ltd. – Heiwado

Heiwado Co., Ltd. procures its stock of products from manufactures and processors and then sells them to customers. It leases land and buildings to related companies and commissions the stores' operations. Specifically, it lends buildings to developers and commissions restaurant companies in the stores. Finally, the company commissions the cleaning of the stores (Fig. 21).

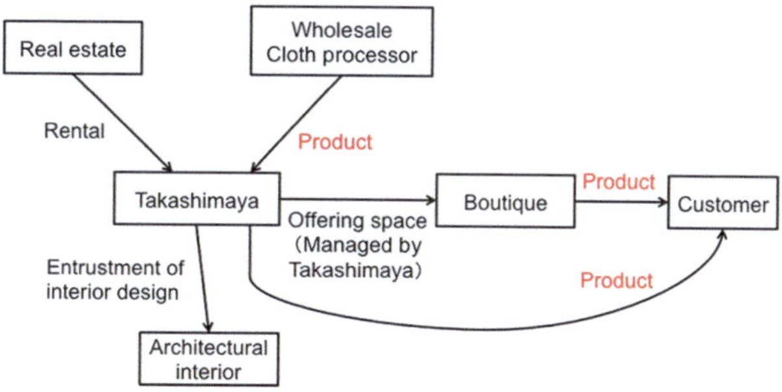

Fig. 19 Service model of Takashimaya Co., Ltd. – Takashimaya

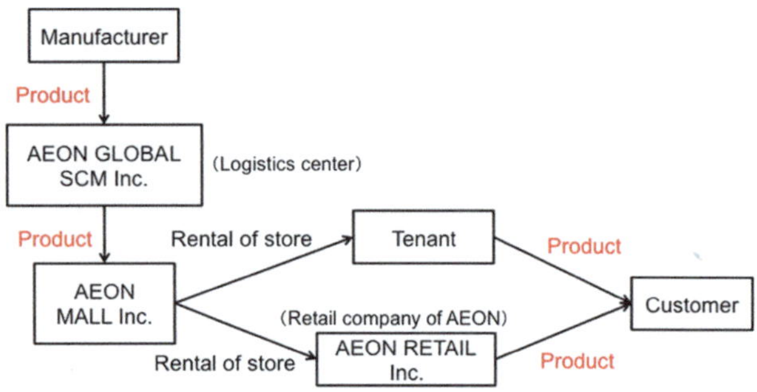

Fig. 20 Service model of AEON MALL Co., Ltd. – AEON MALL

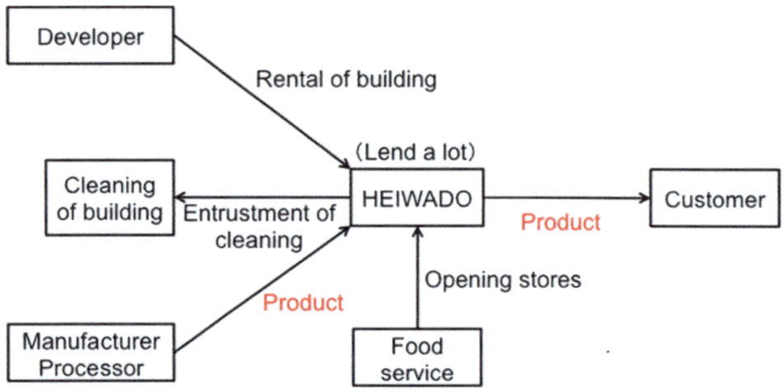

Fig. 21 Service model of Heiwado Co., Ltd. – Heiwado

4.2.4 Don Quijote Co., Ltd. – Don Quijote

Don Quijote Co., Ltd. procures its stock of products from manufacturers and wholesale dealers and then sells them to customers (Fig. 22).

4.2.5 Fast Retailing Co., Ltd. – UNIQLO

Fast Retailing CO., Ltd. is the parent company of UNIQLO Co., Ltd. It is a specialty store retailer of private label apparel and overseas the design, production, and sale of its products [23]. This system is different from those of other retail companies, which procure their stock of products from other companies. Fast Retailing Co., Ltd. is able to maintain flexible production and low costs because it produces original products.

The products are transported from the factory to the warehouse and are then stocked in stores through a control system. The stores sell the products to customers and order more products through the control system [18]. The customer center relays customer requests to store staff (Fig. 23).

Fig. 22 Service model of Don Quijote Co., Ltd. – Don Quijote

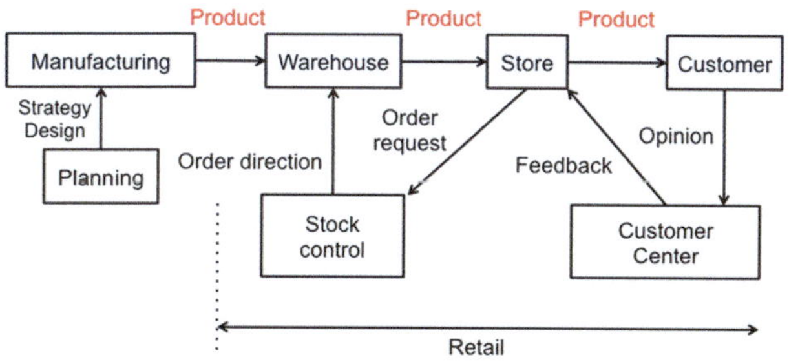

Fig. 23 Service model of Fast Retailing Co., Ltd. – UNIQLO

4.2.6 The Maruetsu, Inc. – Maruetsu

The Maruetsu, Inc. procures inventory from related companies and then sells them provided to customers. The company entrusts the lending and management of its real estate to the development sector and the processing of fresh food to the food processing sector (Fig. 24).

4.2.7 Yamada Denki Co., Ltd. – Yamada Denki

Yamada Denki Co., Ltd. sells products in three ways. First, it sells products directly to customers. Second, it supplies products to a subsidiary, which then sells the products directly to customers. Third, it supplies products to a subsidiary, which then transfers the products to franchise stores, which then sells the products directly to customers. The subsidiaries include the subsidiaries in the local areas as well as electronics retailer Best Denki Co., Ltd. [24], which was acquired by Yamada Denki Co., Ltd. In 2012. The franchise stores are community-based stores, which are also called "electronics centers in the town."

The company procures most of its products directly from manufacturers. It entrusts the transportation and distribution of products to the subsidiaries (Fig. 25).

4.2.8 Shimachu Co., Ltd. – Shimachu

Shimachu Co., Ltd. procures its products from suppliers and then sells them to customers (Fig. 26).

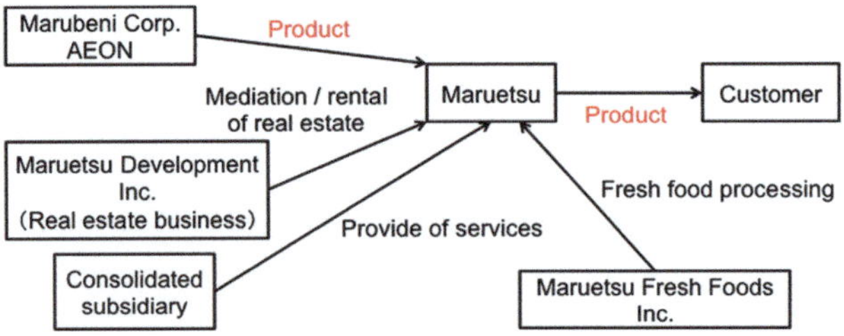

Fig. 24 Service model of The Maruetsu, Inc. – Maruetsu

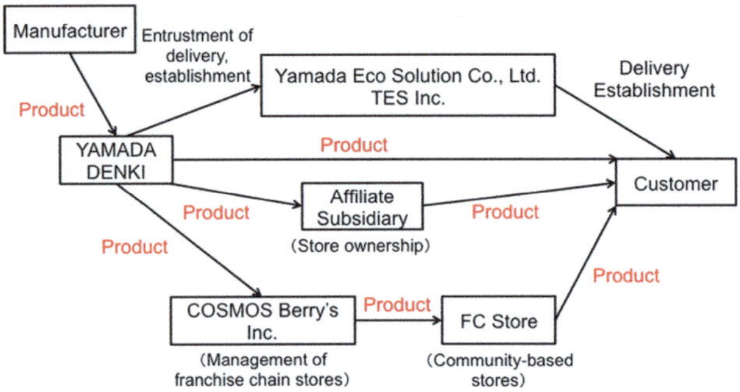

Fig. 25 Service model of Yamada Denki Co., Ltd. – Yamada

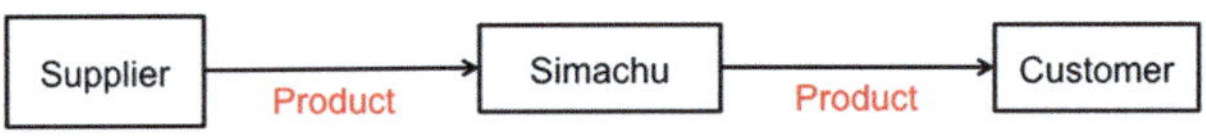

Fig. 26 Service model of SHIMACHU Co., Ltd. – Shimachu

4.3 Comparison Retail Chain Specifications

Takashimaya and AEON MALL have similar business models. However, while AEON MALL provides retail spaces to its tenants in exchange for rent payments, Takashimaya provides retails spaces to its tenants in exchange for a percentage of the sales of the tenants. [25].

Most of the retail chains in this study have the same supply chain model, that is, a streamlined model that involves buying products from implementer wholesalers, and processors and then selling the products to customers. AEON MALL and Fast retailing, however, are exceptions. Their unique models promote flexibility and efficient service, which is ideal for meeting customers' demands.

5 Conclusions

This study explores the various service business models of restaurant and retail chains in Japan. Among the restaurant chains, McDonald's runs an efficient supply chain through the use of IT systems, while Saizeriya operates a centralized kitchen that reduces the cooking times at its restaurants, which in turn, decreases the price of foods and increases the quality of services. Among the retail chains, the traditional department store Takashimaya and general products store chains like AEON MALL and Fast Retailing have different supply chain models than the other chains. The latter two companies own the original supply chain used by their affiliated companies, which is then adapted to the affiliated companies' circumstances such as customer demands and economic cycles.

This study covers only the service models of restaurant and retail chains. Therefore, future studies could examine the service models in other industries or sectors and to classify them according to specific criteria. In the aspect of classification, Ueda et al. [26] suggest the following basic classifications of services: Class I — Providing value model, Class II — Adaptive value model, and Class III — Co-creative value model. They will link the difference in the method of creating value between products and services.

Acknowledgments This study was partially supported by Service Science, Solutions and Foundation Integrated Research Program (S3FIRE), Research Institute of Science and Technology for Society (RISTEX), and Japan Science and Technology Agency (JST).

References

1. OECD (2012) http://stats.oecd.org/
2. Cook DP, Goh CH, Chung CH (1999) Service typologies: a state of the art survey. Prod Oper Manage 8(3):318–338
3. Judd RC (1964) The case for redefining services. J Mark 28:58–59
4. Lovelock CH (1983) Classifying services to gain strategic marketing insights. J Mark 47 (3):9–20
5. Shostack GL (1987) Service positioning through structure change. J Mark 51(1):34–43
6. Bowen J (1990) Development of a taxonomy of services to gain strategic marketing insights. J Acad Mark Sci 18(1):43–49
7. Liu CH, Wang CC (2009) Integrative service taxonomy and business strategies formulation: 4P model with open system and modularization approaches. Int J Inf Manag Sci 20:331–349
8. EDINET – Financial Services Agency. http://disclosure.edinet-fsa.go.jp/
9. KEIEI COM. Independent analysis of Saizeriya's management strategy and advantage (in Japanese). http://www.00keiei.com/kigyou-senryaku/saizeriya.html
10. Commutative Weblog 2. Saizeriya (in Japanese). http://commutative.world.coocan.jp/blog2/2009/11/post-384.html
11. Sukiya. Safety of SUKIYA – Effort of SUKIYA (in Japanese). http://www.sukiya.jp/about/safety.html
12. PRESIDENT Online. Maximum utility – why do people stand in a line of 'Gyoza no Ohsho'? (in Japanese). http://president.jp/articles/-/3338

13. Ichibanya's Bloom System. Characteristics of bloom system (in Japanese). https://www.ichibanya.co.jp/comp/bs/chara/index.html
14. TBS – Gacchiri Monday. Archives – CoCo Ichibanya (in Japanese). http://www.tbs.co.jp/gacchiri/archives/20070225/3.html
15. Black Gold Movie. Economics of coffee (in Japanese). http://blackgoldmovie.com/economics-of-coffee
16. Starbucks Coffee Japan. Sustainable supply model (in Japanese). http://www.starbucks.co.jp/csr/ethicalsourcing/sustainable_model.html
17. Kensnews. The reason why Starbucks dislikes a franchise system and sticks to a direct management scheme (in Japanese). http://kensnews.net/?p=423
18. Fagansusanto. Value chain and competitive advantage of McDonald's. http://fagansusanto.wordpress.com/2012/11/22/value-chain-and-competitive-advantage-of-mcdonalds/
19. Gartner. The Gartner supply chain top 25 for 2013
20. Shared Research (2014) Research report – Takashimaya Co., Ltd. (in Japanese). https://www.gartner.com/doc/2493115?plc=ddr
21. J-marketing.net (2006) Strategies of AEON aiming at improving profit by direct stoking (in Japanese). http://www.jmrlsi.co.jp/sc100/case/2008/ct01-aeon.html. June 2006
22. AEON MALL (2013) Annual report, "Image of profit structure – business model" (in Japanese). http://www.aeonmall.com/ir/annual/business-model/earnings-structure/
23. FAST RETAILING Co., Ltd. Business model of UNIQLO (in Japanese). http://www.fastretailing.com/jp/group/strategy/uniqlobusiness.html
24. J-CAST News (2012) YAMADA DENKI Co., Ltd. has completed acquisition of BEST DENKI Co., Ltd. – account for 30 % of domestic market, more reorganization is estimated (in Japanese). http://www.j-cast.com/2012/12/22158847.html. December 2012
25. J-Net 21. Q926. Please tell me the difference of opening a store in Dpt. and in GMS (in Japanese). http://j-net21.smrj.go.jp/well/qa/entry/926.html
26. Ueda K, Takenaka T, Váncza J, Monostori L (2009) Value creation and decision-making in sustainable society. CIRP Ann Manuf Technol 58(2):681–700

Employee Satisfaction Analysis in Food Service Industry – Resultant of Questionnaire to the Restaurant Staff

Tomomi Nonaka, Toshiya Kaihara, Nobutada Fujii, Fang Yu, Takeshi Shimmura, Yoshihiro Hisano, and Tomoyuki Asakawa

Abstract This study analyzed employee satisfaction in the food service industry by looking at the results of a questionnaire delivered to the restaurant staff. Question items were divided into seven question categories: work environment, work efficiency and service quality, relationship with bosses, rules, education system, attitude and motivation toward work, and interest in multi-skills development. A Japanese restaurant chain located in Japan is selected as an analysis target. Satisfaction structures and the differences among the attributes of work position, employee pattern, age group, and length of continuous employment were analyzed with correlation analysis and covariance structure analysis.

Keywords Service management • Employee satisfaction • Customer satisfaction • Food service industry • Questionnaire • Covariance structure analysis

1 Introduction

The economic activity of the service industry is increasing, and this industry creates an important amount of employment in developed as well as developing countries. The service industry is responsible for about 70 % of Japan's GDP and two-thirds of its employment. In restaurant service, which is categorized as a food service industry, production is based on intensive labour. Staff members who work on the service floor of a restaurant provide their service face-to-face with customers. In a kitchen, service operations still have many hand-made processes that create value

T. Nonaka (✉)
Department of Industrial and Systems Engineering, Aoyama Gakuin University, 5-10-1 Fuchinobe, Chuo-ku Sagamihara-shi, Kanagawa 252-5258, Japan
e-mail: nonaka@ise.aoyama.ac.jp

T. Kaihara • N. Fujii • F. Yu
Graduate School of System Informatics, Kobe University, Rokkodai 1-1, Nada, Kobe 657-8501, Japan

T. Shimmura • Y. Hisano • T. Asakawa
Ganko Food Service Co. Ltd, 1-2-13 Shinkitano, Yodogawa-ku, Osaka 532-0025, Japan

T. Maeno et al. (eds.), *Serviceology for Designing the Future*,
DOI 10.1007/978-4-431-55861-3_2

[1, 2]. Hence, the working conditions of the staff may possibly influence their provision of service. The importance of employee satisfaction for the delivery of good service has been pointed out [3–5]. An earlier study [3] shows a substantial relationship between individual job satisfaction and individual performance. When the performance of employees improves, it can be expected that service qualities will be enhanced. From the viewpoint of service satisfaction [6], it has been indicated that the relationship between service quality and customer satisfaction is independent but closely related [7]. The importance of linkage effects and improving overall satisfaction, including customer satisfaction, employee satisfaction, and management satisfaction, is described by [6, 8].

To enhance both employee satisfaction and service quality, investigating employee satisfaction and revealing its structure is required. This paper analyzes employee satisfaction by looking at the results of a questionnaire distributed to restaurant staff. It first develops a structural model of employee satisfaction. The question items are divided into seven question categories: as work environment, work efficiency and service quality, relationship with bosses, rules, education system, attitude and motivation toward work, and interest in multi-skills development. The results of questions are investigated from multiple points of view in terms of relationships between employee satisfaction and the seven categories. Hence, the satisfaction structures and differences among attributes of work position, employment pattern, age group, and length of continuous employment are analyzed from several perspectives. The results are analyzed with the two approaches of correlation analysis and covariance structure analysis.

In this paper, a Japanese cuisine restaurant chain located in Japan was selected as an analysis target. The restaurant chain hires various types of staff. Many part-time and full-time workers falling into a wide age distribution are employed. Many full-time staff working in the kitchen have traditional cooking customs and behaviours as professional chefs [1]. On the other hand, there are many students and married women working part-time. Hence, it is to be expected that the structures of employee satisfaction for several workers have different characteristics. Thus, the questionnaire was directed at all segments of store staff: both part-time and full-time staff working in kitchens, service floors, washing places, and pantries.

In the remainder of this paper, Sect. 2 describes the structure of the survey and questionnaire. Section 3 provides the results of the questionnaire. A correlation analysis is performed in Sect. 4, and a covariance structure analysis is executed in Sect. 5. Final discussion and conclusions are provided in Sect. 6.

2 Questionnaire for Employee Satisfaction

This section describes the questionnaire.

2.1 An Outline of the Question

Table 1 outlines the question categories and question items. The questionnaire consisted of two parts. The first part investigated employees' attributes, and the second part was for questions. The questionnaires were conducted by Likert question items and free descriptions. The Likert question items were demonstrated with six-level Likert items: (1) strongly agree, (2) agree, (3) slightly agree, (4) slightly disagree, (5) disagree, and (6) strongly disagree. The question sheets were provided by paper, and respondents filled in the question sheet. The questionnaire was conducted as an anonymous survey.

The question categories were divided into two parts exploring the respondents' attributes and the seven-question item categories as follows: Q1. Work environment, viewpoint about work, Q2. Work efficiency, service quality, Q3. Relationship with bosses, Q4. Philosophy, rules, and personal system, Q5. Education system, Q6. Attitude and motivation toward work, and Q7. Interest in multi-skills development.

2.2 Targets

The questionnaire was distributed at 6 stores, and 128 employees answered it, as shown in Table 2. All segments of the restaurants' staff were investigated, including both part-time and full-time workers in kitchens, service floors, washing places, and pantries.

The percentage of respondents widely differed in each age group from the teens to the sixties: 9 % of people in their teens responded, 23 % of people in their twenties, 9 % of people in their thirties, 19 % of people in their forties, 23 % of people in their fifties, and 16 % of people in their sixties. All of the teenaged staff members were part-time workers. Thirty-one per cent of the respondents were men, and 68 % were women. Twenty-six per cent of the staff worked in the kitchen, 18 % in the washing place and pantry, and 50 % on the service floor.

Table 1 Outline of the questionnaire

Question categories	Question items	Question type
Respondent attributes	Gender, Age group, Employee pattern, Length of working period at the current store, Length of continuous employment, Current working position, Experimented working position at the company	Multiple choice question Free descriptions
Q1: Work environment, viewpoint about work	*23 items [Q1-1. ~ Q1-23.] regarding;* Understanding of the company and workplace, Expectation of bosses and colleagues, Opportunity for challenging, Congenially employed, Good relationship with colleagues and bosses, Satisfaction with salary, Cooperation and collaboration with colleagues and bosses, Discretion, Satisfactory work, Pride in work, Understanding of their jobs by family (part of the items)	Six level Likert question items
Q2: Work efficiency, service quality	*7 items [Q2-1. ~ Q2-7] regarding;* Quality of dishes, Food preparation speed, Customer's reaction and satisfaction, Efficiency of work, Teamwork and collaboration, Staff assignment, Work environment (part of the items)	Six level Likert question items
Q3: Relationship with bosses	*8 items [Q3-1. ~ Q3-8] regarding;* Exact direction for work assignment, Advance directive, Work planning, Support, Complying with a request about shift schedule, Giving a goal for skill improvement, Listening to employee's voice (part of the items)	Six level Likert question items
Q4: Philosophy, rules, and personal system	*5 items [Q4-1. ~ Q4-5] regarding;* Understanding of the personal system, Understanding of the work assignment and job rotation, Understanding of the batch production system, Understanding of the demand prediction, Understanding of the importance of collaborations	Six level Likert question items
Q5: Education system	*2 items [Q5-1. ~ Q5-2];* Satisfaction for education system and its reason	Six level Likert question items free descriptions
Q6: Attitude and motivation toward work	*8 items [Q6-1. ~ Q6-8] regarding;* Interest in working for long period at the current company, Work fun, Desire to be more helpful at work place, Desire to enhance customer satisfaction, Interest in improving job skill, Continual awareness of providing good quality service (part of the items)	Six level Likert question items
Q7: Interest in multi-skills development	1 item *[Q7.]* about interests in increasing multi-skills	Six level Likert question items

Table 2 Number of respondents

Store	Number of persons
Store A	35
Store B	18
Store C	27
Store D	14
Store E	15
Store F	19
Total	128

3 Results of the Questionnaire

In this section, the results are analyzed in two parts. The first one analyzes all segments of the staff collectively, and the second one analyzes the employees in specific segments.

3.1 All Segments

This section shows the overall results. In question category Q1. 'Work environment, viewpoint about work', positive answers include (1) strongly agree, (2) agree and (3) slightly agree wholly. Q1-10. 'Have a feeling of attachment for the job, or doing a job that you like' directly represents employee satisfaction. The percentage of workers who answered (1) strongly agree or (2) agree is 49.2 %, and sum of those who answered (1), (2), or (3) slightly agree is 81.3 %. Of the answers that Q1-22. 'Feel that my work leads to customer satisfaction' received, 79.7 % were positive responses of (1), (2), or (3). In question items regarding job growth (Q1-3., Q1-6., and Q1-12.), expectations from bosses and colleagues (Q1-2. and Q1-13.) and self-reliance (Q1-7., Q1-17., and Q1-19.), the positive answers, including the levels of (1), (2) and (3), totalled 60 % or over.

Next, question Q4-2. 'Deeply understanding the purpose of work assignment and job rotation' in the category of Q4. 'Philosophy, rule and personal system' yielded a positive answer rate of 69.7 %, which includes (1), (2) and (3). The results of Q4-5. 'Deep understanding of the importance of cooperation beyond assigned positions and hierarchical relations' show a positive result rate of 75.7 %. From these results, the high ratio of understanding for the company's philosophy and strategies is confirmed; otherwise, there is a certain amount of workers working short periods, part-time workers as well as full-time ones.

With questions Q6-1. and Q6-2, which may be influential trackers of employee satisfaction, many positive answers are observed. Q6-1. 'Interest in working for long period at current company' yielded a positive answer rate of 75.0 %, including answers of (1), (2), and (3). Q6-2. 'Work fun' yielded a positive answer rate of 76.5 %. Additionally, one of the remarkable question items Q6-4. 'Desire to enhance customer satisfaction' had the highest ratio of positive answers. The sum of answers (1) and (2) is 77.3 %, and the sum of answers (1), (2), and (3) is 95.3 %. It is possible that this item provides the most support for the idea that high employee satisfaction leads to high customer satisfaction.

Many positive answers, suggesting a high motivation for improving job skills, are observed in the results of Q6-6. 'Interest in improving job skill' and Q6-7. 'Interest in increasing tasks which can be handled'. Both Q6-6. and Q6-7. received 89 % of answers that were positive, a categorization determined by the sum of (1), (2), and (3).

3.2 Segment of Employee Patterns

This sub-section analyzes separate employee patterns that emerge when the staff is divided into different segments. Many positive answers, including (1), (2), and (3), are observed for both the results of part-time workers and full-time ones. Similar trends with results across all segments are described in Sect. 3.1.

In question category Q4. 'Philosophy and rules', full-time workers answer more positively compared to part-time ones. In question Q4-5. 'Deep understanding for importance of cooperation beyond assigned positions and hierarchical relations', more and stronger positive answers are observed from full-time workers than from part-time ones. The results of Q4-3. 'Understanding of batch production system' and Q4-4. 'Understanding of demand prediction' yield a 40 % medium and negative answer rate, and (4) slightly disagree was included in this tally. It is confirmed that the results are derived mainly from part-time workers and indicate the different level of understanding between full-time workers and part-time ones. In addition, the results of Q1-17. 'Discretion to take a decision as necessary' show more positive answers from the full-time workers than the part-time ones.

As a result, one can see similar trends for both full-time workers and part-time ones in terms of motivation and feelings. Moreover, full-time workers have a stronger motivation and deeper understanding of strategies and rules than part-time ones do.

4 Correlation Analysis

4.1 Overview

A correlation analysis was performed to evaluate the association between the two variables which represent each question item. In this paper, the analysis is performed for all combinations of question items, which are analysed separately by the respondent attributes. The questionnaire results of the six levels of Likert

Table 3 Correlation coefficients with among Q1-22 and variables

Q1-22. Do you feel that your work leads to customer satisfaction?

		Employment pattern		Current work position			Age group		
	Overall	Full time	Part time	Kitchen	Washing place and pantry	Service floor	10s and 20s	30s and 40s	50s and 60s
Q1-10. Is your job one that you like? Do you like your job?	0.373**	0.393*	0.368**	0.270	0.164	0.455**	0.396*	0.372*	0.380**
Q1-15. Do you perform a satisfying assingment?	0.567**	0.558**	0.575**	0.436*	0.472*	0.673**	0.540**	0.467**	0.622**
Q1-17. Can you make a decision according to your own judgement as necessary?	0.283**	0.209	0.313**	0.293	0.020	0.352**	0.436**	0.053	0.291*
Q1-19. Do you propose or implement improvement plans and problem resolutions actively?	0.347**	0.261	0.387**	0.103	0.034	0.520**	0.421**	0.253	0.305*
Q6-6. Do you want to improve your skills and acquire new skills?	0.444**	0.403*	0.464**	0.546**	0.074	0.472**	0.366*	0.503**	0.402**
Q6-7. Do you want to make progress in acquiring more skills?	0.372**	0.309	0.404**	0.500**	−0.241	0.442**	0.225	0.468**	0.399**
Q6-8. Do you always consciously to provide good service?	0.428**	0.207	0.518**	0.401*	0.308	0.507**	0.381*	0.498**	0.445**

$*P < 0.05$, $**P < 0.01$

Table 4 Correlation coefficients with Q6-2. and variables

Q6-2. Do you enjoy your work?		Employment pattern		Current work position			Age group		
	Overall	Full time	Part time	Kitchen	Washing place and pantry	Service floor	10s and 20s	30s and 40s	50s and 60s
Q1-3. Do you have challenging opportunities?	0.339**	0.471**	0.338**	0.269	0.493*	0.280*	−0.066	0.255	0.581**
Q1-8. Do you feel that the company's mission and target give recognition to the importance of your own task?	0.253**	0.607**	0.415**	0.237	0.663**	0.456**	0.222	0.400*	0.587**
Q1-13. Do you feel the expectations of bosses and colleagues?	0.327**	0.509**	0.428**	0.431*	0.352	0.425**	0.275	0.376*	0.583**
Q1-18. Are common rules arranged for job processes?	0.278**	0.444*	0.386**	0.564**	0.356	0.428**	0.116	0.568**	0.486**
Q1-19. Do you propose or implement improvement plans and problem resolutions actively?	0.180*	0.330	0.288**	0.005	0.581**	0.284*	0.126	−0.076	0.586**
Q1-21. Do you cooperate and collaborate with colleagues?	0.239**	0.389*	0.497**	0.209	0.405	0.538**	0.373*	0.575**	0.491**
Q1-22. Do you feel that your work leads to customer satisfaction?	0.366**	0.613**	0.409**	0.462**	−0.029	0.559**	0.299	0.613**	0.443**
Q6-1. Do you want to work for a long period at the current company?	0.328**	0.792**	0.785**	0.812**	0.723**	0.800**	0.560**	0.875**	0.816**
Q6-3. Do you deire to be more helpful in the work place?	0.530**	0.709**	0.740**	0.598**	0.752**	0.706**	0.535**	0.659**	0.799**

*$P < 0.05$, **$P < 0.01$

Table 5 Correlation coefficients with Q6-4 and variables

Q6-4. Do you want to enhance customer satisfaction?

		Employment pattern		Current work position			Age group		
	Overall	Full time	Part time	Kitchen	Washing place and pantry	Service floor	10s and 20s	30s and 40s	50s and 60s
Q1-19. Do you propose or implement improvement plans and problem resolutions actively?	0.180*	0.318	0.137	0.059	0.115	0.238	0.075	0.138	0.294*
Q1-21. Do you cooperate and collaborate with colleagues?	0.239**	0.141	0.305**	0.109	0.581**	0.227	0.141	−0.062	0.476**
Q1-22. Do you feel that your work leads to customer satisfaction?	0.366**	0.327	0.382**	0.353*	0.354	0.489**	0.320*	0.504**	0.318*
Q2-4. Is the work environment efficient?	0.109	0.115	0.093	0.012	−0.058	0.149	−0.021	0.241	0.106
Q2-5. Do you work in a good team setting?	0.248**	0.262	0.257*	0.299	0.129	0.257*	0.050	0.310	0.333*
Q2-6. Is the staff assignment efficient?	0.209*	0.217	0.223*	0.131	0.177	0.169	0.104	−0.002	0.381**
Q2-7. Are the working processes efficient?	0.135	0.248	0.118	0.144	−0.116	0.132	0.082	0.097	0.193
Q6-1. Do you want to work for a long period at the current company?	0.328**	0.520**	0.266**	0.317	0.162	0.372**	0.333*	0.269	0.402**
Q6-3. Do you desire to be more helpful in the workplace?	0.530**	0.716**	0.462**	0.556**	0.240	0.689**	0.728**	0.595**	0.438**
Q6-6. Do you want to improve your skills and acquire new skills?	0.548**	0.606**	0.521**	0.678**	0.212	0.679**	0.642**	0.781**	0.415**
Q6-7. Do you want to make progress to acquire more skills?	0.495**	0.584**	0.462**	0.651**	0.014	0.718**	0.507**	0.704**	0.385**

*$P < 0.05$, **$P < 0.01$

items are converted from the value of '1' to '6' which stand for (1) strongly agree, (2) agree, (3) slightly agree, (4) slightly disagree, (5) disagree, and (6) strongly disagree. A value of $P < 0.05$ is considered to indicate statistical significance.

In this section, representative results are shown with particular reference to three specific question items. The correlation coefficients among questions Q1-22., Q6-2., and Q6-4. and variables are shown in Tables 3, 4, and 5, respectively. Three question items are considered in reference to whether they may directly represent or influence employee satisfaction and customer satisfaction, respectively:

Q1-22. Do you feel that your work leads to customer satisfaction?
Q6-2. Do you enjoy your work?
Q6-4. Do you want to enhance customer satisfaction?

4.2 *Results and Discussion*

As shown in Table 3, Q1-22. shows fair correlation between Q1-10. ($r = 0.455$, $P < 0.01$) answered by the staff working on service floors. This correlation is only observed for this group. Results from the answers of kitchen, washing place, and pantry staff show no significant differences. In addition, fair condition is shown between Q1-19 ($r = 0.520$, $P < 0.01$), which were only answered by the staff working at service floors similarly. This might be because service floor staff are required to make many judgments on their own because they have to adopt customer's requirements flexibly. On the other hand, the kitchen staff works under the hierarchies of traditional chef systems. The other fair correlations between Q1-22. are shown in Q6-6. and Q6-7 for both the kitchen and service floor staff. The fair correlation levels are stronger for the kitchen staff (Q6-6: $r = 0.546$, $P < 0.01$), (Q6-7: $r = 0.500$, $P < 0.01$) than the service floor staff (Q6-6: $r = 0.472$, $P < 0.01$), (Q6-7: $r = 0.442$, $P < 0.01$).

As shown in Table 4, which represents the correlation with Q6-2., Q1-3. ($r = 0.581$, $P < 0.01$), and Q1-19. ($r = 0.586$, $P < 0.01$), there is only fair correlation in the segment of ages for workers who are in their fifties and sixties. Aggressive policies providing the opportunity to propose, challenge, or implement improvement plans and problem resolutions have fair correlation in particular for high-aged workers. These policies may have more influence on the employee satisfaction of workers in their fifties and sixties than of other age groups. Q1-22. shows fair correlation with the kitchen staff ($r = 0.462$, $P < 0.01$) and service floor staff ($r = 0.559$, $P < 0.01$). The results show that they may enjoy feelings toward contribution for customer satisfaction.

In the Table 5, the correlations between Q6-4. and several other questions are described. The fair and strong correlations are shown with Q6-6. and Q6-7 among the kitchen staff, service floor staff, and the staff in their thirties and forties, respectively. In the segmentation of current working positions, the results of kitchen and service floor staffs show fair correlations, and there are no significant differences shown for the washing place and pantry staffs. In the segmentation of ages, each age has fair correlations, and the categories of workers in their thirties and forties show particularly strong correlations. The results show that they may feel that more skill development leads to higher customer satisfaction. Thus, the results may represent the relationship between employee satisfaction and its realization of effects on customer satisfaction.

From the above-mentioned results, correlation differences can be observed according to the segmentations. In particular, the segment of current working position shows different trends of results.

5 Covariance Structure Analysis

5.1 Overview

This section analyzes employee satisfaction structures with covariance structure analysis. The purpose of the analysis is to capture the structure of employee satisfaction with factors related to customer satisfaction. According to the results described in Sect. 4, the category of current working position is selected for evaluation in order to outline the different structures between employees.

5.2 Results and Discussion

Using the question items which have fair or strong correlations with Q1-22., Q6-2., or Q6-4. evaluated in Sect. 4, combinations of question items were tested in a sampling study to investigate the cross-validity of model modifications. As a result, the three models were acquired by verification with fit indexes. Figures 1a, 1b, and 1c show the path diagrams of models tracing employee satisfaction to customer satisfaction by relating factors for the kitchen, washing place and pantry, and

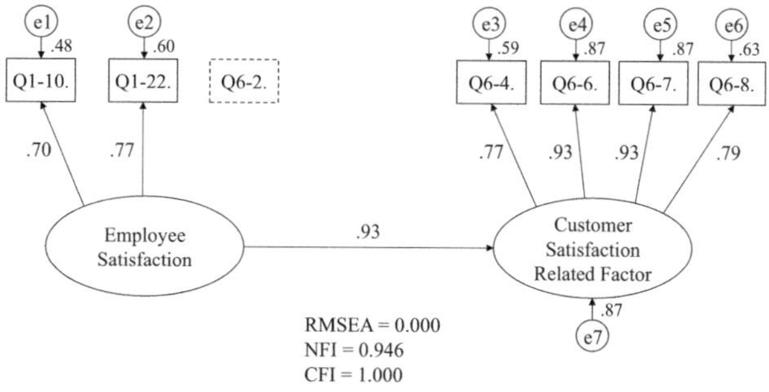

Fig. 1a Path diagram of model of employee satisfaction to customer satisfaction related factors for kitchen staff

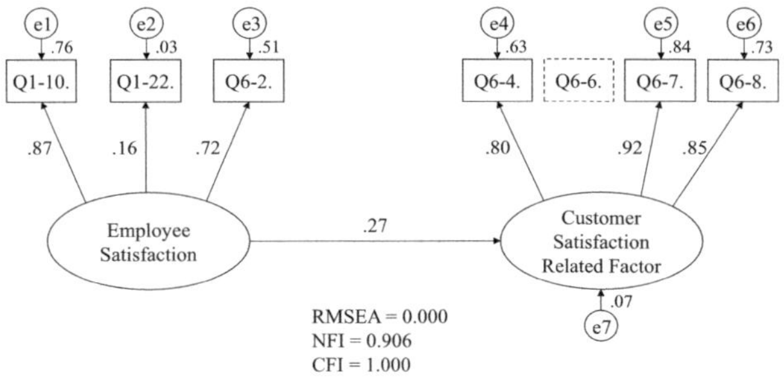

Fig. 1b Path diagram of model of employee satisfaction to customer satisfaction related factors for washing place and pantry staffs

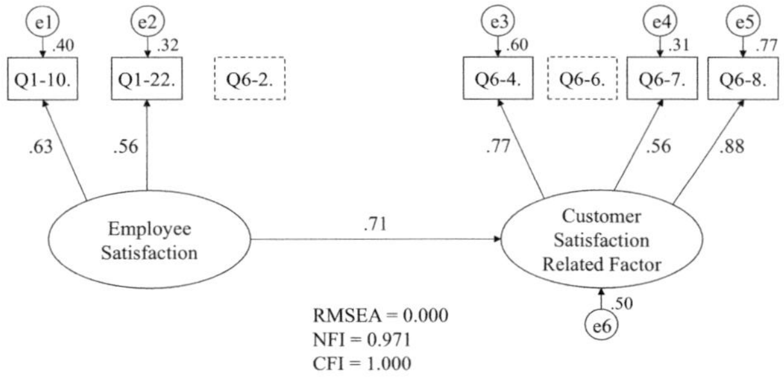

Fig. 1c Path diagram of model of employee satisfaction to customer satisfaction related factors for service floor staff

service floor staffs, respectively. The models are tested by fit indexes of the *P* value from a chi-square score, RMSEA, NFI, and CFI. The *P* values from chi-square scores for the three models are 0.485, 0.792, and 0.686, respectively.

The two independent latent variables in the models are employee satisfaction and customer satisfaction related factor. The latent variables are defined as follows:

Employee satisfaction: Employee satisfaction that staff members feel when they are satisfied with their job and work conditions.

Customer satisfaction related factor: The viewpoint required of staff members that allows their service provisions to enhance customer satisfaction.

Observed variables which represent each question item were tested to represent various hypotheses about the relationship between employee satisfaction and the factors which may lead to customer satisfaction. As a result, seven sets of the observed variables are chosen for structure modelling. The question items as observed variables are:

Q1-10. Is your job one that you like? Do you like your job?
Q1-22. Do you feel that your work leads to customer satisfaction?
Q6-2. Do you enjoy your work?
Q6-4. Do you want to enhance customer satisfaction?
Q6-6. Do you want to improve your skills and acquire new skills?
Q6-7. Do you want to make progress in acquiring more skills?
Q6-8. Do you always consciously provide good service?

When compared to the results of the pass analysis between the observed variable Q1-22. and the latent variable employee satisfaction, the results of kitchen and service floor staffs show higher standard partial regression coefficients. The values are 0.77 for kitchen staff, 0.56 for service floor staff, and 0.16 for washing place and pantry staff. The viewpoint that their own work leads to customer satisfaction is present in the kitchen staff most and next in the service floor staff. The value for washing place and pantry staffs shows a low coefficient.

In examining the structures between the observed variables represented by Q6-6. and Q6-7. and the latent variable of the customer satisfaction related factor, higher coefficient values are shown for the kitchen staff. They show high values of 0.93 for both Q6-6. and Q6-7. For the service floor staff, the coefficient value is observed only for Q6-7., with a lower value of 0.56. Hence, the members of the kitchen staff consider the relationship between customer satisfaction and skill development to be more direct than staff in the other positions. Additionally, service floor staff may feel that acquiring skills leads to customer satisfaction but may see some differences between the acquiring new skills and developing skill levels. For the washing place and pantry staffs, a high coefficient value is observed for Q6-7.: 0.92. However, the pass to Q6-6. is not modelled by fit value verification.

In the passes between the independent values of employee satisfaction and the customer satisfaction related factor, the highest coefficient value is shown in the model for the kitchen staff (0.93), and the next highest is for the service floor staff (0.71). The value for the washing place and pantry staffs is 0.27.

As above-mentioned results demonstrate, the different structures of employee satisfaction are revealed by comparing the results of segmentation by current working positions. The results indicate the different levels of understanding of the relationship between their own tasks and customer satisfaction. It is expected that the findings can be utilized for daily instruction, operations, and education to enhance the quality of service for customer satisfaction.

6 Conclusions

This paper analyzes employee satisfaction through the results of a questionnaire delivered to the staff of six restaurants. The results of these questions were investigated with correlation analysis and covariance structure analysis to capture the structure of employee satisfaction. The results indicate the different trends among the staffs' attributes and the different structures of employee satisfaction which depend on the categorization of current working positions. Developing a more detailed model of employee satisfaction will be a future step taken with further research.

Acknowledgements This work was partially supported by the project of Service Science, Solutions and Foundation Integrated Research Program from JST-RISTEX, and JSPS KAKENHI Grant-in-Aid for Young Scientists (B) Number 26730159.

References

1. Shimmura T, Takenaka T, Ohura S (2013) Improving labor productivity and labor elasticity at multiproduct Japanese cuisine restaurant introducing cell-production system. In: Proceedings of the international conference advances in production management systems (APMS2013), State College, USA, 9–12 September, II11–17
2. Takenaka T, Shimmura T, Ishigaki T, Motomura Y, Ohura S (2011) Process management in restaurant service – a case study of a Japanese restaurant chain. In: Proceedings of the international symposium of scheduling 2011 (ISS2011), Osaka, Japan, 2–4 July, pp 191–194
3. Judge TA, Thoresen CJ, Bono JE, Patton GK (2001) The job satisfaction-job performance relationship: a qualitative and quantitative review. Psychol Bull 127:379–407
4. Iaffaldano MT, Muchinsky PM (1985) Job satisfaction and job performance: a meta-analysis. Psychol Bull 97:251–273
5. Harter JK, Schmidt FL, Hayes TL (2002) Business-unit-level relationship between employee satisfaction, employee engagement, and business outcomes: a meta-analysis. J Appl Psychol 87 (2):268–279
6. Fujii N, Toshiya K, Nonaka T, Oda J, Shimmura T (2013) A study on staff shift scheduling improving service satisfaction in restaurant business – proposal of planning method considering difference in employee abilities. In: Proceedings of the 23rd conference on the Japan society of mechanical engineers, design and systems division (in Japanese) (USB), 2314
7. Sureshchandar GS, Rajendran C, Anantharaman RN (2002) The relationship between service quality and customer satisfaction – a factor specific approach. J Serv Mark 16(4):363–379
8. Fujii N, Toshiya K, Uemura M, Nonaka T, Shimmura T (2013) Facility layout planning of central kitchen in food service industry – application to the real-scale problem. In: Proceedings of the international conference advances in production management systems (APMS2013), State College, USA, 9–12 September, pp 33–40

Exploration of Service System and Value Co-creation Mechanism in Islamic Banking in Pakistan

Amna Javed, Youji Kohda, and Hisashi Masuda

Abstract This study examines service system and value co-creation mechanism in Islamic banking. This paper combines the results from literature review, press release data, interviews and questionnaire analysis conducted in Pakistan. The results revealed that in Islamic banking the backbone of the mechanism of value co-creation is contract-based projects that are regulated by sharia board, to enforce profit and loss sharing that refers to a partnership approach. Islamic banks do not have independence to perform transactions without the supervision of sharia board that ensures the secure investment. These research findings show better understanding of value co-creation in the modern services.

Keywords Islamic banking • Service system • Value co-creation

1 Introduction

Banking industry has gained importance to promote business operations to boost economic situation and leads towards an established economy [1]. In today's global market, there is a progressive aspect of banking system that is known as Islamic banking system [2]. Islamic banks play an important role for the economic growth and well-being of humanity by mobilizing the funds and execution of social activities [3]. As Islamic banking is generally based on a business relationship and partnership approach [1], so it helps to eradicate unjustified operations of different stakeholders that are created by interest-based banking system [4].

Islamic banking system is based on the principles of Islamic law and conducts their activities in accordance with the principles of Islam [5]. It is an interest free banking and promotes partnership concept. As Islamic banking is based on Islamic principles that are more concerned with non-economic values (well-being, partnership, fairness, trust worthy, and justice) than economic values as money in terms of interest. Many scholars have focused on co-creation of values in services mainly on economic dimension but not on non-economic dimension. The core objective of

A. Javed (✉) • Y. Kohda • H. Masuda
Japan Advanced Institute of Science and Technology, 923-1292 Nomi, Japan
e-mail: amna.javed@jaist.ac.jp

T. Maeno et al. (eds.), *Serviceology for Designing the Future*,
DOI 10.1007/978-4-431-55861-3_3

this research can be expressed as to assess team work or sharing experiences among the banks and the customers to analyze the strength of value co-creation.

On the contrary conventional banking is based on a pure financial model, and follows the theory of capitalism with the purpose of profit maximization [6].

This study is about to explore the concept of service systems that combines the created values and organizational networks designed to deliver services that satisfy the requirements and hopes of the customers in Islamic banking system [7].

Following Sharia (Islamic principles that are based on Qur'an and Sunnah) traditions, Islamic banking is more concerned with commodity-based transactions than simply lending loans, they have more check and balance in terms of audit than conventional banks to ensure fairness and transparency. The results also show that Islamic banks deal in secure investments; because of this customers are more satisfied with Islamic banks than conventional banks. The key features of Islamic banks are to promote profit and loss sharing and charity concept for human well-being by following Sharia principles and business ethical standards [8]. This study will also depict that Islamic banking promotes non-economic values used for the betterment of the society by the discussion of the results generated from the press release data of Islamic and conventional banks by using text mining technique [9]. Islamic banks deal in many financing products and services that directly relate to the economic values accompanied with the non-economic values. The economic values are normally monetary values, defined in terms of money that relates to the market such as profit, service charges etc. However the non-economic values are the non-monetary values or standards to regulate human activities like partnership, transparency, fairness and teamwork. The most famous Islamic banking financing products are Musharakah, Mudarabah, Murabaha, and Ijarah [9].

2 Research Goals

This study examines all the perspectives of Islamic banking together with the value co-creation process and co-created values to explore the concept of value co-creation in service science.

This research will extend the services by having Islamic banking cases, because it shows an emerging trend in the banking world even though there are some investment constraints, some prohibitions like interest. It is conducted to know why new services are emerged. And what kind of values customers can get from them?

3 Research Target

This research is basically about Islamic banking in Pakistan which is a developing country having both Islamic and conventional banking systems but Islamic banking is rapidly expanding in Islamic and even in western countries. This research is basically conducted to answer some of the questions like: Why are the western people interested in Islamic banking products? Is there any new type of value that cannot be achieved from conventional banking?

4 Research Significance

This study shows that conventional service system implicitly focuses on the western developed countries value criteria [10] but this research will extend the coverage to the developing countries as well and will contributes to understand more about value system towards value co-creation.

This research provides better understanding of value co-creation in the modern services by using the findings from Islamic Banking case study.

5 Research Methodology

This paper combines the results from literature review to make clear the differences between Islamic and conventional banking, press release data available on the websites of the banks to find out the unique characteristics of the banks related to economic and non-economic or social perspective, and a qualitative study was made through four in-depth interviews in Pakistan from two Islamic banks to know the banking policies and business models used to narrate the co-creation process in Islamic banking system. Interviews were made from top management and middle management positions in two different Islamic banks. Questions asked during interview were based on business activities, customer satisfaction, and social activities.

The interviews of Islamic banking management people show that Islamic banking is strongly focused on fairness and transparency of projects because of benevolent watch of sharia board on all the banking activities. There are many products focused on profit and loss sharing concept that shows a higher sense of involvement because of the contracts established between the parties involved. However, in conventional banking there are no such characteristics found precisely.

Finally analysis of the questionnaire data collected from the customers of Islamic and conventional banks in Pakistan to know the customer perspective regarding their respective banking system.

6 Results and Discussion

6.1 Literature Review

From literature this study covers many concepts and differences and shows some of the interesting differences in Islamic and conventional banking and their dealings with their customers to co-create values. Individual bank's roles and processes contrast in their presentation. These differences follow several motives including the fundamental laws in the country, practices, aims and the manner of interaction between the banks and the customers as well [3].

Following Sharia (Islamic principles that are based on Qur'an and Sunnah) traditions, Islamic banking is more concerned with commodity-based transactions than simply lending loans [8], they have more check and balance in terms of audit than conventional banks to ensure fairness and transparency [11].

The results also show that Islamic banks deal in secure investments; because of this customers are more satisfied with Islamic banks than conventional banks [12]. The main objectives of Islamic banks are to promote profit and loss sharing and charity concept for human well-being by following Sharia principles and business ethical standards.

Table 1 explains the summarized differences between Islamic and conventional banking as a bird view that is conducted from literature review in detail.

6.2 Press Release Data Analysis

In case of press release data analysis the data were collected from the press releases published on the websites of 4 banks working in Pakistan (2 Islamic banks and 2 conventional banks) for the years 2007–2013 and we got 478 press releases from all 4 banks. The data were analyzed as follows.

Firstly the data were organized and entered into SPSS text mining tool and we got 8000 words. We observed 116 financial values, 433 social values and 7451 other words. Secondly we developed the dictionary by analyzing the papers about

Table 1 Difference between Islamic banking and conventional banking

Islamic banking	Conventional banking
Commodity-based transactions than simply lending loans	Interest-based mobilization of funds and lending
More check and balance in terms of sharia audit to ensure fairness and transparency	Less check and balance and customers do not know everything about the transaction
Deal in restricted and secure investments	Deal in all kind of investments with no restriction
Promote profit and loss sharing and charity concept for human well-being	Promote interest for profit maximization

economic, financial and social science aspects of the banking sector explaining all the economic and non-economic values and terms related to Islamic and conventional banking, here we got 192 values after detailed analysis of the terms by using research papers. Thirdly, we divided the press release data from different banks by using the developed dictionary and divided the data into three main groups, 100 % occurrences in Islamic banking, these values do not exist in conventional banking press releases, 100 % occurrences in conventional banking, not exist in Islamic banking press releases, and the others that exist in both banking system. Finally, we compared the values and observed the unique characteristics of Islamic and conventional banking based on economic and non-economic values [9]. Figure 1 explains the whole procedure of press release data analysis at a glance.

Figure 2 demonstrates the detail of the two main groups of values constructed from the press release data. 100 % occurrences in Islamic banks have a focus towards non-economic values and it has unique characteristics including team work, Islamic spirit, trust worthiness that includes transparency of projects and finally practicability that relates to the co-creation between the provider and the customer. Contrary to the Islamic banking, 100 % occurrences in conventional banking has shown a great trend towards economic values and deals in the monetary transactions mainly, as it refers to capitalism with the purpose of profit maximization [9].

The results generated in the figure, shows that firstly, Islamic banking has a great focus on sharing services like teamwork and participation that ensure higher involvement of the concerned parties into the transaction because it promotes cooperation. Secondly, it represents Islamic spirit that explains more about Sharia involvement to have secure investment.

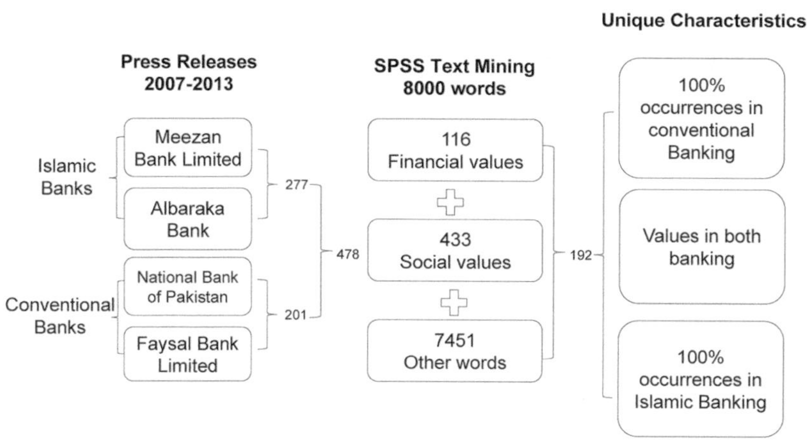

Fig. 1 Procedure of press releases data analysis

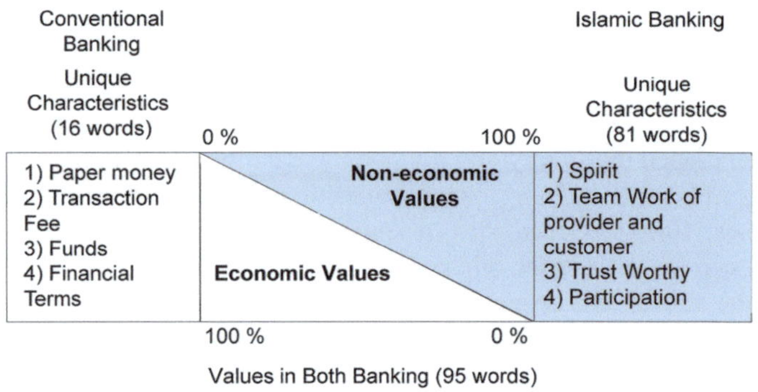

Fig. 2 Unique characteristics of Islamic and conventional banking

6.3 Interview and Questionnaire Analysis

From the pre interview analysis conducted in Pakistan, it is concluded that Sharia board has proper check and balance on each transaction to be carried out, and there is a restriction for making investment (Halal products only). The bank cannot perform the activities independently.

As one interviewee said:

> Behind Islamic banking dealings there is a Sharia advisory board. Before introducing and implementing any product or service, there is an approval from sharia board to make sure that it is as per Islamic principles or not.

6.3.1 Service System in Islamic Banking

On the basis of interview in this study a framework is developed about the service system in Islamic banking as shown in Fig. 3.

As Service science is the study of value co-creation and service systems that have a great focus on the organizational network designed to provide services to fulfil the needs and expectations of the customers and finally to co-create values [13]. In Islamic banking a service system involves sharia advisory, bank itself, customers and either supplier or manufacturer or seller (Fig. 3). Islamic banks do not have independence to perform any kind of transaction without the supervision of sharia board which ensures the secure investment by having extra audit that makes it different from conventional banks.

Any amount of penalty during the transaction cannot be included in the income of the Islamic banks; it must be consumed for charity for the establishment of the community and the well-being of mankind [11]. From interview analysis it is ensured that in case of conventional banks that amount of penalty becomes the profit of the bank.

Figure 3 depicts the Islamic banking service system to explain that sharia advisory is the source of Islamic banking, and charity is the destination of the transaction in case of the penalty charged to the bank. The doted area in the figure

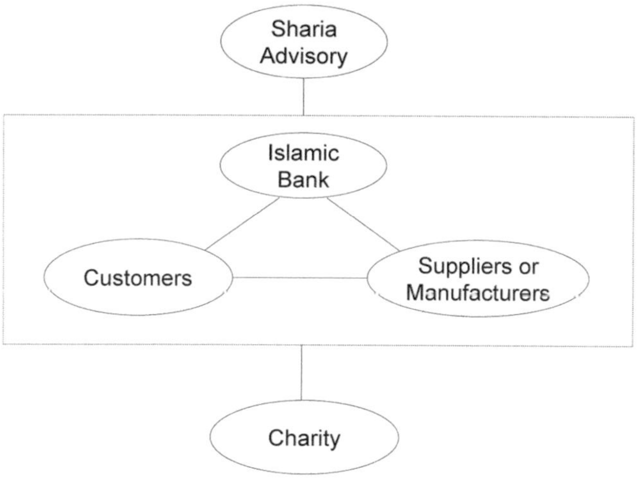

Fig. 3 Islamic banking service system

shows Islamic banking as a sub-system that further explains the transaction carried out among the concerned parties means stakeholders like bank itself, customers, suppliers or manufacturers.

6.3.2 Value Co-creation in Islamic Banking

Islamic banking works according to the principles of sharing of profit and loss. All the transactions are contract based and designed and approved by sharia board (partnership contract), that shows higher involvement of the parties involved.

As one interviewee said:

> Islamic banking works according to the principles of sharing of profit and loss on post facto basic arising from deferred trading and profit margin from the seller.

Another interviewee said:

> When a customer enters into a transaction with Islamic bank, it is usually contract based. We give him a form to explain the detail of the transaction like profit and loss sharing ratios and service charges to make him understand fully about the transaction to be carried out.

On the basis of the results this study proposed a value co-creation process model as explained in Fig. 4. Islamic banking is generally based on a business relationship and partnership approach; such approaches are now attaining popularity with the passage of time [1]. As a financing practice, it implicates a request by the customer to the bank to purchase some goods for him from the supplier. The bank performs this activity for a certain profit over the cost, which is specified in advance. Recovery could be decided in instalments or full payment; it depends on the contract [14]. Here bank enters into two types of contract one with customer and another with the supplier or the manufacturer. So, values co-creation occurs at both

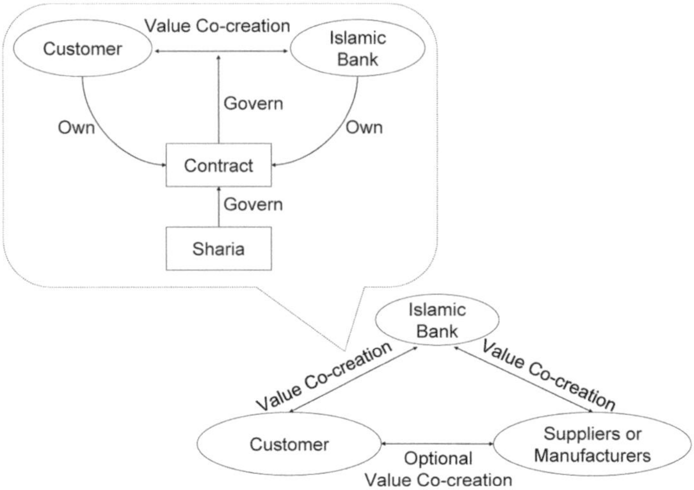

Fig. 4 Value co-creation in Islamic banking

sides, because they have the freedom to revise the contract under the supervision of the sharia board.

Figure 4 shows that in Islamic banking system all of the transactions are based on the contract developed among the involved parties. A project includes three stakeholders, a bank, a customer and a provider. There are two contacts, one between the bank and the customer, and second between the bank and the provider. The bank and the customer enter into a contract that is further governed by the sharia to make the verification of the transaction as per Islamic law. And after making contract they become the partners and work as a team to co-create values like co-operation, transparency and fairness etc. Moreover, customer can also access the supplier anytime for his satisfaction and to see the feasibility of the project, because they know each and everything about the transaction to be carried out. Same process is followed on the supplier side that is not shown in Fig. 4.

6.3.3 Co-created Values in Islamic Banking

Figures 3 and 4 demonstrate the values created as a result of co-creation process and the values are: First, a sense of involvement, as Islamic banking is based on contract based projects and concerned with sharing of profit and loss on partnership basis, they promote sharing of services that leads towards higher involvement of the parties into the transaction. Second, a sense of benevolent watch, in terms of supervisory check of sharia advisory to make sure the feasibility of the contract for secure investment and the bank does not have independence to perform any kind of activity without the approval of sharia board.

Table 2 Co-created values in Islamic banking

Service value embedded in Islamic banking products	Implementation of the service value	Implemented features	Also found in the other modern service	Customer's perceived value
Sharing of profit and loss	Contract – based projects between customers and providers	Identification of customer need	Money-sharing	A sense of involvement
		Consensus building between customers and providers	Crowd- funding	
Compliance with Sharia	Seeking directions from Sharia board	Audit of contents of the contracts	Fair trade mark	A sense of benevolent watch
		Anomaly detection of performance of contracts	Call center service operated by the sharing service provider	

Table 2 shows that there are basically two types of service values embedded in the Islamic banking products and those values are profit and loss sharing that promotes partnership concept and the other one is compliance with the sharia which shows the supervisory of the sharia board in all matters of Islamic banks. These values are extracted from the interviews conducted in Pakistan.

Firstly, profit and loss sharing concept can be implemented by making some contract between the customers and providers. It clarifies the contents of the contract including profit and loss sharing ratios. So Islamic banking has a great focus on contract based projects as this contract considers the identification of the customer's need because of the direct communication between the customer and the provider, which resembles to the money sharing concept [15], where lenders offer their money after deep understanding of real needs of the borrowers. And also there is a consensus building between the customers and the providers that promotes co-operative working, and this feature is similar to crowd-funding [16], where a crowd of people could express their needs to proposals and proposers could reflect the needs on their projects on a flexible basis. In those modern services, the customer's perceived value is a higher sense of involvement because the customer knows everything about the product and work under partnership deed.

Secondly, compliance with sharia ensures the supervision of the sharia board and work as per Islamic principles. Islamic banks seek directions from sharia board before entering into any type of contract with the customers that guarantees pre check and audit of the contents of the contracts to confirm secure investment. It is similar to fair trade mark as in this case before issuing anything pre checking of the information is compulsory. And also there is an anomaly detection of performance of the contracts by sharia board that ensures real time check, similar to call center service operated by the sharing service provider [17]. The customer can make the complaint directly to the managing body in case of emergency, like "car not there" or "a dirty car unexpectedly" in the car sharing service. In those modern services, the customer's perceived value is a higher sense of benevolent watch because the customer knows the transactions are watched for security.

For the purpose of conducting some evidence regarding the values generated, we analyzed the questionnaire data collected from the 25 customers from each bank and in total 100 customers from two Islamic (Meezan bank limited and Alfalah bank limited-Islamic banking) and two conventional banks (National bank of Pakistan and Allied bank limited), by using a statistical software R. We used the Pearson's chi-squared test to narrate the differences of two banking customers. We assessed the significant differences by re-structuring the questionnaire items to the following list for judgment of the significances. This study shows the contingency tables and the results of chi-squared test from the questionnaire items having significant differences. The level of significance is 5 %.

Table 3 illustrates that "This bank do not exploit its customers in any way?" In case of Islamic banks, 30 out of 50 customers gave the support to this question because of the involvement of the customers is higher in all transactions than conventional banks and they do not exploit their customers in any way. Islamic banks give priority to the customer's needs and choices, and they usually work by consensus building between customers and the providers. In case of conventional banks only 10 out of 50 customers were agree to state that these banks do not exploit its customers, and their priority is profit maximization instead of customer satisfaction.

Table 4 demonstrates the contribution of the banks towards social welfare, where 37 out of 50 customers in Islamic banks were agreed that they have major contribution to social welfare in terms of charity, zakat, donations and benevolent loans, however in conventional banks 23 out of 50 customers were agreed and the results show that Islamic banks work for the betterment of the economy and for the well-being of the mankind [14]. Therefore the created values in Islamic banks do not depict only economic values, but they also represent social values.

Table 5 is about to depict the results to explain that "This bank contributes more in removing society's inequalities and improving general standard of living?" And the results show that Islamic banks are more concerned with the reduction in

Table 3 Difference of customer satisfaction and involvement

	Strongly agree	Agree	Neutral	Disagree	Strongly disagree	Sum
Customers of conventional banking	2	8	20	19	1	50
Customers of Islamic banking	5	25	18	1	1	50
Sum	7	33	38	20	2	100
Pearson's Chi-squared test						
X-squared	26.3486					
df	4					
p-value	2.69E-05					

Table 4 Difference of social welfare

	Strongly agree	Agree	Neutral	Disagree	Strongly disagree	Sum
Customers of conventional Banking	4	19	14	10	3	50
Customers of Islamic banking	4	33	12	1	0	50
Sum	8	52	26	11	3	100
Pearson's Chi-squared test						
X-squared	14.2867					
df	4					
p-value	6.43E-03					

Table 5 Difference of economic standards

	Strongly agree	Agree	Neutral	Disagree	Strongly disagree	Sum
Customers of conventional banking	2	11	23	9	5	50
Customers of Islamic banking	4	28	11	7	0	50
Sum	6	39	34	16	5	100
Pearson's Chi-squared test						
X-squared	17.5622					
df	4					
p-value	1.50 E-03					

society's inequalities and improvement of living standards because they consider customer as a partner not just a buyer or seller. The results generated from the questionnaire analysis was, 32 out of 50 customers showed agreement to the statement than conventional banks where only 13 out of 50 customers were agreed.

Tables 3, 4, and 5 express the first concept of Islamic banking that is sharing of profit and loss to lead towards higher sense of involvement that certifies cooperative work among the parties involved (Table 2).

Table 6 represents the differences between two baking systems to explain that "This banking system encourages secure investment projects?", where 36 out of 50 customers of Islamic banks were agreed to state that Islamic banks promote secure projects, but in conventional banks 32 out of 50 customers were disagree with the statement that conventional banks promote secure investment. From the results it is clear that Islamic banks work as per Islamic principles under sharia to promote secure investment that ensure transparency, fairness and project feasibility because of sharia supervisory check and higher involvement that guarantees a sense of benevolent watch.

And finally Table 7 describes that "This bank is governed by cultural norms or not?". And from results, it is observed that Islamic banks follow religious norms

Table 6 Difference of investment projects

	Strongly agree	Agree	Neutral	Disagree	Strongly disagree	Sum
Customers of conventional banking	3	6	9	21	11	50
Customers of Islamic banking	9	27	10	4	0	50
Sum	12	33	19	25	11	100
Pearson's Chi-squared test						
X-squared	38.9763					
df	4					
p-value	7.05 E-08					

Table 7 Difference of cultural norms

	Strongly agree	Agree	Neutral	Disagree	Strongly disagree	Sum
Customers of conventional banking	2	6	10	15	17	50
Customers of Islamic banking	11	27	9	3	0	50
Sum	13	33	19	18	17	100
Pearson's Chi-squared test						
X-squared	44.647					
df	4					
p-value	4.71E-09					

and work as per teachings of Islam (Quran and Sunnah) and these principles cannot be changed by human. And according to these principles Islamic banks address humanity and sense of justice. They do not have independence to perform any kind of activity and for each transaction they need the approval from the sharia advisory. 38 out of 50 customers showed agreement to the statement. It is the fact that makes it different from conventional banks where 32 out of 50 customers were agreed at this point. It is the feature that relates Islamic banks to the modernized services like fair trade mark and call centre services operated by sharing services and like these services Islamic banks have proper check and balance of Sharia in terms of audit to make feasible the terms of contracts established among the parties.

Tables 6 and 7 show the second concept of Islamic banking that is compliance with sharia and the ultimate perceived value is a sense of benevolent watch (Table 2).

From the questionnaire analysis it can be assured that Islamic banks have more customized strategies that are consumer centric than conventional banks. The created values in Islamic banks like a sense of involvement and a sense of benevolent watch, these are the new types of values, and the main success factors of the modern services like Islamic banking in the banking world that resembles to crowd funding, money sharing, sharing services, fair trade mark, etc. And it is the main factor of success that is promoting Islamic banking day by day even in the western culture as well.

7 Conclusion

This paper enlightens value co-creation process and the co-created values in Islamic banking from the literature reviews, press release data analysis, questionnaire analysis and finally the initial analysis of interviews from the management people in Islamic banking in Pakistan.

The core objective of this research can be expressed as to assess created values. Firstly, higher sense of involvement among the bank and the customers because they are dealing in a partnership contract and all the parties are well aware of the contents of the contract before starting the transaction. Secondly, a sense of benevolent watch because of sharia audit and supervision that ensures secure investment, and expected as a growing trend in the business world that provide a great opportunity to broaden the view of value co-creation with respect to non-economic values in service science. These values are the main source of attraction towards Islamic banking system even for the view point of western people.

Our findings from Islamic banking cases show similarities to the emerging modern services like crowd funding, money sharing, fair trade mark and call center service operated by the sharing service provider. So the results generated from this research are not only beneficial to developing and Islamic countries but also for developed and western countries as well.

Acknowledgement This work was supported by JSPS KAKENHI Grant Number 25240050.

References

1. Shahid H, Rehman R, Niazi GSK, Raoof A (2010) Efficiencies comparison of Islamic and conventional banks of Pakistan. Int Res J Financ Econ 1(49):24–42
2. Fazlan S, Mohammad A (2007) The efficiency of Islamic banks: empirical evidence from the MENA and Asian countries Islamic banking sectors. Paper presented at the fifth international Islamic finance conference, Kuala Lumpur, 3–4th September
3. Javed A, Umemoto K, Kohda Y (2013) Co-creation of non-economic values in Islamic banking: a new frontier in service science. Int J Soc Hum Sci Eng 7(9):331–338
4. Masruki R, Ibrahim N, Osman E, Wahab HA (2011) Financial performance of Malaysian founder Islamic banks versus conventional banks. J Bus Policy Res 6:67–79
5. Iqbal Z, Mirakhor A (2007) An introduction to Islamic finance: theory and practice. Willey, Chichester

6. Parashar SP, Venkatesh J (2010) How did Islamic banks do during global financial crisis? Banks Bank Syst 5:54–62
7. Maglio P, Srinivasan S, Kreulen J, Spohrer J (2006) Service systems, service scientists, SSME, and innovation. Commun ACM 49(7):81–85
8. Usmani TS (2002) An introduction to Islamic finance. Maktaba Ma'arif Al Quran, Karachi
9. Javed A, Masuda H, Kohda Y (2014) Value analysis of Islamic banking and conventional banking to measure value co-creation. Int J Soc Hum Sci Eng 8(2):85–93
10. Spohrer J, Maglio PP (2008) The emergence of service science: toward systematic service innovations to accelerate co-creation of value. Prod Oper Manag 17(3):1–9
11. Hanif M (2011) Difference and similarities in Islamic and conventional banking. Int J Bus Soc Sci 2(2):166–175
12. Khan F (2010) How 'Islamic' is Islamic banking. J Econ Behav Organ 76:805–820
13. Stephen LV, Robert FL (2004) Evolving to a new dominant logic for marketing. J Mark 68 (1):1–17
14. Pervez A (2012) Principles of Islamic interest free banking in Pakistan: study focusing on three Islamic banks in Pakistan. LAP LAMBERT Academic Publishing, Germany
15. Robert RR, Waller C (2006) Money and risk sharing. J Money Credit Bank 38(6):1599–1618
16. Mollick E (2014) The dynamics of crowd funding: an exploratory study. J Bus Ventur 29:1–16
17. Kohda Y, Masuda K (2013) How do sharing service providers create value? ACIS2013, 2013/10/31, Phuket, Thailand

The Ordering of Fast Food Using Menu

Kiyoteru Kitano, Yutaka Yamauchi, and Takeshi Hiramoto

Abstract Fast-food services are highly routinized and standardized. Previous studies have insisted that service routinization and standardization was treated as a top-down or automatic process in which interactants master their respective roles through learning scripts. However, we suppose that service routinization is achieved through the contingent, situated actual interaction between service providers and consumers. Therefore, we focused on the process of actual service encounters and analyzed how routinization and standardization standardized are achieved. Indeed, we videotaped ten actual customer interactions at a certain hamburger restaurant, and analyzed them by using conversation analysis based on ethnomethodology.

Keywords Service encounter • Ethnomethodology • Value co-creation • Boundary object

1 Study Outline

In this study, we used conversation analysis [1], a method based on the sociological approach of ethnomethodology [2], to examine service encounters (interactions between service providers and consumers) in a fast food (hamburger) restaurant. It is said that the service provided in fast food restaurants is highly standardized and routinized [3]. Previous studies have tended to view this standardization and routinization as a top-down or automatic process in which interactants master their respective roles through learning scripts [4]. However, what in fact occurs during service encounters in fast food restaurants is the achievement of an exchange between mutual strangers who are meeting in a chance encounter. In this study, we analyzed how standardization and routinization are achieved and how fast food service is achieved in context-dependent service encounters. For this purpose, we

---

K. Kitano (✉) • Y. Yamauchi • T. Hiramoto
Graduate School of Management, Kyoto University, Yoshida-Honmachi, Sakyo-ku, Kyoto 606-8501, Japan
e-mail: kitano.kiyoteru.43z@st.kyoto-u.ac.jp

T. Maeno et al. (eds.), *Serviceology for Designing the Future*,
DOI 10.1007/978-4-431-55861-3_4

used conversation analysis, a method that places importance on the process of actual mutual interactions.

2 Study Background

2.1 *The Standardization and Routinization of Service and the Role of the Customer*

Part of the value of the service provided at fast food restaurants lies in each restaurant's characteristics (i.e., its streamlining, standardization, and routinization of service provision and consumption). If this is the case, then it follows that the creation of this value cannot be achieved by the service provider alone but by the consumer's involvement in the process [5]. In other words, the restaurant's provision of standardized and routinized service and the customer's consumption of the service in a standardized and routinized manner must come together as a set.

What is important here is that in a service encounter, the service provider and the customer are typically mutual strangers [6]. The question that must be asked is how mutual strangers can achieve standardized and routinized service. To enable the standardization and routinization of service, the service provider and the consumer must be able to easily predict each other's ensuing actions [7]. Previous studies have identified script learning and role mastery as the means by which predictability can be increased [4]. It is argued that the predictability of the mutual interaction will be increased if the service provider and consumer both learn the script for the encounter and learn their respective roles (that of staff member and of customer).

However, in interactions between mutual strangers, how is it possible for the interactants to know from the outset whether their counterparts have learned the scripts and mastered their roles? There is of course no need to retreat into agnosticism and skepticism. In most fast food restaurant encounters, the service provider and consumer provide/consume the service in a standardized and routinized manner with the greatest of ease. However, we consider the above argument to be overly dependent on the premise that script and role mastery are common among all interactants.

In this study, we focused on the process of actual service encounters and analyzed how the provision and consumption of standardized and routinized service are enabled through this process.

2.2 *The Menu as a Boundary Object*

As we argued in the previous chapter, the staff member and the customer will not know at the outset whether their counterparts have learned the scripts and mastered their roles for each encounter. Therefore, in many service encounters, including but not limited to hamburger restaurants, the first exchange will involve investigating what kind of customer (or staff member) one's counterpart is. For example, at Edomae-sushi bar, even at those outlets that do not have menus, the service provider will first ask the customer, "What drink would you like?" and ascertain the type of customer based on his or her reaction to the question [8]. Alternatively, in certain types of restaurants, i.e., Izakaya, the service provider will first ask, "Is this your first time here?" In relation to this, we have focused on how menus are used as the first step in hamburger restaurant service encounters.

Menus are boundary objects [9] that exist between the shop and the customer. Boundary objects exist on the boundary between contrasting communities; they are semantic constructions (objects, codes, events, etc.) whose existences form boundaries, and they link communities together.

Menus are not merely information media—they are intrinsic cultural objects that instruct customers on how they should behave. In the case of the hamburger restaurant we studied (Fig. 1), the design of the menu itself was used to structure the flow of decisions, such that customers select a hamburger from the main menu and then move to the combination menu or the drink menu. In other words, it is not the case that customers select products from information provided to them in a completely unbiased state. Instead, customers select the products to order through an interaction with the cultural instructions that are presented on the menu itself.

We focused on menus because we believe that the body of cultural knowledge presented in a menu can be a resource for indicating a particular type of customer. To state it another way, it is because the menu presents the shop's body of cultural knowledge that it can be used as a benchmark to determine the extent to which a customer is familiar with the shop's culture. However, the menu can only be a boundary object between the shop and customers when it is positioned in the midst of the shop-customer interactions in an appropriate way. The presence of the menu inside the shop does not itself guarantee that the menu will be used as the first step

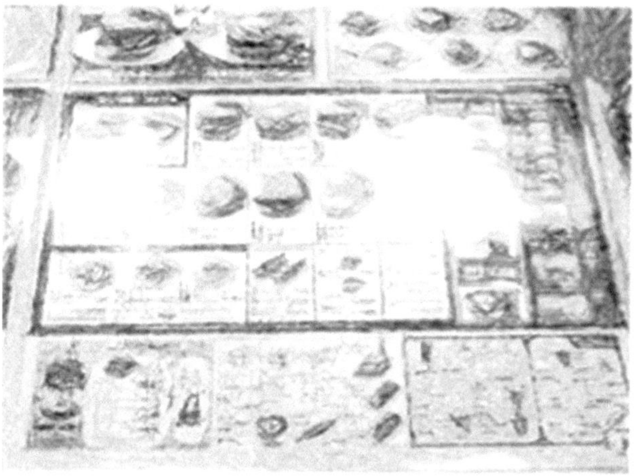

Fig. 1 Menus at the service counter

of the service encounter and that customer orders will be based on the information the menu displays.

In view of the above, we conducted our analysis of service encounters in a hamburger restaurant with a focus on the following two questions:

1. How does using the menu to place an order become an appropriate action?
2. Assuming that customers indicate the kind of people they are through using the menu, how do they do this?

3 Research Overview and Analysis Method

3.1 Video-Recording of Scenes of Orders

The site we selected for research was the storefront of a large hamburger chain that is developing nationwide. We selected this site because it is a setting in which standardization and routinization of service encounter is being advanced. The research overview is shown in Table 1.

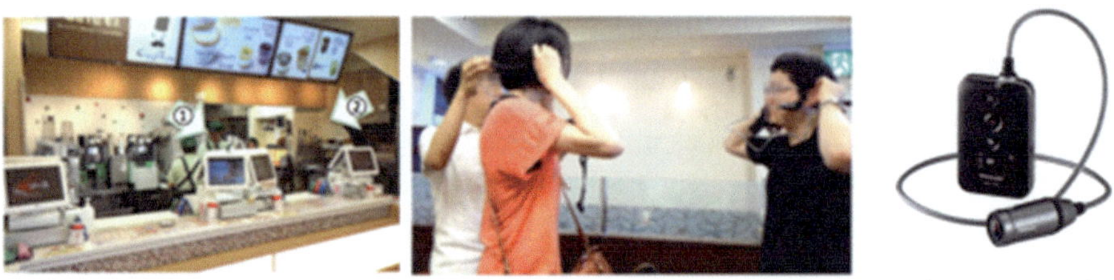

fixed cameras   wearable camera HX-A100   (Panasonic)

3.2 Analysis Method

In this study, we conducted conversation analysis with a focus on the scene of the participants' placing orders at the store counter. It involves recording data on an interaction that actually took place and making a detailed transcription of the

Table 1 Research overview

Date	13 June 2013 (Thu), 2:00–5:00 pm
Place	An outlet of a certain hamburger chain, Shijo Kawaramachi
Participants	Twelve persons (ten students, two housewives)
Method	The participants equipped themselves with a wearable camera (video camera attached to the head) and recorded the process from entering the shop to placing their orders. Stationary recording of the interactions was carried out with two fixed cameras.

interaction. When transcribing the data, we used the Jefferson system,[1] a widely used notation system in conversation analysis.

4 Results of the Analysis of the Research Data

4.1 Using the Menu to Place an Order

We first investigated (1): How does using the menu to place an order become an appropriate action?

Service encounters are generally considered to be scenes of interactions between shop staff and customers, but interaction between the two does not always begin at the moment the customer enters the shop. Even if it does begin at this point, it does

[1] Hereunder is a list of the notations used in the transcription of the data:

[Indicates the start points of overlapping speech.
] Indicates the end points of overlapping speech.
= When attached to the end of one line and the beginning of another, equal signs indicate that the later talk was "latched onto" the earlier talk with no hesitation.
(#) A number in parentheses indicates the time, in seconds, of a pause in speech.
(.) An extremely brief pause, typically 1 second or so.
: Colons indicate an elongated syllable; the more colons, the more the syllable or sound is stretched.
- Indicates an abrupt halt or interruption in an utterance.
. Indicates falling pitch or intonation.
? Indicates rising pitch or intonation.
, Indicates a temporary rise or fall in intonation.
↑↓ Indicates a sharp rise or fall in the pitch or intonation of a subsequent utterance.
text Indicates that the speaker is emphasizing or stressing the speech.
°Text° Text within degree symbols indicates whispering, reduced volume, or quiet speech.
hh Audible exhalation; the more h's, the longer the exhalation.
text (h) text (h) When there are multiple audible exhalations, (h)'s are inserted in the utterance.
.hh Audible inhalation; the more h's, the longer the inhalation.
<text> Indicates that the enclosed speech was delivered more slowly than usual for the speaker.
>text< Indicates that the enclosed speech was delivered more rapidly than usual for the speaker.
(text) Speech that was unclear or in doubt in the transcript.
((text)) Annotation of nonverbal activity.

Fig. 2 Menu on the counter

not necessarily follow that the action of placing an order will be conducted in an appropriate manner. Existing studies of service counter interactions tend to begin their discussions from the point at which interactions (placing an order, etc.) have already begun [10]. In this study, we examined how a common interactional space [11] that renders the action of placing an order appropriate is created. The results revealed that the common interactional space is created in a way that renders the action of placing an order while referring to the menu appropriate.

In Extract 1 below, at the moment the customer enters the shop, he is greeted with the words, "Welcome" (Line 01) and directed toward the counter, where a staff member is waiting. As shown in Fig. 2, at the counter, there is a menu that is positioned close to and facing the customer. We used the words "positioned close to the customer" and "facing the customer," and it should be noted that this phenomenon is itself achieved through an interactive process. To explain this in more detail, at the moment of entry into the shop, the customer is beckoned toward the counter. The presence of the counter—a physical structure that prevents the customer from advancing any farther—and the fact that it is positioned in such a way as to ensure that the distance between the customer and the staff member is appropriate for face-to-face conversation require that the customer come to a halt in the position shown in Fig. 3. The resulting "positions" of the customer and staff member and the "postures" they adopt within these positions in turn create a condition in which the menu is positioned close to the customer and facing the customer.

Extract 1

Mr. AO (uses the shop approximately once a month)

((Mr. AO enters the shop))
01 Staff: ↑いらっしゃいませ:
↑irasshai mase
↑Welcome:
((the moment Mr. AO enters the shop))

02 (1.2)
((Mr. AO approaches the counter on his right))
03 Staff: 店内でおめしあがりですか¿=
tennnai de omeshi agari desu ka¿=
Will you be eating in¿=
04 Mr. AO: =はい
=hai
=Yes

In Figures 2 and 3, we see that the staff members' gazes are facing away from the customers. The staff prompt the customers to place orders while waiting in postures that allow them to enter information on their display panels. Instead of interacting with customers face to face, the staff shows a certain degree of disengagement [12] from the customers by averting their stances and gazes. As far as we observed, staff, while displaying such physical disengagement, prompted customers to place orders and created a situation in which the menu was positioned close to the customer and facing the customer, thereby encouraging customers to order from the menu. In response, the customers indicated their understanding of how they should order from the menu by placing their hands on the counter and perusing the menu. During the process of placing orders, customers would often point to the menu and state the name of the product.

Menu-based orders are by no means automatic. They are achieved through an interactional process, specifically, by the staff member's action of directing the customer and by the placement of an object (the menu).

Fig. 3 Service encounter

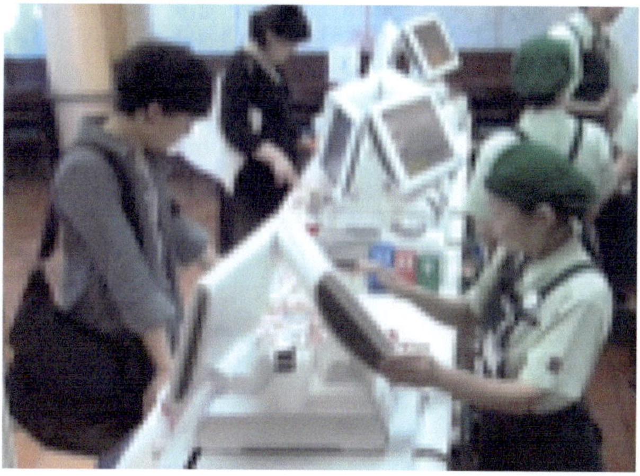

4.2 Using the Menu to Indicate Degree of Understanding

Next, we investigated (2): Assuming that customers indicate the kinds of people they are through using the menu, how do they do this?

If it is the case that the menu presents the shop's body of cultural knowledge and that it can therefore be used as a benchmark to determine the extent to which a customer is familiarized, how does this happen specifically?

First, as argued in (1), the menu is presented to the customer as a boundary object. In other words, in the position of the first interaction with the staff member after entering the shop, the customer is instructed to undertake what is a core action in catering establishments, namely, using the menu to place an order, and this is precisely why the menu can be used as an object that links the customer to the shop's cultural knowledge. At this point, the customer, through the action of placing an order, shows his or her degree of "distance" from the body of cultural knowledge reflected in the menu and thereby conveys the extent to which he or she is familiar with that body of knowledge.

An examination of the data revealed that in cases in which customers were not familiar with the shop's system, they would use the menu to indicate their levels of understanding of the system with one of the three methods described hereunder. We will explain these methods in order.

4.2.1 Using the Product Number (Mr. IS)

Customers can convey their unfamiliarity with the shop's system by placing an order in such a way that they deliberately refer to the product number despite being able to refer to the product name.

In Extract 2 below, after initially saying "XX Cheese" (Line 02), Mr. IS refers to the product again—this time by its number, saying "the number three" (Line 03). It is likely that by referring to the number printed on the menu, Mr. IS indicated his level of understanding (i.e., his lack of familiarity with the shop's system).

Extract 2

Mr. IS (a first-time customer to this hamburger outlet)

01 Staff: ご注文お決まりでしたらどうぞ
gochuumonn okimari deshitara douzo
Are you ready to place an order

02 Mr.IS: ん::(6.0)え: XX<チーズ>:
nn::(6.0)e: XX<chi-zu>:
Hmm:: (0.6) We:ll Can I have the XX<Cheese>:

03 Staff: はい[XX]
hai [XX]
Yes [XX]

04 Mr.IS: [三番]はい
[sannbann] hai
[Number three] right.
05 Staff: はいXXチーズ(.)三番で
hai XX chi-zu(.)sannbann de
OK XX Cheese (.) Number three

What is important here is not simply the fact that he referred to the product number but the fact that despite initially referring to the product by its name, saying "XX <Cheese>" (Line 02), he deliberately switched to the product number ("number three"). It was actually the case that product numbers "3" and "11" were both included in the section of the menu that displayed the "XX Cheese" product. Number 11 was the "WXX Burger." The customer most likely quoted the number so as not to confuse these two products. However, once the customer deliberately switched his statement to "number three" (XX Cheese), the staff member was able to infer the customer's uncertainty regarding the shop's ordering system.

4.2.2 Using Excessive Referring Expressions (Ms.GA)

When placing orders, customers frequently point to the menu and use demonstrative pronouns. There are cases when these referring expressions are used multiple times in a clearly redundant way.

In Extract 3 below, the customer and the staff member are focusing their attention together on the menu (joint attention.) The customer points to the menu and uses the referring expression "this" before stating the product name (Line 12.)

Extract 3

Ms.GA (comes to the shop approximately twice a year)

10 GA: ど>しょっかな<
do>shokkana<
Hmm>What shall I have<
11: (5.0)
12 GA: じゃあ(.)この¿(0.2)<XXバーガー>(0.2)の::
jaa(.)kono¿(0.2)<XX ba-ga->(0.2)no::
OK (.) This¿ (0.2)<XX Burger>(0.2) with::
((GA says this while pointing to XX Burger in the menu))

The action of pointing is generally undertaken to establish joint attention. For example, if someone points to the moon while saying, "Isn't the moon beautiful tonight?" the listener will look at where the speaker is pointing and establish joint attention upon the "moon." At this point (in the conversation extract), the customer and staff member are both looking at the menu. Therefore, joint attention to the menu itself has already been established. When the customer placed her order while pointing to the menu, the staff member would have understood not that the customer was pointing to the menu itself but that she was pointing to a particular product displayed on the menu. However, the essential point here is that the staff member originally instructed the customer to place an order by using the menu (Sect. 1). It can therefore be argued that the staff member behaved as someone who should as a matter of course know not only that the menu is placed where it is but also what products are displayed on the menu and in what positions they are displayed. Therefore, the act of pointing to the product on the menu under such a condition would suggest that the staff member should also know the whereabouts of what is being pointed at. Moreover, the customer uses speech to indicate the product by using the demonstrative pronoun "this." At this point, with joint attention to the desired product having already been established, the customer supplements his pointing action by the referring expression "this," thus making multiple referring expressions. It can therefore be understood that the referring expressions are excessive.

The use of "excessive" referring expressions can be a means by which customers convey their uncertainty, similar to the case shown in 4.2.1.

4.2.3 Using Expressions that Indicate Uncertainty (Ms. IH)

To this point, we have seen how customers indicate their uncertainty regarding the act of placing an order as a means to convey their lack of familiarity with the shop's system. Occasionally when placing orders, customers express their uncertainty directly. In Extract 4 below, the customer refers to the product name with the referring expression "this" and then adds the words "...or whatever it's called" to indicate his uncertainty (Line 09). This direct expression of uncertainty can be a means for a customer to indicate his or her level of understanding regarding the shop's system.

Extract 4

Ms. IH (a first-time customer to this hamburger outlet)

07 Ms.IH:	°ん:::::° °nn:::::° °Hmm:::::°
08 Ms.IH:	(7.5) ((During this pause, the staff member takes out a tray))
09 Ms.IH:	えと(.)このなつなの照り焼きチキンって[いう]のを: eto(.)kono natsuna no teriyakichikinn tte[iu]nowo: Uh (.) This Natsuna Teriyaki [Chicken] or whatever it's called: ((IH says this while pointing at the menu))
10 Staff:	[はい] [hai] [OK]

なつみの照り焼き[チキンが]おひとつ
natsumi no teriyaki[chikinn ga] ohitotsu
One Natsumi Teriyaki [Chicken]
11 Ms.IH: [なつみか]
[natsumika]
[Natsumi]
12 Ms.IH: て(.)いうのと:(0.2)あと(.)このシュ::ポットシエイク[の:]
te(.)iunoto:(0.2)ato(.)konoshu::pottosheiku [no:]
That (.)and: (0.2) and I'll also have (.) this Choux:: Pot Shake
13 Staff: [はい]
[hai]
[OK]
14 Ms.IH: バニ[ラ(0.5)]ひとつ
bani[ra(0.5)] hitotsu
Vani [lla (0.5)] one please
15 Staff: [バニラで]
[Vanilla]
16 Staff: はい(0.2)ありがとうございます
hai(0.2)arigatou gozaimasu
OK (0.2) Thank you very much sir
((The staff member is looking at the monitor in order to register information))

Attention should be drawn to how the expression "or whatever it's called" can be understood as an expression of uncertainty because the action of placing an order by using the menu is established as an appropriate action. If the customer had recited "Natsuna Teriyaki Chicken" without the menu being present, then adding the words "or whatever it's called" would have had little to do with his level of understanding regarding the shop's system. Rather, the reason it is possible to understand the customer's uncertainty regarding the shop's system is that the customer used the qualification "or whatever it's called" *while looking at the menu and placing an order*, and *under a condition in which the product name was printed on the menu*.

4.3 How Staff Respond When Customers Indicate their Levels of Understanding

Heretofore, we have examined the ways by which customers indicate their uncertainty in a manner that is embedded in the action of placing an order. How does staff respond to these methods? The responses of staff confirm that the analysis we have so far conducted of customer behavior is not simply the result of the analyzers' own interpretations but is instead based on the comprehension of the participants at the scene.

The data show that when customers convey their uncertainty using the menu, staff may add a thorough explanation that makes reference to the menu. In the case of Mr. IS, the staff member says, “You can select the drink you want from the section on the right” (Extract 5, Line 10), and in the case of Ms. GA, the staff member says, “There are four combinations available” (Extract 6, Line 16). The staff treat their counterparts as unfamiliarized customers and thoroughly instruct the customers in how to use the menu.

Extract 5

Mr. IS

08 IS: はい,のポテトSセット:
hai,no potato esu setto:
Uh, the potato combination S:
09 Staff: ポテトSセット,かしこまりました.
potato esu setto,kashikomarimashita
Potato combination S, no problem
10 Staff: お飲み物を右の欄からお選びください
onomimono wo migi no ran kara oerabi kudasai
For your drink, you can select the drink you want from the section on the right.
((staff member runs a finger down the drink section))

Extract 6

Ms.GA

14 GA: えっと(0.5)セット?(0.5)[で::]
etto(0.5)setto?(0.5)[de::]
Erm (0.5) can I have it as a combination? (0.5) [please::]
15 Staff: [セット]
[setto]
[combination]
16 Staff: 4種類ございますが[°どうな]さいますか°
yonn shurui gozaimasu ga [°douna]saimasuka°
There are four combinations available, °Which would you like°
((staff member runs a finger down the combination section))

Naturally, the fact that these explanations are thorough means that there must also be explanations that are not thorough. When customers order a product from a combination menu and it is apparent that they have not finished ordering, staff will normally prompt the customer to continue the order with such utterances as, “Please choose a combination” or “What drink would you like with that?” Compared with these utterances, the staff in Extracts 5 and 6 offer more detailed explanations, with utterances such as, “You can select the drink you want from the section on the right” and “There are four combinations available.” Moreover, when prompting the customers to continue their orders, the staff ran their fingers down through the drinks (Extract 5) and combinations (Extract 6) sections. Through these actions, the staff explained in a thorough manner about selecting products from the menu. It may be considered that the staff members behaved in this way to indicate that they understood the customers' uncertainty regarding the shop's system. Undertaking such thorough explanations will enable even unfamiliarized customers to behave in a manner appropriate to the scene.

In Extract 4 however, after the customer conveys his uncertainty, the staff member does not particularly demonstrate that he has recognized the customer's uncertainty. The reason for this discrepancy is found in the environment of the sequence of utterances in which the customer was being prompted. In Extracts 2 and 3, once the customer in Extract 2 selected the combination (Line 08) and once the customer in Extract 3 conveyed that she wanted the combination (the 0.5-s pause in Line 14), the conversation ceased temporarily, forming a break in the order of utterances [13], which is formed from a number of utterance sequence units. Therefore, the staff were able to use these breaks as opportunities to prompt the customers to order a drink or a combination. In Extract 4, in contrast, after ordering the "Natsuna Teriyaki Chicken," the customer promptly continued his order by adding the connective expression "that and" (Line 12). Therefore, the staff member had no opportunity to add an explanation by prompting the customer's order. In other words, staff can recognize the customer's uncertainty and add a thorough explanation only in cases in which the interaction is appropriate for adding such an explanation.

5 Discussion: The "Fast Food Characteristics" of a Fast Food Restaurant

The purpose of this study is to illustrate how it is possible for the interactants to know from the outset whether their counterparts have learned the scripts and mastered their roles, in interactions between mutual strangers. In this study, we focused on the process of menu-based ordering in actual service encounters. On the assumption that menu-based ordering is achieved interactionally by the staff member's directing of the customer and the placing of the menu. Inasmuch as it presents a body of cultural knowledge, the menu is a resource that instructs customers on the appropriate behavior for placing an order, and it functions as a boundary object that links the shop to the customer.

Analysis result shows, in cases in which customers were unfamiliar with the shop's system, we observed them using the menu to indicate their levels of understanding of the system. We observed three methods of doing this: using the product number, using excessive referring expressions, and using expressions that indicated uncertainty.

What is noteworthy about the responses of staff members is that once they understood the kinds of customer they were serving, they always indicated their understanding embedded in the course of the action of placing an order. In other words, it was not the case that staff halted the flow of the action of placing an order and then issued explanations of the shop's system. To state it differently, both the customer and the staff member formed, embedded in the course of the action of placing an order, an understanding of the "kind of customer" each customer was. In this way, the customers and staff appeared to place priority on advancing the flow of the action of placing an order and, concomitant with this flow, negotiating an understanding of each kind of customer.

It is likely that fast food outlets prioritize the fluidity of the action of placing an order in this way in order to highlight the "fast food characteristics" of their restaurants. In other words, customers and staff (inasmuch as they are concerned with progressing the order) use subtle methods to negotiate an understanding of the type of customer, and this will be reflected in what on the surface appears to be a problem-free progression of the action of placing an order. In cases in which staff did not have an opportunity to give thorough explanations, the fact that they did not recognize the customers' uncertainty may have been related to the fast food characteristics of this fast food outlet.

6 Summary

In this study, we used conversation analysis [1], a method based on the sociological approach of ethnomethodology [2], to examine service encounters (interactions between service providers and consumers) in a fast food restaurant. Previous studies have tended to view this standardization and routinization as a top-down or automatic process in which interactants master their respective roles through learning scripts [4]. However, what in fact occurs during service encounters in fast food restaurants is the achievement of an exchange between mutual strangers who are meeting in a chance encounter. The results revealed that both the customer and the staff member formed, embedded in the course of the action of placing an order, an understanding of the type of customer each one was. It is likely that fast food outlets prioritize the fluidity of the action of placing an order in this way in order to highlight the "fast food characteristics" of their fast food restaurants.

The number of these participants is not enough, therefore we are going to plan additional research to raise analysis precision.

References

1. Sidnell J, Stivers T (eds) (2012) The handbook of conversation analysis, Vol. 121. John Wiley & Sons

2. Garfinkel H (1967) Studies in ethnomethodology. Printice-Hall, Englewood Cliffs
3. Ritzer G (1996) The McDonaldization of society (Japanese translation: Masaoka K. Waseda University Press)
4. Solomon MR, Surprenant C, Czepiel JA, Gutman EG (1985) A role theory perspective on dyadic interactions: the service encounter. J Mark 49(1):99–111
5. Vargo SL, Lusch RF (2004) Evolving to a new dominant logic for marketing. J Mark 68 (1):1–17
6. Czepiel JA (1990) Managing relationships with customers: a differentiating philosophy of marketing. In: Bowen DE, Chase RB, Cummings TG (eds) Service management effectiveness. Jossey-Bass, San Francisco, pp 299–323
7. Surprenant CF, Solomon MR (1987) Predictability and personalization in the service encounter. J Mark 51(2):86–96
8. Yamauchi Y, Hiramoto T (2013) Sushiya ni Okeru Chumon no Tekisetsusei o Meguru Yaritori – Kaiwa Bunseki o Mochiite—(Exchanges in a sushi restaurant that revolve around the appropriateness of placing an order: a study using conversation analysis). The first international conference on serviceology
9. Star SL, Griesemer JR (1989) Institutional ecology, 'Translations' and boundary objects: amateurs and professionals in Berkeley's museum of vertebrate zoology, 1907–39. Soc Stud Sci 19:387–420
10. Clark C, Pinch T (2010) Some 'Major' organizational consequences of some 'Minor' organized behaviour: a video analysis of pre-verbal service encounters in a showroom retail store. In: Llewellyn N, Hindmarsh J (eds) Organisation, interaction and practice. Cambridge University Press, Cambridge, pp 140–171
11. Mondada L (2009) Emergent focused interactions in public places: a systematic analysis of the multimodal achievement of a common interactional space. J Pragmat 41:1977–1997
12. Goodwin C (1981) Conversational organization: interaction between speakers and hearers. Academic, New York
13. Schegloff EA (1982) Discourse as an interactional achievement: some uses of "Uh Huh" and other things that come between sentences. In: Tannen D (ed) Analyzing discourse: text and talk, Georgetown university roundtable on languages and linguistics. Georgetown University Press, Washington, DC, pp 71–93

Evaluation of Taxiing at a Large Airport Considering Customer Satisfaction

Hiroya Daimaru, Satoshi Shimada, Shinsuke Shimizu, Jun Ota, and Tatsunori Hara

Abstract At large airports, aircraft take-off and land simultaneously on multiple runways. As a result, taxiing is a complex and lengthy process. To support in planning aircraft taxiing, we have developed a taxiing model and simulated the taxiing time at Narita International Airport so far. On the other hand, such objective taxiing time should be evaluated from the customer's (passenger's) perspective. This paper constructs mixed distribution of perceived waiting time of customers based on a customer survey. Distribution of customer satisfaction is estimated through simulation by application of a method of evaluating customer satisfaction. As a result, possible taxiing strategies and issue on synthesis from the viewpoint of customer satisfaction are discussed.

Keywords Customer satisfaction • Simulation • Design

1 Introduction

Today, the aviation industry is promoting the increase of flights at large airports, aiming at progress for further globalization. This trend is expected to bring about delays of aircraft taxiing in taking-off and landing. Aircraft taxiing means a movement of aircraft from a runway to a gate or from a gate to a runway (Fig. 1). Airport planning and scheduling activities is to optimize the use of infrastructure and the cost of operation [1]. Improving an aircraft taxiing system is a complex synthesis problem that includes both issues of design optimization [2] and adaptation to dynamic environment [3]. Dealing with this problem is indispensable to relieve delays of aircraft taxiing and to enable additional flight increases. Resolving delays of aircraft taxiing does not simply mean the reduction of total taxiing time: the degree of satisfaction or dissatisfaction felt by passengers because of delays in

---

H. Daimaru (✉) • S. Shimada • J. Ota • T. Hara
RACE(Research into Artifacts, Center for Engineering), The University of Tokyo, Chiba 277-8568, Japan
e-mail: daimaru@race.u-tokyo.ac.jp

S. Shimizu
ANA Strategic Research Institute Co., Ltd, Tokyo, Japan

T. Maeno et al. (eds.), *Serviceology for Designing the Future*,
DOI 10.1007/978-4-431-55861-3_5

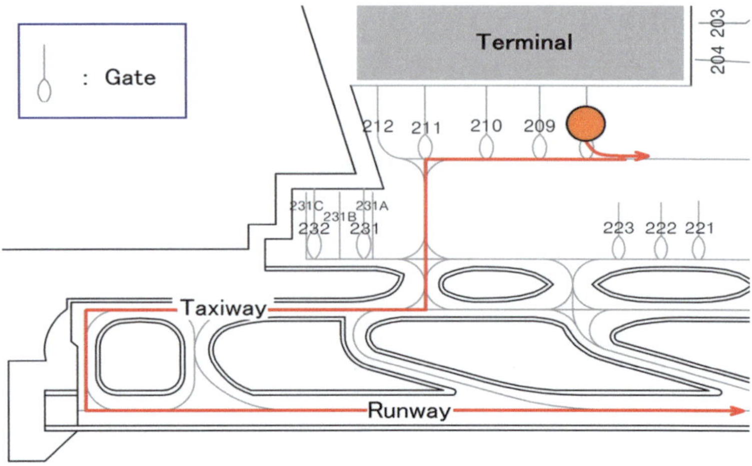

Fig. 1 Aircraft taxiing route (case of departure)

taking-off and landing is an important performance index. Accordingly, this study is conducted to evaluate aircraft taxiing according to customer satisfaction.

The rest of this paper is organized as follows. Section 2 reviews previous studies on taxiing simulation and design of aircraft. Section 3 refers problem settings and illustrates a method of estimating customer's satisfaction. In Sect. 4, we employ the method to large-scale airport. Section 5 discusses its result, and possible taxiing strategies, and issue on synthesis. Section 6 concludes the paper.

2 Preceding Studies on Aircraft Taxiing Simulation and Design

2.1 Preceding Studies on Operations in Airport

Olaru et al. reviews simulation and design studies on airports [1]. Simulation represents a standard tool for modeling stochastic queuing systems. According to [1], simulation studies on airports focus on either particular sub-system in the airport operation (traffic control, passenger vs baggage flows, modeling of airspace, airfield, or terminal operations) or treat the whole airport as an entity. However, a review of the publications in airport modelling reveals very few applications combining simulation with optimization occurs for enhancing the modelling results.

2.2 Preceding Studies on Aircraft Taxiing

Compared to studies on other operations in airport, less preceding studies have examined the aircraft taxiing system. Jung et al. specifically examined the number of times and duration of stopping caused by a take-off wait for runway clearance to reduce the number of average stops of outbound aircraft from 7 to 4 and to shorten the stopping time [4]. Kariya et al. reduced the waiting time in front of the runway by equalizing the arrival intervals of outbound aircraft to a runway for departure using queuing theory, thereby shortening the average and maximum taxiing times of outbound aircraft by 16.5 % and 7.8 %, respectively [5]. Those earlier studies were conducted from the viewpoint of increased efficiency of taxiing, such as the number and duration of stops and the average and maximum taxiing time attributable to take-off waiting for runway clearance. Regarding other evaluation indices and more algorithmic design approach, Yahagi et al. examined simulation-based optimization to obtain simple and robust rules of taxiing operations by representing the taxiing system as a multi-agent system [6].

This paper investigates taxiing time more from customer's point of view. It is extremely important for implementing further globalization of the aviation industry to evaluate the degree of satisfaction or dissatisfaction of passengers with delay in taking-off or landing.

3 Approach and Method

3.1 Approach

This study examines a proposal of an aircraft taxiing system that incorporates passenger satisfaction and dissatisfaction as well as improvement in the efficiency of taxiing. The objective of this paper is to evaluate aircraft taxiing at a large airport by passenger satisfaction with take-off delay, which is fundamental for the target of this study: this study specifically examines take-off delays among delays in taking-off and landing of aircraft. Issues to be addressed for accomplishing this objective are the following.

- How can the degree of satisfaction of a passenger with taxiing be assessed?
- Individual needs of each passenger cannot always be fulfilled under a constraint that passengers with diverse needs are obliged to board the same aircraft. How can this point be addressed?
- How can the aircraft taxiing and the take-off delay be represented?

In the rest of this paper, the term "passenger" is described as the term "customer". Customer satisfaction (CS in short) is studied as an indicator of manufacturing system or service. In general, CS is evaluated in two approaches: to do questionnaires directly to customers as a part of marketing activity, and to estimate CS based on service performance as an engineering activity. Based ideas in the latter approach are represented as nonlinear satisfaction functions. In this study, such a nonlinear satisfaction function that presumes CS with delay is adopted for the first issue, although a satisfaction function for each aircraft is computed according to the mixed distribution of passengers' expectation with take-off delay. Moreover, this study uses the take-off delay of each aircraft computed by the aircraft-taxiing simulator of Narita International Airport [5] as input data.

3.2 Satisfaction Function with Delay of Service Element

Methods of estimating CS include a satisfaction function with each element of service [7]. This approach expresses a satisfaction function with Eq. (1):

$$S_v = a\left\{1 - exp\left(-\frac{b}{a}(v - c)\right)\right\} \tag{1}$$

Varying the sign of a in the ranges of $v > c$ and $v < c$ yields a graph as Fig. 2b, which exhibits a case in which CS decreases monotonously according to the value of the factor. Application of the Prospect Theory [8] in behavioral economics is assumed at the decision of the value of a, in which the absolute value is generally set greater for a- of negative satisfaction than for $a+$ of positive satisfaction. It is noteworthy that this makes Fig. 2b asymmetrical.

The variable v representing the value of the element (factor) of service is the input of this function, and S_v, the degree of satisfaction of a customer to that factor, is computed as an output. Parameters a, b, and c respectively denote the maximum/minimum of the degree of satisfaction with the factor, the slope of the function at an inflection point, and the expectation value of a customer over a factor. A satisfaction function with a certain service factor is computed by determining these parameters to the factor. CS is determined by difference (v-c) between the quality of actual service and the expectation value thereof, and becomes ± 0 when the value of a factor is in agreement with expectations, as expressed by parameter c.

Yoshimitsu, Hara et al. [9, 10] address the correspondence of the satisfaction function described above with the Kano model [11], in which the degree of satisfaction is defined in both positive and negative regions when a service factor as an object for evaluation is categorized into the one-dimensional quality in the Kano model (Fig. 2b), although it is defined in the negative region only in case of the must-be quality (Fig. 2a), and in the positive region only in case of the attractive quality in the Kano model (Fig. 2c).

This study introduces a satisfaction function with take-off delay in aircraft taxiing as a service factor (i.e., td_{act} in Eq. (2)). This take-off delay is assumed to have a must-be quality in the Kano model, so that Eq. (1) is applied for $v > c$, although the degree of satisfaction is always 0 in $v < c$ as in Eq. (2). The procedure

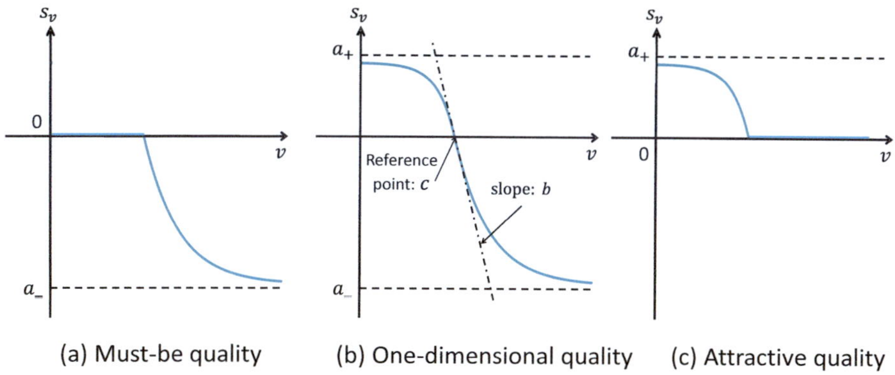

Fig. 2 Graph of satisfaction function according to the Kano model (in the case of monotonously decreasing and $v > 0$)

for determining the expectation value of a customer c (i.e. td_{expt} in Eq. (2)) is discussed in Sect. 4.2.

$$S_{td} = \begin{cases} a\left\{1 - exp\left(-\frac{b}{a}\left(td_{act} - td_{expt}\right)\right)\right\} & , td_{act} \geq td_{expt} \\ 0 & , td_{act} < td_{expt} \end{cases} \tag{2}$$

3.3 *Computation Procedure of Satisfaction Function for Each Aircraft*

A satisfaction function should be computed not for every customer but for every aircraft in actual evaluation of a taxiing system because take-off delays take place not per customer but per aircraft. Customers with different expectations c are mixed in an aircraft, and satisfaction functions also vary among customers. Therefore, the satisfaction function of an aircraft is determined by the linear combination with probability weight of satisfaction functions of customers under the premise that the distribution of expectation values of customers c are obtainable for each aircraft in this study. Let the satisfaction function of customers with expectation value td_{expt} on a certain aircraft be $S_{td}\left(td_{act}, td_{expt}\right)$, and the fraction of such customers be $P\left(td_{expt}\right)$, where $\sum P\left(td_{expt}\right) = 1$. Then S_{mix}, the satisfaction function of the aircraft as the linear combination with probability weight of the satisfaction function of customers given by Eq. (3):

$$S_{mix}(td_{act}, td_{mix}) = \sum_{td_{expt}} P(td_{expt}) * S_{td}(td_{act}, td_{expt}) \quad (3)$$

4 Case Study for Large-Scale Airport

4.1 Computation of Take-Off Delay by Aircraft Taxiing Simulator

This study employs the aircraft taxiing simulator of Narita International Airport [5] to compute the take-off delay of an aircraft. This simulator models the relation between movement and spacing of aircraft during taxiing. It has been applied to Narita International Airport. It uses a real or assumed service diagram as an input, and visualizes a departing aircraft taxiing from its parking spot to a runway and taking off thereafter. An arriving aircraft reaches its spot from a runway on a map (Fig. 3). Furthermore, the taxiing time of each aircraft is computed. This simulation has been verified as capable of reproducing actual taxiing with accuracy parameter of 1 minute, so that the congestion of taxiing aircraft is apparent at rush hours of the diagram, just as at real Narita Airport. This study defines the take-off delay of each aircraft as the taxiing time under an environment with absolutely no congestion subtracted from that at taking-off computed with this simulator.

4.2 Estimation of Customer's Expectation with Take-Off Delay

Data summarization was conducted to obtain the distribution of td_{expt}, the expectation values of customers for each aircraft. Employed data were collected from

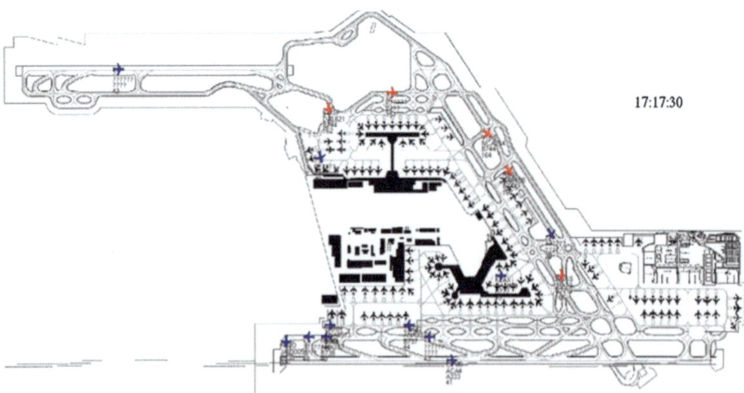

Fig. 3 Aircraft taxiing simulation in Narita International Airport [5]

international airline users in April–June 2010 (irrespective of carrier) in questionnaire format on a web site after travel. All questions were fixed-choice. In all, 1753 replies were collected, from which 1074 datasets of international airline users who departed from Narita International Airport were used in this study. Some questionnaire items are described below.

- Customer data of users
- Information on use, such as departing airport, arriving airport, carrier, and seat class
- Questions on delay in taking-off and landing, etc.

One question on delay in taking-off and landing asks a user to "answer on allowable departure delay in an international flight" by selecting one of seven choices of "less than 5 min, less than 10 min, less than 15 min, less than 30 min, less than 45 min, less than 60 min, and can not determine." This study adopts users' replies to this item as the expectation value td_{expt} of customers on take-off delay: a user is considered not dissatisfied if the delay is within the answered limit (degree of satisfaction = 0), although the customer is assumed to be dissatisfied with an excess delay over the limit in monotonous increase.

4.3 Computation of Satisfaction Function of Each Aircraft

The satisfaction function of each aircraft is computed using the expectation values of customers on take-off delay td_{expt} defined in Sect. 4.2. Aircraft were classified according to the destination airport. The degree of satisfaction was determined for each class in this study. Specifically, the distribution of expectation value of customers td_{expt} according to the destination airport was computed based on the questionnaire described in Sect. 4.2. The satisfaction function of aircraft was computed according to the destination airport using Eq. (3) in Sect. 4.2. Regarding "aircraft bound for a certain destination that exists in a service diagram as a destination but does not exist in the choices of the questionnaire", a satisfaction function was computed in the following fashion: distribution of the expectation value of customers c is computed according to flight time using the data given in Sect. 4.2. As a result, the former fashion covers about 70 % of aircraft, while the latter covers about 30 %.

The distribution of the expectation value td_{expt} of a close flight time was applied. a and b in Eq. (2) were set as -1 and -0.1, respectively here. Figure 4 presents results of cross tabulation of the expectation values of customers according to distribution airport. As a measure of the degree of satisfaction S_{td}, when take-off is more delayed by 10 min than a certain expectation (e.g., $S_{td}(20, 10)$), $S_{td} = -0.632$ ($-1 \leq S_{td} \leq 0$; the greater S_{td} is, i.e., the smaller its absolute value is, the higher the degree of satisfaction is). Figure 5 overviews variations of satisfaction functions on take-off delay according to destination airport.

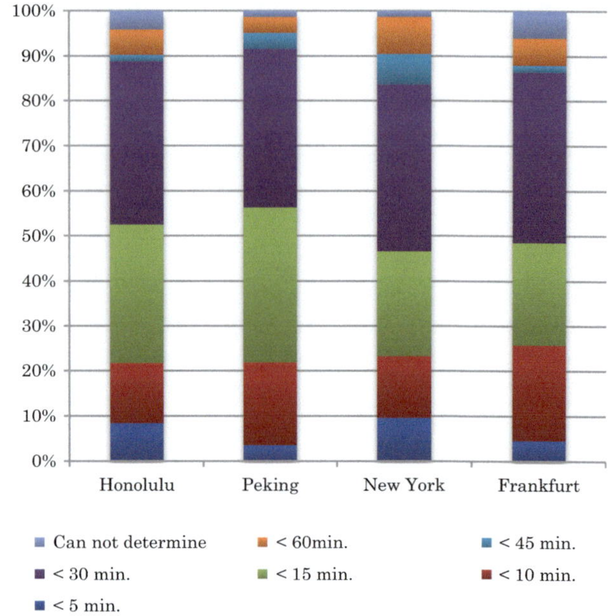

Fig. 4 Cross tabulation of expectation value of passengers on take-off delay according to destination airport

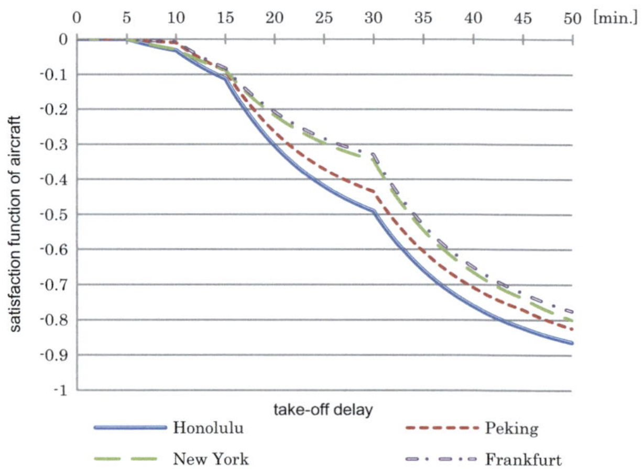

Fig. 5 Satisfaction functions on take-offf delay according to destination airport

Superposition of satisfaction functions that have different expectation parameters can be seen.

4.4 Result

The satisfaction function of aircraft with take-off delay S_{mix} was applied in this study to two types of service diagrams with congestion at a different level in Narita

International Airport; a case crowded rather than normal (109 departing aircraft in 14:00–19:00, hereinafter designated as Diagram 1), and a very crowded case (147 aircraft in the same conditions, hereinafter designated as Diagram 2). Diagram 1 corresponds to a situation in 2012, while Diagram 2 a situation in 2013 after expansion of the airport.

Table 1 shows the maximum, third quartile, median, first quartile, minimum, mean, and standard deviation of taxiing time and degree of satisfaction S_{mix} in service diagrams of two types. Figure 6 present scatter plot of taxiing time and the degree of satisfaction with take-off delay S_{mix}, covering partial data ranged from minimum to 1st Quartile satisfaction. Size of circles represents flight time of aircraft

Table 1 Taxiing time and degree of satisfaction

	Taxiing time (old evaluation index in previous study [5])		Degree of satisfaction for each aircraft (S_{mix}) (new evaluation index in this study)	
	Diagram no. 1	Diagram no. 2	Diagram no. 1	Diagram no. 2
Maximum	2565	3499	0	0
3rd quartile	1259	1527	0	0
Median	1071	1245	0	−0.007
1st quartile	870	1047	−0.018	−0.045
Minimum	550	550	0.62	−0.82
Mean	1119	1317	−0.035	−0.068
Standard deviation	351	445	0.098	0.14

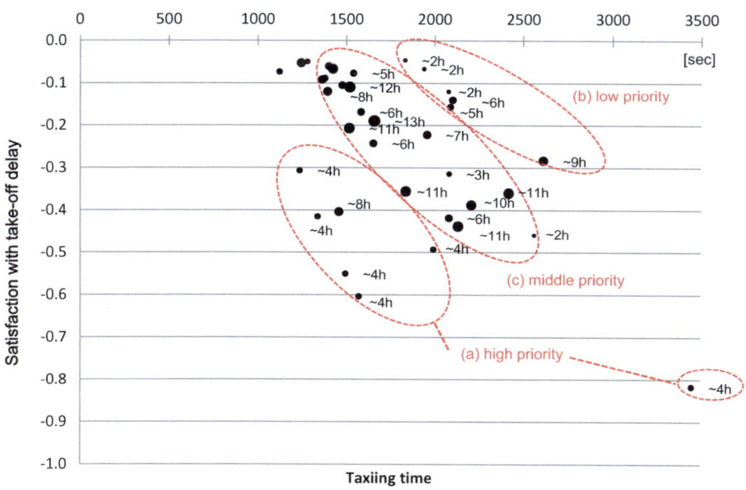

Fig. 6 Scatter prot of taxiing time and satisfaction of take-off delay (only minimum to 1st Quartile satisfaction data)

5 Discussion

5.1 Possible Taxiing Strategies

Regarding the simulation result described in Sect. 4.3, both median and mean of taxiing time in Diagram 2 are about 3 min. longer than those of Diagram 1. However, the approximate upper half of aircraft in terms of satisfaction in both Diagram 1 and Diagram 2 do not feel dissatisfaction (i.e., $S_{td} = 0$), while a marked increase of dissatisfaction is seen in lower 25 % of aircraft. Aircraft plotted in Fig. 6a shows a marked increase of dissatisfaction compared to that of taxiing time, so that urgent attention on these aircraft's taxiing would be necessary. Furthermore, 3–4 h aircraft are dominant in this group. On the other hand, aircraft plotted in Fig. 6b would have low priority because they show less increase of dissatisfaction according to that of taxiing time. Aircraft plotted in Fig. 6c are regarded as middle-priority targets to improve. This is because sensitivity of their taxiing time to dissatisfaction is between that of (a) and (c).

As discussed in the above, we obtained some priority strategies for further improvement of aircraft taxiing by introducing CS as an evaluation index instead of objective taxiing time. These priority strategies can be used for rules generations of taxiing operations in simulation-based optimization described in Sect. 2.2.

5.2 Issue on Synthesis from the Viewpoint of CS

This study determined a mixed distribution of customer expectation based a certain customer survey for analysing CS. However, customer's expectation is not static in nature, and it is formed and likely to be updated according to product/service contents customer has experienced. This means that satisfaction functions also may changes overtime. To address issue on synthesis driven by satisfaction function beyond just analysis, it is necessary to incorporate such dynamics of customer expectation and satisfaction function as a result.

Among class problems of synthesis discussed in emergent synthesis approaches [3, 12], Class II type problem (complete specification and incomplete environment) is to cope with the dynamic properties of the unknown environment, while in Class I type problem (complete specification and complete environment) the emergence of the solution is not time related. In other words, Class II type problem is a dynamic Class I type problem [12].

In the above context, Fig. 7 examines an approach to Class II problem of synthesis driven by the viewpoint of CS. What this paper studied is evaluation of CS based on customer expectation based on the solution for current transaction as shown in the upper part of the figure. Update of customer expectation as a part of environment change based on the solution for further transaction. Furthermore, for example, it is expected to develop simulation-based optimization (or generation)

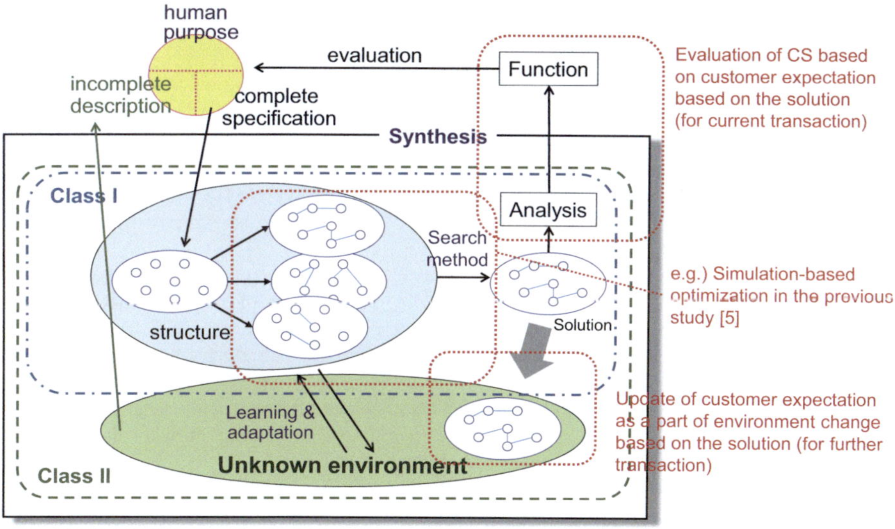

Fig. 7 An approach to Class II problem of synthesis from the viewpoint of CS (Modified from Ueda et al. [12])

that includes learning of- and adaptation to dynamic customer expectation as environmental information.

6 Conclusion

This study was conducted to evaluate aircraft taxiing at a large-scale airport by customer (i.e., passenger) satisfaction with take-off delay. The satisfaction function of customers with delay and distribution of customers' expectation value in each aircraft were adopted, and the satisfaction function with a take-off delay of each aircraft was computed. Furthermore, the application of satisfaction function to service diagrams of two types in Narita International Airport demonstrated that the estimated relation between two service diagrams varies with taxiing time and the degree of satisfaction. As a result, we obtained some priority strategies for further improvement of aircraft taxiing by introducing customer satisfaction instead of taxiing time.

Our future research will be aimed at more dynamic and adaptive design of coordination rules on aircraft taxiing by introducing update of customer's expectations as an environment change.

References

1. Olaru D, Emery S (2007) Simulation and GA-optimisation for modeling the operation of airport passenger terminals. In: Proceedings of the 29th conference of Australian institutes of transport research (CAITR). Adelaide, Australia
2. Roy R, Hinduja S, Teti R (2008) Recent advances in engineering design optimisation: challenges and future trends. CIRP Ann Manuf Technol 57(2):697–715
3. Ueda K, Markus A, Monostori L, Kals HJJ, Arai T (2001) Emergent synthesis methodologies for manufacturing. CIRP Ann Manuf Technol 50(2):535–551
4. Jung YC, Hoang T, Gupta G, Malik W, Tobias L (2010) A concept and implementation of optimized operations of airport surface traffic. In: Proceedings of the 19th AIAA aviation technology, integration, and operations conference. Fort Worth, Texas, USA
5. Kariya Y et al (2013) Modeling and designing aircraft taxiing patterns for a large airport. Adv Robot 27(14):1059–1072
6. Yahagi H et al (2013) Simulation-based simple and robust rule generation for motion coordination of multi-agent system. In: Proceedings of the 2013 I.E. international conference on systems, man, and cybernetics, pp 421–426. Manchester, UK
7. Kimita K, Shimomura Y, Arai T (2009) Evaluation of customer satisfaction for PSS design. J Manuf Technol Manage 20(5):654–673
8. Kahneman D, Tversky A (1979) Prospect theory: an analysis of decision under risk. Econometrica XVLII:263–291
9. Yoshimitsu Y, Kimita K, Arai T, Shimomura Y (2008) Analysis of service using an evaluation model of customer satisfaction. In: Proceedings of the 15th CIRP life cycle engineering seminar. Sydney, Australia
10. Hara T, Arai T (2011) Simulation of product lead time in design customization service for better customer satisfaction. CIRP Ann Manuf Technol 60(1):179–182
11. Kano N et al (1984) Attractive quality and must-be quality. Quality 14(2):39–48
12. Ueda K et al (2004) Emergent synthesis approaches to control and planning in make to order manufacturing environments. CIRP Ann Manuf Technol 53(1):385–388

Research of the Social New Transportation Service on Electric Full Flat Floor Bus

Toshiki Nishiyama

Abstract Author has developed a prototype large sized "Electric Full Flat Floor Bus" (ELFB). Taking the field test results into consideration as well, it is also revealed that ELFB has the specifications to meet most of the existing bus service requirements. Introducing the concept and technology of the integrated platform that motors, inverters, batteries and controllers are installed under the ELFB's floor can lower the minimum height from ground to floor, construct a full flat cabin, and improve the universal design performance. So it turned out that trial ride monitor 380 persons show high concern and a willingness to pay for introduction of the service which reduces resistance of a change or movement.

Keywords Electric Low and Full Flat Floor Bus (ELFB) • Universal design • Ecological design • Integrated platform • Public transportation services

1 Introduction

The application of electric car technology to public transportation is a short cut to the spread of electric cars. Above all, with the application of the technology to a big size city bus it is possible to protect the environment to give new service to passengers, and to make the quality of an electric bus gain wide publicity. An electric low and full flat floor bus is a vehicle which has made it possible to lower a floor, to get rid of exhaust gas, and to prevent noises by using electric car technology. Therefore, it is significant to carry out research on the application of electric car technology to a big size city bus. This research has revealed what citizens and bus companies in Japan think of the application of the electric car technology to a big size bus.

T. Nishiyama (✉)
Faculty of Urban Life Studies, Tokyo City University, Tokyo, Japan
e-mail: nishibus@tcu.ac.jp

T. Maeno et al. (eds.), *Serviceology for Designing the Future*,
DOI 10.1007/978-4-431-55861-3_6

2 Back Ground of This Research

2.1 The Development of a Grand-Up Type of Electric Car with Basic Auto Parts Concentrated Under a Floor

We have developed an electric car based on the "Integrated Platform" called "KAZ" and "Eliica" with basic auto parts, for example a lithium ion battery, a tandem wheel suspension system, concentrated under a floor. And a motor is kept in each of eight wheels (Figs. 1, 2, and 3). As a result, an electric car has more room and passengers can enjoy roominess. A previous sport car which can do more than 300 km an hour can accommodate only two passengers because of a big size engine. But "Eliica" can accommodate four passengers. Mere alteration of the form or size of an electric car makes it easier to apply the technology to buses, trucks, and so on. We have certified that a grand-up type of electric car rides better than an electric car now in use. We have tried to make the maximum speed of "Eliica" 400 km an hour. At present "Eliica" can do 370 km an hour. The quality of "Eliica" is very high speed and very high level of function of acceleration.

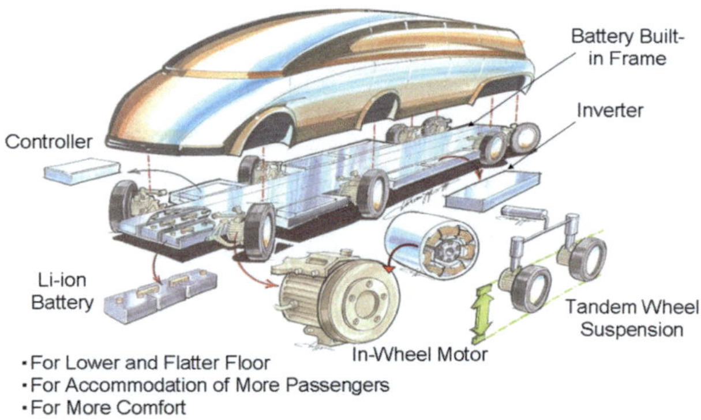

Fig. 1 Concept of integrated platform by Keio University

Fig. 2 "KAZ" (Based on integrated platform)

Fig. 3 "Eliica" (Based on integrated platform, next generation of "KAZ")

2.2 *More Interest in the Next Generation's City Buses Designed Universally and Ecologically*

Nowadays Japan is facing an aging society. We predict that people aged over 65 years will account for 25 % of the population of Japan by 2025. The number of physically-challenged people is increasing. In this aging society we are more and more interested in city buses. But recently buses have been facing serious problems about exhaust gas, noises and so on. But Electric Low and Full Flat Floor Buses are highly appreciated in accessibility and low emission. To solve these problems, we have developed an electric city bus. It is a means of transportation designed universally and ecologically. There is a growing need for a big size electric city bus which can accommodate more passengers [1–3].

2.3 *Greater Interest in Big Size City Buses Useful for Promoting the Spread of Electric Motor Buses in Japan*

Japanese people are getting more and more interested in the development of emission-low vehicles to cut down on carbon dioxide. Nowadays, Japanese municipalities are adopting a policy about assuming a part of the rent for a parking lot, which is paid by a driver of an emission-low car and are trying to spread emission-low vehicles all over Japan. Under the present condition of Japan, people are not very interested in an electric car because of its low efficiency involved in going up a slope. But electric cars are gaining wide publicity because a lithium ion battery can be charged more efficiently now. In spite of its publicity, electric cars have not come into wide use because it costs more to produce a lithium ion battery. But we have proved the high speed and the high acceleration of "Eliica". So, Japanese municipalities and various bus companies are showing greater interest in applying "Eliica" technology to city buses. Scientific societies are showing interest in "Eliica" technology, too. As a trigger for popularizing electric cars, applying the quality of a grand-up type of electric car to a big-sized city bus is worth noticing.

3 Development of Electric Low and Full Flat Floor Bus (ELFB) and the Purpose of this Research

3.1 R&D's Core Technology

The Electric Vehicle Laboratory of Keio University has been working on the development of EVs for years. The laboratory's basic concept is to build a dedicated platform for EVs from scratch, instead of the conversion type that involves retrofitting an engine with a motor. This innovative design technology, called integrated platform, stores all equipments required for operating a vehicle, such as batteries, motors (in-wheel motors), and inverters, beneath the vehicle floor. The application of this technology makes it possible to concurrently achieve expansion of usable cabin space in EVs, extension of the mileage on a single charge by using in-wheel motors, and a greater number of lithium-ion batteries, as well as improvement in universal design performance, which has a brisk demand (Fig. 1). Based on the assumption of using the above concept, the Electric Vehicle Laboratory of Keio University has developed a prototype ELFB (The Trial Electric Low and Full Flat Floor Bus) (Figs. 4, 5, and 6).

Fig. 4 The front-view of "ELFB"

Fig. 5 The rear-view of "ELFB"

Fig. 6 The interior of electric bus

3.2 Prototype ELFB by Keio University

Keio University has developed a prototype large sized ELFB. The development of ELFB was pushed forward based on a framework of the industrial sector (Isuzu Motors Limited), public sector (Kanagawa Prefecture and Kanagawa Bus Association, along its member bus companies that operate public transportation services), and academic sector (Keio University). In the development project, Keio University has worked on creating a practical bus by expanding the opportunities to exchange ideas with industrial and public sectors. In manufacturing a route bus to be used as a public transportation vehicle, it is important to construct a safe, secure, and strong body. Aluminum and polycarbonate are used for the body in order to reduce as much weight as possible. The newly developed direct drive in-wheel motors are mounted on the chassis in order to achieve high-efficiency vehicle operation without loss. The specifications of the large-size ELFB that was developed through this project are summarized in Table 1. The major concern of bus service companies regarding ELFB is the mileage on a single charge. As the results of a field test conducted in Yokohama City, Kanagawa Prefecture in Japan, mileage per charge

Table 1 Specification of the electric bus

Items	Specifications
Overall width	2490 mm
Overall length	10,050 mm
Overall height	2730 mm; 300–400 mm lower than conventional
Capacity	49 persons; 21 seated and 28 standing
Vehicle weight	11,800 kg; Gloss VW, 8600 kg; Vehicle Weight
Floor height	270 mm; barrier free
Range / charge	121 km > Average route bus range 120 km/day

was 121 km. The author of this article conducted an interview survey with 12 bus service companies that are the members of Kanagawa Bus Association. The survey revealed that vehicles of the member bus service companies run an average distance of about 120 km per day between departure from and return to the bus garage. Taking the field test results into consideration as well, it is also revealed that ELFB has the specifications to meet most of the existing bus service requirements (vehicle driving range required for a normal bus service and various other services provided by bus service companies). Introducing the concept and technology of the integrated platform can lower the minimum height from ground to floor, construct a full flat cabin, and improve the universal design performance. According to the Kanagawa Bus Association, the fuel cost per kilometer for the existing large size non-step bus (10.5 m overall length and 2.5 m overall width standard) is ¥38 (US $0.47). On the other hand, the prototype ELFB of the same large size can run on ¥8 (US$0.1) per kilometer (when using nighttime electric power). In other words, a fuel economy merit of ¥30 (US$0.38) per kilometer can be achieved by replacing the existing non-step buses with electric ones. The route buses run a distance of about 120 km per day and 300 days per year, which means using electric buses can result in an annual fuel cost reduction effect of ¥1.08 million (US$13,500) per vehicle. In addition, electric buses contain less number of components and thus, about 50 % of the remaining running costs (mainly components-related maintenance costs) after subtracting the fuel cost from the total running cost can be reduced from the present expenses. Calculations made based on this fact indicate that the amount of components-related maintenance costs that can be reduced annually is about ¥770,000 (US$9625) for a large bus. Therefore, based on the above calculations, approximately ¥1.85 million (US$23,125) can be reduced annually for each bus. The electric bus offers a great advantage to Japanese bus service companies because about 80 % of them are presently operating in the red. Furthermore, the amount of carbon dioxide emissions from existing large-sized internal combustion engine buses is 0.61 kg/km (FY2009 data of Ministry of Land, Infrastructure, Transport and Tourism). EVs do not emit carbon dioxide during operation. EVs can reduce 90 % of carbon dioxide emissions even when emissions during power generation are taken into consideration. In addition, EVs do not produce any noise when they are operated. Therefore, they are better for the environment in areas along the bus route.

3.3 The Purpose of Our Research Study

We made a demonstrated evaluation of running performance of an electric full-flat and low-flat floor bus (ELFB) which was made as a trial bus in a project carried out by the Ministry of Environment in fiscal year 2009. With this research in the 2013 fiscal year, the main point towards the validity and spread strategies of an electric low and full flat bus was summarized based on the test run from the 2011 fiscal year to the 2012 fiscal year. In this research, the following knowledge was able to be acquired as a result. In this research, the possibility of the new public transportation service by an electric bus is verified.

4 Experiment Run and Evaluating of "ELFB"

4.1 Progress of an Experiment Run

We performed the actual proof run of the electric full-flat floor bus (ELFB) in the following six routes from 2011 to 2012 (Figs 7, 8, 9, 10 and 11).

Fig. 7 The scene of ELFB's trial run (FY2011)

Fig. 8 The scene of electric supply by professional drivers

Fig. 9 The overly rapid electric supply machine tried this time (ELFB becomes full in about 30 min from SOC 50 %)

Fig. 10 Getting on and off at a mimic bus stop

Fig. 11 The situation of the experiment to which load was applied in imitation

Route A: ODAKYU Shonandai Station – Keio University, 4.0 km of one way. This route is a bus route in an urban area and university students mainly use it for attending the school.
Route B: JR Kamata Station – Haneda Airport, 7.0 km of one way. This route is a bus route in an urban area and the users of Haneda Airport mainly use it for going the airport.
Route C: JR Oomori Station – Haneda Airport, 7.5 km of one way. This route is a bus route in an urban area and the users of Haneda Airport mainly use it for going the airport.
Route D: JR Isogo Station – Oppama Bus Terminal, 12.7 km of one way. This route is a bus route in a suburban area and the users of it is mainly residents along the bus route.
Route E: ODAKYU Shonandai Station – Ayase Bus Terminal, 8.7 km of one way. This route is a bus route in a suburban area and the users of it is mainly residents along the bus route.
Route F: JR Tsujido Station – Ayase Bus Terminal, 14.7 km of one way. This route is a bus route in a suburban area and the users of it is mainly residents along the bus route.

All the above-mentioned routes are routes in which a bus company does commercial service. About operation of ELFB, we requested the professional drivers which is always running the same route. We carried out the test run of ELFB for 10 days in each route. And we acquired and analyzed the test run data of ELFB in all the test runs. We got 380 citizen monitors through the Internet, and got them to test-ride ELFB and it was evaluated by the citizen monitors. The overly rapid electric supply machine tried this time is a product of JFE engineering. ELFB becomes full in about 30 min from SOC 50 %.

We conducted interview investigation also to the bus transit company's management and drivers who cooperated in the actual proof run of ELFB.

4.2 A Result of the Actual Proof Run for Two Years

(1) We have confirmed that the use of electric buses in towns or cities will lead to the reduction of a minimum of 24 % and a maximum of 49 % in carbon dioxide in using an air conditioner system and the reduction of a minimum of 35 % and a maximum of 52 % in carbon dioxide in not using an air conditioner system, compared with non-electric buses.
(2) When the operation cost data of the existing bus on a regular route was compared with the operation cost data of this electric full-flat and low-floor bus, it has checked that the electric bus was running at cost lower 10 % than the existing bus.
(3) In general expected earliness could be attained, and when there were about 30 min on the occasion of the rest after 1 round trip, it was able to compensate sufficient electricity for the next employment. However, since the connector for

electric supply was large and heavy, there is much voice made into a subject and the driver showed the improvement necessity for future.

(4) When a large-sized electric full flat bus was compared with the conventional bus, 70 % of the trial ride monitor 380 persons in the 2011 fiscal year did high evaluation as a bus which the former takes easily. Citizens are pleased with a full-flat floor, quietness, and the ease of riding of ELFB [4].

(5) Drivers rated electric full-flat and low-floor buses higher than non-electric buses in present use as environmentally-friendly and barrier-free vehicles because of much room taken within a bus. These many advantages show that an electric full-flat and low-floor bus has easy access to other areas and practicality. Driving interface of drivers who participated in the demonstrative experiment was rated high.

(6) Bus companies have added that electric buses will be able to take middle or long distance routes and to have access to higher and colder areas if technological improvements are made in butteries, motors and wheel drives. Qualitative improvement in them will promote the prevalence of electric buses in Japan without fail. And, in Japan, we should disseminate large size EV buses with a door in center and front and in rear and front.

5 The Interview to the Bus Transit Companies Towards the Spread of ELFB

We held an interview to the bus transit companies towards the spread of ELFB from October 2013 to November 2013. Concretely, we went to each bus association of Hokkaido, Chiba Prefecture, Kanagawa Prefecture, Yamanashi Prefecture, Kanagawa Prefecture, Hyogo Prefecture, Hiroshima Prefecture, Fukuoka Prefecture, and Okinawa Prefecture, and interviewed the bus companies. We explained the result of the actual proof experiment. And we interviewed the ideal form of ELFB needed in bus companies. As a result, the demand of buses with the following 2 appearance was high.

The following results became the almost same tendency as the nationwide survey to the bus transit companies in FY 2010 under ELFB manufacture done in our related paper.

5.1 Model A: An Urban Type of Electric Full-Flat Large Size Bus (The Front Door – The Middle Door, an Amount to Be Borne by a Bus Company Is 23,600,000 Yen)

Mileage on a single charge is 200 km (including electric power for air conditioners, equipments indispensable to a one-man-operated bus and subsidiary ones). This type of bus is designed to leave enough standing room by decreasing seats in

Fig. 12 The exterior of large-size urban use electric full-flat bus with a low floor (Width = 2.5 m, Length = 10.5 m, Height = 2.8 m, The front-door – the middle-door type, Mileage on a single charge is 200 km) designed by Seijiro NODA

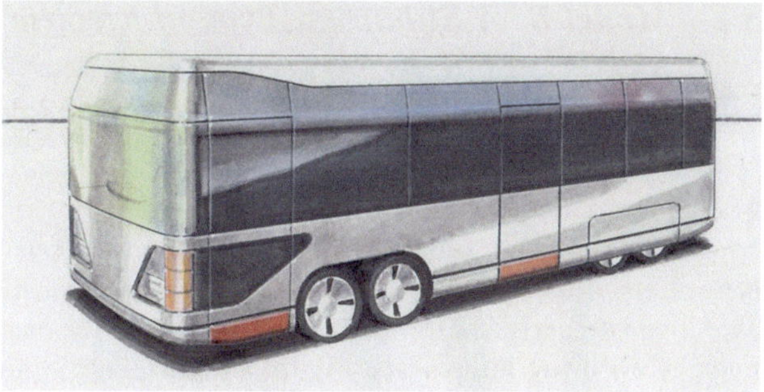

Fig. 13 The interior of an large-size electric full-flat bus with a low floor designed by Seijiro NODA

number in order to accommodate many passengers. Moreover, to satisfy passengers' needs it has some space for baggage. Figure 12 represents the exterior of a model A. Figure 13 represents the interior. A large size electric full-flat non-step bus can supply passengers with enough space for baggage, large chairs, desks, and so on, with an integrated platform installed. Moreover, it is possible to get power source for welfare equipments under a wheelchair. In this way this type of electric bus which runs in cities makes it possible to combine universal design with eco-design because of enough space.

5.2 *Model B: A Suburban Type of Electric Full-Flat Large Size Bus (The Front Door – The Rear Door, an Amount to Be Borne by a Bus Company Is 23,600,000 Yen)*

Mileage on a single charge is 200 km (including electric power for air conditioners, equipments indispensable to a one-man-operated bus and subsidiary ones). This type of bus is designed to leave as many seats as possible, with little standing room. Moreover it is possible to secure enough space for baggage and to make room for bicycles by using jump seats. Figure 14 represents a model B. of an electric bus. In our survey many bus companies hoped that a bus with the front door and the rear door would survive in suburban cities. Concerning a non-step bus with a rear engine installed, the rear engine makes it possible to adopt "the front door and the rear door" system. But many bus companies like a bus with the front door and the rear door better than a bus with the front door and the middle door because a bus with the rear door can accommodate more passengers and passengers are used to taking the bus. We visualized a bus in Fig. 14 as a model of a suburban type of bus. As represented in Fig. 13. the seats in half of the forward part of the bus are arranged in two rows. The position of space for baggage, bicycles, children, and so on is subject to change according to what characteristics of buses with a regular route have.

5.3 *Model C: An Small Size Electric Full-Flat Bus (The Front Door – The Middle Door, an Amount to Be Borne by a Bus Company Is 12,800,000 Yen)*

Mileage on a single charge is 200 km (including electric power for air conditioners, equipments indispensable to a one-man-operated bus and subsidiary ones). This type increase seats in number with little standing room by using larger seats and by holding subsidiary equipments under seats because many elderly people will take the small bus. We have confirmed that electric buses will be classified into a large-size buses and small-size buses through interview with bus companies. On the other

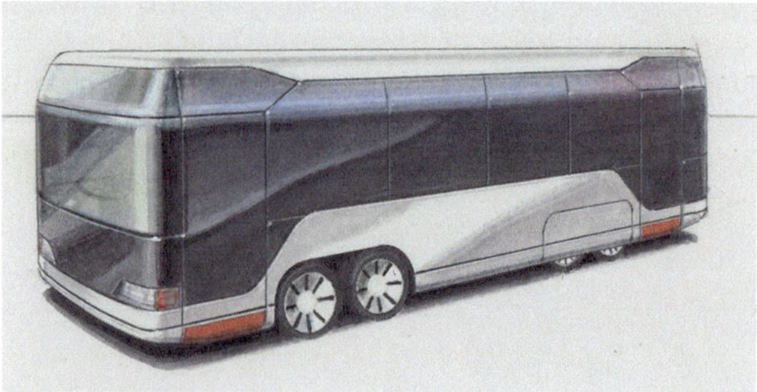

Fig. 14 The exterior of a suburban type of large-size electric full-flat bus with a low floor (Width = 2.5 m, Length = 10.5 m, Height = 2.8 m, The front-door – rear-door type, Mileage on a single charge is 200 km) designed by Seijiro NODA

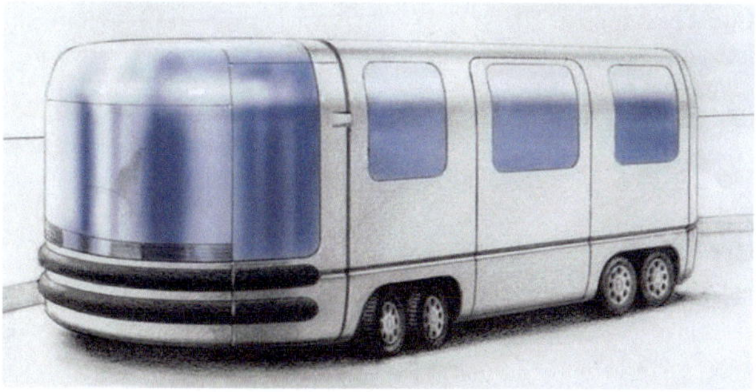

Fig. 15 The exterior design of a grand-up style community bus designed by Seijiro NODA

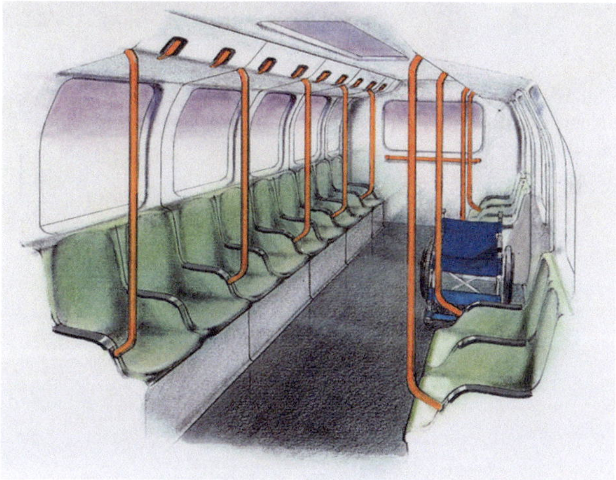

Fig. 16 The interior design of a grand-up style community bus designed by Seijiro NODA

hand, middle-size buses will cease to be used from now on because of halfway transportation. So we visualized a small-size electric full-flat bus which elderly people and physically challenged people can take easily in order to any places with poor access, as represented Figs. 15 and 16. Moreover the bus can be loaded with wheelchairs. This community bus can contribute to combining eco-design with universal design.

To others, although there was also expectation for an interurban coach (Fig. 17) and a long-distance night high speed bus (Fig. 18), many bus companies wanted to introduce when development of a battery with many high-density are done [5].

Fig. 17 Image of interurban coach on integrated platform designed by Kikuo EMOTO

Fig. 18 Image of a long-distance night high speed bus on integrated platform designed by Kikuo EMOTO

6 New Social Service Based on ELFB

ELFB does not emit an exhaust gas, noise, and carbon dioxide. So, ELFB is put in public buildings. If ELFB goes into the station and airport, and a harbour especially, the burden and resistance of citizens' change can be reduced. If ELFB rides into a hospital and a supermarket, resistance of movement of citizen like elderly people or physically challenged people can be reduced greatly 80 % of 380 monitors want ELFB to ride into a station, a hospital, and a shopping center, and it also turned out that increase of 60 Japanese yen is permitted on the bus fare per time towards the environmental improvement in a station, a hospital and a shopping center. In the car of ELFB, the floor is a full flat. Moreover, since ELFB can also cut an exhaust gas and sound, the public service of the universal design of a high level and ecological design can be provided for citizens (Figs. 19 and 20).

The technology of ELFB is applicable also to an ambulance or an inspection car like an X-ray car which puts much medical equipment in the car. The technology of ELFB has the application possibility to large-sized vehicles other than a bus (Figs. 21 and 22).

Fig. 19 The image which an electric bus rides into a shopping center designed by Kikuo EMOTO

Fig. 20 The image which an electric bus rides into a city park designed by Kikuo EMOTO

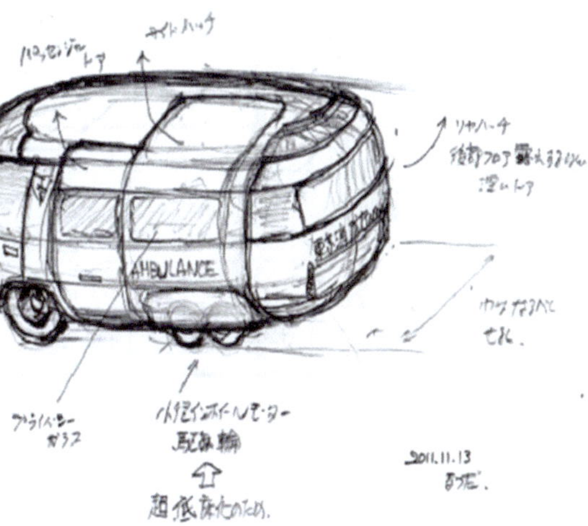

Fig. 21 The appearance image of the electric ambulance using an in-wheel motor designed by Atsushi MATSUDA

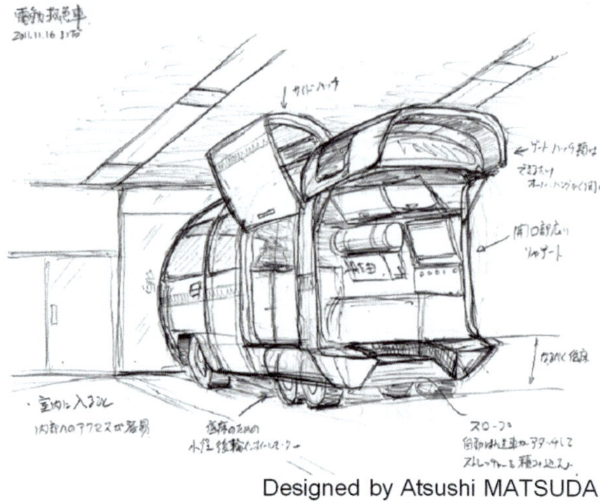

Fig. 22 The practical use image of the electric ambulance using an in-wheel motor designed by Atsushi MATSUDA

Fig. 23 The image to which the patient is turning around the hospital with the small electric vehicle with an automatic operation function

In particular, the monitor urged it strongly to apply the technology of ELFB to an ambulance by the free answer column. If an electric ambulance rides in into a hospital, time of a patient's emergency step can be shortened on the whole.

Furthermore, if a patient enables it to turn around the inside of a hospital freely by the small electric vehicle of automatic operation, level of the universal design and an ecological design will go up markedly in a hospital. The demand of the

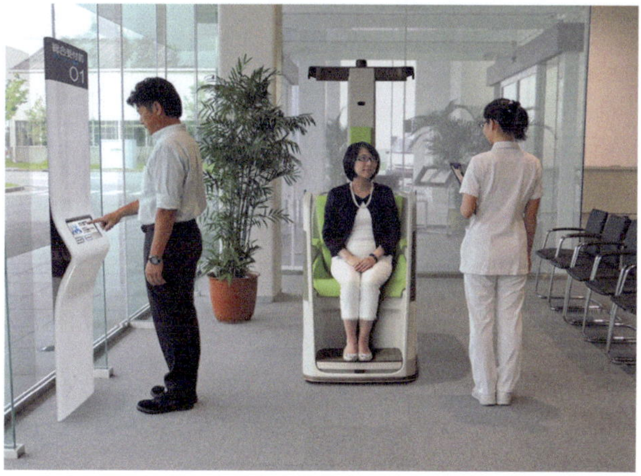

Fig. 24 The service robot for hospitals which operate automatically a author, Nishiyama, has developed (Securement of citizen's mobility service in the hospital by an electric car will be important from now on)

service which raises the level of a universal design and an ecological design with the technology of such an electric vehicle increases from now on.

A writer names this MOBITALITY SERVICE and is trying hard towards the spread (Figs. 23 and 24).

7 Conclusion

It is urgently necessary to enhance further practical utility of EVs through research by evaluating the prototype vehicles. One of the reasons why EVs are being spotlighted in recent years is that they can be driven right to the inside of a building. In the case of electric route buses, they can be driven right up to the front of the ticket gate of a station, medical treatment facilities, or shopping malls, which can reduce the moving distance for the elderly or persons with disabilities. In the future, the research would like to take the moving distance into consideration while studying scenarios of low-carbon town development that utilizes EVs. The research also intends to continue studies on the construction of future environments that are created through a fusion of eco design and universal design. In addition, the plan to embark on the development of electric trucks by making the most of the integrated platform structure is underway as well. Furthermore, we put in a building the electric buses which an exhaust gas and noise do not come, and also advance research to reduce resistance of the change by traffic environments. If it puts in the station of a railroad, or the building of an airport, resistance of a change will decrease and be pleased with elderly people. ELFB is useful in the field of city planning.

References

1. Ballabio E (1998) The European Commission R&D initiatives to promote design for all. In: Proceedings of international workshop on universal design, building research institute, Ministry of Construction and Japan International Science and Technology Exchange Center, Tokyo, Japan (CD-ROM)
2. Benkzton M (1993) Designing for our future selves: the Swedish experience. Appl Ergon 24:19–27
3. Bicknell J, McQuiston L (eds) (1997) Design for need: the social contribution of design. In: Conference proceedings, Pergamon Press & Royal College of Art, London, UK (CD-ROM)
4. Kahneman D, Knetsch JL (1992) Contingent valuation and the value of public goods: reply. J Environ Econ Manag 22:90–94
5. Nishiyama T, Noda S, Ishikawa M, Shimizu H (2012) Electric Full Flat Floor bus's trial production and research of the spread strategy in a public transportation system. In: Proceedings of the third international engineering systems symposium, on Internet

Analysis of Business Process Innovation Using Outsourcing

Takeshi Takenaka, Naoki Tomotake, Rui Suzuki, Masumi Yoshida, Taiki Yamada, and Shigeaki Shiraishi

Abstract Innovative change of business processes is an important concern for both service and manufacturing industries to achieve more sustainable businesses corresponding to rapid changes and globalization of the world economy. This paper presents a discussion of how companies can use outsourcing for business process innovation. We conducted a questionnaire survey targeting business managers in Japan and the U.S. It also analyzes best practices and challenges related to this issue through interviews with BPO vendor companies and user companies in both countries. Finally, the classification of business process innovation using outsourcing it discussed from the viewpoint of a value creation model.

Keywords Business model • Innovation • Outsourcing • Value creation model

1 Introduction

In an increasingly globalized and rapidly changing world economy, worldwide competition compels companies not only to dominate an existing market but also to expand their businesses by creating new value for customers. Actually, the boundaries of business sectors such as manufacturing and service industries have become more blurred in terms of value creation in society. Therefore mutual learning and cooperation among industries has become increasingly important for sustainable growth at an industry level. At the firm level, the manner in which a company can incorporate external knowledge and expertise is important.

In recent years, some global companies have grown rapidly through increased merger and acquisition (M&A) activities. In Japan, for instance, the total transaction value of M&A acquisitions of foreign companies surpassed that of domestic M&A in 2006 [1]. This phenomenon suggests that Japanese markets are highly

T. Takenaka (✉)
National Institute of Advanced Industrial Science and Technology, 135-0064 Tokyo, Japan

Ministry of Economy, Trade and Industry, Tokyo, Japan
e-mail: Takenaka-t@aist.go.jp

N. Tomotake • R. Suzuki • M. Yoshida • T. Yamada • S. Shiraishi
Ministry of Economy, Trade and Industry, Tokyo, Japan

T. Maeno et al. (eds.), *Serviceology for Designing the Future*,
DOI 10.1007/978-4-431-55861-3_7

mature and saturated. Those companies that implemented international M&A might require external sources of new knowledge and business functions to enhance their respective competitive positions domestically and abroad. Nevertheless, it is not always easy for an acquiring company to achieve desirable results through M&A, even if they are in the same industry, because both the acquiring and acquired companies have their own distinct business cultures or implicit processes. Therefore, companies that require business restructuring should marshal their business model or operational processes in advance of M&A.

Recently, business services such as outsourcing play more important roles for companies to make their business processes more efficient and effective. Although the salient purpose of introducing outsourcing for companies is cost reduction, it can also be beneficial for the reconstitution of their business models or operational processes for future growth. However, companies cannot always use outsourcing effectively. For effective outsourcing, both outsourcing vendors and user companies need mutual understanding to co-create greater value.

This paper presents a discussion of how companies can use outsourcing for the innovative restructuring of their business processes. First, this discussion presents overviews of recent trends prevailing in outsourcing industries. Then, it presents results of a questionnaire survey that aims to clarify business managers' motivation or attitudes for introducing outsourcing in Japan and the U.S. Then it introduces best practices for introducing outsourcing for innovative change of business processes. Finally, it proposes a classification of business process innovation using outsourcing from the viewpoint of a value creation model.

2 Outsourcing Industry

Although many definitions of outsourcing have been presented in the literature, Gereff and Fernandez-Stark [2], for example, classified outsourcing into ITO, KPO, BPO, and other industry-specific outsourcings from the viewpoint of their global value chain framework (see Fig. 1). They refer to knowledge process outsourcing (KPO) as higher value services across all industries, such as market intelligence, business analytics, and legal services. However, business process outsourcing (BPO), which includes enterprise resource management (ERP), human resource management (HRM), and customer relationship management (CRM) are positioned lower than KPO in terms of value added.

In practice, BPO and KPO are often inseparable for the improvement of business processes, although KPO usually includes professional services provided by consultants with public qualification. Moreover, information technology outsourcing (ITO) is playing a crucial role in KPO and BPO. Because of this situation, this paper mainly specifically examines BPO as outsourcing including descriptions of some functions of KPO and ITO.

HfS Research [3] estimated the global market size for both ITO and BPO, which are gradually expanding, as presented in Fig. 2. Table 1 presents the estimated

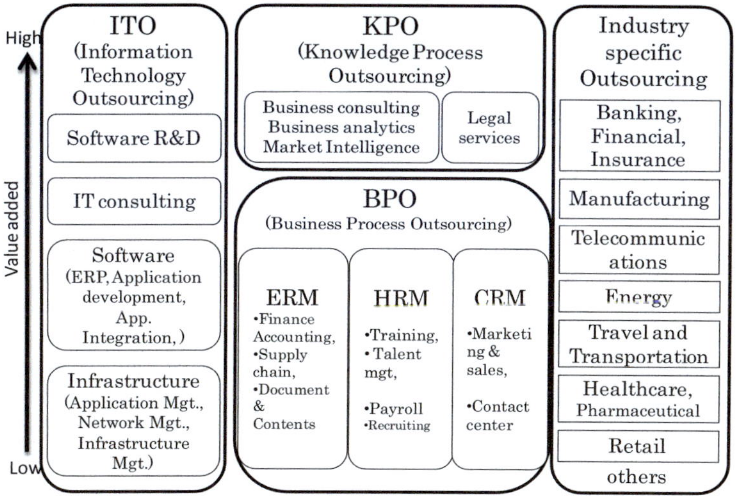

Fig. 1 Types of outsourcing (Based on Gereffi and Fernandez-Stark [2])

Fig. 2 Estimated world ITO and BPO markets (Ref. [3])

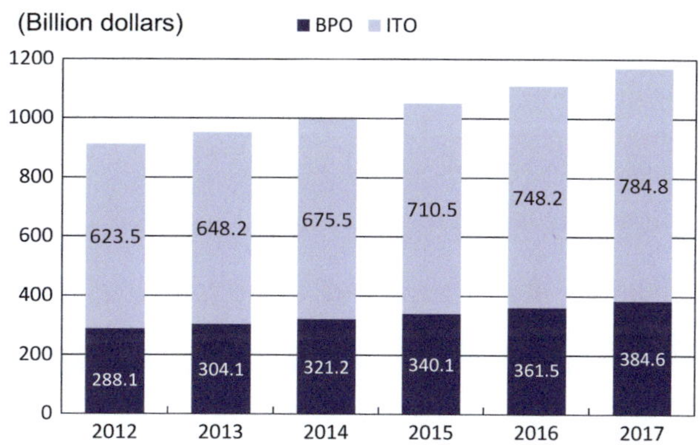

Table 1 Estimated market size of BPO and ITO in the U.S. and Japan, expressed as percentages of those countries' GDP[a]

	Country	2012 (%)	2013 (%)	2014 (%)	2015 (%)
BPO	US	0.75	0.76	0.77	0.77
	JP	0.14	0.17	0.17	0.18
ITO	US	1.18	1.18	1.18	1.20
	JP	1.40	1.76	1.74	1.73

[a]Based on HfS Research "State of the Outsourcing Industry 2013" and IMF "World Economic Outlook Database, October 2013"

shares of GDP of BPO and ITO in the U.S. and Japan. The major difference between the U.S. and Japan is that the BPO market in Japan is much smaller than that in the U.S. in terms of GDP. The authors' fundamental research question is why Japanese companies can use BPOs less than foreign companies can. Lower productivity of Japanese service industries has remained a key issue in Japan [4]. Moreover, we are interested in how BPO is useful for the creation of added value and for cost reduction. Japanese service industries exhibit higher ratios of small and medium-sized enterprises (SME's). Therefore, the benefit of BPO cost reductions cannot be anticipated to be high for SMEs. Additionally, cheaper offshore outsourcing presents limited opportunities for Japanese companies because Japan is a non-English-speaking country. The following section introduces the results of a questionnaire survey that was administered to clarify companies' current situations and problems for introducing outsourcing in Japan and the U.S.

3 Questionnaire Survey of BPO in the U.S. and Japan

The Japanese Ministry of Economy, Trade and Industry (METI) [4] created a study group for the creation of higher value-added service industries, which was established to assess new policies for increasing high added value in the service industry, a key sector for Japan's successful economic growth in January 2014. The major topics to be discussed were the following.

1. Fostering human resources to lead approaches to increasing high added value in businesses
2. Shifting to a new approach called "Proactive Investment in IT," including the acquisition of new customers and the development of new services
3. Stimulating business-supporting service industries and using such services
4. Innovation of a business model aimed at increasing high added value in service industries, etc.

A web-based questionnaire was administered during February 27 – March 14 in 2014 to clarify current problems of the third topic above, comparing the U.S and Japan, and also comparing manufacturing and service industries. The survey targeted executives and managers in the four industries presented in Table 2.

Table 2 Number of questionnaire survey respondents

Nation	Co. size (Number of emps.)	Retail, Whlsale.	Restnt., Hotel	Finance, Insur.	Mfg.
Japan	>500 employees	50	25[a]	50	50
	≤500 employees	50	50	50	50
U.S.	>500 employees	50	50	50	50
	≤500 employees	50	50	50	50

[a]Only 25 respondents answered with data for the decade

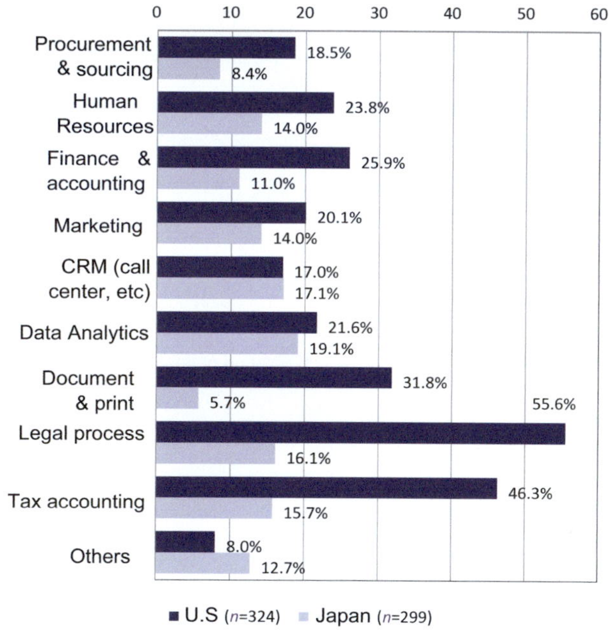

Fig. 3 Currently used outsourcing services in the U.S. and Japan

Figure 3 presents responses to question No. 4 comparing U.S. and Japan (Q4: Which kind of outsourcing has your company used?). In many operations, U.S. companies use outsourcing more than Japanese companies do.

Japanese companies use outsourcing much less in the fields of human resource management, finance and accounting, document and print operations, legal processes, and tax accounting. Additional investigations reveal that 23.2 % of large companies (>500 employees) in service industries use data analysis outsourcing, although only 16.4 % of smaller companies offering services (≤500 employees) use such outsourcing. However, small service companies in the service sector use customer relationship management (CRM) including call-centre services and marketing outsourcing more often than large service companies and manufacturing companies do.

Figure 4 portrays companies' expected results for outsourcing (Q8: Which effects did you expect for outsourcing when you decided to introduce it?). Many companies reported expecting the effects of reduced operational costs and process efficiencies in both Japan and the U.S. because those effects were expected to be the first or direct expected effects of outsourcing. Further investigation reveals that those tendencies are more apparent in large companies irrespective of service or manufacturing industries. However, there are apparently interesting differences that prevail between U.S. and Japanese companies in other expected effects. U.S. companies apparently expect outsourcing to make their business processes more robust against changes in business environment such as business expansion, globalization, and satisfaction of compliance requirements. Moreover, they expect

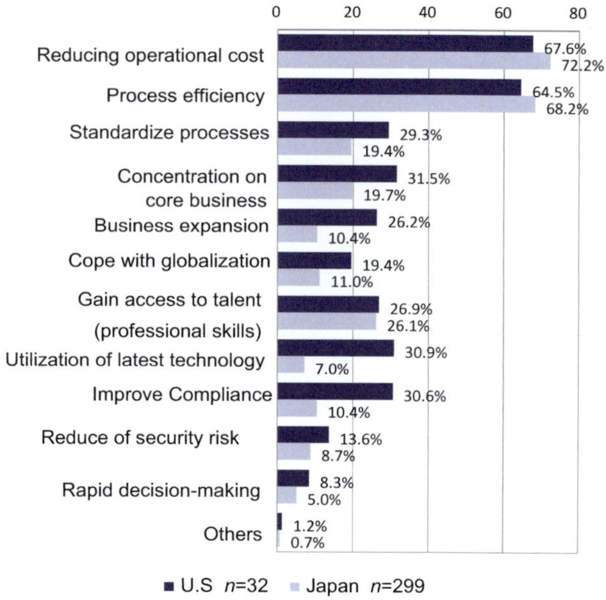

Fig. 4 Expectations related to outsourcing

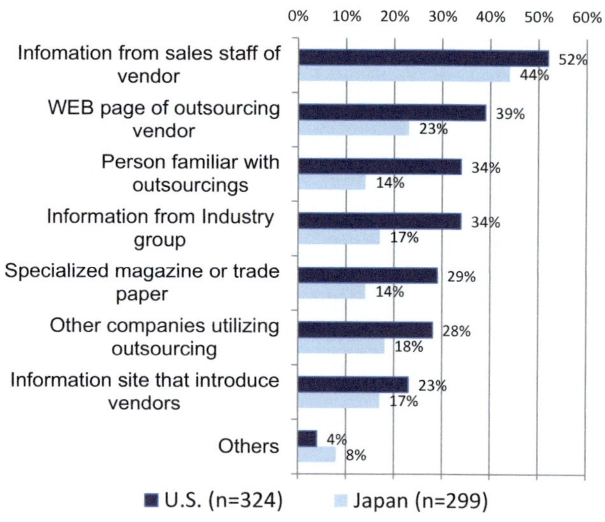

Fig. 5 Information resources of outsourcing vendors

secondary or indirect effects such as standardization of processes or use of the latest technologies. Companies that anticipate those effects might react positively to changes in their business processes.

Those differences in expectations might be related to the degree of familiarity of using outsourcing. Figure 5 presents a comparison of information sources when they choose outsourcing vendor companies. Fewer information sources are expected to exist in Japan. Throughout our interviews with a BPO vendor company in the U.S., a manager remarked that an extremely reliable information resource for user companies is other companies' experiences. Therefore, the company often introduces their existing client companies to a new client who is considering outsourcing. Such user networks constitute an important resource for user companies. Moreover, industrial groups can play important roles for the promotion of effective utilization of outsourcing.

4 Analysis of Outsourcing Best Practices

We conducted interviews with outsourcing vendor companies, user companies and researchers familiar with the outsourcing industry in Japan and the U.S. from December 2013 through April 2014. As described herein, it briefly introduces a case study of Prudential Financial Inc. Subsequently, it proposes some important topics for the success of outsourcing.

4.1 Case of Prudential Financial Inc

The authors conducted an interview of two vice presidents and key persons of the outsourcing project in Prudential Financial Inc. on March 13, 2014 at their headquarters in Newark, New Jersey, U.S. This company is the world's largest provider of insurance, investment management, and other financial products in more than 30 countries. Their use of outsourcing, especially of human resources, is well known because of its extremely large BPO deals in service industries [5]. This interview ascertains the factors contributing to success of BPO of this company because they introduced BPO not merely for cost reduction or streamlining but for bold reconstruction of their business processes.

The vice president has remarked that they decided to outsource human resources management (HRM) to an external BPO vendor with the opportunity to go public in 2001. They strongly drove globalization through numerous M&A projects while planning to reduce many employees in the U.S. simultaneously. Actually, U.S. domestic employees decreased from 48,000 in 2000 to 22,500 in 2002, although employees overseas increased from 9000 in 2000 to 22,500 in 2002. Results show that they cut over 250 jobs related to human resource management. However, their true goal might be to make all HRM processes more robust against various risks or uncertain environmental changes. The interview revealed some important factors to use outsourcing for business model innovation, some of which are listed below.

- They started a project team including employees and external experts to re-examine and describe all existing HRM-related processes.

- They started a shared service on HRM in their company before outsource processes.
- They chose an outsourcing vendor prepared to manage dozens of HRM processes as an important partner with a more than 10-year long-term contract.
- IT systems play a crucially important role in business processes related to re-engineering.
- Key performance indicators were prepared from diverse perspectives.

Although the case of Prudential Financial Inc. should be regarded as an outstanding example of best practices, the authors regard some factors above as common in successful best practices that use outsourcing as an opportunity for business process innovation. However, interviews with more than 10 BPO vendor companies indicate disappointments in BPO outcomes arising from short-sighted expectations or poor relations between vendors and user companies.

The author's investigations to date remain limited. Additional studies conducting detailed analyses for the success of outsourcing considering industrial background or company sizes.

5 Classification of Business Process Innovation Using Outsourcing

This section presents discussion of business model innovation using outsourcing from the viewpoint of value creation.

5.1 *Utilization Strategy of Outsourcing*

Although the main purpose of outsourcing is expected to be cost reduction, additional development is another possible goal. Results of this study suggest that differences in ranges of expected goals of outsourcing among companies (clients). Figure 6 portrays three scenarios of using outsourcing according to the expected outcomes with business processes.

As one example, a client company might want to outsource a single business process (process B in Fig. 6) such as payroll or accounting management because it is regarded as inefficient. In scenario 1, the company merely seeks to outsource process B with no changes. Actually, many companies expect this type of scenario for outsourcing vendors. Although the purpose of outsourcing is clear, some difficulty is expected to exist in examining existing processes. When existing process B is not well described, the expected effect of outsourcing will be smaller. In this case, a possible solution is to standardize the process according to the vendor recommendation. However, the relation between the client and vendor is fixed in the outsourcing of process B.

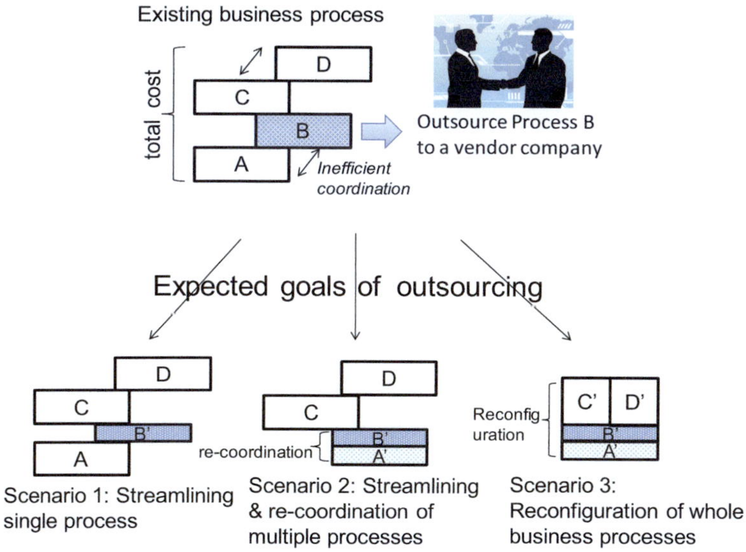

Fig. 6 Expected results of outsourcing for companies

A second scenario is that the client outsources process B, thereby influencing other processes related to B. In this case, the vendor must consider the coordination of multiple processes. Through client cooperation, a synergetic effect of re-coordination of multiple processes might be achieved.

A third scenario is that the total business process of the client might change through outsourcing of process B. In this case, the outsourcing purpose is not limited to the streamlining of process B. The reconfiguration of all business processes is expected. A new business process can be co-created through mutual interaction between the client and vendor. However, this scenario might require long-term and mutual cooperation based on trust. The case of Prudential Financial Inc. might correspond to this scenario because they positively seek a more effective business processes as a system.

5.2 *Problem Structure of Outsourcing from the Viewpoint of Value Creation Model*

According to the value creation class model proposed by Ueda et al. [6], relationships among the outsourcing vendor, client, and client's business processes can be described as portrayed in Fig. 7. As discussed with three scenarios in Sect. 5.1, expected results of outsourcing could vary according to the completeness of description of business processes and information about environment. Therefore outsourcing vendor and its client company should discuss and determine the best

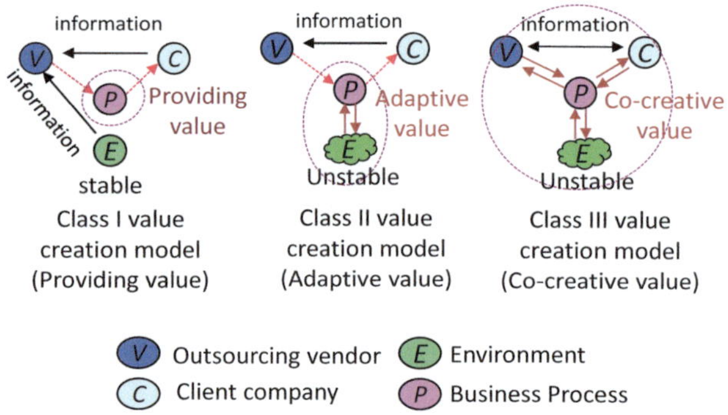

Fig. 7 Value creation model in outsourcing

strategy based on verification of existing business processes and predictability of business enviroments. In the changing environment and increasing uncertainness of economic situation, mutual interaction between outsourcing vendor companies and client companies will be more important to achieve sustainable growth of the world economy. Future work will explore relationships among business stakeholders and consumers for the co-creation of business models in a sustainable society.

6 Conclusion and Future Work

This paper presents a discussion of how companies can use outsourcing for business process innovation. Through results of questionnaire survey and analysis of best practices, the authors can assess the possible policies or measures for assistance using outsourcing, especially in service industries. A quality certification system discussed in the industry group of BPO is anticipated as a useful tool for user companies. Additionally, a user network and introduction of best practices introducing BPO will help user companies.

References

1. Marr online (2014) Statistics on M&A, RECOF data corporation. http://www.recofdata.co.jp/english.do
2. Gereffi G, Fernandez-Stark K (2010) The offshore services value chain: developing countries and the crisis, Policy research working paper 5262. The World Bank, pp 1–31. https://openknowledge.worldbank.org/handle/10986/3751
3. HfS Research, State of the Outsourcing Industry (2013). http://www.hfsresearch.com/State-of-the-Outsourcing-Industry-2013-Executive-Findings
4. The Ministry of Economy, Trade and Industry. http://www.meti.go.jp/english/press/2014/0116_02.html
5. Cooke FL, Shen J, McBride A (2005) Outsourcing HR as a competitive strategy? A literature review and an assessment of implications. Hum Resour Manag 44:413–432
6. Ueda K, Takenaka T, Vancza J, Monostori L (2009) Value creation and decision-making in collaborative society. Ann CIRP 58(2):681–700

Part II
Context Based Service and Technology

Mixed Reality Navigation on a Tablet Computer for Supporting Machine Maintenance in Wide-Area Indoor Environment

Koji Makita, Thomas Vincent, Soichi Ebisuno, Masakatsu Kourogi, Tomoya Ishikawa, Takashi Okuma, Minoru Yoshida, Laurence Nigay, and Takeshi Kurata

Abstract This paper describes a maintenance service support system for wide-area indoor environment, such as a factory and a hospital. In maintenance services, operators often have to check a map to find out a way to a target machine, and also have to refer documents to get information about check-up and repair of the machine. In order to reduce working load of operators, information technology can help operators carry out additional but important operations during maintenance, such as referring documents and maps, recording maintenance logs and so on. In this paper, we propose mixed reality navigation on a tablet computer composed of augmented virtuality mode and augmented reality mode. Augmented virtuality mode performs map-based navigation shows positions of the user and the target machine. Augmented reality mode performs intuitive visualization of information about the machine by overlaying annotations on camera images. The proposed system is based on a hybrid localization technique realized with pedestrian dead reckoning (PDR) and 3D model-based image processing for the purpose of covering wide-area indoor environment. Experimental results using our

K. Makita (✉) • T. Okuma • T. Kurata
Center for Service Research, National Institute of Advanced Industrial Science and Technology, AIST Tsukuba Central 2, 1-1-1, Umezono, Tsukuba, Ibaraki 305-8568, Japan
e-mail: k.makita@aist.go.jp

T. Vincent • L. Nigay
Université Joseph Fourier, Saint-Martin-d'Hères, France

S. Ebisuno • T. Ishikawa • M. Yoshida
KODO Lab Inc, Tsukuba, Ibaraki 305-0872, Japan

M. Kourogi
Center for Service Research, National Institute of Advanced Industrial Science and Technology, AIST Tsukuba Central 2, 1-1-1, Umezono, Tsukuba, Ibaraki 305-8568, Japan

KODO Lab Inc, Tsukuba, Ibaraki 305-0872, Japan

T. Maeno et al. (eds.), *Serviceology for Designing the Future*,
DOI 10.1007/978-4-431-55861-3_8

prototype with a mock-up model of a machine are also described for showing feasibility of our concept in the paper.

Keywords Maintenance support • Human navigation • Mixed reality • Mobile computing

1 Introduction

Machine maintenance services are essential for workers to safely and efficiently use machines. For maintenance operators, main activities of maintenance services are to go in front of the target machine, and check-up of the machine. Moreover, in case failure parts are detected, additional activities such as doing repairs, component replacement and orders for components are occurred.

Because maintenance services are not routine, the activities can be usually inexperienced works. Therefore, maintenance services can be difficult and time-consuming because they require lots of information. Operators often have to check a map to find out a way to a target machine, and also have to refer documents to get information about check-up and repair of the machine. Therefore, especially in wide-area indoor environment, such as a factory and a hospital, working load of operators should be reduced.

In this paper, we propose a maintenance service support system for wide-area indoor environment to reduce working load of operators. We focus on working load of obtaining information. In order to realize the system to reduce working load of obtaining information, we introduce mixed reality (MR) concept. MR is an inclusive term of techniques to merge virtual and real world. Since MR has a possibility to realize intuitive information presentation, we propose mixed reality navigation on a tablet computer composed of augmented virtuality mode and augmented reality mode. Augmented virtuality (AV) and Augmented Reality (AR) are components of MR. Augmented virtuality mode performs map-based navigation shows positions of the user and the target machine. Augmented reality mode performs intuitive visualization of information about the machine by overlaying annotations on camera images.

2 Related Works

How MR can assist in reducing the time and effort in maintenance is evaluated in previous works [1, 2]. The majority of related work focuses on MR applications work in front of the target object. On the contrary, we focus on how to realize localization in wide-area indoor environment for MR.

To achieve robust and global localization in wide areas for mobile MR, one solution is to combine a method works constantly and the visual tracking method, because visual tracking can be used for precisely estimating position and posture [3] only in case a camera of mobile device is active.

GPS is often used for initialization [4, 5]. While GPS can provide global position and posture without any previous knowledge, it works only in outdoor environments. On the other hand, for indoor MR, several approaches requiring previous knowledge have been proposed, one of which is the marker based localization method. This method can be used in environments with marker infrastructure already in place [6]. Therefore, these methods are appropriate for constructing a special environment or desktop application that works in small spaces.

Marker based methods can be used for initialization only when markers are captured by a camera. For more efficient construction of a wide MR space, there exist devices that can be applied to construct positioning infrastructure. For example, the radio frequency identification RFID and infrared data association IrDA based methods [7] have been proposed. On the other hand, a positioning system using wireless LAN has also been proposed [8]. These methods are effective for acquiring an approximate global position in large indoor environments. However, the accuracy of these methods is not normally adequate for the initialization of visual tracking methods. Image based matching methods [9–11] can also be applied for initialization. However, because these methods use 2D image based matching, many reference images are needed to cover large environments and for accurate initialization. On the other hand, in existing tracking methods using 3D models [12, 13], a matching between a captured image and an image generated from models is used to estimate relative position and posture.

Recently, 3D reconstruction methods have been proposed. For large scale accurate reconstruction, modeling methods using robots with range sensors are effective [14, 15]. Moreover, interactive image-based modeling methods have been proposed [16–19].

These methods encourage more convenient 3D reconstruction, and can be applied to various environments. For example, Fig. 1 shows a sample generated image from 3D models generated using an interactive 3D modeler [18]. Once models are generated, images from any views can be obtained. Therefore, models can be applied various applications. Figure 2 shows a sample image of augmented virtuality (AV) application. In a scene 3D dataset (point features, edgelets, etc.) can be obtained, global position and posture of a device are estimated with the dataset [20–22]. But normally generation cost of the 3D dataset is higher than the one of 3D models in terms of a number of images needed to be generated. Therefore, the estimation using 3D models has possibility to efficiently extend areas for mobile MR.

This study focuses on the ability to merge pedestrian dead reckoning (PDR) and 3D model-based image processing for the purpose of covering wide-area indoor environment. PDR is a technique for the measurement of position and orientation based on dead reckoning that focuses on human walking motion detection. Because the PDR is normally realized with self-contained wearable or mobile sensors, localization can always work. The PDR works well for application systems such as human-navigation systems which user is mainly walking during using the

Fig. 1 An example of an input image and a corresponding image from models (*Left*: Input image taken in indoor environment. *Right*: Corresponding image from models that are generated using an interactive 3D modeler [18])

Fig. 2 *Left*: Input image taken in indoor environment. *Right*: Screen shot of augmented virtuality (*AV*) application. This generated image is composed of an image from models and navigation information

system. However, the positioning accuracy of the PDR is often getting down in working situation, and includes accumulative errors. On the contrary, image processing can work only in case a camera of mobile device is active and quality of camera images are well. For instance, blur of the images and less image features affect accuracy of the localization. However, normally the accuracy of the localization is high, and accumulative errors can be reduced with key-frame based initialization.

3 Prototype

Figure 3 shows an overview of the proposed localization method. Position and orientation of the user of a tablet computer can be always estimated with PDR. However, the positioning accuracy is getting down with the accumulative errors

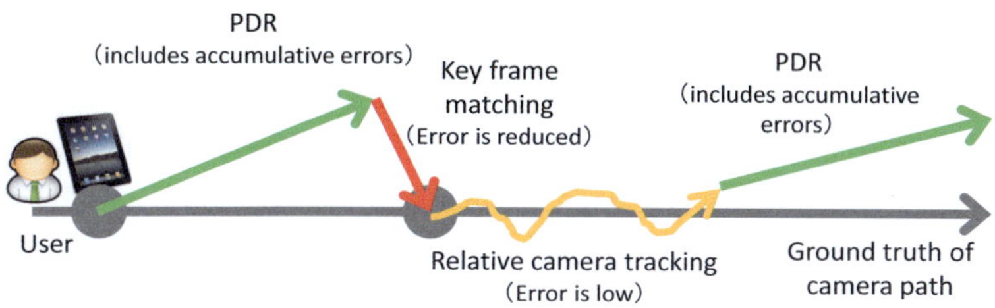

Fig. 3 Overview of the localization method

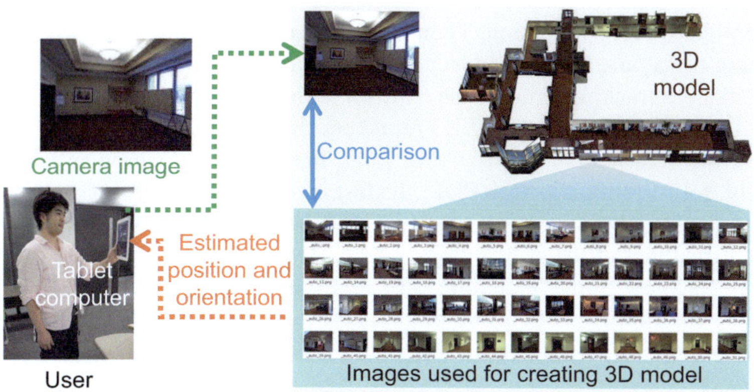

Fig. 4 Overview of the key frame matching

because of various kinds of actions except for walking. For creating a prototype system in this paper, we implemented a PDR application based on a method shown in [23] on a smart phone (Samsung RZ Galaxy). Acceleration sensors, angular velocity sensors, and magnetic sensors in the smart phone are used for the estimation procedure of PDR. The smart phone is attached to the waist of the user, and position and orientation data estimated by the PDR application is always sent to the tablet computer via wireless network. Only with PDR, the accumulative errors are not able to be reduced.

On the other hand, when the camera of the tablet computer is active, comparisons of camera image and images used for creating 3D model are conducted. Figure 4 shows an overview of the comparisons. Afterward, we call this comparison "key frame matching". Images used for creating 3D model are photos with known photo-shoot positions and orientations, and depth data. For creating a prototype system in this paper, we implemented an application for (re-)initialization based on a method shown in [24] on a tablet computer (Apple iPad2). Because of applying secondary product of modeling process, additional works are not necessary for set up. In the method, an image of real environment taken by a mobile camera of the tablet computer is compared to images used for creating 3D model. Because images used for creating 3D model are linked with photo-shoot position, orientation, and depth data, 3D coordinates of each pixel on the images is available in the model coordinate system. Therefore, various types of camera tracking method using image features, such as feature points and edges, are able to be applied to the localization. For the prototype, we applied an estimation technique using feature point-matching method, and selected Speeded Up Robust Features (SURF) [25] for the detection of feature points. Normally, the calculation time of the matching procedure is too long to be applied to MR application. Therefore, after the matching is succeeded, image processing procedure is changed to a relative camera tracking. Since normally the relative camera tracking can also be realized with image features, in this paper, we applied Lucas-Kanade-Tomasi (LKT) trackers shown in [26].

When the relative camera tracking fails, the PDR application is used again. Currently, for re-initializing the PDR, the last result of the relative camera tracking is simply applied, because the accuracy of relative camera tracking is supposed to be higher than that of the PDR in most cases. However, since there is a case to get errors of the key frame matching and relative camera tracking, we have the possibility of improvement in re-initialization of the PDR. After that, the key frame matching also work again.

4 Experiments with the Prototype

We conducted an experiment in our office building in order to check the behavior of the prototype system. In the experiment, we evaluated a moving distance when the key frame matching is succeeded. Motivation of the evaluation is below.

Figure 5 shows samples of moving distances when the key frame matching is succeeded. As shown in Fig. 4, when the key frame matching is conducted, a camera image captured by the tablet computer is compared to images used for creating 3D model. Generally, we have lots of images used for creating 3D model, and retrieve the images in distance order from the position estimated by PDR. Therefore, the computational time of the key frame matching is proportional to the moving distance. Fundamentally, lots of images should be used, and the computational time should be measured. However, since currently we do not have any speeding up method in the retrieval, there is a high possibility of exceedingly-long computational time. Therefore, in this experiment, we preliminarily set only one image as a target of the key frame matching, and measured the moving distance.

Experimental set up is below. Figure 6 shows appearances of the user and the AV/AR application. First, we created a virtualized reality model of our office

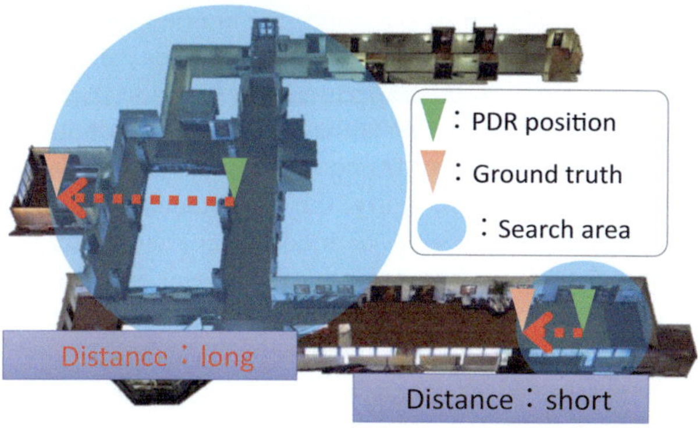

Fig. 5 Moving distances when the key frame matching is succeeded

Fig. 6 Appearances of the user and the AV/AR application

building with the method shown in [18], and prepare one image for the key frame matching. Next, we set a round course in the building whose distance is about 60 m, and set one user as a subject. In the experiment, at first, the user stands at the start position of the course, and hold the tablet computer to conduct the key frame matching with AR mode. After key frame matching is succeeded, the user walks along the course with AV mode. Finally, the user stands at the start position and conduct the key frame matching with AR mode again, and the moving distance of latter matching is measured. In order to check difference between walking distance and the moving distance of latter matching, we conducted experiments five times with one cycle (about 60 m) and two cycles (about 120 m).

Figure 7 shows examples of localization results. In these figures, blue lines indicate the moving distance of former matching occurred because the difference of manual initialization and the start point. Red lines indicate the pass when the method is changed from the relative camera tracking to PDR. Sky blue lines indicate the pass estimated with PDR. Finally, green lines indicate the moving distance of latter matching we focus on in the experiments.

Figure 8 shows all results of the experiments. In the results, average moving distance of latter matching was about 1.44 m with one cycle, and about 2.03 m with two cycles.

As a result, we have successfully checked the behaviour of the prototype system. Because we have the variability of walking motion, the moving distances of latter matching also have the variability. In terms of the average, however, the results shown in Fig. 8 are reasonable because basically the error of PDR is proportional to

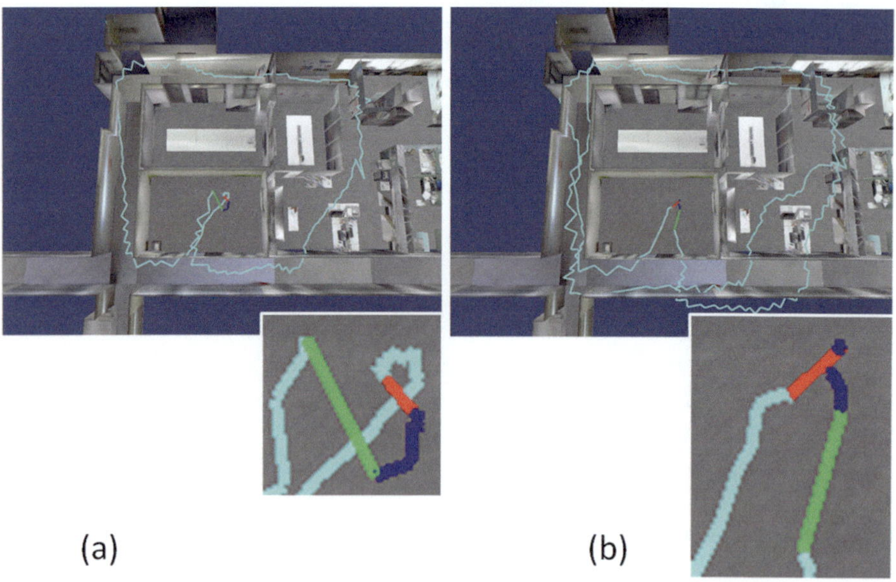

Fig. 7 Examples of localization results. (**a**) one cycle, (**b**) two cycles. Each *green line* indicates the moving distance of latter matching

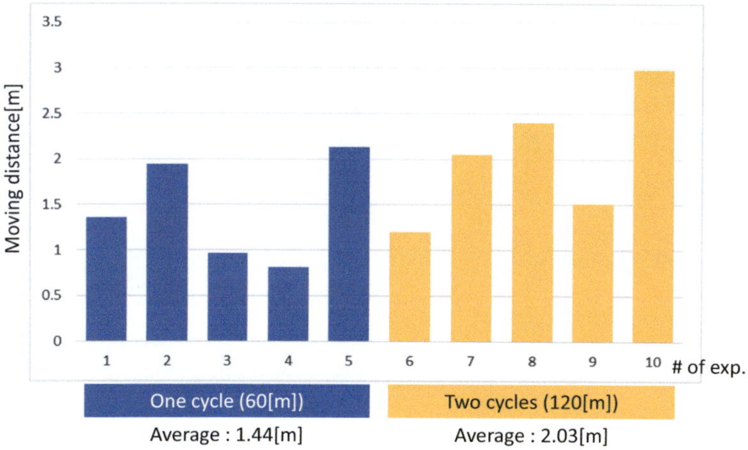

Fig. 8 Moving distances of latter matching

the walking distance. In future, in case we use lots of images for the key frame matching, we have to consider about the trade-off between the number of chances of the matching and the computational time. Specifically, spatial configuration and resolution of the images for the matching should be optimized. In the experiment, we set the resolution at 240 * 180, and the computational time was about 1–2 s.

Moreover in the future, we plan to conduct subjective experiments with a mock-up model of a machine. Figure 9 shows appearances of the tablet computer (Toshiba dynabook Tab for weight saving) and the mock-up. In order to study the effectiveness, we have been implementing both AV and AR applications to effectively create and show annotations with precise pointing techniques shown in [27]. In the experiments of this section, AV mode was only used for the map based navigation. However, we already have implemented both AR and AV mode to showing annotations. In experiments described in Sect. 5, we apply AV mode to show annotations, and conduct subjective experiments.

5 In-Field Evaluation

We conducted customer interest check (CIC) as in-field evaluations. The goal of the evaluations is to survey the acceptability, usefulness and usability of the prototype system for engineers and operators working in a factory. The evaluations were conducted with six maintenance engineers and operators working in France and six maintenance engineers and operators working in Japan. Procedures of evaluations are below.

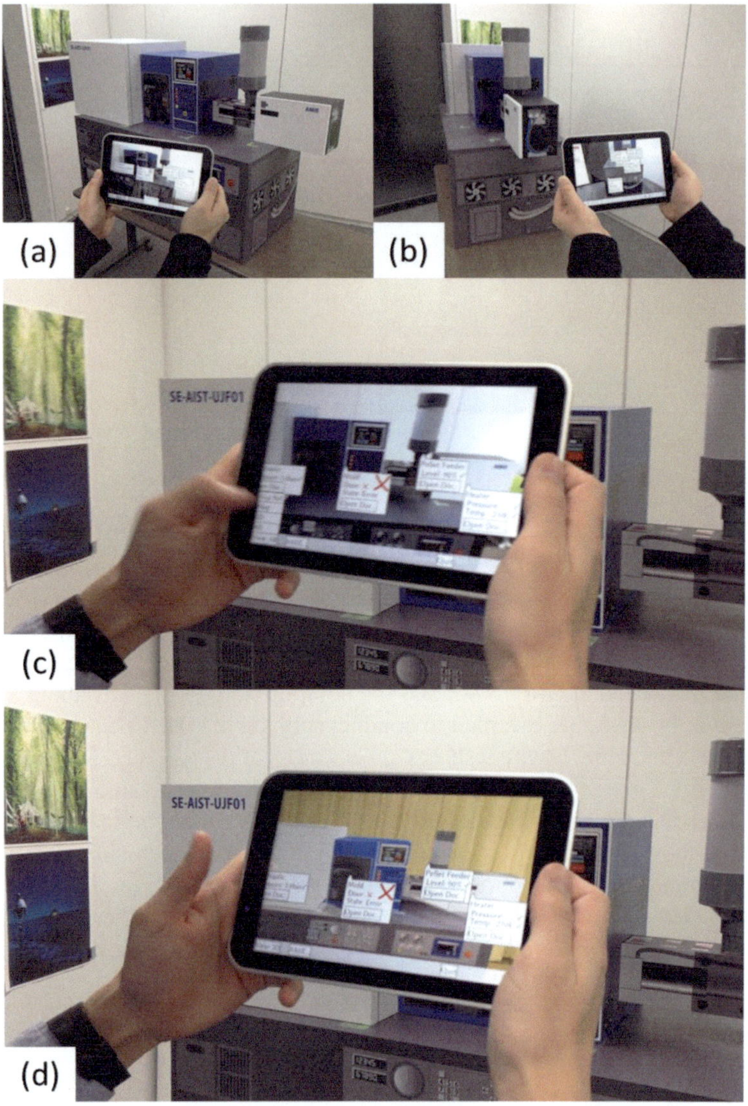

Fig. 9 Appearances of the tablet computer and the mock-up. (**a**) and (**b**): Overviews of the mock-up. (**c**): Annotation overlay with AR mode. (**d**): Annotation overlay with AV mode

5.1 Presentation of the Concept and the Movie

Each interviewee watches a movie for showing the concept of the proposed system. Interviewers supplement the explanation of the movie with their talking when the interviewee has questions about the proposed system.

5.2 *Evaluation of the Acceptability*

Each interviewee is asked 20 questions of CAUTIC [28] about the proposed system, and choose the one answer from "Acceptable", "Acceptable under conditions" and "Not acceptable". CAUTIC is an interview method for evaluating an acceptability of a product. CAUTIC has scientifically identified 20 criteria at 4 levels of analysis (technical, practical, user identity and user environment) that must be validated in order to permit to establish, beyond doubt, whether or not the compelling reason to buy exists. If the 20 criteria are not or only partially validated, the method identifies the problem areas from each of the 4 levels that obstruct acceptance of the innovation in question. These areas can be subsequently given remedial treatment, which in turn can be verified by the CAUTIC method if seemed necessary.

5.3 *Evaluation of Usefulness*

Each interviewee is asked a couple of questions about each function of the proposed system from KANO method [29]. In KANO method, each question is composed of a functional and dysfunctional questions, and the interviewee choose the one answer from "Like", "Must be", "Neutral", "Live with" and "Dislike". By combining the two answers, quality attributes are grouped into six categories (Attractive, One-dimensional, Must be, Questionable, Reverse, Indifferent) with a different impact on customer satisfaction [30].

Results of the evaluations are shown below.

Acceptability

Table 1 indicates CAUTIC's questions and of results of six respondents in France and Japan. The number of stars (* in the right side of Table 1) is decided with an average score M of the answers. We set scores (2,1,0) for "Acceptable", "Acceptable under conditions" and "Not acceptable" respectively to calculate M ($M >=$ 1.33: ***, $0.67 <= M < 1.33$: **, $M < 0.67$: *). Totally, the evaluation of the acceptability by CAUTIC reveals that the proposed system is well received by maintenance technicians and managers. In the results of France, average score M of 1.2 and 2.5 are comparatively low. In 1.2, there were several questions about how to connecting the machine to get information of the machine, and about scalability of the system. In 2.5, there were several questions about capability of false recognition of the machine, and about capacity to update technical documentations. In the results of Japan, average score M of 3.2 and 4.3 are comparatively low. The main reason of these results were interviewees' impression of the proposed system as their private role. Two interviewees answered "Not acceptable" for 3.2 and 4.3 because they do not have any ideas of using the proposed system privately.

As was pointed out in the results of France, scalability and the capability of false recognition are very important in terms of the comfort of the system. Experiments

Table 1 Results of CAUTIC

		France	Japan
Assimilation to know-how	1.1. Does the concept interest the user in principle?	***	***
	1.2. Is the concept's technical principle understandable enough for the user?	**	***
	1.3. Can the user connect / compare the concept techniques to existing techniques?	***	***
	1.4. Can the user easily identify and use the function (s) which interest him in the concept?	***	***
	1.5. Is the technical principle of the concept simple enough to be used at anytime?	***	***
Association to current practices	2.1. Does the concept seem simple to use for the user?	***	***
	2.2. Does the concept allow to complete current practices, without interfering with them?	***	***
	2.3. Does the comparison with the current practices validate and make credible the new suggested practices?	***	***
	2.4. Does the concept solve at least one practical problem that the user is facing?	***	***
	2.5. Can the user gradually foresee a new personal organization of his practices by using the concept?	**	***
Appropriation to identity	3.1. Does the user recognize himself as the targeted user and know who else the concept concerns?	***	***
	3.2. Can the user use the concept to act on his private / professional role?	***	**
	3.3. Is the concept compatible with the user's personal / professional values?	***	***
	3.4. Does the user appropriate the concept?	***	***
	3.5. Does the user imagine alternative or extended uses to the concept?	***	***
Adaptation to environment	4.1. Is the concept arriving on the market at the right time?	***	***
	4.2. Is the concept adapted to the user's client-supplier / family relations evolution?	***	***
	4.3. Is the concept adapted to the user's position in his/her professional / private circle?	***	**
	4.4. Is the concept adapted to the user's working organization/ way of living and its evolution?	***	***
	4.5. Does the user agree to pay for the concept?	***	***

in wide area with multiple machines are next steps of this study. Required time and false recognition rate of the key frame matching are supposed to become progressively higher with an increase in the number of key frames. In future, we plan to study fast search of the key frames. Human-computer interaction technique is one way to realize the fast search. For example, showing candidates of the key frames is supposed to be worth to shorten the required time of the image matching.

Table 2 Results of KANO

Functions of the proposed system	France		Japan	
	Dim1	Dim2	Dim1	Dim2
1: Visualization of the machine in Augmented Reality	Q(5)	R(1)	O(3)	A(2)
2: Guidance toward the machine	I(3)	A(2)	A(4)	O(2)
3: Maintenance history	I(2)	A(2)	O(3)	A(2)
4: Access to the diagnosis	A(4)	Q(2)	A(3)	O(2)
5: Technical documentation	A(3)	Q(2)	O(2)	A(2)
6: Access to parts inventory replacement	O(2)	N	O(4)	N
7: Contents of the electric cabinet without having to open the machine	A(3)	I(3)	O(3)	N
8: Creation of maintenance report	A(4)	N	O(4)	N
9: Freeze function	–	–	O(4)	N
10: Visualization of the machine in Augmented Virtuality	–	–	A(3)	I(2)

Usefulness
Table 2 indicates functions of the proposed system, and KANO's evaluation table of six respondents in France and Japan. First, in France, we applied eight questions for the KANO method. Subsequently, in Japan, we added two functions: "Freeze function" and "Visualization of the machine in Augmented Virtuality" to the system, and applied ten questions for the KANO method. "Freeze functions" is a function to temporarily stop updating the displayed image of AR/AV mode. We assume "Freeze functions" is helpful as hands-free capability. "Visualization of the machine in Augmented Virtuality " is a function to overlay annotations with AV mode as shown in Fig. 9d.

Dim1 and Dim2 indicate first and second components of the category of the answers (A: Attractive, M: Must-be, R: Reverse, O: One-dimensional, Q: Questionable, I: Indifferent, N: None) with the number of the answers. For example, "Q (5)" indicates that there are five answers categorised in "Questionable". In case the number of the category of multiple answers is one, we applied "None" as the component of the category. In the results of France, functions 4(Access to the diagnosis), 5(Technical documentation), 7(Contents of the electric cabinet without having to open the machine) and 8(Creation of maintenance report) were categorised in " Attractive" as first component. In the results of Japan, 2(Guidance toward the machine), 4(Access to the diagnosis) and 10(Visualization of the machine in Augmented Virtuality) were categorised in " Attractive" as first component.

Throughout the whole experiments, interviewees in Japan had more previous knowledge of AR than interviewees in France, and there were several answers categorised in "Questionable" only in France. 1(Visualization of the machine in Augmented Reality) was categorised in "Questionable" as first component in France. This is due to the difficulty of the users to imagine AR function before using it. Five interviewees could not understand the difference of the way of provision of information. Therefore, they selected the same answer for the both questions. Moreover, in the results in France, 4(Access to the diagnosis) and 5 (Technical documentation) were also categorised in "Questionable" as second component. These are also probably due to the difficulty to imagine AR function. On the contrary, in Japan, there were not any answers categorised in "Questionable".

6 Conclusion

This paper has proposed a maintenance service support system for wide-area indoor environment. In order to realize MR maintenance service support system, we proposed a hybrid localization technique realized with pedestrian dead reckoning (PDR) and 3D model-based image processing for the purpose of covering wide-area indoor environment. As a first step of the realization, this paper presents experimental results that show the relation between the walking distance and the moving distance of the key frame matching that is strongly related to the computational time of the key frame matching. As regards future works, we plan to study the optimization of the key frame matching to realize appropriate accuracy and computational time of the localization, and subjective experiments with both mock-up and real machine in wide indoor environments.

Acknowledgements This work was supported by Strategic Japanese-French Cooperative Program on Information and Communication Technology Including Computer Science (ANR in France and JST in Japan).

References

1. Henderson SJ, Feiner S (2009) Evaluating the benefits of augmented reality for task localization in maintenance of an armored personnel carrier turret. In: Proceedings of the IEEE international symposium on mixed and augmented reality 2009, Science and technology, pp 135–144
2. Platonov J, Heibe H, Meier P, Grollmann B (2006) A mobile markerless AR system for maintenance and repair. In: Proceedings of the 5th IEEE and ACM international symposium on mixed and augmented reality, pp 105–108
3. Klein G, Murray D (2009) Parallel tracking and mapping on a camera phone. In: Proceedings of the international symposium on mixed and augmented reality, pp 83–86
4. Reitmayr G, Drummond T (2007) Initialisation for visual tracking in urban environments. In: Proceedings of the international symposium on mixed and augmented reality, pp 161–172
5. Höllerer T, Feiner S, Terauchi T, Rashid G, Hallaway D (1999) Exploring MARS: developing indoor and outdooruser interfaces to a mobile augmented reality system. Comput Graph 23 (6):779–785
6. Saito S, Hiyama A, Tanikawa T, Hirose M (2007) Indoor marker-based localization using coded seamless pattern for interior decoration. In: Proceedings of the IEEE virtual reality 2007, pp 67–74

7. Tenmoku R, Kanbara M, Yokoya N (2003) A wearable augmented reality system using positioning infrastructures and a pedometer. In: Proceedings of the IEEE international symposium on wearable computers, pp 110–117
8. Yoshida H, Ito S, Kawaguchi N (2006) Evaluation of pre- acquisition methods for position estimation system using wireless LAN. The third International Conference on Mobile Computing and Ubiquitous networking (ICMU 2006), pp 148–155
9. Cipolla R, Robertson D, Tordoff B (2004) Image-based localization. In: Proceedings of the international conference on Virtual System and Multimedia (VSMM2004), pp 22–29
10. Kourogi M, Kurata T, Sakaue K (2001) A panorama-based method of personal positioning and orientation and its realtime applications for wearable computers. In: Proceedings of the fifth International Symposium on Wearable Computers (ISWC'01), pp 107–114
11. Torr P, Zisserman A (1999) Feature based methods for structure and motion estimation. International workshop on vision algorithms, pp 278–295
12. Bleser G, Wuest H, Stricker D (2006) Online camera pose estimation in partially known and dynamic scenes. In: Proceedings of the international symposium on mixed and augmented reality, pp 56–65
13. Simon G, Berger M (2002) Pose estimation from planar structures. IEEE Comput Graph Appl 22(6):46–53
14. Hähnel D, Burgard W, Thrun S (2003) Learning compact 3D models of indoor and outdoor environments with a mobile robot. Robot Auton Syst 44(1):15–27
15. Jensen B, Weingarten J, Kolski S, Siegwart R (2005) Laser range imaging using mobile robots: from pose estimation to 3D-models. In: Proceedings of the of 1st range imaging research day, pp 129–144
16. Neubert J, Pretlove J, Drummond T (2007) Semi-autonomous generation of appearance-based edge models from image sequences. In: Proceedings of the international symposium on mixed and augmented reality, pp 79–89
17. SketchUp. http://www.sketchup.com/
18. Ishikawa T, Thangamani K, Kourogi M, Gee AP, Mayol W, Jung K, Kurata T (2009) In-situ 3D indoor modeler with a camera and self-contained sensors. In: Proceedings of the HCII2009, LNCS 5622, pp 454–464
19. Langer M, BenHimane S (2010) An interactive vision-based 3D reconstruction workflow for industrial AR applications, online. In: Proceedings of the ISMAR 2010 workshop on augmented reality super models
20. Arth C, Wagner D, Klopschitz M, Irschara A, Schmalstieg D (2009) Wide area localization on mobile phones. In: Proceedings of the international symposium on mixed and augmented reality, pp 73–82
21. Klein G, Murray D (2008) Improving the agility of keyframe-based SLAM. In: Proceedings of the 10th European Conference on Computer Vision (ECCV'08)
22. Irschara A, Zach C, Frahm JM, Bischof H (2009) From structure-from-motion point clouds to fast location recognition. Conference on Computer Vision and Pattern Recognition (CVPR)
23. Kourogi M, Ishikawa T, Kurata T (2010) A method of pedestrian dead reckoning using action recognition. In: Proceedings of the IEEE/ION Position, Location and Navigation Symposium (PLANS), pp 85–89
24. Makita K et al (2013) Photo-shoot localization of a mobile camera based on registered frame data of virtualized reality models. In: Proceedings of the S&T ISMAR2013, pp 273–274
25. Bay H, Tuytelaars T, Gool LV (2006) SURF: speeded up robust features. In: Proceedings of the 9th European Conference on Computer Vision (ECCV'2006), Springer LNCS volume 3951, part 1, pp 404–417
26. Tomasi C, Kanade T (1991) Detection and tracking of point features. Technical Report CMU-CS-91-132, Carnegie Mellon University, April 1991
27. Vincent T, Nigay L, Kurata T (2013) Precise pointing techniques for handheld augmented reality. In: Proceedings of the INTERACT 2013, IFIP-Springer (2013), pp 122–139

28. CAUTIC (Conception Assistée par l'Usage pour les Technologies l'Innovation et le Changement). http://innovacs.upmf-grenoble.fr/innovacs/wp-content/uploads/2011/12/PPT-CAUTIC-Janv-2012.pdf
29. Kano N (1984) Attractive quality and must-be quality [in Japanese]. J Jpn Soc Qual Control 14 (2):147–156
30. Sauerwein E, Bailon F, Matzler K, Hinterhuber H (1996) The Kano model: how to delight your customers. In: Preprints volume I of the IX: international working seminar on production economics, pp 313–327

Business Structure of e-Book Service as a Product Service System: A Game Theoretic Approach

Nariaki Nishino and Keisuke Okuda

Abstract The study demonstrates a theoretic mechanism of e-book business using game theory. Recently, the e-book business has spread. Many e-book titles are now readily available through smart phones or dedicated devices. The e-book business incorporates aspects of service and products into one system, which can be regarded as a product service system. Thereby, stakeholders of various kinds are intertwined. Designing an appropriate product service system requires elucidation of its theoretic mechanism. Against this background, we first investigated the book industry in Japan and categorized its business structures into four cases: (1) paper-made books, (2) CD-ROM books, (3) multidevice-readable e-books, and (4) cellphone-readable e-books. Then e-book service models of four kinds were constructed, including several players such as publishers, device manufacturers, e-book stores, telecommunication firms, and consumers. Results show that, in a theoretical equilibrium, consumers are separable into two segments: consumers purchasing all e-books and consumers with no purchase. Its border is determined by conditions of production cost, commission of e-book sales, the number of available e-books, etc. Results of our study can contribute to clarification of the fundamental mechanisms underlying the e-book business.

Keywords Game theory • Business model • Categorization • Decision-making

1 Introduction

Recently e-books have spread widely along with the development of information communication technologies. Consumers can readily read e-books using several devices such as smart phones, tablets, and PCs. However, in contrast to paper-made books, e-books cannot be provided as an independent physical product because e-books require a device for reading. For that reason, an e-marketplace for e-book purchases is necessary. Content developers for e-books must be involved.

N. Nishino (✉) • K. Okuda
School of Engineering, The University of Tokyo, 7-3-1 Hongo, Bunkyo-ku, Tokyo 113-8656, Japan
e-mail: nishino@tmi.t.u-tokyo.ac.jp

T. Maeno et al. (eds.), *Serviceology for Designing the Future*,
DOI 10.1007/978-4-431-55861-3_9

Therefore, it is a typical product service system (PSS). Nowadays, numerous devices are available and e-book stores of many kinds have opened. However, compatibility among e-books is often lacking. Therefore, it is important to clarify its mechanism as a PSS to realize an appropriate business structure. Our study first investigates current e-book situations in Japan and categorizes them into four classes. Then, based on the categorization, we construct game theoretic models of four kinds comprising several players such as publishers, publications distributors, device manufacturers, and consumers. Using an approach of theoretical analysis with game theory, our study elucidates the fundamental mechanisms underlying the e-book business.

2 Industry Structures of Books/e-Books in Japan

Through the ages, paper-based books have been used and read by many readers worldwide. In the 1980s, CD-ROM type books first appeared in Japan. At that time, the CD-ROM format had been used for dictionaries because the format is extremely helpful for searching for words, especially in dictionaries. E-books first appeared in the 1990s. They started as e-books for PCs. After about 2002, e-books for cell phones came to be sold. Now e-books have come to be available in many media.

Fundamental differences of PSS structures suggest four structure categories: (1) paper-made books, (2) CD-ROM books, (3) multidevice-readable e-books, and (4) cellphone-readable e-books. According to previous studies [1, 2], each structure can be re-depicted as Fig. 1. However, it is important to ascertain a fundamental structure to consider it as a decision-making problem having an interdependent relation in e-book PSS. In subsequent sections, we simplify the structure of each business and clarify its relation.

2.1 Paper-Made Book Structure

In the business of paper-made books, publishers, publications distributors (called "toritsugi" in Japanese), consumers can be regarded as the main players. Publishers produce books and sell them wholesale to distributors. Then distributors generally choose how many books should be purchased from publishers. Books are then distributed to book stores by distributors. Consumers finally find the printed books there. That relation is a one-way directional flow, as depicted in Fig. 2a.

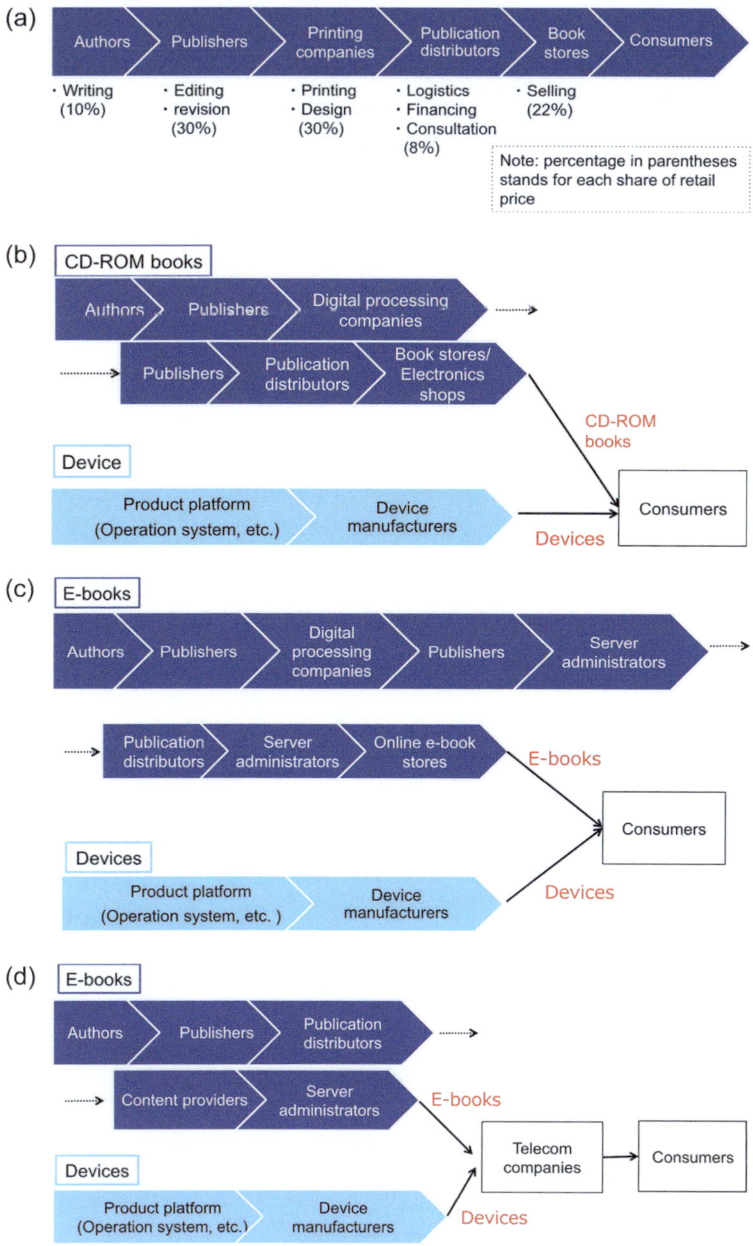

Fig. 1 Business structures of four kinds. (**a**) Paper-made books. (**b**) CD-ROM books. (**c**) Multidevice-readable e-books. (**d**) Cellphone-readable e-books

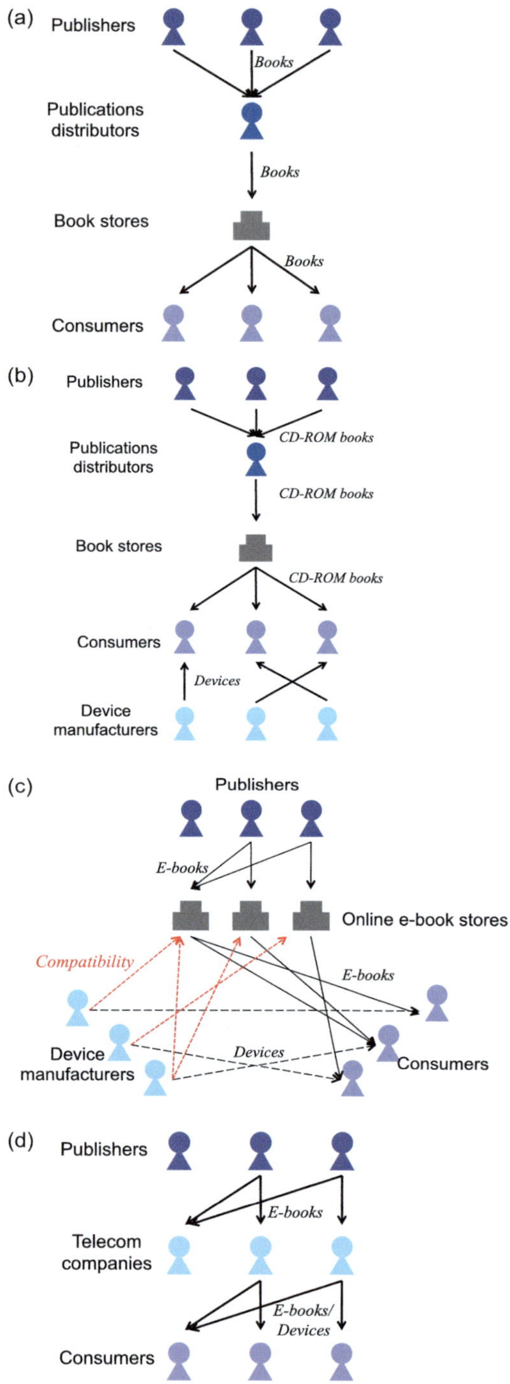

Fig. 2 Relation among main players. (**a**) Paper-made books. (**b**) CD-ROM books. (**c**) Multidevice-readable e-books. (**d**) Cellphone-readable e-books

2.2 CD-ROM Book Structure

The CD-ROM book business structure resembles that of paper-made books: publishers, publications distributors, and consumers are the main players. In addition to them, device manufacturers are important players because a CD-ROM cannot be read without some device for reading. In the mid-1980s, when CD-ROM books first appeared, several reading formats existed. Now CD-ROMs can be read by personal computers. Therefore, it is reasonable to infer that no compatibility problem exists. A business structure of this type is depicted in Fig. 2b.

2.3 Multi-Device-Readable e-Book Structure

To read e-books, the e-book data and proprietary devices are required. Several players are involved in this business structure. The main players are publishers, device manufacturers, online stores for e-books, and consumers. The device manufacturers produce devices for reading e-books. Publishers produce the contents of e-books and digitize them; occasionally they commission its digital processing. Consumers must purchase devices and e-books independently. In addition, compatibility problems arise. Many device manufacturers and many online e-book stores have different standards for formatting e-book digital data. Therefore, accessible e-books are limited depending on the kind of device. The business structure is presented in Fig. 2c.

2.4 Cellphone-Readable e-Book Structure

Cellphone-readable e-books are characteristic of the Japanese e-book business. This type of e-book service is generally provided by telecom companies that sell cellphones. In this business, a telecom company usually certifies some e-book stores as official stores so that cellphone users can access and purchase them only from certified stores. Especially in Japan, telecom companies do a business of selling cellphones. Device manufacturers do not generally sell cellphones directly to consumers. Therefore, telecom companies can be regarded as a unified entity, including the role of manufacturer. This structure is shown in Fig. 2d.

3 Game Theoretic Modeling

In this section, the four business structures presented previously are modeled based on the game theory framework.

3.1 Some Assumptions for Modeling

In modeling, we use the following assumptions for simplicity and clarity of analysis.

- Each model has only one type of e-book. We do not consider a case with sales of paper-made books and cellphone-readable e-books, for example.
- The models incorporate no qualitative difference between devices, such as usability or familiarity with a device interface.
- If book titles differ, they are regarded as heterogeneous goods. Book quality is not considered.
- A consumer cannot buy books of the same title and cannot buy more than one device.

4 Model 1: Paper-Made Book

4.1 Publishers

The model includes multiple publishers. A set of publishers is defined as $Q = \left\{q_{(1)}, q_{(2)}, \ldots\right\}$,where $q_{(i)}$stands for the i-th publisher.[1] A publisher, $q \in Q$, has a set of books, B^q, which can be published. Now a set of all books that all publishers can publish is defined as shown below.

$$B = \bigcup_{q \in Q} B^q$$

Each publisher makes a decision related to the book price p_1^q. In our model, each publisher has a different book price, but prices are equal for the same publisher.

4.1.1 Publications Distributor

The model has only one publication distributor. The distributor decides how many books are sold wholesale from respective publishers. The distributor chooses the tuple, $m_1^a = \left(m_{(1)}, m_{(2)}, \ldots\right)$, where $m_{(i)}$ stands for the number of books sold

[1] In our model, the following description rule for variables is introduced: superscript stands for player kind and subscript stands for model number. For example, Π_1^q means the publisher's profit in model 1. For superscript, we use q as a publisher, a as a publications distributor, c as a consumer, d as a device manufacturer, and t as a telecom company. On the other hand, if we use a number with parentheses for subscript like $q_{(1)}$, then it presents the index for players.

wholesale from the i-th publisher. M_1^a is the total number of books the distributor purchased.

4.1.2 Consumers

A number of consumers are described as a set, $C = \{c_{(1)}, c_{(2)}, \ldots\}$. Each consumer makes a decision of purchasing books. We define a set of books purchased by a consumer c as $Y_1^c \subseteq B$; a set of consumers who purchased a book $b_{(i)}$ is $N_{(i)}$, which is formulated as

$$N_{(i)} = \{c \mid b_{(i)} \in Y_1^c,\ c \in C\}.$$

It is noteworthy that $b_{(i)}$ stands for the i-th book in all books. For any i, the following inequality must hold.

$$|N_{(i)}| \leq m_{(i)}.$$

4.1.3 Profit Function for Each Player

The profit function of a publisher, q, is described as

$$\Pi_1^q(p_1^q) = \alpha p_1^q |N_1^q| - f_1^q - c_1^q M_1^a - \omega\left(M_1^a - |N_1^q|\right),$$

where the first term means the book sales; α is the commission rate that a publisher takes from the total sales; f_1^q and c_1^q means a fixed cost and variable cost for book production; $\omega(\cdot)$ stands for a disposal cost. Here, N_1^q stands for a set of books that publisher q sold and is defined as the following formula.

$$N_1^q = \bigcup_{c \in C} N_1^{q|c}$$

In that equation, $N_1^{q|c}$ means that a set of books purchased by consumer c and which are published by publisher q; and $N_1^{q|c}$ can be formulated using a mapping from a book to its publisher, $g : B \rightarrow Q$, as shown below.

$$N_1^{q|c} = \{y \mid g(y) = q,\ y \in Y_1^c\}.$$

The distributor's profit is defined as

$$\Pi_1^a(M_1^a) = \beta p_1^q |N_1^q| - t(M_1^a),$$

where the first term represents the sales, β is the commission rate that the distributor takes, and the second term, $t(\cdot)$, denotes the transportation cost.

Consumer utility is described as

$$U_1^c = u_1^c |Y_1^c| - \sum_{q \in Q} p_1^q |N_1^{q|c}|.$$

The first term denotes a sort of satisfaction obtained by purchasing books. Here, u_1^c stands for the utility for one book. In other words, it can be regarded as a reservation price for one book. The second term signifies a payment for purchase. Because of consistency with models 3 and 4, we assume that the following formula is satisfied.

$$u_1^c = \frac{v^{c\prime}}{|B|}$$

Herein, $v^{c\prime}$ means a consumer type and is assumed to be distributed uniformly in [0,1].

4.2 *Model 2: CD-ROM Book*

4.2.1 Publishers

It is the same as that in model 1. Each publisher chooses a book price p_2^q.

4.2.2 Publications Distributor

Similarly to model 1, the distributor chooses m_2^a.

4.2.3 Device Manufacturers

The model includes multiple manufacturers. Each produces one kind of device that is differentiated from the other manufacturers' devices. A set of manufacturers is defined as $D = \{d_{(1)}, d_{(2)}, \ldots\}$. Each device manufacturer chooses a device price p_2^d. CD-ROM is standardized and is compatible with many devices. Therefore, we assume that CD-ROM books can be read by any device.

4.2.4 Consumers

In the model, each consumer makes a decision to purchase a device and CD-ROM books. Specifically, they respectively decide a tuple of $\left(s_2^c, Y_2^c\right)$, where $s_2^c \in D$ is a device purchased and $Y_2^c \subseteq B$ is a set of CD-ROM books purchased. A consumer who does not purchase any device cannot read any CD-ROM book.

4.2.5 Profit Function for Each Player

The profit function of a publisher q is

$$\Pi_2^q\left(p_2^q\right) = \alpha p_2^q \left|N_2^q\right| - f_2^q - c_2^q M_2^a - \omega\left(M_2^a - \left|N_2^q\right|\right).$$

This function is fundamentally the same as that in model 1. The distributor's profit is defined as

$$\Pi_2^a\left(M_2^a\right) = \beta p_2^q \left|N_2^q\right| - t\left(M_2^a\right).$$

Then the device manufacturer's profit function is

$$\Pi_2^d\left(p_2^d\right) = \left(p_2^d - c_2^d\right)\left|N_2^d\right| - f_2^d,$$

where c_2^d and f_2^d respectively stand for a variable cost and fixed cost for device manufacturing. N_2^d represents a set of consumers who purchased devices produced by a manufacturer, d, which is defined as

$$N_2^d = \{\, c \mid s_2^c = d,\ c \in C\}.$$

A consumer's utility is calculated as shown below.

$$U_2^c = \begin{cases} m_2\left|X_2^d\right| + v^c \dfrac{\left|Y_2^c\right|}{|B|} - \sum\limits_{q \in Q} p_2^q \left|N_2^{q|c}\right| - p_2^d & \text{(if purchasing)} \\ 0 & \text{(if no purchasing)} \end{cases}$$

The first term represents the indirect effect of network externality, meaning that the utility increases as the number of books that are read by the device purchased increases X_2^d means a set of all CD-ROM books. The effect is assumed here as a linear function using a coefficient m_2. v^c represents a type for consumer c and is distributed uniformly in [0,1]. The second term represents the direct satisfaction derived from one PSS including some books. The third term stands for the total payment for purchased CD-ROM books. The fourth term is the price for the purchased device.

4.3 Model 3: Multidevice-Readable e-Book

4.3.1 Publishers

Similarly to models 1 and 2, each publisher respectively has an exogenous variable of B^q as a set of e-books. However, unlike models 1 and 2, a publisher must make a decision about which online e-book stores should be contracted with. Therefore, they choose a tuple of $\left(p_3^q, S_3^q\right)$, where p_3^q is the e-book price and S_3^q stands for a set of online e-book stores to be distributed. Now we define a set of all online e-book stores as $R = \{r_{(1)}, r_{(2)}, \ldots\}$. Thereby S_3^q can be described as a subset of R. For our model, we assume that the same title of e-books is not sold from another store.

4.3.2 Device Manufacturers

Unlike model 2, each device manufacturer chooses a set of online stores that is accessible from their own device. Accordingly, a manufacturer decides a tuple of $\left(p_3^d, S_3^d\right)$, where p_3^d is the device price and S_3^d is a subset of R. Here, a set of e-books that is accessible from the device produced by manufacturer d is

$$X_3^d = \bigcup_{r \in S_3^d} f\left(S_3^q\right)(r),$$

where $f\left(S_3^q\right)$ is defined as a correspondence, $f\left(S_3^q\right) : R \rightarrow 2^B$, which means that under the decision of S_3^q, the correspondence returns the set of available e-books in online store r.

4.3.3 Consumers

Similarly to model 2, each consumer chooses a tuple of $\left(s_3^c, Y_3^c\right)$, where s_3^c stands for a device purchased and Y_3^c is a set of e-books purchased. Consumers can purchase e-books only from the set $X_3^{s_3^c}$.

4.3.4 Profit Function for Each Player

Publisher's profit function is described as

$$\Pi_3^q\left(p_3^q, S_3^q\right) = (1 - \varepsilon) p_3^q \left|N_3^q\right| - f_3^q,$$

where ε stands for the commission rate that publishers must pay to online stores. f_3^q denotes a fixed cost for e-book production. Here, we assume that it requires a fixed cost for e-book production but it requires no variable cost for additional production because digital data can be copied with very little expense.

The device manufacturer's profit is described as follows in the same manner as that used for model 2:

$$\Pi_3^d(p_3^d) = (p_3^d - c_3^d)|N_3^d| - f_3^d.$$

The consumer utility is also the same.

$$U_3^c = \begin{cases} m_3|X_3^d| + v^c \dfrac{|Y_3^c|}{|B|} - \sum\limits_{q \in Q} p_3^q |N_3^{q|c}| - p_3^d & \text{(if purchasing)} \\ 0 & \text{(if no purchasing)} \end{cases}$$

4.4 *Model 4: Cellphone-Readable e-Book*

4.4.1 Publishers

Each publisher chooses the e-book price as p_4^q.

4.4.2 Telecom Companies

The model includes multiple telecom companies. They are, respectively, selling one kind of cellphone. A set of telecom companies is defined as $T = \{t_{(1)}, t_{(2)}, \ldots\}$. Each telecom company makes a decision of (p_4^t, S_4^t), where p_4^t is cellphone price and S_4^t is a set of publishers that a telecom company t admits to be official publishers for that company. A set of e-books that is accessible from a certain cellphone is described as presented below.

$$X_4^t = \bigcup_{q \in S_4^t} B^q$$

4.4.3 Consumers

Similarly to models 2 an 3, each consumer makes a decision of a tuple (s_4^c, Y_4^c), where s_4^c stands for a cellphone purchased and where Y_4^c is a set of e-books purchased.

4.4.4 Profit Function for Each Player

The publisher's profit function is

$$\Pi_4^q(p_4^q, S_4^q) = (1-\zeta)p_4^q|N_4^q| - f_4^q,$$

where ζ stands for the commission rate that telecom companies take from the total sales. f_4^q represents a fixed cost for e-book production.

The telecom company's profit function is defined as

$$\Pi_4^t(p_4^t) = \zeta p_4^q|N_4^q| + (p_4^t - c_4^t)|N_4^t| - f_4^t,$$

where the first term denotes revenues derived from the commission of e-books sales. Also, c_4^t and f_4^t respectively denote the variable and fixed costs for cellphone production.

The consumer utility function is shown below.

$$U_4^c = \begin{cases} m_4|X_4^t| + v^c\dfrac{|Y_4^c|}{|B|} - \sum\limits_{q\in Q} p_4^q|N_4^{q|c}| - p_4^t & \text{(if purchasing)} \\ 0 & \text{(if no purchasing)} \end{cases}$$

5 Theoretical Analysis

For theoretical analysis, a Nash equilibrium is derived. A Nash equilibrium is the state of theoretical predictable outcome under the assumption that all players follow perfectly rational behavior. Unrealistic assumptions are often given, but it is important to understand the theoretical and fundamental mechanism of PSS.

For simplicity, we derive a Nash equilibrium in which there is only one player of each kind. The equilibrium is presented in Tables 1, 2, 3, and 4. In model 1, two states are obtained depending on the conditions of costs, the number of books, and the commission rate. In case 1-1, most variables are zero, meaning the state in which no one purchases any books. In case 1-2, it means the state in which about half of consumers purchased books. Figure 3 presents such a situation and presents an intuitive explanation for it. Figures 4, 5 and 6 present respective situations for cases 2, 3 and 4.

Regarding model 2, it is fundamentally the same as that in model 1. In all, cases of four kinds are obtained depending on the conditions of costs and etc. Cases 2-1 and 2-2 correspond respectively to cases 1-1 and 1-2. Case 2-4 presents the state in which some consumers have a low reservation price to purchase devices but who purchase no e-book. In case 2-3, both states of cases 2-2 and 2-4 become the equilibrium state. Conditions of models 3 and 4 differ from those of model 2, but the fundamental properties resemble those of model 2.

Table 1 Theoretical equilibrium in Model 1

(a) Case 1-1		
Condition of $\alpha < c_1^q\lvert B^q\rvert$		
Publisher	$p_1^{q^*} > \frac{1}{\lvert B^q\rvert}$	$\prod_1^{q^*} = -f_1^q$
Distributor	$M_1^{a^*} = 0$	$\Pi_1^{a^*} = 0$
Consumers	$Y_1^{c^*} = \phi$	$U_1^{c^*} = 0$
(b) Case 1-2		
Condition of $\alpha \geq c_1^q\lvert B^q\rvert$, $\beta\frac{\{(\alpha\bar{v}'^c)^2-(c_1^q\lvert B^q\rvert)^2\}}{4\alpha^2} \geq t(M_1^{a^*})$		
Publisher	$p_1^{q^*} = \frac{\alpha\bar{v}'^c + c_1^q\lvert B^q\rvert}{2\alpha\lvert B^q\rvert}$	$\prod_1^{q^*} = \frac{(\alpha\bar{v}'^c - c_1^q\lvert B^q\rvert)^2}{4\alpha} - f_1^q$
Distributor	$M_1^{a^*} = \frac{(\alpha\bar{v}'^c - c_1^q\lvert B^q\rvert)\lvert B^q\rvert}{2\alpha}$	$\prod_1^{a^*} = \beta\frac{\{(\alpha\bar{v}'^c)^2-(c_1^q\lvert B^q\rvert)^2\}}{4\alpha^2} - t(M_1^{a^*})$
Consumers	$Y_1^{c^*} = \begin{cases} \phi & \text{if } 0 \leq v^c < v^* \\ B^q & \text{if } v^* \leq v^c \leq \bar{v}'^c \end{cases}$	$U_1^{c^*} = \begin{cases} 0 & \text{if } 0 \leq v'^c < v'^* \\ v'^c - v'^* & \text{if } v'^* \leq v'^c \leq \bar{v}'^c \end{cases}$

6 Conclusion

Our study addresses e-book business as one example of product service systems and investigates actual business situation in Japan. Classifying e-book business into four types, we construct preliminary game theoretic models. Theoretical analysis presents that e-book business has a theoretical equilibrium state that all consumers purchase devices. However, paper-made book business has no such tendency. This is one of remarkable points.

The current analysis focuses on the simple situation that the number of each player is only one. But in general there exist competition among players. Different characteristics might be observed in case of competitive players. Therefore, we should extend the model to more competitive or complicated situations.

Table 2 Theoretical equilibrium in Model 2

(a) Case 2-1		
Condition of $\gamma\overline{v}^c + \gamma m_2\|B^q\| - \gamma c_2^d - c_2^q\|B^q\| < 0$		
Publisher	$p_2^{d^*} > m_2\|B^q\|$ and $p_2^{d^*} - m_2\|B^q\| + p_2^{q^*}\|B^q\| > \overline{v}^c$	$\prod_2^{q^*} = -f_2^q$
Manufacturer		$\prod_2^{d^*} = -f_2^d$
Distributor	$M_2^{a^*} = 0$	$\prod_2^{a^*} = 0$
Consumers	$Y_2^{c^*} = \phi$	$U_2^{c^*} = 0$
(b) Case 2-2		
Condition of $\gamma\overline{v}^c + \gamma m_2\|B^q\| - \gamma c_2^d - c_2^q\|B^q\| \geq 0$ and $2\gamma(m_2\|B^q\| - c_2^d) \leq 3\gamma\overline{v}^c + c_2^q\|B^q\| - 2\sqrt{2\gamma\overline{v}^c(\gamma\overline{v}^c + c_2^q\|B^q\|)}$		
Publisher	$p_2^{q^*} = \frac{\gamma(\overline{v}^c + m_2\|B^q\|) - \gamma c_2^d + 2c_2^q\|B^q\|}{3\gamma\|B^q\|}$	$\Pi_2^{q^*} = \frac{(\gamma\overline{v}^c + \gamma m_2\|B^q\| - \gamma c_2^d - c_2^q\|B^q\|)^2}{9\gamma} - f_2^q$
Manufacturer	$p_2^{d^*} = \frac{\gamma(\overline{v}^c + m_2\|B^q\|) + 2\gamma c_2^d - c_2^q\|B^q\|}{3\gamma}$	$\Pi_2^{d^*} = \frac{(\gamma\overline{v}^c + \gamma m_2\|B^q\| - \gamma c_2^d - c_2^q\|B^q\|)^2}{9\gamma^2} - f_2^d$
Distributor	$M_2^{d^*} = \left(\frac{\gamma\overline{v}^c + \gamma m_2\|B^q\| - \gamma c_2^d - c_2^q\|B^q\|}{3\gamma}\right)\|B^q\|$	$\Pi_2^{a^*} = \delta p_2^{q^*} M_2^{a^*} - t(M_2^{a^*})$
Consumers	$(Y_2^{c^*}, s_2^{c^*}) = \begin{cases} (\phi,\ NoPurchase) & if\ 0 \leq v^c < v^{c}* \\ (B^q, d) & if\ v^{c}* \leq v^c \leq \overline{v}^c \end{cases}$	$U_2^{c^*} = \begin{cases} 0 & if\ 0 \leq v^c < v^{c}* \\ v^c - v^{c}* & if\ v^{c}* \leq v^c \leq \overline{v}^c \end{cases}$
(c) Case 2-3-1		
Condition of $2\gamma(m_2\|B^q\| - c_2^d) > 3\gamma\overline{v}^c + c_2^q\|B^q\| - 2\sqrt{2\gamma\overline{v}^c(\gamma\overline{v}^c + c_2^q\|B^q\|)}$ and $\gamma\overline{v}^c - 2\gamma m_2\|B^q\| + 2\gamma c_2^d - c_2^q\|B^q\| \geq 0$		
Publisher	$p_2^{q^*} = \frac{\gamma(\overline{v}^c + m_2\|B^q\|) - \gamma c_2^d + 2c_2^q\|B^q\|}{3\gamma\|B^q\|}$	$\Pi_2^{q^*} = \frac{(\gamma\overline{v}^c + \gamma m_2\|B^q\| - \gamma c_2^d - c_2^q\|B^q\|)^2}{9\gamma} - f_2^q$
Manufacturer	$p_2^{d^*} = \frac{\gamma(\overline{v}^c + m_2\|B^q\|) + 2\gamma c_2^d - c_2^q\|B^q\|}{3\gamma}$	$\Pi_2^{d^*} = \frac{(\gamma\overline{v}^c + \gamma m_2\|B^q\| - \gamma c_2^d - c_2^q\|B^q\|)^2}{9\gamma^2} - f_2^d$
Distributor	$M_2^{a^*} = \left(\frac{\gamma\overline{v}^c + \gamma m_2\|B^q\| - \gamma c_2^d - c_2^q\|B^q\|}{3\gamma}\right)\|B^q\|$	$\Pi_2^{a^*} = \delta p_2^{q^*} M_2^{a^*} - t(M_2^{a^*})$
Consumers	$(Y_2^{c^*}, S_2^{c^*}) = \begin{cases} (\phi,\ NoPurchase) & if\ 0 \leq v^c < v^{c}* \\ (B^q, d) & if\ v^{c}* \leq v^c \leq \overline{v}^c \end{cases}$	$U_2^{c^*} = \begin{cases} 0 & if\ 0 \leq v^c < v^{c}* \\ v^c - v^{c}* & if\ v^{c}* \leq v^c \leq \overline{v}^c \end{cases}$

(d) Case 2-3-2		
Condition of $2\gamma(m_2\|B^q\| - c_2^d) > 3\gamma\overline{v}^c + c_2^q\|B^q\| - 2\sqrt{2\gamma\overline{v}^c(\gamma\overline{v}^c + c_2^q\|B^q\|)}$ and $\gamma\overline{v}^c - 2\gamma m_2\|B^q\| + 2\gamma c_2^d - c_2^q\|B^q\| \geq 0$		
Publisher	$p_2^{q^*} = \frac{\gamma\overline{v}^c + c_2^q\|B^q\|}{2\gamma\|B^q\|}$	$\Pi_2^{q^*} = \frac{(\gamma\overline{v}^c - c_2^q\|B^q\|)^2}{4\gamma} - f_2^q$
Manufacturer	$p_2^{d^*} = m_2\|B^q\|$	$\Pi_2^{d^*} = (m_2\|B^q\| - c_2^d)\overline{v}^c - f_2^d$
Distributor	$M_2^{a^*} = \left(\frac{\gamma\overline{v}^c - c_2^q\|B^q\|}{2\gamma}\right)\|B^q\|$	$\Pi_2^{a^*} = \delta p_2^{q^*} M_2^{a^*} - t(M_2^{a^*})$
Consumers	$(Y_2^{c^*}, s_2^{c^*}) = \begin{cases} (\phi,\ NoPurchase) & if\ 0 \leq v^c < v^{c*} \\ (B^q, d) & if\ v^{c*} \leq v^c \leq \overline{v}^c \end{cases}$	$U_2^{c^*} = \begin{cases} 0 & if\ 0 \leq v^c < v^{c*} \\ v^c - v^{c*} & if\ v^{c*} \leq v^c \leq \overline{v}^c \end{cases}$
(e) Case 2-4		
Condition of $\gamma\overline{v}^c - 2\gamma m_2\|B^q\| + 2\gamma c_2^d - c_2^q\|B^q\| < 0$		
Publisher	$p_2^{q^*} = \frac{\gamma\overline{v}^c + c_2^q\|B^q\|}{2\gamma\|B^q\|}$	$\Pi_2^{q^*} = \frac{(\gamma\overline{v}^c - c_2^q\|B^q\|)^2}{4\gamma} - f_2^q$
Manufacturer	$p_2^{d^*} = m_2\|B^q\|$	$\Pi_2^{d^*} = (m_2\|B^q\| - c_2^d)\overline{v}^c - f_2^d$
Distributor	$M_2^{a^*} = \left(\frac{\gamma\overline{v}^c - c_2^q\|B^q\|}{2\gamma}\right)\|B^q\|$	$\Pi_2^{a^*} = \delta p_2^{q^*} M_2^{a^*} - t(M_2^{a^*})$
Consumers	$(Y_2^{c^*}, S_2^{c^*}) = \begin{cases} (\phi,\ NoPurchase) & if\ 0 \leq v^c < v^{c*} \\ (B^q, d) & if\ v^{c*} \leq v^c \leq \overline{v}^c \end{cases}$	$U_2^{c^*} = \begin{cases} 0 & if\ 0 \leq v^c < v^{c*} \\ v^c - v^{c*} & if\ v^{c*} \leq v^c \leq \overline{v}^c \end{cases}$

Table 3 Theoretical equilibrium in Model 3

(a) Case 3-1-1		
Condition of $m_3\lvert B^q\rvert + \overline{v}^c < c_3^d$		
Publisher	$p_3^{q^*} = any,\ \left(S_3^{q^*},\ S_3^{d^*}\right) = (\phi, \phi)$	$\Pi_3^{q^*} = -f_3^q$
Manufacturer	$p_3^{d^*} \neq 0$	$\Pi_3^{d^*} = -f_3^d$
Consumers	$\left(Y_3^{c^*}, s_3^{c^*}\right) = (\phi,\ NoPurchase)$	$U_3^{c^*} = 0$
(b) Case 3-1-2		
Condition of $m_3\lvert B^q\rvert + \overline{v}^c < c_3^d$		
Publisher	$p_3^{d^*} > m_3\lvert B^q\rvert$ and $p_3^{d^*} - m_3\lvert B^q\rvert + p_3^{q^*}\lvert B^q\rvert > \overline{v}^c$	$\Pi_3^{q^*} = -f_3^q$
Manufacturer		$\Pi_3^{d^*} = -f_3^d$
Consumers	$\left(Y_3^{c^*}, s_3^{c^*}\right) = (\phi,\ NoPurchase)$	$U_3^{c^*} = 0$
(c) Case 3-2		
Condition of $m_3\lvert B^q\rvert - \frac{3-2\sqrt{2}}{2}\overline{v}^c < c_3^d \leq m_3\lvert B^q\rvert + \overline{v}^c$		
Publisher	$p_3^{q^*} = \frac{\overline{v}^c + m_3\lvert B^q\rvert - c_3^d}{3\lvert B^q\rvert},\ S_3^{q^*} = R$	$\Pi_3^{q^*} = \frac{(1-\varepsilon)(\overline{v}^c + m_3\lvert B^q\rvert - c_3^d)^2}{9} - f_3^q$
Manufacturer	$p_3^{d^*} = \frac{\overline{v}^c + m_3\lvert B^q\rvert + 2c_3^d}{3},\ S_3^{d^*} = R$	$\Pi_3^{d^*} = \frac{(\overline{v}^c + m_3\lvert B^q\rvert - c_3^d)^2}{9} - f_3^d$
Consumers	$(Y_3^{c^*}, s_3^{c^*}) = \begin{cases} (\phi, NoPurchase) & if\ 0 \leq v^c < v^{c*} \\ (B^q, d) & if\ v^{c*} \leq v^c \leq \overline{v}^c \end{cases}$	$U_3^{c^*} = \begin{cases} 0 & if\ 0 \leq v^c < v^{c*} \\ v^c - v^{c*} & if\ v^{c*} \leq v^c \leq \overline{v}^c \end{cases}$
(d) Case 3-3-1		
Condition of $m_3\lvert B^q\rvert - \frac{1}{2}\overline{v}^c < c_3^d \leq m_3\lvert B^q\rvert - \frac{3-2\sqrt{2}}{2}\overline{v}^c$		
Publisher	$p_3^{q^*} = \frac{\overline{v}^c + m_3\lvert B^q\rvert - c_3^d}{3\lvert B^q\rvert},\ S_3^{q^*} = R$	$\Pi_3^{q^*} = \frac{(1-\varepsilon)(\overline{v}^c + m_3\lvert B^q\rvert - c_3^d)^2}{9} - f_3^q$
Manufacturer	$p_3^{d^*} = \frac{\overline{v}^c + m_3\lvert B^q\rvert + 2c_3^d}{3},\ S_3^{d^*} = R$	$\Pi_3^{d^*} = \frac{(\overline{v}^c + m_3\lvert B^q\rvert - c_3^d)^2}{9} - f_3^d$
Consumers	$(Y_3^{c^*}, s_3^{c^*}) = \begin{cases} (\phi, NoPurchase) & if\ 0 \leq v^c < v^{c*} \\ (B^q, d) & if\ v^{c*} \leq v^c \leq \overline{v}^c \end{cases}$	$U_3^{c^*} = \begin{cases} 0 & if\ 0 \leq v^c < v^{c*} \\ v^c - v^{c*} & if\ v^{c*} \leq v^c \leq \overline{v}^c \end{cases}$

(e) Case 3-3-2		
Condition of $m_3\lvert B^q\rvert - \frac{1}{2}\overline{v}^c < c_3^d \le m_3\lvert B^q\rvert - \frac{3-2\sqrt{2}}{2}\overline{v}^c$		
Publisher	${p_3^q}^* = \frac{\overline{v}^c}{2\lvert B^q\rvert}$, ${S_3^q}^* = R$	${\Pi_3^q}^* = \frac{(1-\varepsilon)}{4}\overline{v}^{c2} - f_3^q$
Manufacturer	${p_3^d}^* = m_3\lvert B^q\rvert$, ${S_3^d}^* = R$	${\Pi_3^d}^* = \left(m_3\lvert B^q\rvert - c_3^d\right)\overline{v}^c - f_3^d$
Consumers	$({Y_3^c}^*, {s_3^c}^*) = \begin{cases} (\phi, d) & \mathit{if}\ 0 \le v^c < \frac{1}{2} \\ (B^q, d) & \mathit{if}\ \frac{1}{2} \le v^c \le \overline{v}^c \end{cases}$	$U_3^c = \begin{cases} 0 & \mathit{if}\ 0 \le v^c < \frac{1}{2} \\ v^c - \frac{1}{2} & \mathit{if}\ \frac{1}{2} \le v^c \le \overline{v}^c \end{cases}$
(f) Case 3-4		
Condition of $c_3^d < m_3\lvert B^q\rvert - \frac{1}{2}\overline{v}^c$		
Publisher	${p_3^q}^* = \frac{\overline{v}^c}{2\lvert B^q\rvert}$, ${S_3^q}^* = R$	${\Pi_3^q}^* = \frac{(1-\varepsilon)}{4}\overline{v}^{c2} - f_3^q$
Manufacturer	${p_3^d}^* = m_3\lvert B^q\rvert$, ${S_3^d}^* = R$	${\Pi_3^d}^* = (m_3\lvert B^q\rvert - c_3^d)\overline{v}^c - f_3^d$
Consumers	$({Y_3^c}^*, {s_3^c}^*) = \begin{cases} (\phi, d) & \mathit{if}\ 0 \le v^c < \frac{1}{2} \\ (B^q, d) & \mathit{if}\ \frac{1}{2} \le v^c \le \overline{v}^c \end{cases}$	$U_3^c = \begin{cases} 0 & \mathit{if}\ 0 \le v^c < \frac{1}{2} \\ v^c - \frac{1}{2} & \mathit{if}\ \frac{1}{2} \le v^c \le \overline{v}^c \end{cases}$

Table 4 Theoretical equilibrium in Model 4

(a) Case 4-1-1		
Condition of $m\|B^q\| + \bar{v}_c < c_t$		
Publisher	$p_4^{q^*} = any$	$\Pi_4^{q^*} = -f_4^q$
Manufacturer	$p_4^{t^*} \neq 0,\ S_4^{t^*} = \phi$	$\Pi_4^{t^*} = -f_4^t$
Consumers	$\left(Y_4^{c^*}, s_4^{c^*}\right) = (\phi, NoPurchase)$	$U_4^{c^*} = 0$
(b) Case 4-1-2		
Condition of $m\|B^q\| + \bar{v}_c < c_t$		
Publisher	$p_4^{r^*} > m_4\|B^q\|$ and $p_4^{r^*} - m_4\|B^q\| + p_4^{q^*}\|B^q\| > \bar{v}^c,\ S_4^{r^*} = Q$	$\Pi_4^{q^*} = -f_4^q$
Telecom company		$\Pi_4^{r^*} = -f_4^r$
Consumers	$\left(Y_4^{c^*}, s_4^{c^*}\right) = (\phi, NoPurchase)$	$U_4^{c^*} = 0$
(c) Case 4-2		
Condition of $m_4\|B^q\| - (7 - 2\sqrt{12+\zeta})\bar{v}^c \leq c_4^t \leq m_4\|B^q\| + \bar{v}^c$		
Publisher	$p_4^{q^*} = \frac{\bar{v}^c + m_4\|B^q\| - c_4^t}{2(1-\zeta)\|B^q\|}$	$\Pi_4^{q^*} = \frac{(\bar{v}^c + m_4\|B^q\| - c_4^t)^2}{8} - f_4^q$
Telecom company	$p_4^{t^*} = \frac{(1-3\zeta)\bar{v}^c + (1-3\zeta)m_4\|B^q\| + (3-\zeta)c_4^t}{4(1-\zeta)},\ S_4^{r^*} = Q$	$\Pi_4^{t^*} = \frac{(\bar{v}^c + m_4\|B^q\| - c_4^t)^2}{16} - f_4^t$
Consumers	$\left(Y_4^{c^*}, s_4^{c^*}\right) = \begin{cases} (\phi, NoPurchase) & if\ 0 \leq v^c < v^{c*} \\ (B^q, t) & if\ v^{c*} \leq v^c \leq \bar{v}^c \end{cases}$	$U_4^{c^*} = \begin{cases} 0 & if\ 0 \leq v^c < v^{c*} \\ v^c - v^{c*} & if\ v^{c*} \leq v^c \leq \bar{v}^c \end{cases}$
(d) Case 4-3		
Condition of $c_4^t \leq m_4\|B^q\| - (7 - 2\sqrt{12+\zeta})\bar{v}^c$		
Publisher	$p_4^{q^*} = \frac{\bar{v}^c}{2\|B^q\|}$	$\Pi_4^{q^*} = \frac{(1-\zeta)}{4}\bar{v}^{c2} - f_4^q$
Telecom company	$p_4^{t^*} = m_4\|B^q\|,\ S_4^{t^*} = Q$	$\Pi_4^{t^*} = \frac{\zeta}{4}\bar{v}^{c2} + \left(m_4\|B^q\| - c_4^t\right)\bar{v}^c - f_4^t$
Consumers	$Y_4^c = \begin{cases} (\phi, t) & if\ 0 \leq v^c < \frac{1}{2} \\ (B^q, t) & if\ \frac{1}{2} \leq v^c \leq \bar{v}^c \end{cases}$	$U_4^c = \begin{cases} 0 & if\ 0 \leq v^c < \frac{1}{2} \\ v^c - \frac{1}{2} & if\ \frac{1}{2} \leq v^c \leq \bar{v}^c \end{cases}$

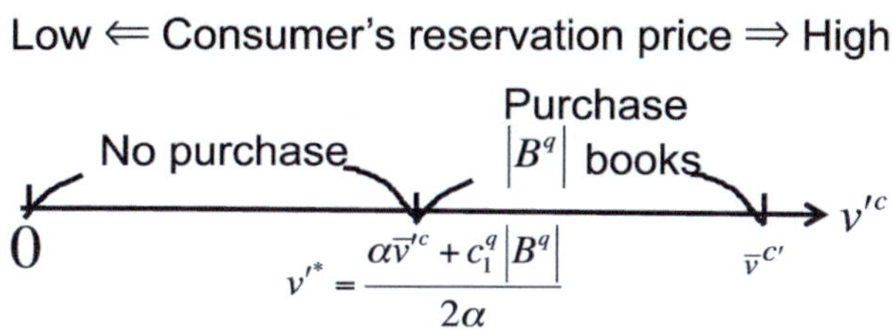

Fig. 3 Consumer decisions at equilibrium in model 1 (Case 1-2)

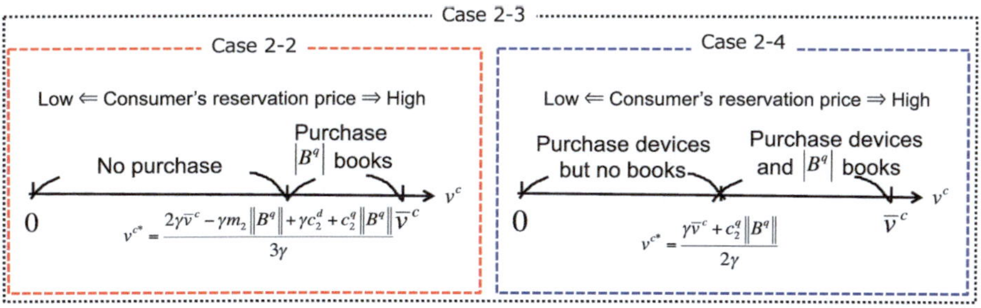

Fig. 4 Consumer decisions at equilibrium in model 2

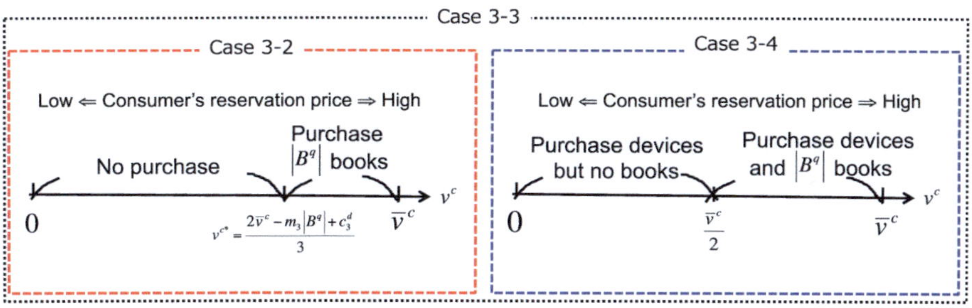

Fig. 5 Consumer decisions at equilibrium in model 3

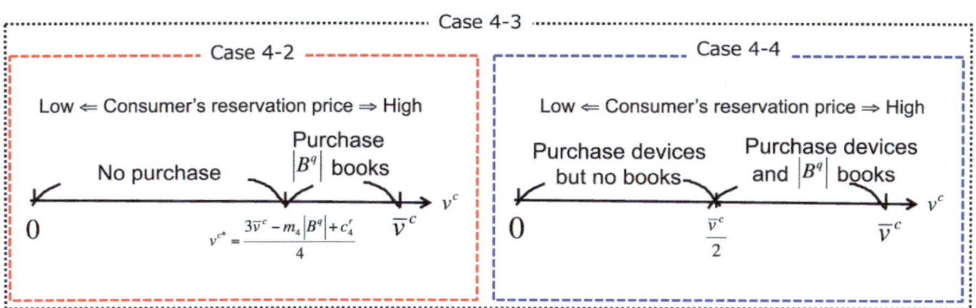

Fig. 6 Consumer decisions at equilibrium in model 4

Acknowledgement This research was supported by Service Science, Solutions and Foundation Integrated Research Program (S3FIRE), Research Institute of Science and Technology for Society (RISTEX), Japan Science and Technology Agency (JST).

References

1. Kitamura Y (2012) Fundamental study on the online publishing. Ota Publishing Company, Tokyo, Japan, in Japanese
2. Impress Business Media, Corp (2013) e-Book marketing report 2013. Impress Business Media, Corp, Tokyo, Japan, in Japanese

Service Field Simulator: Virtual Environment Display System for Analyzing Human Behavior in Service Fields

Takashi Okuma and Takeshi Kurata

Abstract This paper describes a virtual reality system for analyzing human behavior in service fields and its two practical use cases. Sensing and analyzing human behavior are very useful for evaluating and improving processes and environment of services. However, a change of service process and environment carries some risks. If we can evaluate them in advance, we can reduce the risks. Therefore, we have been developing a service field simulator using virtual reality technologies. A person who experiences the virtual service field is surrounded by displays in 360° not for losing sense of direction during 'walking around' the environment. He or she can walk around the virtual environment by walking-in-place motion. The direction of travel is decided with the body direction. In this paper, we introduce our new implementation of the simulator. We replaced four image projectors to 40 liquid crystal displays for improving resolution of virtual environment. New setting up can provide sufficient resolution to evaluate package design of products in the virtual retail store. We also replaced a wearable sensor module for detecting walking-in-place motion into RGB-D sensors set on a frame of the simulator. It reduces number of the sensors attached to the person and makes it easier to experience the virtual environment. In order to show feasibility of our concept of the service field simulator, this paper also describes two use cases of the simulator. In the first use case, we use the simulator with eye-tracking device for analyzing buying behavior in a drag store. In the second use case, we use the simulator with EEG for studying a method for evaluating human interest to an environment.

Keywords Service engineering • Virtual environment • Human activity sensing

T. Okuma (✉) • T. Kurata
Center for Service Research, National Institute of Advanced Industrial Science and Technology, Tsukuba Central 2, 1-1-1, Umezono, Tsukuba, Ibaraki 305-8568, Japan
e-mail: takashi-okuma@aist.go.jp; t.kurata@aist.go.jp

T. Maeno et al. (eds.), *Serviceology for Designing the Future*,
DOI 10.1007/978-4-431-55861-3_10

1 Introduction

Sensing and analyzing human behavior are very useful for evaluating and improving processes and environment of services. Fukuhara et al. showed that service providers could improve their service process using their own behavior data [1]. In other cases, eye-tracking devices are used for analyzing monitor customers' attention in service environment such as a retail store. This kind of analysis is conducted for improving the service environment. However, a change of service process and environment carries some risks. If we can evaluate them in advance, we can reduce the risks.

A simulator that provides virtual environment of the service field might be useful for this kind of evaluations in advance. In addition, evaluation with the simulator has two advantages. One advantage is that we can control conditions in the virtual environment easier than those in the real field. The other advantage is that we can acquire more detail data in the simulator, because we can use sensors that is difficult to be used in the real field, such as electroencephalograph (EEG). Another reason why we can acquire more detail data is the limited sensing area. When the sensing area is limited, many kinds of sensing devices are available including motion capture devices.

Therefore, we have been developing service field simulators using virtual reality technologies [2]. The aim of this research is to develop a method for evaluating service environment and process in advance by sensing and analyzing actual human behavior.

This paper describes improvements upon the service field simulator for making it more practical. In addition, our recent activities for substantiating our hypothesis that the service field simulator is useful for improvement of service process and environment are also described.

2 Related Research and Use Cases

In the fields of retail and related B2B service such as market research services, some advanced companies such as consumer product manufacturer and retail companies put "virtual store," which is reconstructed store in virtual environment, to practical use on marketing research. The in-store marketing institute reports investigation with virtual store has begun to attract notice because this kind of investigation is possible to deliver a more accurate representation of at-shelf product selection and other shopping behaviors than traditional methods of consumer research if it is conducted properly [3]. In addition, they also report marketing research using virtual store has some benefit on cost and flexibility and can prevent to leak information such as new package designs or research targets during in-store marketing.

Burger et al. proposed a test bed called ServLab [4] for designing and evaluating service operation and environment using virtual reality technologies. They combined 3-D interactive stereo projection display with the real stage called service theatre where professional actors play some roles of customers and employees to review possible situations. They also showed some case studies such as the redesign of the check-in processes of the Accor hotels group and testing a new service concept on customer training and consulting that are taken place over the Internet.

Our service field simulator and the ServLab have some common goals including redesign and evaluation of service concept, service process and service environment without actual reproduction of real situations. However, the two simulators are developed on different concepts. Service field simulator mainly focuses on sensing of natural human behavior and biological data in the simulated service field. Therefore, it is important for us to keep human's sense of direction and provide natural way to walk around in virtual environments. A person who experiences the virtual service field is surrounded by displays in 360° not for losing sense of direction during 'walking around' the environment. He or she can walk around the virtual environment by walking-in-place motion. The direction of travel is decided with the body direction. This traveling method is designed as similar as possible to that in the real environment. Hyun et al. showed some experimental results on the service field simulator. They confirmed that subjects felt sense of presence and sense of involvement in the simulator. They also confirmed that subjects could preserve sense of absolute direction [2]. However, the horizontal resolving power of the simulator was approximately equivalent to 20/100 vision (about 0.190 pixels per minute). This limitation led to low experienced realism and high task load, because subjects watched screen with worse eyesight than their real eyesight. Figure 1 shows a scene a person is observing a virtual shopping mall in the service field simulator.

Fig. 1 Inside of the service field simulator

3 Service Field Simulator

3.1 Concept

In many service fields such as shopping mall, retail stores, restaurants, and hotels, service processes are often constituted by the repetition of small activities such as walking, simple work, and gaining some information. By providing virtual environment where we can experience these activities, we could sense and record human reaction to some possible service process, layouts of the service environment, design of the real tools to be used in the service process and so on. We are developing the service field simulator according to this hypothesis.

Therefore, we set the following three design policy for developing the service field simulator; (1) virtual environment is displayed through 360° Omni-directional screen, (2) virtual viewpoint is controlled by walking-in-place motion, (3) photo-realistic avatars can be displayed in the virtual environment for communication with other person. Figure 2 shows a whole concept of the service field simulator with the first version implementation. Because new contributions of this paper are concerned with the first and the second policy, we describe these two policies in this section.

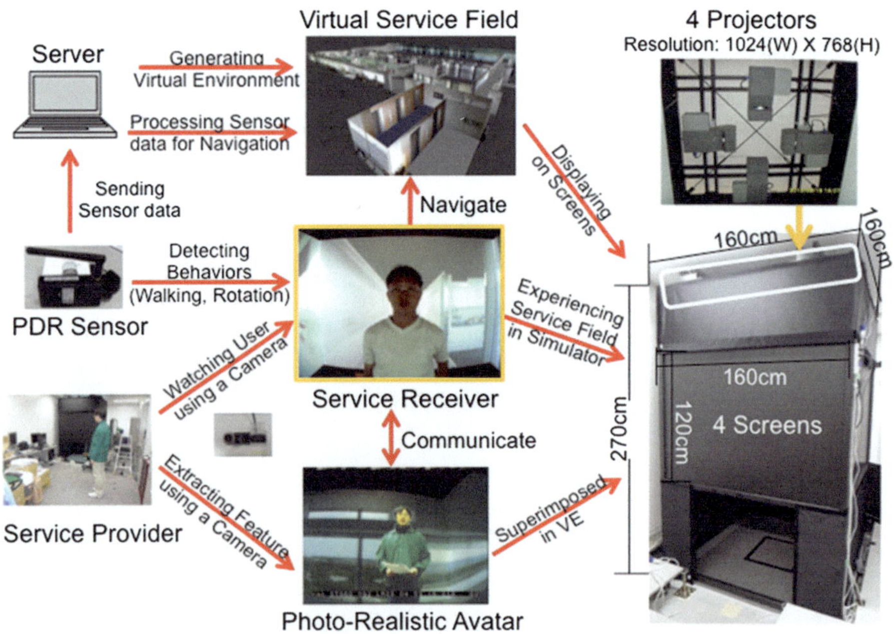

Fig. 2 Service field simulator version 1.0

3.1.1 Screen Configuration

For providing immersive virtual environment, various immersive displays have been investigated. Head-mounted displays (HMDs) are one of the most popular devices and have benefits that they are small and easy to provide immersiveness relative to other devices. However, they are always required to be improved on their narrow field of view, resolution, eye-fatigue and how they feel fitted. In addition, they make users feel unnatural to see holding real objects such as mobile information devices and paper maps. They sometime make users feels virtual motion sickness because of delay between when a user moves head and when the appropriate CG scene has been rendered. It is also a demerit for our application that they cannot be used with eye-tracking device that is used for in-store marketing and so on.

The CAVE [5] like immersive display room can solve problems of HMDs. However, it requires large space to be constructed. In addition, it is very difficult for a user to keep sense of direction when the user walks around in virtual environment with partially Omni-directional setting because system has to rotate virtual environment to show all direction of the virtual environment.

It is important for us to keep sense of absolute direction as well as the real environment because we would like to sense and record human activities in virtual environment as similar to the real environment as possible. Full solid angle display such as the COSMOS [6] and the GarnetVision [7] would be ideal for our purpose, but they require huge space to be set or very complex optical system that have to be calibrated very strictly. Because of limitation of the cost and the room, we adopt a fully Omni-directional screen setting to our system in order to keep sense of horizontal direction as a minimum requirement to our purpose.

3.1.2 Virtual Viewpoint Control Method

In order to imitate the way to move in service fields, we decided to provide a function to control virtual viewpoint by walking motion. This function allows a person who experiences the virtual service field to be hands-free, so that we can test service process with real tools used in the real field. Another important point is that he or she is required physical load to move around in the virtual environment as much as in the real environment. Therefore, the function is suitable for evaluating the service process and layout settings based on their embodiment.

We have two possible implementations to realize the function in the simulator; one is implementation using Omni-directional treadmill and the other is implementation with methods for detecting walking-in-place motion and body direction. Walking on the Omni-directional treadmill would be very similar to that in the real environment. However, considering safety, difficulty to change walking direction and cost limitation, we adopt a method for detecting walk-in-place motion and body direction to our system because it would be more practical than the treadmill setting.

3.2 Implementation of Version 2.0

The first implementation of the service field simulator used four image projectors to construct Omni-directional screen. As mentioned before, the resolving power of the simulator was about 0.190 pixels per minute, and this limitation led to low experienced realism and high task load. For evaluating service environment, at least, letters in the environment should be read easily. Therefore, we replaced four image projectors to 24 liquid crystal displays (LCDs) for improving resolution of virtual environment. Each LCD has the full-HD resolution (1920 $\times$ 1080 pixels), and they are placed along shape of octagonal prism (Figs. 3 and 4). As a result, horizontal angle resolution was improved to about 0.733 pixels per minute so it can represent eyesight of 20/30 vision. In addition, this configuration reduces required room space and height (about 2300 mm). So this version of the service field simulator can be placed on Japanese popular office building floor.

We also replaced a wearable sensor module for detecting walking-in-place motion into two RGB-D sensors set on the frame of the simulator. It reduces number of the sensors attached to a person who experiences the virtual environment and makes it easier. So the physical load of the person was reduced. It also reduces time and the number of the operation for preparing virtual in-store investigation.

Our system can detect the person's walking-in-place motion and the waling direction using the shape of the distribution of measured point cloud (See Fig. 5). We don't describe the detail of the algorithm for detecting waling-in-place motion and walking direction because it is out of scope of this paper. Actually, we don't think the current algorithm is the optimal for detecting walk-in-place motion because it is simple and straightforward implementation and there exists many threshold parameters to be adjusted in it.

We describe two use cases for verifying the efficiency of the service field simulator version 2.0 in the following subsections.

3.2.1 Gazing Point Analysis Using Combination of an Eye-Tracking Device and the Simulator Ver. 2.0

Our new hypothesis is that we can do the same investigation using an eye tracker and the Ver. 2.0 as real in-store marketing. We did preliminary investigation using a virtual drag store with project members that includes people from a consumer product manufacturer. In this project, a real drug store is reconstructed as a computer graphics model in the Ver. 2.0 (Fig. 6). We gave subjects a task that is to find and decide a product that they want to buy from one product category.

Fig. 3 Appearance of the simulator Ver. 2.0

Fig. 4 Inside of the simulator Ver. 2.0

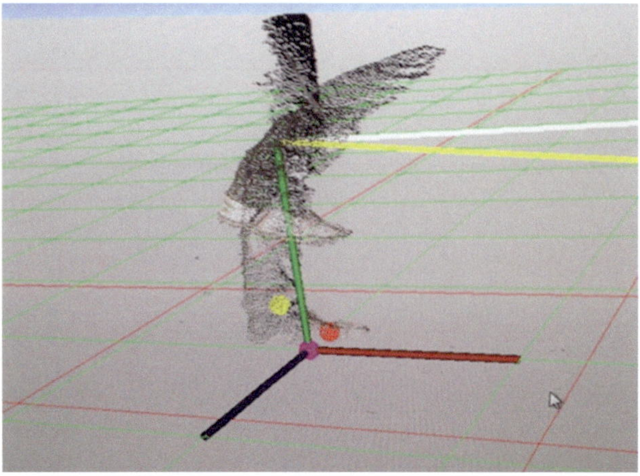

Fig. 5 Captured point cloud of legs and detected walking direction (*yellow line*)

Fig. 6 Constructed virtual drag store

During the task, position of the subject in the virtual store and eye tracking data is recorded (Fig. 7).

In the preliminary investigation stage, five project members played a role of subjects and did the task for testing how it works. Then, we could get some positive

Fig. 7 Examples of recorded eye-tracking data

opinion, which the motion of the gazing point in the virtual environment is similar to that in the real store especially from the entrance to in front of the shelf where target products are layout, from a person who sometimes does in-store marketing investigation using eye-tracker as a part of his business.

On the other hand, after subjects stood in front of the shelf, the gazing point moved different way with that in the real store. Based on opinions of people who played a role of subjects in the preliminary investigation, two possible reasons were found out. One reason is that the shelves in the virtual store were modeled roughly and not enough to see the detail of the products packages. The other reason is that fine adjustment of standing point was difficult for them with current implementation of the virtual viewpoint control function based on walking-in-place motion detection.

3.2.2 Investigation for a Method to Measure Human Interest Using EEG and the Simulator Ver. 2.0

Recent few years, marketing activities based on brain and neuroscience called 'neuro-marketing' has been attracting attention. Our institute also launched a feasibility study project named 'neuro-aided-design' for designing service process and products. As a part of the project, we started a feasibility study on an investigation method using an EEG in the simulator. This study is placed as a verification of the merit that we can acquire more detail data in the simulator than in the real situation.

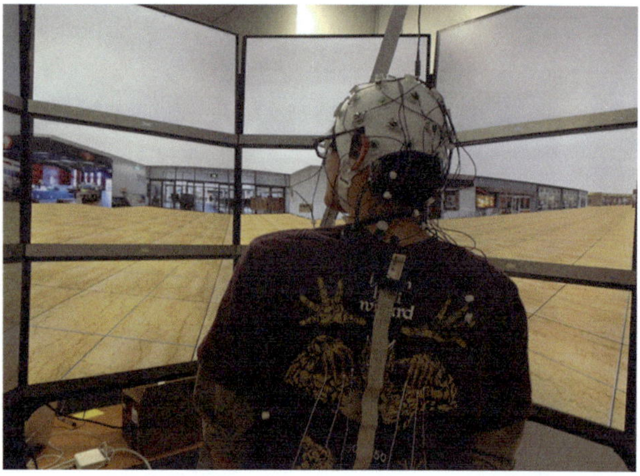

Fig. 8 A participant wearing EEG in the simulator Ver.2.0

Therefore, we reconstruct a shopping mall in the simulator and measured brain waves of subjects while the subjects walking around in the virtual shopping mall. Experimental variable was states of the shop appearances, open and close. The project members explored electrophysiological measures that reflect an observer's interest in the simulator (See Fig. 8). The results from the brain wave analysis could be reasonably explained by supposing that participants allocated more attentional resources to visual information in an interesting environment than in a boring environment. The detail of the experimental results is described in other submitting paper and to be appeared before long [8].

Currently, we are investigating a method for measuring potential interest of customers using EEG and the simulator for designing interesting service environment.

3.3 Implementation of Ver. 2.1

During some investigations with the simulator Ver. 2.0, lack of vertical field of view angle was pointed out again. Actually, that was one of the problems from the first version, but had not been resolved in Ver. 2.0 too. Especially in the investigation using a virtual drag store, because we set LCDs at low position to show products on the lowest shelf, we could not show some direction board hanged from the ceiling. As a result, we could not confirm if this kind of direction board is useful or not for customers to find products.

Therefore, we increased the number of LCDs to 40 and increased vertical viewing angle. The upper vertical viewing angle and the lower vertical viewing angle of the new setting are about 35° and 58.5° each (Figs. 9 and 10).

During subjects are experiencing virtual service field, experimenters could not see what is occurring inside from outside. Therefore, we developed a program that

Fig. 9 Appearance of the simulator version 2.1. The entrance door can be closed(*left*) and opened (*right*)

Fig. 10 Inside of the simulator Ver. 2.1

Fig. 11 Visualized virtual position of the subject as a teddy bear on the tiled wall dispay

visualizes virtual service field and virtual position of the subject. We display the visualizing program on another tiled wall displays to see the position of the subjects in real time (Fig. 11). Currently, we use the program for observing current situation. However, it might be possible to use as 'virtual stage' just like the ServLab.

An experiment has been conducted with this version. We did quantitative evaluation of task complete time, virtual movement distance and objective evaluation values, and qualitative evaluation of gazing points. Although we cannot show the detail result in this paper, we could have done 'in-virtual-store' investigation using the service field simulator just like the in-store investigation.

4 Conclusion

We have described improvements upon the service field simulator for making it more practical. We also described our recent activities for substantiating our hypothesis that the service field simulator is useful in improvement of service process and service environment. The practicability and advantages of the service field simulator is being confirmed through our investigation activities.

This paper mainly focuses on verifying the practicability of the service field simulator. However, in order to make it clear how much investigation results with the simulator are applicable to the real service field, it is still very important that verifying the duplicability of the human activities in the simulator through comparison with that in the real environment.

In addition, it is also very important to represent human communications in the simulator for redesigning and evaluating actual service processes. We did not describe the detail about it in this paper, but the original concept of the service field simulator includes the function using photo-realistic avatars. The practicability of this function also should be confirmed through experiments in future.

References

1. Fukuhara T, Tenmoku R, Okuma T, Ueoka R, Takehara M, Kurata T (2013) Improving service processes based on visualization of human-behavior and POS data: a case study in a Japanese restaurant. The 1st international conference on serviceology, Japan, 16–18 October, pp 1–8
2. Hyun J, Park A, Ishikawa T, Kourogi M, Habuchi Y, Okuma T, Kurata T (2011) Hands-free omni-directional walkthrough simulator for service-field pre-evaluation and its user study. J Virtual Real Soc Jpn 16(1):45–56
3. In-Store Marketing Institute. Shaping retail: the use of virtual store simulations in marketing research and beyond. http://www.kelley.iu.edu/CERR/files/09ISMI_virtualretailing.pdf
4. Burger T, Kim KJ, Meiren T (2009) Visualizing and testing service concepts. In: The 1st international symposium on services science, Logos, Berlin, pp 149–159
5. Cruz-Neira C, Sandin DJ, DeFanti TA (1993) Surround-screen projection-based virtual reality: the design and implementation of the CAVE. In: The 20th annual conference on computer graphics and interactive techniques, ACM SIGGRAPH 93, pp 135–142
6. Yamada T, Tanahashi H, Ogi T, Hirose M (1999) Development of fully immersive 6-screen display "COSMOS" and evaluation of virtual space navigation. J Virtual Real Soc Jpn 4 (3):531–538
7. Iwata H (1996) Rear-projection-based full solid angle display. In: The sixth international conference on artificial reality and tele-existence, Japan, pp 59–64
8. Yuji T, Takashi O, Motohiro K, Takeshi K, Takeshi T, Sunao I (in press) Electrophysiological measurement of interest during walking in a simulated environment. Int J Psychophysiol, Available online 1 June 2014

Part III
Healthcare Services

Improvement of Sharing of Observations and Awareness in Nursing and Caregiving by Voice Tweets

Kentaro Torii, Naoshi Uchihira, Yuji Hirabayashi, Testuro Chino, Takanori Yamamoto, and Satoko Tsuru

Abstract Nurses and caregivers provide care for patients using their hands and move around hospital wards, staff stations, operating rooms, residents' rooms and so on. They also read and write, sending and receiving a lot of information of various types in the course of their work. Therefore, nursing and the provision of care could be described as "physical and adaptive intelligent services." Several types of information systems and communication systems are used in "physical and adaptive intelligent services." But there is a striking mismatch between "physical and adaptive intelligent services" and conventional information systems whose interfaces are generally designed for deskwork. The mismatch is one of the significant causes of the fact that the quality and efficiency of information processing in nursing has not been improved as much as people would expect. In this paper, we describe how a voice interface improves the quality of the information processing in nursing, focusing on care recording. The smart voice messaging system for nursing makes it possible for nurses to input care records, take notes for themselves, and organize voice messages to other staffs in a unified way during care. The experimental evaluation performed in a nursing home for the elderly shows that the smart voice messaging system can contribute to improvement in the quality of care because care staffs can easily retain various observations and concerns that are often lost in the case of conventional care records.

K. Torii (✉) • T. Chino • T. Yamamoto
Corporate R&D Center, Toshiba Corporation, 1 Komukai-Toshibacho, Kawasaki 212-8582, Japan
e-mail: kentaro.torii@toshiba.co.jp

N. Uchihira
School of Knowledge Science, Japan Advanced Institute of Science and Technology, Nomi, Japan

Y. Hirabayashi
Institute of Technology, Shimizu Corporation, Tokyo, Japan

S. Tsuru
Health Social System Engineering Laboratory, School of Engineering, The University of Tokyo, Tokyo, Japan

T. Maeno et al. (eds.), *Serviceology for Designing the Future*,
DOI 10.1007/978-4-431-55861-3_11

Keywords Healthcare service • Awareness • Speech recognition • Nursing record • Smart voice messaging

1 Introduction

More and more countries are experiencing population aging and a rise in the number of patients, and more and more time and money will be devoted to healthcare services. Therefore, improvement of the efficiency of healthcare processes is an important issue. Nursing and caregiving are an important aspect of health services. We focus on nursing care processes in this paper.

Nurses and caregivers provide care for patients using their hands and move around hospital and care facility wards, staff stations, operating rooms, residents' rooms and so on. They also read and write, sending and receiving a lot of information of various types in the course of their work. Since their work involves both physical action and information processing, it could be described as "physical and adaptive intelligent services."

Several types of information systems and communication systems are used in nursing such as telephones, mobile phones, PCs and papers. But there is a striking mismatch between "physical and adaptive intelligent services" and conventional information systems that are generally developed as tools for deskwork. Since care staffs' hands are occupied during the patient care and care staffs move around hospital wards, staff stations, operating rooms and so on, it is difficult for them to record by keyboard, to compose messages, to telephone other staffs and to take telephone calls from other staffs at the right moment.

Lemonidou, et al. show that care staffs spend much of their time on indirect care such as recording the provision of nursing, accepting medication orders, and messaging to other staffs [9]. Hollingsworth, et al. show that emergency physicians and care staffs spent almost half of their time on indirect care [6]. Hendrich, et al. show that care staffs devote 35.3 % of their time to documentation according to their time study of 767 care staffs [4]. The mismatch is one of the significant causes of the inefficiency of indirect care. Smartphones and tablet PCs have recently come into use in nursing and caregiving, but it is difficult to type messages during "physical and adaptive intelligent services" other than fixed inputs such as multiple-choice, autocompletion, and numeric inputs.

Uchihira, et al. [13] and Torii, et al. [11] have proposed a "smart voice messaging system" to overcome this mismatch. The smart voice messaging system provides a single semi-hands-free voice interface and enables nurses and caregivers to input nursing records, take notes for themselves, and compose voice messages to other staffs during the provision of care.

In this paper, we describe an experimental application of the smart voice messaging system in actual nursing processes in a nursing home for the elderly, and evaluate effects of the system on recording and sharing of observations and awareness of care staffs. The remainder of this paper is organized as follows. We describe related works in the second section. The third section describes the overview of the smart voice messaging system. The fourth section describes the experiments using the smart voice messaging system in actual care processes in a nursing home for the elderly in Japan, and analysis of collected voice tweets is shown in the section six. In the last section, we present our conclusions. This research was approved by JAIST Life Science Committee (#25-001).

2 Related Works

Much research has been done on voice input interfaces and automatic speech recognition (ASR) for medical reporting [1, 3, 7, 8, 10, 12]. ASR technology was first adopted for radiology more than 25 years ago [1, 8]. The most common objective of introducing speech recognition in a radiology department is to reduce report turnaround time (RTT), i.e., the time interval between the examination and the finalized report digitally available in the hospital information systems (HIS) [1]. Kang, et.al. also show reduction in turnaround time of reporting about surgical pathology using speech recognition software [7]. Devine, et.al. evaluate accuracy of speech recognition softwares about dictation of medical records by physicians [3]. In these researches, it is assumed that users sit at their desks and use applications via laptop PC interfaces, typically a laptop screen, a keyboard and a mouse. So these systems are not necessarily applicable to "physical and adaptive intelligent services" such as nursing and caregiving.

3 The Architecture of the Smart Voice Messaging System

Figure 1 adapted from [11] shows the outline of the architecture of the smart voice messaging system, which consists of a smartphone application and a Web server application. The voice interface of the smartphone application enables semi-hands-free messaging. Several sensors on the smartphone continuously collect data around the user in a background process, and the application periodically sends the data to the server of the system. Several estimators on the server estimate the state (context) of each staff who carries a smartphone. The states estimated by each estimator are called "tags". Tags are, for example, specific types of keywords included in a recognized voice, location and time when the staff makes a voice tweet, the task which the staff is executing at the moment of the voice tweet, and the staff ID.

Tags have two important functions. First, the tags improve comprehensibility of voice tweets for recipients. So, care staffs need not say everything about a voice tweet. For example, they can leave out context information (WHO, WHEN, WHERE) in their voice tweets. Secondly, the tags enable the system to decide the type of delivery of voice tweets and to deliver them to an appropriate destination

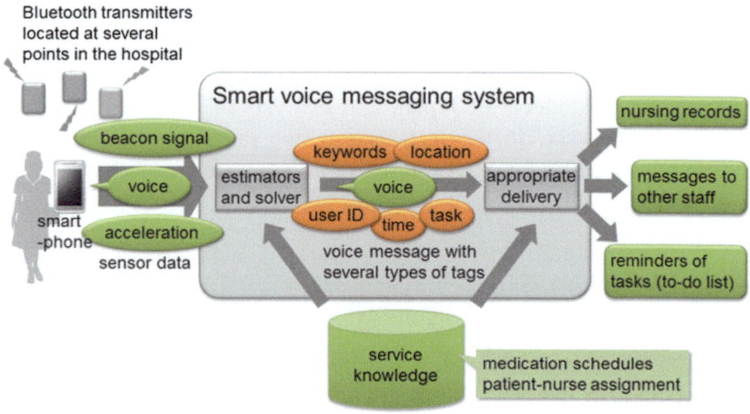

Fig. 1 Architecture of the smart voice messaging system (Adapted from [11])

automatically, for example a message to other staffs, a reminder of one's task, and inputting to an appropriate database of nursing records.

3.1 Automatic Tagging to Voice Tweets

Figure 2 illustrates the structure of "estimators and solver" in Fig. 1. There are three estimators in the server: voice recognition and keyword extraction, location estimator, and task estimator. The voice recognition in the server recognizes the voice and extracts keywords included in the recognized voice. For example, patient names and names of cares for the patients may be included in the keywords. To collect indoor location data with smartphones, Torii et al. developed a small Bluetooth transmitter that emits a beacon signal including ID of the Bluetooth device on the transmitter [11]. The smartphone receives beacon signals from the transmitters around it, and the application on the smartphone sends ID and RSSI (Received Signal Strength Indicator) of each signal to the server. The location estimator estimates the location of each staff at each time from the Bluetooth beacon signal. The task estimator estimates what task was executed at each time from acceleration sensor data and so on. These estimators generate candidates of tags with degrees of confidence. The tag solver in the server searches an optimal set of tags from the candidates using service knowledge such as medication schedules for patients and patient-nurse assignments. An optimal set of tags is a set of tags whose sum of degrees of confidence of tags is maximal among sets that satisfy the constraints derived from service knowledge. The server attaches the optimal set of tags to the voice.

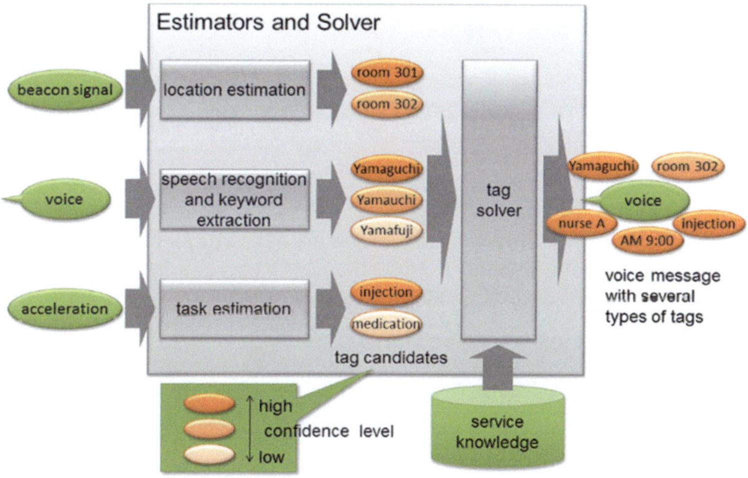

Fig. 2 Architecture of the module for automatic tagging to voice tweets (Adapted from [11])

3.2 Delivery of Voice Tweets to Appropriate Destinations

The system estimates the types of delivery of each tweet according to the attached tags, and decides appropriate destinations of voice tweets (Fig. 1). Messages are delivered to the staffs who actually need the messages. Reminders by a care staff are sent to herself/himself at the right time. Nursing records are reported to an appropriate database. Details of the estimators are explained in [11].

Figure 3 illustrates typical use cases of the smart voice messaging system. Care staffs are able to compose voice tweets such as a patient's pain complaints, a patient's questions about care processes and several observations of concern about a patient only by one hand at a patient's bedside. Tags are automatically attached to the voice tweets. For example, a voice tweet about a patient's pain complaint is delivered to the care staff who cares for the patient in the bathroom, and then the care staff is able to easily grasp the patient's pain, and can appropriately assist the patient to bathe. Since the tweets with tags are categorized automatically, care staffs can efficiently hand over information about patients.

4 Field Studies in a Nursing Home for the Elderly

We performed field studies at a nursing home for the elderly in Japan. Several "physical and adaptive intelligent services" are provided in the nursing home as shown below.

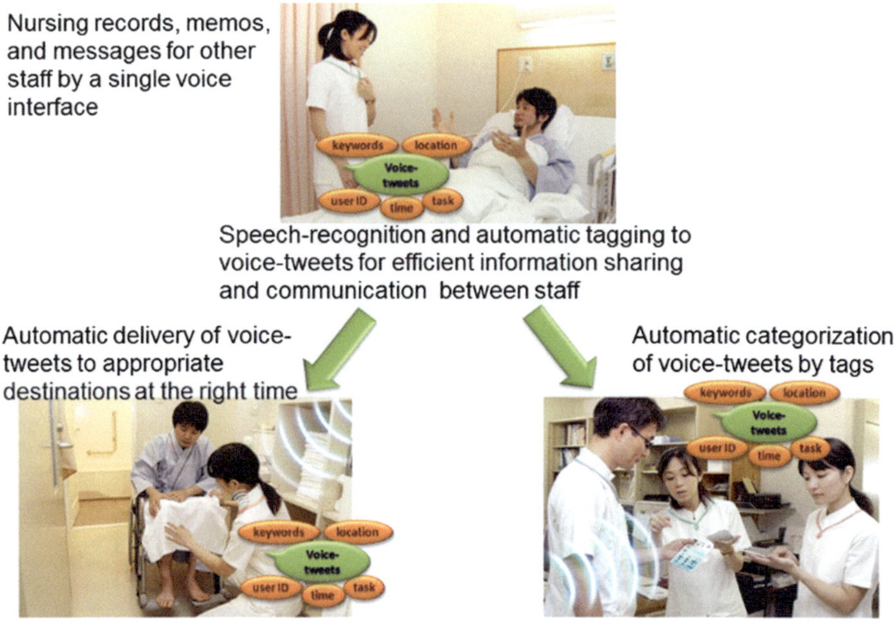

Fig. 3 Use cases of the smart voice messaging system (Adapted from [11])

4.1 *Overview of the Nursing Home for the Elderly*

Figure 4 shows the sketch map of the nursing home where we performed field studies. The nursing home is a four-story building. Thirty residents live in the nursing home. Each resident has his/her private room. There are two dining rooms, one on the first floor and the other on the second floor. The care staff station is next to the dining room on the second floor. PCs for accessing care records are in the care station.

4.2 *"Action Oriented Intellectual Service" Provided in the Nursing Home for the Elderly*

Several "physical and adaptive intelligent services" are provided in the nursing home. Feeding assistance care is a typical one as shown below. All the residents take meals together at a dining room on the first floor or at a dining room on the second floor. During the provision of feeding assistance care, nurses and caregivers not only assist residents in taking meals, but also give them medicines and help them brush their teeth and use the lavatory. Furthermore, since many of the residents have physical or cognitive difficulties with walking, caregivers attend residents moving between their rooms and the dining rooms. Care staffs also

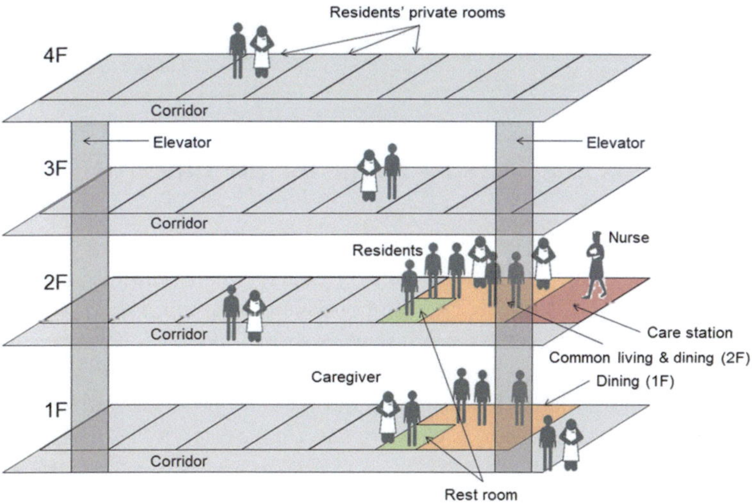

Fig. 4 Sketch map of the nursing home for the elderly

process several types of information. For example, during the provision of feeding assistance care, they check whether the meals for each resident are served correctly, check which residents have been prescribed which medicine after a meal or before a meal, and record how much each resident ate of each dish. They also observe behaviors and conditions of residents.

Experienced care staffs are aware of slight behavioral abnormalities of residents, and slight but significant changes of physical and mental conditions of residents. Such observations often include information helpful for evaluating and revising care plans for residents. For example, in the feeding assistance care, if a resident spills food from her/his mouth during a meal, the resident might have a problem concerning the swallowing function. Or the behavior might be symptomatic of dementia. Then other nutritional strategies such as jelly to ease deglutition might be examined and training in swallowing might be planned. In bathing assistance, care staffs observe the skin condition of each resident while washing his/her body, and a care staff assesses whether bedsores are developing on his/her skin and examines whether pressure ulcer care is appropriate for the resident. Some care staff may become aware of problematic states of facilities during care. For example, they may find an obstacle on a corridor that could cause residents to fall.

These observations and awareness should be recorded and shared with other staffs and managers. But it is difficult for them even to take notes on slips of paper, much less record the events on a PC because both hands are occupied in providing care or they are moving around in the nursing home. Therefore, they must remember all those important events. But no small part of the memory is lost before they have the opportunity to sit and record at a desk. The smart voice messaging system provides care staffs with a single semi-hands-free voice interface for both recording and immediate messaging to other staffs. The system enables care staffs to take notes of such awareness and observations during care.

5 Experiments in a Nursing Home for the Elderly

We performed experiments using the smart voice messaging system in the nursing home 7 times for feeding assistance care on 4 days in September, October, and November 2012, and 10 times for feeding assistance care on 5 days in May 2013. The system configuration in the experiments is shown in Fig. 5. Twenty caregivers and nurses used this system in the provision of care at lunch and dinner on the 9 days. About 6 staffs cared for residents during each provision of feeding assistance care. Each care staff wore a wired headset connected to a smartphone when they started attending residents moving to dining rooms. The care staffs push the switch of the headset and articulate their observations or messages whenever they observe events to be recorded about residents or they have messages for other staffs during the provision of feeding assistance care. Input voice tweets are sent to the managing server on the internet via Wi-Fi or 3G. The ASR server recognizes the voice of each tweet and returns the result to the managing server. The managing server adds tags to each tweet and processes it appropriately. The voice and other data are transferred through a secure channel.

Each care staff usually fills in important observations for the nursing records after the completion of feeding assistance care for all residents. In the experiments,

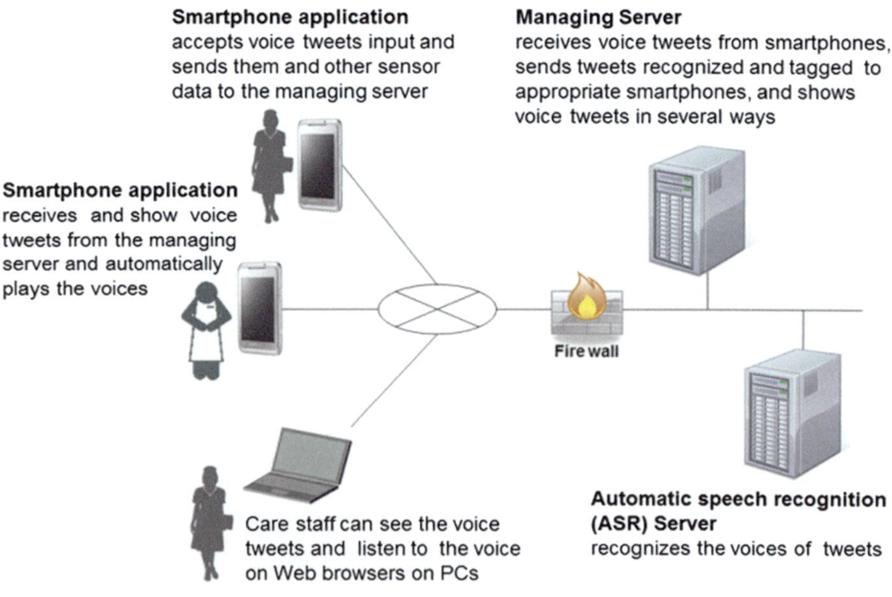

Fig. 5 The experimental voice messaging system

the care staffs made nursing records as usual without reviewing their voice tweets spoken during the provision of feeding assistance care, in order to evaluate how many tweeted observations are recorded or lost in the case of the conventional provision of feeding assistance care and recording.

6 Experimental Results

As shown in Tables 1 and 2, 330 voice tweets were collected in the experiments in 2012, and 720 voice tweets were collected in the experiments in 2013. Thus, 1050 voice tweets were collected during the provision of feeding assistance care at lunch and dinner 17 times on 9 days.

6.1 Types of Voice Tweets in the Feeding Assistance Care

All the voice tweets are classified in two types of "record" and "message" according to their contents. Two hundred and fifteen voice tweets (74 of 2012 and 141 of 2013) are of record and 785 voice tweets (256 of 2012 and 579 of 2013) are messages to other staffs. Voice tweets of recording are for example "Mrs. A leans to left side when she eats," "Mr. B exhibits agitation behaviour after dinner," and "Mrs. C is putting too much food in her mouth at once." And voice tweets of messaging are for example "Mrs. A has finished dinner, now I take her to her room," and "Some residents are getting agitated. Please come to the dining room to help."

Table 1 The results of the count of voice tweets in the care assistance in the 4 days experiments in 2012

	Record	Message	Total
Lunch	46	180	226
Dinner	28	76	104
Total	74	256	330

Table 2 The results of the count of voice tweets in the care assistance in the 5 days experiments in 2013

	Record	Message	Total
Lunch	90	337	427
Dinner	51	242	293
Total	141	579	720

6.2 The Occurrence of Voice Tweets in the Feeding Assistance Care

Figures 6 and 7 show the occurrence of voice tweets in the experiments in 2013. Usually lunch care starts at about 10:00 or 10:30, and finishes at about 13:00. Dinner care starts at about 17:00 and finishes at about 19:30 or 20:00. For lunch, care staffs start tweeting from about 10:00 and finish at about 13:00. For dinner, they start tweeting from about 17:00 and finish at about 20:00. It can be seen from these figures that voice tweets of records occur throughout the provision of care. Since care staffs usually can finally input nursing records after all the feeding assistance care is completed, the data on the time at which nursing records input into the current information system without using the smart voice messaging system might be distributed to the right end of these figures.

Figures 8 and 9 show cumulative relative frequency of voice tweets of records during the provision of care assistance at lunch and dinner over time axis in all the experiments. Fifty-four percent of voice tweets of records for lunch care occur before 11:50. And 52 % of voice tweets of records for dinner care occur before 18:15. These results mean that about half of observations and events that could be recorded occur 1 or 2 h before all the feeding assistance care is completed and care staffs have the opportunity to make records at desks.

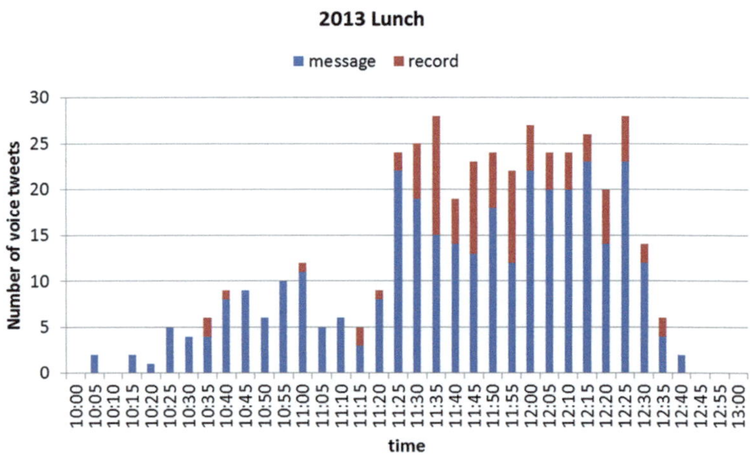

Fig. 6 The occurrence of voice tweets in the lunch care assistance in the 5 days experiments in 2013

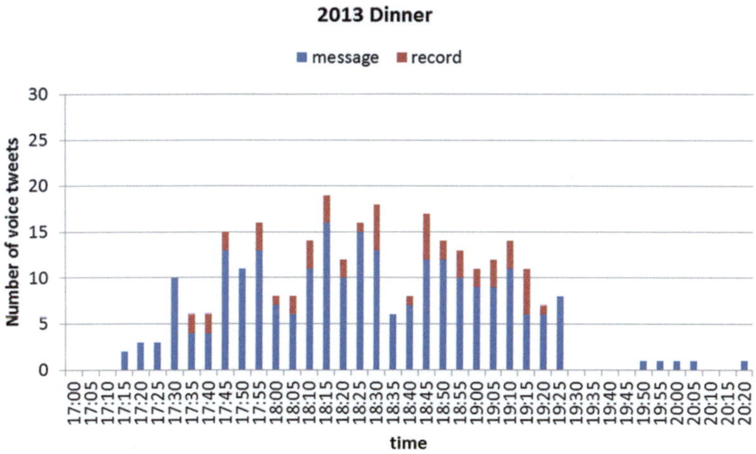

Fig. 7 The occurrence of voice tweets in the dinner care assistance in the 5 days experiments in 2013

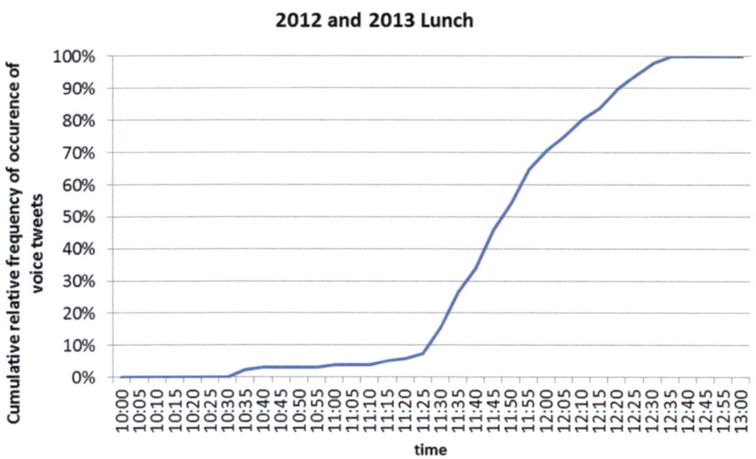

Fig. 8 Cumulative relative frequency of voice tweets of records in the lunch care assistance over time axis

6.3 Analysis of Voice Tweets of Records

Seventy-four voice tweets of records in the experiments in 2012 and 141 voice tweets of records in the experiments in 2013 were classified into four categories according to the following two criteria.

1. Every voice tweet was classified into either of two groups, "recorded" and "unrecorded," depending on whether a corresponding topic was written for the care record.

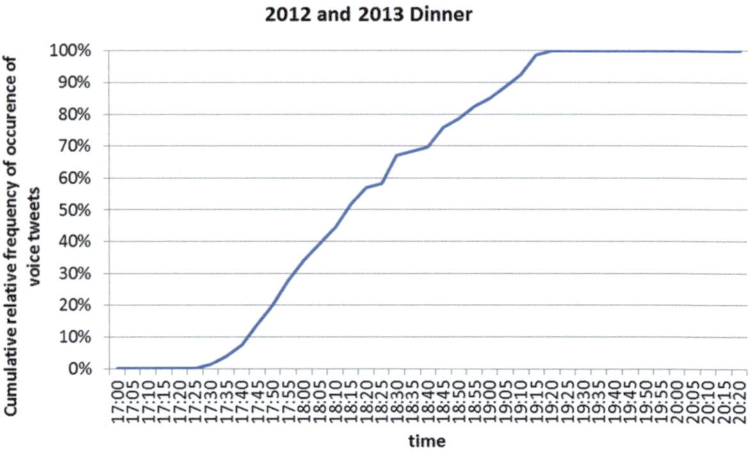

Fig. 9 Cumulative relative frequency of voice tweets of records in the dinner care assistance over time axis

Table 3 The result of the classification of tweets of records in the experiments in 2012 by care staff A

	Important	Unimportant	Total
Recorded	11	5	16
Unrecorded	41	17	58
Total	52	22	74

Table 4 The result of the classification of tweets of records in the experiments in 2012 by care staff B

	Important	Unimportant	Total
Recorded	10	6	16
Unrecorded	36	22	58
Total	46	28	74

Table 5 The result of the classification of tweets of records in the experiments in 2013 by care staff A

	Important	Unimportant	Total
Recorded	53	6	59
Unrecorded	60	22	82
Total	113	28	141

2. Two administrative care staffs (A, B) of the nursing home classified every voice tweet into either of two groups, "important" and "unimportant," according to the effectiveness for care planning. The voice tweets that include information helpful for assessing the care levels of some residents or for revising the care plans for some residents were classified as "important" voice tweets, and the others were labelled as "unimportant".

The results of the classification by the two administrative care staffs are shown in Tables 3, 4, 5 and 6.

Table 6 The result of the classification of tweets of records in the experiments in 2013 by care staff B

	Important	Unimportant	Total
Recorded	48	11	59
Unrecorded	56	26	82
Total	104	37	141

Fifty-two voice tweets out of a total of 74 voice tweets of records in 2012 were classified as "important" by care staff A, but 79 % of them (41/52 in Table 3) were not recorded in the care records. Also 78 % of the voice tweets classified as "important" by care staff B (36/46 in Table 4) were not recorded.

With regard to 2013, care staff A classified 113 voice tweets out of a total of 141 voice tweets of records as "important", but 53 % of them (60/113 in Table 5) were not recorded in the care records. In addition, 54 % of the voice tweets classified as "important" by care staff B (56/104 in Table 6) were not recorded.

These results show that important observations made by care staffs and events seen by them during the provision of care are often lost and not shared with other staffs and managers. According to the interviews of care staffs who participated in the experiments, they observe many residents' behaviours while providing care that are a cause of concern. It is impossible for them to make records at a desk during the provision of care. They try to take notes of the events when they can use their hands. But they often become involved in providing care for other residents before they are able to take notes about their observations. Some care staff omits recording of observations according to the subjective criteria.

Even in the case that these observations remain in the memories of care staffs, it is difficult for them to remember the exact time when each event occurred, and so nursing records tend to be inaccurate. For example, residents with dementia often sleep during the provision of feeding assistance care. The duration of the time when the resident's eyes are closed could be important for revising the care for the residents to provide better nutrition.

Too much tweets could be easily made and disturb care staffs. But the system prints the time and the location where each tweets is made, and also automatically extracts keywords such as patient's names and terms related to care or nursing from the tweet. Such meta-information is displayed attached to each tweet. The meta-information would enable care staffs to grasp the summary of each tweet and to process piles of tweets efficiently.

7 Conclusions and Future Works

In this paper, we describe a smart voice messaging system and its application to "physical and adaptive intelligent services." The smart voice messaging system makes it possible for care staffs to input care records, take notes for themselves, and deliver voice messages to other staffs with a single voice interface of a smartphone application during the provision of care.

We performed experiments in an actual nursing home for the elderly, and collected voice tweets by care staffs during the provision of feeding assistance care. The evaluations of the voice tweets shows that important observations and awareness during the provision of care are often lost and not shared. Recording and sharing of these observations and awareness are indispensable for assessing the care levels of residents accurately or for revising the care plans for residents appropriately. The smart voice messaging system can contribute to improvement of the quality of care records because staffs can easily retain various observations and awareness during the provision of care.

The limitation of the paper is that experiment in a nursing home was done for a short period and it is too short to evaluate care quality improvement in addition to evaluation that a smart voice messaging system can catch more awareness shown in this paper.

Subjects for future work will include:

- Complete stress-free interface
- Speech recognition with higher accuracy
- Reduction of the time required to make nursing records using a smart voice messaging system
- Long term experiment and evaluation using refined smart voice messaging system
- Improvement of the quality of contents of voice tweets. Standardization of observations during the provision of care and corresponding voice tweets between care staffs is indispensable for this issue. This will increase the ratio of important tweets and improve the quality of nursing records.
- Assessment of physical and mental states of residents by utilizing accumulated voice tweets about residents. And evidence-based care planning according to the assessment.

Although we focused on nursing records by voice tweets in this paper, the smart voice messaging system has another important aspect: messaging to other staffs [2, 5, 14]. Improvement of collaboration among care staffs by means of the voice tweet messaging is also included in the research targets.

Acknowledgments This research project has been supported by JST RISTEX S3FIRE (Service Science, Solutions and Foundation Integrated Research Program).

References

1. Borowitz SM (2001) Computer-based speech recognition as an alternative to medical transcription. J Am Med Inform Assoc 8(1):101–102
2. Chino T, Torii K, Uchihira N, Hirabayashi Y (2012) Work and speech interactions among staff at an elderly care facility, in book of human work interaction design. Work analysis and HCI, third IFIP 13.6 Working conference, HWID 2012, Copenhagen, Denmark, December 5–6,

2012, Revised selected papers, Series: IFIP Advances in Information and Communication Technology, vol 407, Campos P, Clemmensen T, Abdelnour Nocera J, Katre D, Lopes A, Ørngreen R (eds) VIII, 199 p 76 illus., Springer, ISBN 978-3-642-41144-12013

3. Devine EG, Gaehde SA, Curtis AC (2000) Comparative evaluation of three continuous speech recognition software packages in the generation of medical reports. J Am Med Inform Assoc 7(5):462–468
4. Hendrich A, Chow M, Skierczynski B, Lu Z (2008) A 36-hospital time and motion study how do medical-surgical nurses spend their time. Perm J 12(3):25–34
5. Hirabayashi Y, Uchihira N, Ikawa Y (2013) Verification of effectiveness of the nursing-care service space visualization and evaluation system: field experiment aiming to improve nursing-care services. In: Proceedings of the Portland international conference on Management Engineering and Technology 2013 (PICMET2013), San Jose, pp 1568–1571
6. Hollingsworth J, Chisholm C, Giles B, Cordell W, Nelson D (1998) How do physicians and nurses spend their time in the emergency department. Ann Emerg Med 31(1):87–91
7. Kang HP, Sirintrapun SJ, Nestler RJ, Parwani AV (2010) Experience with voice recognition in surgical pathology at a large academic multi-institutional center. Am J Clin Pathol 133(1): 156–159
8. Leeming BW, Porter D, Jackson JD, Bleich HL, Simon M (1981) Computerized radiologic reporting with voice data-entry. Radiology 138(3):585–588
9. Lemonidou C, Plati C, Brokalaki H, Mantas J, Lanara V (1996) Allocation of nursing time. Scand J Caring Sci 10(3):131–136
10. Ramaswamy MR, Chaljub G, Esch O, Fanning DD, vanSonnenberg E (2000) Continuous speech recognition in MR imaging reporting, advantages, disadvantages, and impact. Am J Roentgenol 174(3):617–622
11. Torii K, Uchihira N, Chino T, Iwata K, Murakami T, Tanaka T (2012) Service space communication by voice tweets in nursing. In: Proceedings of the AHFE2012, San Francisco, July 2012
12. Trumm CG, Morhard D, Ertl-Wagner B, Glaser C, Reiser MF (2008) Impact of RIS/PACS integrated speech recognition on report availability. Radiol Manage 30(6):16–23
13. Uchihira N, Torii K, Hirabayashi Y, Sugihara T, Hiraishi K (2011) Innovation for service space communication by voice tweets in nursing and caring: concept and approach in Japanese national project. In: 20th annual frontiers in service conference, Columbus, 2011
14. Uchihira N, Choe S, Hiraishi K, Torii K, Chino T, Hirabayashi Y, Sugihara T (2013) Collaboration management by smart voice messaging for physical and adaptive intelligent services. In: Proceedings of the PICMET'13, 2013, pp 251–258

A System Promoting Cooperation Between Medicine and Dentistry Using Key Performance Indicators and Importance-Performance Analysis

Shuichiro Nagaosa, Hironobu Matsushita, Jun Yaeda, Takashi Shinagawa, Norihiro Sonoi, Hiroyuki Nakamura, Hiromi Ohta, Masatoshi Usubuchi, Yasuhisa Arai, and Yasunori Sumi

Abstract With the advent of an aging society and on-going cutbacks of medical reimbursement systems targeting long-term care, "management of oral function" is becoming more salient than ever. The authors devised a system that facilitates cooperation between medicine and dentistry (multi-occupational cooperation) by identifying and using key performance indicators (KPIs) with importance-performance analysis (IPA). As such, the purpose of this study is to empirically

S. Nagaosa (✉)
Division of Dental Surgery, Tokyo Metropolitan Rehabilitation Hospital, Tokyo 131-0034, Japan
e-mail: nagaosa.reha@gmail.com

H. Matsushita
Kanagawa University of Human Services, Yokohama, Japan

J. Yaeda
Graduate School of Comprehensive Human Sciences, University of Tsukuba, Tsukuba, Japan

T. Shinagawa
Division of Oral Surgery, Heisei Yokohama Hospital, Yokohama, Japan

N. Sonoi
Division of Dental Surgery, National Sanatorium Kuriurakusenen, Gunma, Japan

H. Nakamura
Nakamura Dental Clinic, Tokyo, Japan

H. Ohta
Ohta Dental Clinic, Tokyo, Japan

M. Usubuchi
Division of Dental Surgery, Miyagi Cancer Center, Natori, Japan

Y. Arai
Tokyo Metropolitan Rehabilitation Hospital, Tokyo, Japan

Y. Sumi
Department for Advanced Dental Research, Center of Advanced Medicine for Dental and Oral Diseases, National Center for Geriatrics and Gerontology, Obu, Japan

T. Maeno et al. (eds.), *Serviceology for Designing the Future*,
DOI 10.1007/978-4-431-55861-3_12

examine the effectiveness of the system. The research was conducted by using KPIs for nurses in an acute care hospital. Further key improvement areas concerning multi-occupational cooperation were then identified from the findings.

Keywords Management of oral function • Multi-occupational cooperation • Importance-performance analysis • Internal marketing • Visualization

1 Introduction

With the advent of an aging society and on-going cutbacks of medical reimbursement systems targeting long-term care, "management of oral function" is becoming more salient than ever. As such cross-occupational cooperation between medicine and dentistry is becoming crucial for further dissemination. "Visualization of the role expectation for strategy management using the Delphi method and importance-performance Analysis (IPA) in dysphagia rehabilitation nursing services" was reported as a role study for professionals in The 1st International Conference of Serviceology [1]. The methodology used to identify the specific role of nurses specializing in dysphagia (swallowing) rehabilitation was examined. Further, a job analysis was conducted to evaluate the importance and performance of relevant behaviors. The methodology used there appears to be valid and may be applied to identify professional roles for other rehabilitation fields. The implementation of the Delphi method was a straightforward and effective tool used to create a questionnaire. IPA is a strategic tool that can quickly enable a manager to understand "identified customer needs" and "satisfied customer needs" and to assess performance rather than rely solely on importance indicators. Moreover, IPA allows for the visualization of the results of a strategic analysis, which facilitates understanding and interpretation of the results. The validity of IPA has been shown. A new job analysis using the Delphi method and IPA demonstrated one way of exploring human resource management in swallowing rehabilitation.

A hospital dentist is responsible for oral function management, and therefore, swallowing rehabilitation, in Japan [2]. Due to the aging society in Japan, the management of oral function is carried out not only under medical insurance but also under long-term care insurance. So the cooperation of medicine and dentistry (multi-occupational cooperation) is essential for further dissemination. The key people cooperating in this multi-occupational scenario are the nurse and the dentist.

In this study, the authors devised a system that facilitates multi-occupational cooperation using key performance indicators (KPIs). KPIs for management of oral function were created by selecting the item for KPIs for swallowing rehabilitation. A system using IPA that promotes cooperation between nurse and dentists by analyzing KPI is suggested. This system is intended to provide management with an internal marketing perspective using KPI. In this approach, the patient is regarded as an external customer and the rehabilitation professional is regarded

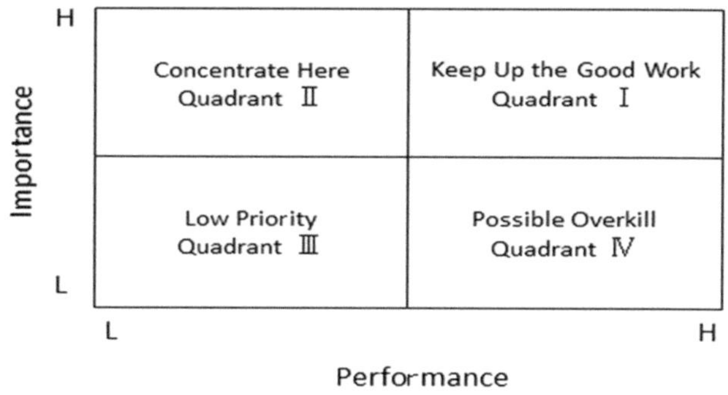

Fig. 1 The original IPA framework [3] (*L* low, *H* high)

as an internal customer. The purpose of this study is to demonstrate the effectiveness of the system.

1.1 Importance-Performance Analysis

IPA was formulated by Martilla and James in 1977 [3]. The IPA model is divided into four quadrants with performance on the x-axis and importance on the y-axis (Fig. 1). The quadrants are arranged according to the average of the importance and performance to differentiate. The IPA model measures customer importance and satisfaction performance. The four-quadrants and implication of IPA are shown in Table 1.

1.2 Internal Marketing for Nursing

The role of nurses in multi-occupational cooperation is expanding, and becoming more important as time progresses. Therefore, the primary concern is to use internal marketing strategies effectively to enhance nurses' organizational commitment. IPA enables management to evaluate and identify the major strengths and weaknesses of key organizational success factors. Therefore, this study used the IPA model to evaluate the importance and performance of internal marketing for nursing.

2 Method

To conduct a job analysis, a questionnaire was developed for swallowing rehabilitation nurses. Since the job description of rehabilitation nurses was not well defined, the Delphi method was used to attain consensus among professionals.

Table 1 Quadrant and implication of IPA [4]

Quadrants	Implication
Quadrant I Keep up the good work	Attributes are perceived to be very important to respondents, and at the same time, the organization seems to have high levels of performance in these activities. The message here is to keep up the good work
Quadrant II Concentrate here	Attributes are perceived to be very important to respondents, but performance levels are fairly low. This suggests that improvement efforts should be concentrated here
Quadrant III Low priority	Attributes here are rated as having low importance and low performance. Although performance levels may be low in this cell, managers should not be overly concerned, since the attributes in this cell are not perceived to be very important. Limited resources should be expended on this low priority cell
Quadrant IV Possible over kill	This cell contains attributes of low importance, but where performance is relatively high. Respondents are satisfied with the performance of the organization, but managers should consider present efforts on the attributes of this cell as being superfluous/unnecessary

After the questionnaire was developed, it was distributed to 45 experts in the field of swallowing nursing.

The operational definition of the term "expected role" is as follows: Expected role is a role that is expected from the patient, family, and other professions. Customer needs are expressed as "job importance," and are measured by a Likert-type scale. Similarly, the satisfaction of customer needs is expressed as "job performance."

The questionnaire survey was distributed to certified nurses. Through IPA, swallowing rehabilitation items, including oral function management items that are expected of nurses were selected. The matrix category "*High Importance and High Performance*" was labeled "*Keep up the Good Work.*" Those 35 job items (45.4 % of all the items), considered important KPIs, were included in the matrix analysis if a certified nurse was assumed to be a role model of high achievement. In order to facilitate multi-occupational cooperation, the KPI for "management of oral function" was created by selecting KPIs from those for swallowing rehabilitation. A pilot study using the KPIs for "management of oral function" was conducted on managerial nurses and hospital dentists. Accordingly, empirical research was conducted by using these KPIs on nurses in an acute care hospital. This study then further identified key improvement areas concerning multi-occupational cooperation using the KPIs.

2.1 A Questionnaire Constructed Using the Delphi Method

2.1.1 The Delphi Method

The Delphi method was developed by the RAND Corporation in the 1950s to forecast the impact of technology on warfare. The method entails a group of experts who anonymously reply to questionnaires and subsequently receive feedback in the form of a statistical representation of the "group response," after which, the process repeats itself. The goal is to reduce the range of responses and arrive at something closer to expert consensus [5].

2.1.2 Round 1

In order to extract the job items, a focus group interview was performed with the 6 chief nurses at the T rehabilitation hospital.

2.1.3 Round 2

After the focus group interview, a semi-structured interview was performed with 2 of the nurses on the K rehabilitation hospital Dysphagia Committee. As a result, some missing job items were added.

2.1.4 Rounds 3 and 4

A group of experts consisting of 24 nurses, 7 medical doctors, 5 dentists, 3 speech therapists, 3 occupational therapists, and 3 physical therapists was selected to constitute the panel ($n = 45$). Nineteen of the 24 nurses were selected from rehabilitation hospitals. Four were the first graduates of the dysphagia rehabilitation certification course. One was the instructor of the certified nursing course.

Seven doctors, 5 dentists, and 9 therapists were selected from rehabilitation hospitals. The 7 doctors were rehabilitation medicine specialists and the 5 dentists were certified in providing care for the elderly.

In Round 3, the questionnaire was handed to each expert panel member, who was asked to rate the validity of the job items using a 5-point Likert-type scale. They could add omitted job items in the space provided. In Round 4, the respondents were given feedback about their Round 3 ratings. Specifically, they were shown the group means and standard deviations. Next, they rated the same job items in the same manner as that described for Round 3. The questionnaires were distributed and collected by e-mail. The items were rank ordered by the mean scores. Items with scores greater than 4 were retained for subsequent analyses. Figure 2 shows a summary of the Delphi method.

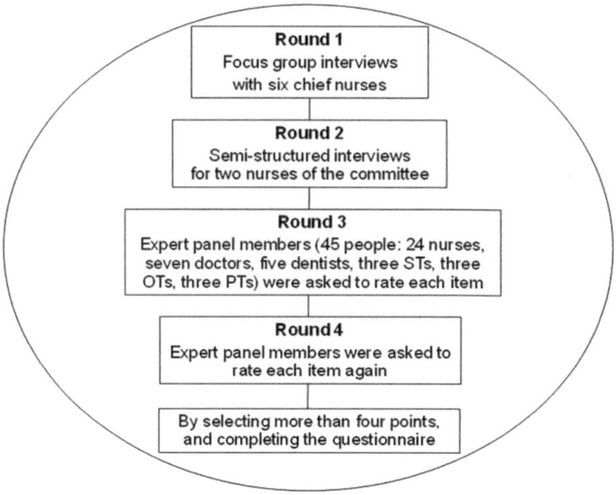

Fig. 2 Summary of the Delphi method

The job items were categorized as risk management, oral care, assessment, instruction to the patient and family, cooperation and collaboration, meal assistance, direct training, environment adjustment, indirect training, correspondence about the meal form, and advocacy.

2.2 Survey of the Job Analysis by Certified Nurses

2.2.1 Subjects

The application of the Delphi method resulted in the final questionnaire. This questionnaire was electronically distributed to 31 nurses who were the first graduates of the swallowing rehabilitation nursing certified nurse education program.

2.2.2 Method

Each professional nurse rated each of the scale items in terms of the job importance and performance in their daily practices of dysphagia rehabilitation. IPA was used for the data analysis. The KPIs for swallowing rehabilitation were taken from Quadrant I (Keep up the good work) in the IPA framework.

2.3 *A Pilot Study Using the KPI for "Management of Oral Function"*

2.3.1 Subjects

The subjects for the pilot study included five managerial nurses and five dentists in five hospitals.

2.3.2 Method

The KPIs for "management of oral function" was created by selecting items such as "oral care," "assessment," and "indirect training" from the KPIs for swallowing rehabilitation. A pilot study using the KPIs for "management of oral function" was conducted on managerial nurses and hospital dentists. In each of the items, both the degrees of expected importance and performance, which had impact on increasing satisfaction, were measured and evaluated by a 7-point Likert-type scale. A ceiling effect was considered, and the marking method was changed from 5 points to 7 points. It should be noted that the score on the dentist side is the expected value of the nurse.

2.4 *Empirical Research Conducted Using KPIs on Nurses in an Acute Care Hospital*

2.4.1 Subjects

The questionnaire on the KPIs for "management of oral function" was distributed to 14 nurses who worked in acute care hospital H and a dentist. The hospital dentist received only one question.

2.4.2 Method

In each of the items, both the degrees of expected importance and performance, which had an impact on increasing satisfaction, were measured and evaluated by a 7-point Likert-type scale. The difference between the importance and performance was calculated. In addition, the improvement factor (IF) was calculated, as shown in Eq. 1 [4].

$$\mathrm{IF} = (\text{performance-importance})/\text{importance} \tag{1}$$

The dentist received a question with possible dental service interventions. These were evaluated by a 7-point Likert-type scale.

3 Results

3.1 *The Delphi Method*

Ninety-four job items were selected using the Delphi method. However, a job list of 77 items with a mean rating greater than 4 points was adopted.

3.2 *Questionnaire Survey on the Job Analysis of the Certified Nurses*

Completed questionnaires were collected from 21 of the 31 certified nurses; thus, the return rate was 68 %. As a result of the IPA, 35 job items (45.4 % of the job items) fell in the "*Keep up the Good Work*" category (Fig. 3, upper right matrix). Ten items fell in the "*Concentrate Here*" category (Fig. 3, upper left matrix). Nine items were included in the "*Possible Overkill*" category (Fig. 3, lower right matrix). Finally, 23 items fell into the "*Low Priority*" category (Fig. 3, lower left matrix). The KPIs for swallowing rehabilitation were job items that fell into the *Keep up the Good Work* category.

3.3 *A Pilot Study Using the KPIs for "Management of Oral Function"*

The KPIs for "management of oral function" (12 job items) were created by selecting items such as "oral care," "assessment," and "indirect training" from the KPIs for swallowing rehabilitation (35 job items), as shown in Table 2. Results of the IPA are shown in Figs. 4 and 5, Tables 3 and 4.

IPA was used to compare the managerial nurses' and hospital dentists' perceptions of attributes in the KPIs for "management of oral function." Each attribute was plotted according to the mean score of its perceived importance and performance, with the importance of attributes on the vertical axis from low (bottom) to high (top), and the performance of attributes on the horizontal axis from low (left) to high (right). The four quadrants are constructed with crosshairs set at the average scores of the importance and performance scores.

For managerial nurses (Fig. 4), the averages for the pooled data were importance, 5.88, and performance, 4.97. For hospital dentists (Fig. 5), they were importance, 5.95, and performance, 4.42. These figures show that managerial nurses and hospital dentists have different perceptions of the 12 factors. In Fig. 4, the *Concentrate Here* quadrant captured a single attribute for nurses: "For oral care, I massage the oral mucosa with a sponge brush (O2)." This attribute also presents a major discrepancy between importance and performance that requires special

Fig. 3 Importance-performance analysis for the certified nurses ($n = 21$)

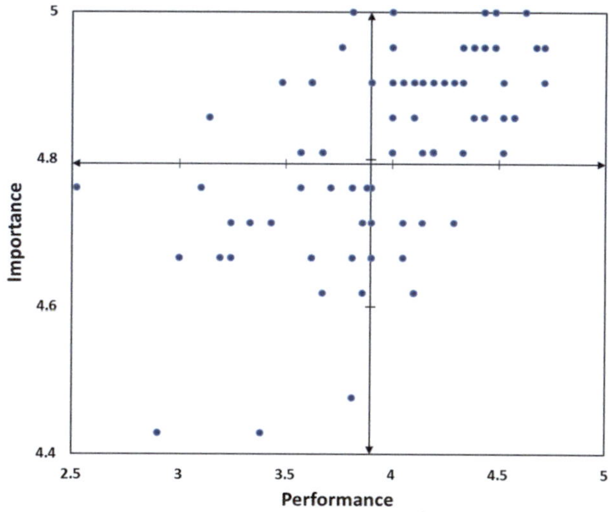

Table 2 Oral function management items

Code	Job items of oral management for nursing	I	P
A1	I asses the patient's oral cavity	5.00	4.62
IT	As indirect training, I train patient on oral function	5.00	4.48
O1	I help patients and their families understand the need for oral care	5.00	4.00
O2	For oral care, I massage the oral mucosa with a sponge brush	4.95	4.67
O3	I confirm the independence of oral care	4.95	4.48
O4	For oral care, I assist the patient with denture cleaning	4.90	4.71
O5	For oral care, I assist the patient with tooth brushing	4.90	4.52
A2	I conduct physical assessments of the cranial nerve system	4.90	3.90
A3	I confirm the functioning of patients' dentures during mealtimes	4.86	4.52
A4	I understand the problems related to feeding, such as in cases of higher brain function disability	4.86	4.00
O6	I confirm oral care according to the ADL	4.81	4.33
A5	While assessing inflammation, malnutrition, and dehydration, I write the individual plan for the patient	4.81	4.00

(a) *I* importance, *P* performance
(b) *O* oral care, *IT* indirect training, *A* assessment
(c) Means above 4.79 on the Importance scale and above 3.89 on the Performance scale in 77 items are shown in the table
(d) A 5-point Likert-type scale was used

attention (Table 3). In Fig. 5, the dentists include two attributes in this quadrant: "I conduct physical assessments of the cranial nerve system (A2)," and "As indirect training, I train patient on oral function (IT)." Nurses identified four attributes in the *Keep up the Good Work* quadrant, which could be considered primary oral care. From the dentists' point of view, 5 attributes are included in this quadrant.

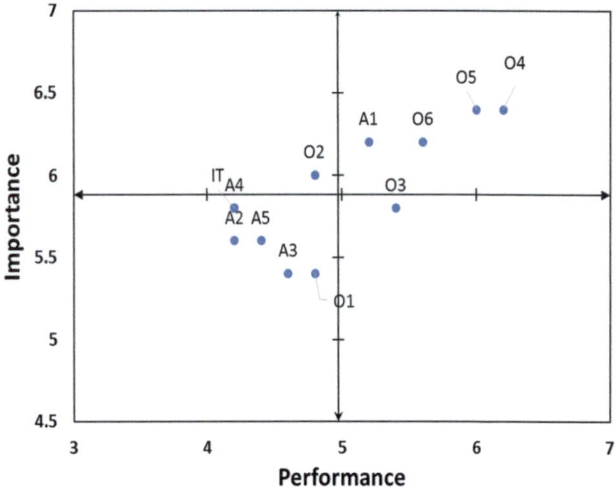

Fig. 4 Importance-performance analysis for managerial nurses ($n=5$)

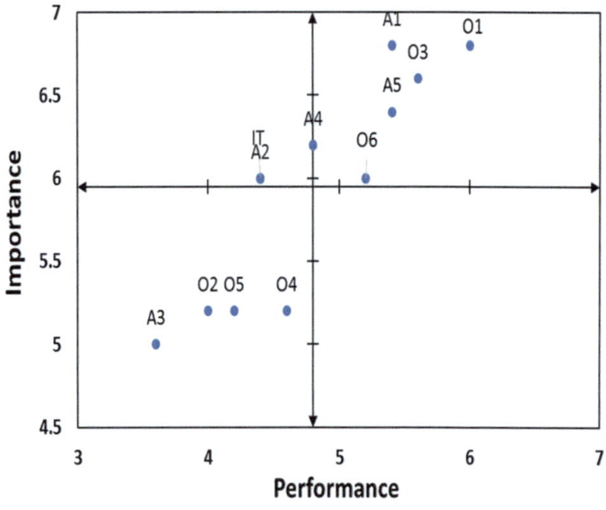

Fig. 5 Importance-performance analysis for hospital dentists ($n=5$)

Table 3 lists the aggregate importance and performance values of each attribute with the difference between the two for nurses and dentists. All the importance scores are higher than the performance scores. It is necessary to analyze the discrepancies between the performance and importance scores so that attributes with greater differences will be given higher priority.

Table 4 presents the results for the difference between the nurses' and the dentists' perceptions in service quality. The primary gap between the nurses' and the dentists' perceptions is found for the job item "I help patients and their families understand the need for oral care (O1)."

Table 3 Aggregate importance and performance scores of each attribute (nurses and dentists)

	Nurses			Dentists		
Code	Importance	Performance	Difference	Importance	Performance	Difference
O1	5.4	4.8	0.6	6.8	6.0	0.8
O2	6.0	4.8	1.2	5.2	4.0	1.2
O3	5.8	5.4	0.4	6.6	5.6	1.0
O4	6.4	6.2	0.2	5.2	4.6	0.6
O5	6.4	6.0	0.4	5.2	4.2	1.0
O6	6.2	5.6	0.6	6.0	5.2	0.8
A1	6.2	5.2	1.0	6.8	5.4	1.4
A2	5.6	4.2	1.4	6.0	4.4	1.6
A3	5.4	4.6	0.8	5.0	3.6	2.4
A4	5.8	4.2	1.6	6.2	4.8	1.4
A5	5.6	4.4	1.2	6.4	5.4	1.2
IT	5.8	4.2	1.6	6.0	4.4	1.6

Table 4 Oral management service quality (Nurses vs. Dentists)

Code	Nurses	Dentists	Gap	Ranking
O1	4.8	6.0	−1.2	1
A5	4.4	5.4	−1.0	2
A4	4.2	4.8	−0.6	3
IT	4.2	4.4	−0.2	4
A2	4.2	4.4	−0.2	4
A1	5.2	5.4	−0.2	4
O3	5.4	5.6	−0.2	4
O6	5.6	5.2	0.2	8
O2	4.8	4.0	0.8	9
A3	4.6	3.6	1.0	10
O4	6.2	4.6	1.6	11
O5	6.0	4.2	1.8	12

3.4 *Empirical Research Conducted Using KPIs on Nurses in an Acute Care Hospital*

Completed questionnaires were collected from 12 of the 14 regular nurses; thus, the return rate was 86 %. IPA was used to assess the acute care hospital nurses' perceptions of attributes in the KPIs for "management of oral function." For nurses (Fig. 6), the averages for the pooled data were importance, 6.42, and performance, 4.15. In Fig. 6, the *Concentrate Here* quadrant captured two attributes: "I asses the patient's oral cavity (O1)," and "While assessing inflammation, malnutrition, and dehydration, I write the individual plan for the patient (A5)." These attributes also present a major discrepancy between importance and performance (Table 5), which also requires special attention. Nurses identified five attributes in the *Keep up the Good Work* quadrant, which could be considered primary oral care.

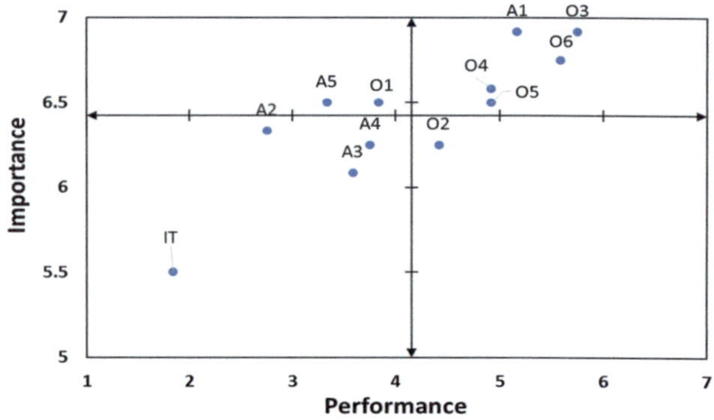

Fig. 6 IPA for nurses in acute care hospital ($n = 12$)

Table 5 Aggregate importance and performance scores and dental intervention scores (Nurses, $n = 12$, Hospital dentist, $n = 1$)

	Nurses			Dentist
Code	Importance	Performance	Difference	Intervention
O1	6.50	3.83	2.67	7
O2	6.25	4.42	1.83	7
O3	6.92	5.75	1.17	7
O4	6.58	4.92	1.66	6
O5	6.50	4.92	1.58	6
O6	6.75	5.58	1.17	7
A1	6.92	5.17	1.75	6
A2	6.33	2.75	3.58	5
A3	6.08	3.58	2.50	6
A4	6.25	3.75	2.50	5
A5	6.50	3.33	3.17	7
IT	5.50	1.83	3.67	7

Table 5 lists the aggregate importance and performance values of each attribute for nurses and dental intervention scores for hospital dentists. All the importance scores are higher than the performance scores.

Figure 7 shows the difference between importance and performance in the IPA matrix; the size of the grid also shows a difference.

Figure 8 shows an IPA matrix applied to dental interventions; the size of the grid shows the intervention value.

Service quality measure factors for oral management services in nursing are shown in Table 6.

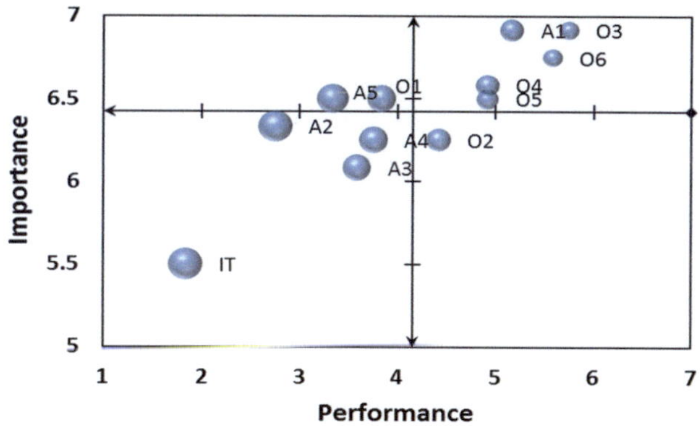

Fig. 7 IPA × difference for nurses in acute care hospital ($n = 12$)

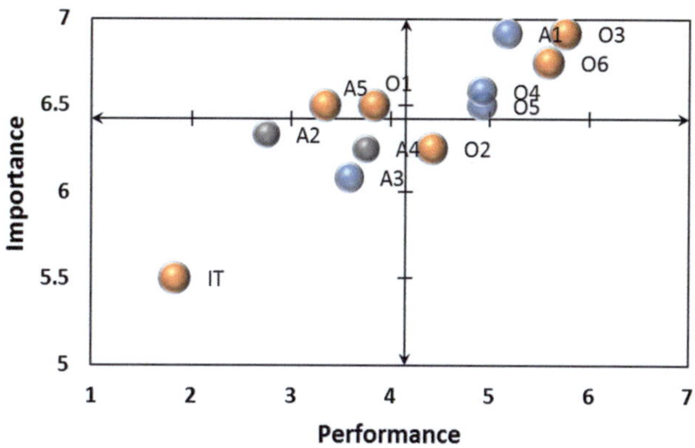

Fig. 8 IPA × dental intervention for nurses in acute care hospital ($n = 12$)

Table 6 Results of IF

Code	Improving factor (IF)	Ranking
IT	−0.667	1
A2	−0.566	2
A5	−0.487	3
A3	−0.411	4
O1	−0.410	5
A4	−0.400	6
O2	−0.293	7
O4	−0.253	8
A1	−0.253	8
O5	−0.244	10
O6	−0.173	11
O3	−0.169	12

(a) IF = (Performance-Importance)/Importance

4 Discussion

The methodology used here appears to be valid and may be applied to identify professional roles for other rehabilitation fields [1]. The implementation of the Delphi method was a straightforward and effective tool used to create a questionnaire. IPA is a strategic tool that can quickly enable a manager to understand "identified customer needs" and "satisfied customer needs," and to assess performance rather than rely solely on importance indicators. Moreover, IPA allows for the visualization of the results of a strategic analysis (see Fig. 3), which facilitates understanding and interpretation of the results.

This study also demonstrated the validity of IPA. A new job analysis using the Delphi method and IPA showed one way of exploring human resource management in swallowing rehabilitation. In addition, as an evaluation tool, KPIs for swallowing rehabilitation (35 job items) were developed.

In this study, the authors devised a system that facilitated cooperation between medicine and dentistry (multi-occupational cooperation) by identifying and using KPIs. As such, one of the purposes of this study was to evaluate the effectiveness of the system. In order to facilitate multi-occupational cooperation, the KPIs for "management of oral function" (12 job items) were created by selecting items such as "oral care," "assessment," and "indirect training" from the KPIs for swallowing rehabilitation (35 job items). A pilot study using the KPIs for "management of oral function" was conducted on managerial nurses and dentists in five hospitals. In each of the items, both the degrees of expected importance and performance, which had an impact on increasing satisfaction, were measured and evaluated. Accordingly the research was empirically conducted by using these KPIs on nurses in an acute care hospital. This study then further identified key improvement areas concerning multi-occupational cooperation by using the KPIs. Note that the KPIs for management of oral function were based on an activity evaluation of the service provider and the structure, process, and outcomes (SBO) classification in the Donabedian Model (1966) [6].

The initial portion of this research was the pilot study. There are importance and performance values of each attribute together with the difference between the two for nurses and dentists in five hospitals. There are the differences between the nurses' and the dentists' perceptions in service quality, and the primary gap between the nurses' and the dentists' perceptions is found for the job item "I help patients and their families understand the need for oral care (O1)." This means that the expected value of the dentist side was high. The results indicated a difference of perspective, in that dentists could not predict the behavior of nurses. Therefore IPA for KPIs is necessary.

The subsequent research was the implementation phase of this study, conducted in an acute care hospital. We can conclude with certainty that three methods (IPA, Difference, and IF) are able to explain service marketing and management. This is more visible by representing the difference by the size of the grid in the IPA. The *Concentrate Here* quadrant captured two attributes; these attributes also present a major discrepancy between importance and performance. However, it was possible to receive dental interventions in these attributes. Thus, this can be solved by coordination. Similarly, it is important to receive the help of doctors and dentists in attributes with high IF.

Acknowledgments We would like to show our greatest appreciation to Ns. Koyama who provided carefully considered feedback and valuable comments. Special thanks also go to nurses who were the first graduates of swallowing rehabilitation nursing certified nurse education program whose comments made enormous contribution to our work.

This study was supported by the Research Funding for Longevity Science (25–7) from National Center for Geriatrics and Gerontology (2014).

References

1. Nagaosa S, Yaeda J, Matsushita H, Abdul MS, Sumi Y (2013) Visualization of the role expectation for strategy management using the Delphi method and importance-performance analysis in dysphagia rehabilitation nursing services. In: The 1st international conference of serviceology, Tokyo, Japan, 16–18 October, pp 217–220
2. Sumi Y, Ozawa N, Miura H, Miura H, Toba K (2011) Community coordination of dental care needs in a home medical care support ward and at home. Jpn J Geriatr 48(4):391–396
3. Martilla J, James J (1977) Importance-performance analysis. J Mark 41(1):77–79
4. Tzeng G, Chang H (2011) Applying importance-performance analysis as a service quality measure in food service industry. J Technol Manag Innov 6(3):106–115
5. Dalkey N, Helmer O (1963) An experimental application of the Delphi method to the use of experts. Manag Sci 9:458–467
6. Donabedian A (1966) Evaluating the quality of medical care. Milbank Q 83(4):691–729

Designing the Amount of Image Delay in Tele-surgery

Iwane Maida, Hisashi Sato, Tetsuya Toma, and Takashi Maeno

Abstract It has become possible enough to take telemedicine by progression of the ICT. However, telesurgery is limited to experimental trials. This reason is that surgeon had been disturbed perceptual motor coordination by visual delay. This study was analyzed to understand about range of work efficiency in telesurgery with the delay. As a result of analysis, it was suggested that pointing task has threshold of the difficulty between 2 mm and 4 mm in diameter. The similar range in the work efficiency was up to approximately 300 ms independent on pointing size.

Keywords Medical service • Tele-surgery • Image delay • Pointing • Perceptual motor coordination • Cognition

1 Introduction

The medical field plays an important role in society of realizing social security. However, many countries and regions around the world lack sufficient healthcare services; one factor for this shortage in medical services is the regional gap, which depends on several parameters such as the size of a country, whether the country is made up of islands, and the number of doctors.

Tele-surgery has attracted attention as a potential solution to this problem, but it has not been established as a common form of medical care from a global standpoint. One reason is that the image delay during tele-surgery has a strong influence on maneuvers. Past studies have not necessarily confirmed its suitability as a medical service. In this study, a psychophysics experiment was carried out using a simplified model for maneuvers. The purpose of this study was to specify that the cognition of the subject changes according to the visual delay.

I. Maida (✉) • T. Toma • T. Maeno
Graduate School of System Design and Management, Keio University, Yokohama 223-0061, Japan
e-mail: i-maida@z6.keio.jp

H. Sato
Visualware Division, Keisoku Giken Co., Ltd, Yokohama 224-0037, Japan

T. Maeno et al. (eds.), *Serviceology for Designing the Future*,
DOI 10.1007/978-4-431-55861-3_13

2 Experimental Tele-surgery and Visual-Delay

In order to provide a fair geographical distribution of healthcare services, the medical system must add not only remote care but also tele-surgery. Surgical robots have seen improved operability relative to actual surgical environments. A major factor is that surgical robots have many degrees of freedom [1]. In other words, while a forceps used for common endoscopic surgical operation has two degrees of freedom, surgical robots such as the da Vinci have seven degrees of freedom. Therefore, surgical robots can faithfully reproduce the movement of the hand during an operation [2]. Surgical robots can also perform the same movements in abdominal operations.

The tele-surgery system has come closer to realization by the development of surgical robots with many degrees of freedom and steady progress in information and communication technology. In one experimental case, Marescaux et al. successfully performed gallbladder extraction remotely with a surgical robot in the United States from France using ZEUS [3]. This surgery had an intermediate degree of difficulty; thus, this group showed that tele-surgery of intermediate difficulty is possible. However, tele-surgery has not been established as a common form of medical care from a global standpoint.

There are two factors. First, the robot and transmission device for tele-surgery requires a transaction, and the transmission data produces an image delay on the screen like a Fig. 1.

Therefore, the visual delay affects perceptual motor coordination. In an example of tele-surgery between Japan and Thailand, the surgical arms controlled by the doctor operated with a delay of 582.4 ms [4]. The transmission speed over the communication line was 10.4 Mb/s, and the total transmission distance was 3700 km. In another example of tele-surgery between the UNIITED STATES and France, the doctor operated remotely on a patient with a delay of 200 ms. The surgical robots were connected by a communication line with a transmission speed of 20 Mb/s at a distance of 6230 km. This delay can have the effect of increasing mistakes by the operator during surgery [3].

Second, surgical operations are very different from other forms of medical treatment. Maneuvers for external invasion have the risk of death for the patient when an operator commits a mistake during surgery. Thus, researchers must increase safety by quantifying the risk factors of tele-surgery.

3 Relationship Between Visual Delay and Perceptual Motor Coordination

The arms of a surgical robot enable complicated and detailed movement with high operational flexibility. Therefore, tele-surgery is in constant demand for minimally invasive surgery requiring accuracy. However, a major structural problem of the

Fig. 1 Difference delay generating structure between tele- surgery and real site surgery

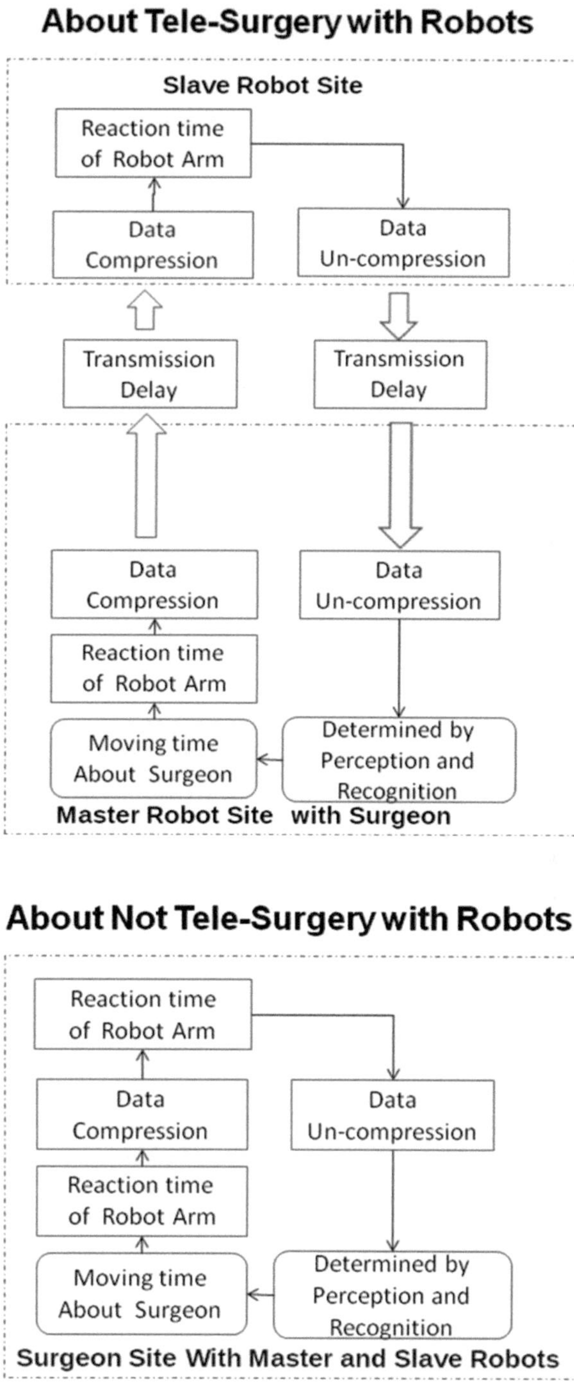

tele-surgery system is the image delay due to various instrumental transactions. This delay comprises delays in transmission and digital signal processing in cameras and monitors.

In general, communication delay is about 200 ms for a person in Japan communicating with someone in the UNIITED STATES using an international communication line. However, a tele-surgery system has a greater delay because of the data size, compression, and prevention of freezing in the camera and monitor. In general, a delay of 0.5 s is very short. However, the time lag caused by the delay is a major problem when the human senses are combined.

During surgery, the operator manipulates the arms based on visual information obtained from the eyes. In other words, the operator moves the arms based on the time order of information from the eyes. Therefore, visual information is always located earlier in the time order of information from each sense. By this principle, perceptual motor coordination works based on visual information. However, during surgery the monitor displays an operating field with a delay because of the nature of tele-surgery. Therefore, the movement of the arm by the operator is displayed on the monitor with a delay. This makes maneuvers very difficult and is representative of the challenges faced in perceptual motor coordination. The difficulty caused to humans by a visual image delay needs to be evaluated. The next section reviews the nuisance value of the image delay for conventional perceptual motor coordination.

4 Work Efficiency of Perceptual Motor Coordination with Delay in Previous Research

Previous studies have shown that exercises with an image delay cause a discordance of the senses, which decreases work efficiency. Past studies have shown that the work efficiency of subjects decreases rapidly when the delay between vision and somatic sensations is 0–500 ms [5]. Maneuvers consist of integrating the somatic sensations of the hand with visual information from the eyes. This delay affects the safety of surgical maneuvers and work efficiency because it suppresses the perceptual motor coordination of the operator. In addition, the transmission delay is not constant and has some fluctuation, and study about delay on transmission has been progressing but the studies cannot suppress a delay enough [6, 7]. Doctors are also restricted in terms of working hours because a wound should be closed as soon as possible to prevent infection and increase the QOL for a patient.

In surgical procedures, the time required to perform each step can be added up to determine the total time necessary for the entire procedure. When the surgeon is operating remotely, a delay occurs with each step. Therefore, surgeons have been demanding the dangerous elements of tele-surgery to be quantified experimentally.

A maneuver basically consists of incision, detachment, suturing, and ligation. In other words, a fundamental element of a maneuver is precisely pointing the fingers based on the sight of the surgeon. The gap between the visual information and somatic sensations makes this pointing difficult. This can lead to accidents, such as excessive operation time or medical mistakes. This is very disadvantageous for patients receiving healthcare. Determining how a difference in image delay affects the work characteristics of a person is necessary. In this study, the following experiments were carried out to identify the effect of delay during a pointing task.

5 Design of Experiments

5.1 Placement of Experiment Machine Parts

Figure 2 shows the machine used for the experiment. The subjects worked on pointing their right hand by fixing their eyes onto a monitor showing an image of their hand taken with a camera. The experimenter carried out the following control to reproduce the transmission delay in tele-surgery. The hand image output from the camera to the monitor was relayed through a delay generator; the delay generator controlled the image delay to up to 500 ms through the use of two frame units (see Table 1).

The experimenter made efforts to remove any sense of incongruity in the subjects with regard to the image on the monitor. These efforts included hand size, color, and brightness on the monitor to remove agitation caused by differences with the subjects' actual senses.

A previous study showed that there is a positioning gap between the somatic sensations to control the body and sight [8].

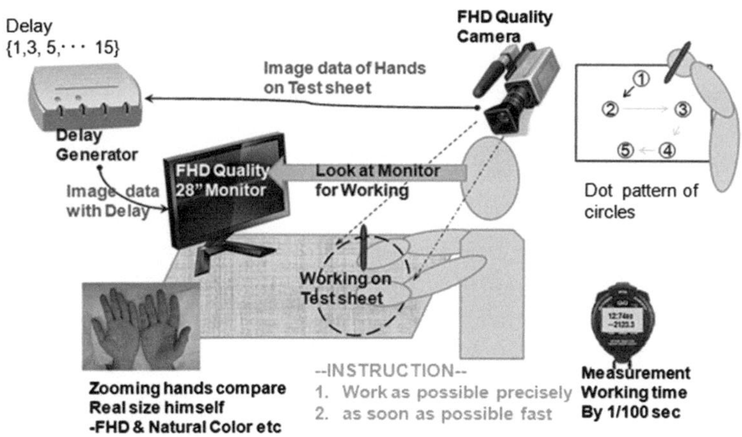

Fig. 2 Outline about a layout plan system in this study

Table 1 Equipment and materials used for experiment

No.	Equipment and materials
1	Full HD camera (30 fps)
2	Full HD28 inch monitor
3	Delay generator (up to 15 frames)
4	Block diagram of experimental equipment and materials

5.2 Subjects

The purpose of this study was to identify the work efficiency when subjects performed a maneuver based on an image with a delay. The subjects were all in their twenties because young people were assumed to have high adaptability to visual delay. They were all right-handed with visual deficits. The ten subjects were randomly placed in groups of three.

5.3 Experiment Plan

The purpose of this study was to determine the work efficiency based on the image delay. The experimenter planned the experimental groups with pointing sizes of 2, 4, and 6 mm.

The analysis considered two points: first, the difference in work efficiency with the amount of image delay; second, the difference in difficulty level with the pointing size. The former was a comparison of performances by a given subject, while the latter was a comparison between subjects.

In the experiment, multiplex analysis was carried out to compare the difference in delay between subjects. When equal variances could not be confirmed through a Bartlett test, a nonparametric Friedman test was performed to examine the work efficiency with regard to differences in delay. Scheffe's method of pair comparisons for multiplex analysis was used when a significant difference was detected in the statistical processing. The work efficiency for surgery was determined by quantifying the degree of difficulty due to the pointing size.

5.4 Experimental Procedure and Instructions

Before the start of the experiment, the subjects were instructed in the ground rules of tele-surgery and to point to a circle without making a mistake. The experiment ethics were explained, and the subjects were told that the experiment would be discontinued if they felt discomfort. The subjects' consent was obtained.

Table 2 Measurement tools used for experiment

No.	Measurement tools
1	Simplified surgery model (five-circle pattern)
2	Stopwatch (1/100)

Using the tools shown in Table 2, the subjects pointed to five circles in order. Pointing was considered to be successful when the finger was inside the frame of each circle. Subjects were not allowed to proceed to the next circle until the task was successfully performed. The time was measured as the working time for one trial when the subject had finished pointing at the fifth circle. The image stimulation with the delay had eight levels. The levels were made by dividing two frames into 1–15 frames at equal intervals.

Combining two of the patterns was found to gradually increase the delay by the subjects. The subjects carried out 16 trials in total. Combinations were used to suppress the learning effect of the subjects. The subjects were given a break of at least 1 min in intervals of each level to remove the fatigue factor as much as possible.

6 Analysis Results

6.1 *Learning Effect*

Figure 3 shows the transition of the working time for each delay level. The working time was found to increase monotonically with the amount of delay. The delay level was not confirmed to have a major effect. Therefore, the experimental data showed the effect of the image delay rather than the learning effect. The data showed that the learning effect could be canceled by comparing the subjects' results statistically.

6.2 *Experiment 1*

Figures 4, 5 and 6 show about the experimental data. These data were analyzed with regard to the range of work efficiency for each pointing size group through the Friedman test. The significance level was set to 5 %. A significant difference of 1 % appeared in the statistical processing. The significant difference allowed the progression to the next step because the results needed multiple comparisons for analysis. Multiplex analysis was performed on the data. The data were analyzed through multiple comparisons because the significant difference validated the advance to the next statistical step.

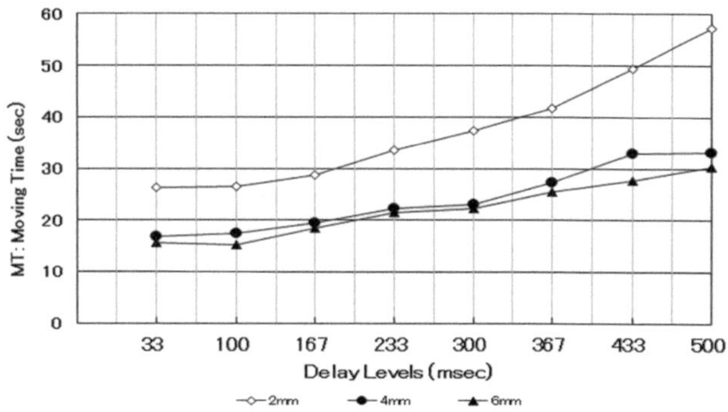

Fig. 3 Examination about the learning effect

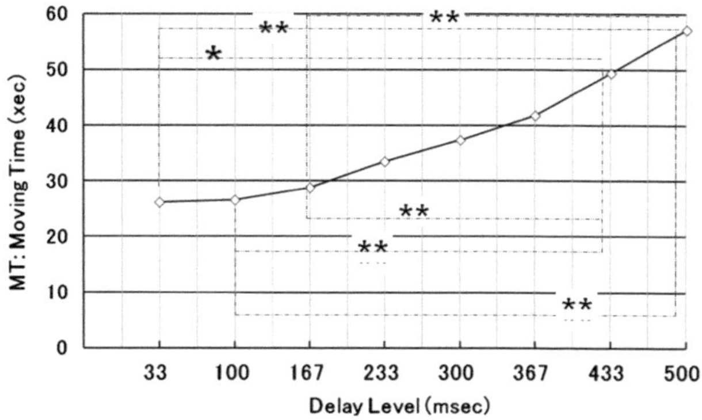

Fig. 4 Range of work efficiency in 2 mm diameter group

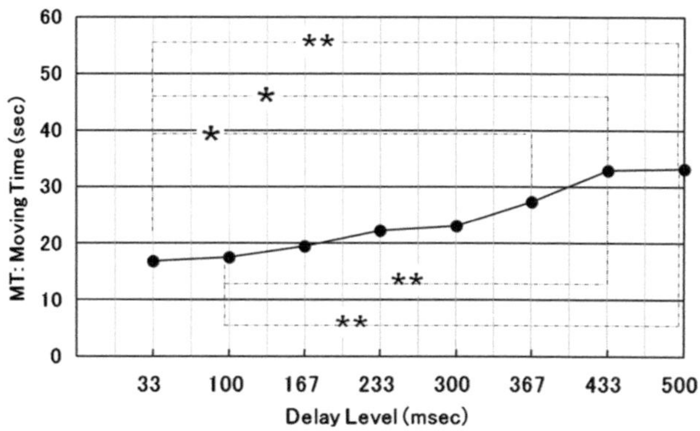

Fig. 5 Range of work efficiency in 4 mm diameter group

In this study, the experiment had many levels of stimulation with eight levels of delay for multiple comparisons. Therefore, Scheffe's method of pair comparisons, which is robust for nonparametric multiple comparison, was used for the analysis.

Fig. 6 Range of work efficiency in 6 mm diameter group

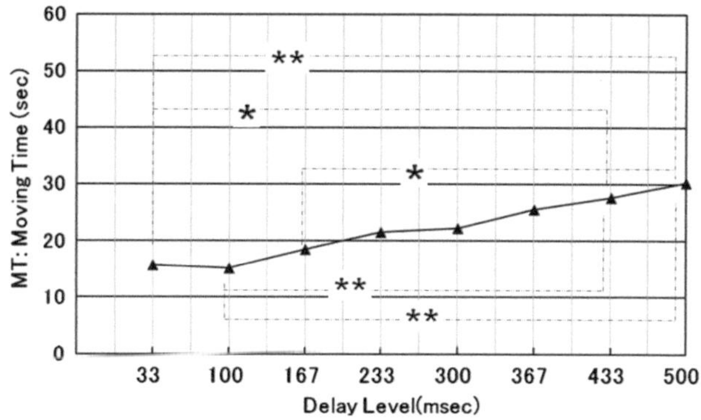

Fig. 7 Comparison of the degree of difficulty in each group

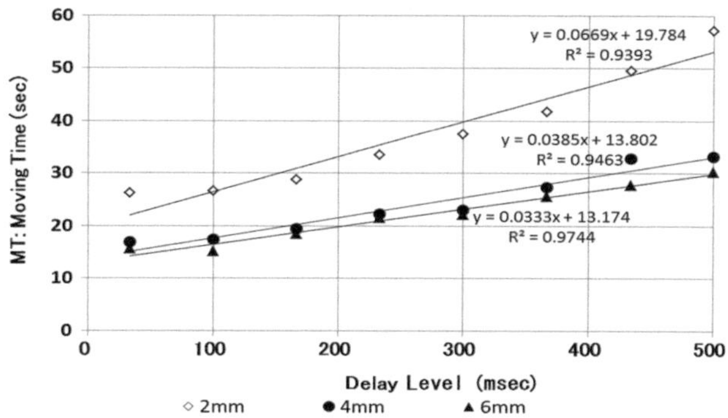

6.3 Experiment 2

This experiment examined the relationship between the pointing size and amount of delay with regard to work efficiency. Figure 6 shows the average working time for each group size. The 2 mm diameter group was found to require more working time than the 4 and 6 mm diameter groups. In this study, the slope of the linear equation was calculated to determine the degree of difficulty for the pointing task of each group and the coefficient of correlation. Thus, the 2 mm diameter group was found to have more than twice the difficulty of other groups. The Pearson test was used to confirm that the correlation was strong with coefficients of 0.9 or more (Fig. 7).

7 Discussion

7.1 Range of Working Efficiency in Each Pointing Size Group in Experiment 1

In experiment 1, the work efficiency did not depend on the degree of difficulty caused by the difference in pointing size. According to Fitts' law, the working time for pointing is given by MT = b(ID) + a. The index of difficulty for pointing is given by $ID = \log_2(A/W + 1)$, where A is the target width (i.e., circle diameter) and W is the distance between points. Therefore, the working time for pointing depends on the variables A and W. The degree of difficulty increases as the circle diameter decreases. This should decrease the range of delay.

A significant difference in the working time was only detected when the delay level difference was more than 10 frames. Therefore, a smaller pointing size makes pointing more difficult. In other words, a smaller pointing size increases the working time.

In the experimental data for the difference in working times for delay levels, the delay range did not depend on the difference in pointing size and was found to be approximately constant. The delay range was detected in more than 8–10 frames for the working time.

This delay range was equivalent to 266–333 ms.

This experiment showed that operators have similar work efficiency when this delay range is selected for communication line quality.

7.2 Relation Between Pointing Size and Delay in Experiment 2

According to Fitts' law, the working time for pointing is given by $MT = b(ID) + a$. In this study, a linear equation was applied to identify the degree of difficulty for a pointing task with regard to the pointing size. The coefficient for the degree of difficulty was found to be about 1.7–2 based on a comparison with other groups. The pointing size of 2 mm was used as the threshold of the difficulty.

8 Conclusion

This study examined the effect of tele-surgery on minute pointing movements by the operator.

The working time of each pointing size group showed a significant statistical difference when the delay level was higher than 8–10 frames regardless of the pointing size.

In other words, reducing the delay is important when a surgeon needs to perform very minute work in tele-surgery.

The threshold of difficulty for a pointing task with visual delay was quantified to a pointing size of 2 mm in diameter.

Future work will involve examining work characteristics of pointing to represent surgical maneuvers under a delay to determine the relationships between the delay and parameters in Fitts' law.

References

1. Intuitive Surgical, Inc. Home. page http://www.intuitivesurgical.com/
2. Yasunaga T, Hashizume M (2006) Development of surgical robotics and open MRI. MEDIX 44:4–7
3. Marescaux J, Leroy J, Gagner M et al (2001) Transatlantic robot-assisted telesurgery. Nature 413:379–380
4. Arata J, Takahashi H, Pitakwachara P, Warisawa S et al (2006) A remote surgery experiment between Japan-Thailand using the minimally invasive surgical system. J Jpn Soc Comput Aided 8(3):174–175
5. Fujisaki W (2012) Effects of delayed visual feedback on grooved pegboard test performance. Front Psychol 3:61: doi: 10.3389
6. Kameyama H, Sato Y, Yoshida Y (2007) Erasure code with small reception overhead and its application to distributed storage. In: IEEE 41st annual conference on information sciences and systems, pp 80–85
7. Kameyama H, Sato Y, Yoshida Y, Yoshida Y, Urita S (2009) High-speed file transfer solution for cloud computing. FUJITSU 60:470–475
8. Kitagawa T, Fukuda H, Fukumura N, Ono Y (2006) Investigation of error in human perception of hand positions during arm movement. IEICE Trans Inf Syst J89-D:1429–1439

Visualization of Muscle Activity During Squat Motion for Skill Education

Koshiro Yanai, Qi An, Yuki Ishikawa, Junki Nakagawa, Wen Wen, Hiroshi Yamakawa, Atsushi Yamashita, and Hajime Asama

Abstract Physical skills are important for various service fields, such as sports, nursing, and manufacturing. Experts show high performance by using specific muscular movements which non-experts are unable to use. It is important to understand expert's characteristic of muscle activity in specific motion and difference between experts and non-experts for skill education. However, people cannot just observe the difference between expert's and non-expert's muscle activity and imitate expert's muscle activity because muscle activities are invisible. In this study, a novel skill education system was developed for non-experts by measuring movement, analysing and visualizing characteristic of the movement in order to improve skill education effect and efficiency. This study focused on squat motion, a kind of strength training. We measured movement of experts with motion capture and surface electromyography, and extract skills for effective and safety training by analysing motion. A method of visualization extracted skills for skill education was proposed. At last, the experiment was carried out in order to verify effectiveness of proposal method for skill education. In the experiment, it was suggested proposal method is effective for physical skill education.

Keywords Visualization • Motion capture • Skill education • Surface electromyography

1 Introduction

The motions which need special physical skills exists in various service fields, such as sports, nursing, and manufacturing. In each field, experts show high performance by using specific physical skills which non-experts are unable to use [1, 2]. In physical skill education, experts teach non-experts their way to use muscle as a kind of physical skill. Experts usually can use only their words and gestures for education in education service fields. However, muscle activity is invisible with a naked

K. Yanai (✉) • Q. An • Y. Ishikawa • J. Nakagawa • W. Wen • H. Yamakawa • A. Yamashita • H. Asama
The University of Tokyo, Tokyo, Japan
e-mail: yanai@robot.t.u-tokyo.ac.jp

T. Maeno et al. (eds.), *Serviceology for Designing the Future*,
DOI 10.1007/978-4-431-55861-3_14

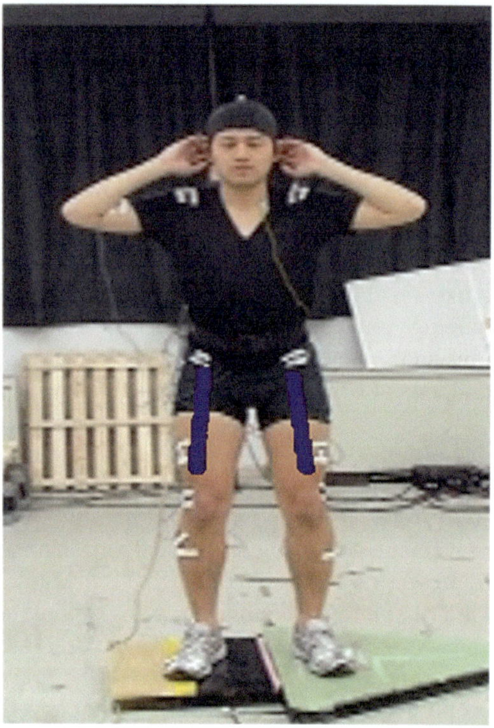

Fig. 1 Image of visualization of muscle activity

eye. Therefore, it is difficult for experts to teach timing and strength of using muscle by their words and gesture.

Several studies have reported that effectiveness and efficiency of physical skill education are increased by feeding back invisible information in different form. Tsubouchi showed that muscle activity was changed by feedback electromyography data as sound [3]. Previous works reported muscle activity was enhanced by visual feedback of electromyography [4]. However, this research did not focused muscle activity during motion as skill. This study focused on visual feedback of muscle activity during motion for skill education.

The significance of this study is to solve problem about inefficiency and ineffective in physical skill education. In this study, we developed physical skill education system by visualization of muscle activity during motion. Figure 1 shows an image of visualization of muscle activity. It is possible that non-experts observe expert's muscle activity directly and understand expert's timing and strength of muscles.

Target of this study was education of squat motion, a kind of strength training. Squat motion is known widely as training of sports. In addition, this motion is noticed as basic of motions in daily life, such as standing up motion and lifting up motion. Squat expert's physical skills were extracted in training books and previous studies. In those, important points of squat motion was suggested in the viewpoint of training and safety. From the viewpoint of training, it is important to use the muscles which should be strengthened. From the viewpoint of safety, it is important to perform squat motion with the posture that is hard to be hurt. Posture is easy to teach to non-experts because posture is always visible. However, muscle activity is invisible and difficult to teach, so this study focused how to use muscle.

The purpose of this paper is to develop physical skill education system by visualization of muscle activity during motion. This paper proposed a method of muscle activity visualization by using motion capture and surface electromyography. In this method, a color was used as visual feedback of muscle activity. Color of visualized muscle was reflected by strength of muscle activity. The experiment was carried out to verified effectiveness of physical skill education by visualizing muscle activity.

2 Visualization of Muscle Activity

There are two sides in physical skill that experts have. One side is kinematic skill, such as posture, trajectory of body movement and speed of movement. The other side is kinetic skill, such as power of muscle. Non-experts need to obtain these skills. However, in skill education, non-experts cannot do performance same as experts by imitating only expert's posture. When non-experts imitate only how to use muscle, non-experts cannot do performance same as experts too. Non-experts can show high performance by imitating both body movement and strength of using muscle of experts. In addition, timing of using muscle against body movement is important too. Therefore, in skill education, it is important for non-expert to understand expert's body movement and how to use muscle at the same time. Development skill education system that teach strength and timing of using muscle with expert's body movement is needed. Measurement body movement and muscle activity at the same time is needed. For the reason, measurement body movement and muscle activity makes quantitative evaluation of those skills possible. Evaluation of difference between experts and non-experts and progress degree of the non-experts make possible by quantitative skill evaluation. Lippold showed that linear relations holds between strength of muscle and electromyography of the muscle [5]. Timing and strength of using muscle are estimated by measuring surface electromyography (sEMG) of muscle. In this study, muscle activity was measured by sEMG. Measuring motion was needed in order to make clear relation between muscle activity and progress of motion. Body movement was measured by motion capture and video camera. Body movement was measured by recording coordination of some body joints continuously by motion capture. How to show of expert's measured physical skill is most important thing in education. It is important for non-experts to imagine expert's body movement more clearly, so actual video during motion was recorded by video camera. The direction to show the motion of the expert is important. Taking video of expert's movement from the plural directions made it possible that non-experts could watch some important parts of motion. This section explained how to measure muscle activity and motion and how to show these data to non-experts in detail.

2.1 Measurement Muscle Activity and Motion

Measurement electrical activity of target muscles was conducted by surface electromyography. sEMG was possible to measure by attaching an electrode on skin on target muscle. Measurement posture during squat motion was conducted by Optical motion capture. Markers was attached on 20 points of subject's body joints and filmed by 8 infrared cameras. It enabled to get the three-dimensional coordinate of the markers. Actual video is recorded by video camera in chronological order same as other data.

2.2 Method of Visualization

Filtering operations to sEMG data fixed base line were performed by using low-pass filter with 300 Hz and high-pass filter with 10 Hz of different cut-off frequency. Filtered sEMG data was rectified and smoothed. Maximum voluntary contraction (MVC) was measured in order to calculate %MVC, ratio of power of the muscle for the maximum power. MVC was obtained by measuring EMG at the position that was easy to perform maximum power. A comparison in others of activity of muscle is enabled by normalizing activity of muscle by MVC.

Three-dimensional coordination of each joints from motion capture were plotted in virtual space and 3D human model was made. This human model was point group and did not have musculoskeletal parts. Virtual muscles was drawn at real muscle position on 3D model. Shape of Virtual muscle was like pole. Parts of muscles drawn as virtual muscle were only parts that wanted to show muscle activity to non-expert. This was to enhance education effect by making it focus in only the activity of the muscle which want to show to non-expert. It was thought that it may cause the perplexity of the non-expert to show the activity of too many muscle.

It is important how to show muscle activity to non-experts. In this study, muscle activity was expressed as color of virtual muscle. Change of color was easily understandable visually and intuitively. Color of virtual muscle was changed by ratio to maximum value of muscle activity during motion. Changing color was performed by changing each elements of RGB color model. The color of muscle nears red so that strength of muscle activity is strong. On the other hand, the color of muscle nears blue so that strength of muscle activity is weak. Figure 2 shows condition of changing virtual muscle color. Virtual muscle on 3D model was projected to 2D plane adjusting direction of actual video and overlaid with actual video data. These were process to make muscle activity visualization system.

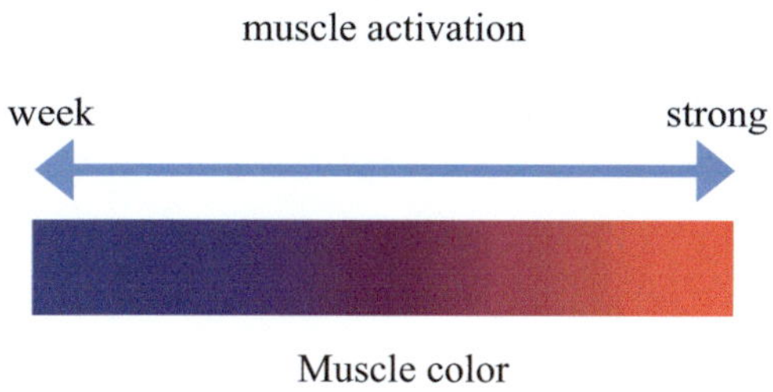

Fig. 2 Color of virtual muscle

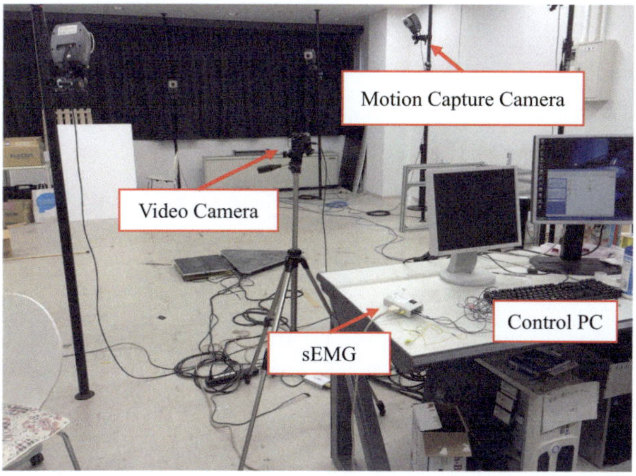

Fig. 3 Measurement devices

3 Visualization of Expert's Motion

3.1 Experimental Set Up

In this experiment, measurement electrical activity of target muscles was conducted by surface electromyography (sEMG) (S & ME DL-721). Sampling frequency of sEMG was 1000 Hz. Measurement posture during squat motion was conducted by Optical motion capture (Motion Analysis EVaRT4.4). Sampling frequency of motion capture was 200 Hz. Video camera (SONY) is connected to control PC via converter (grass valley ADVC55) in order to get a video data in chronological order same as other data. Figure 3 shows condition of measurement devices.

3.2 Subject

Expert's muscle activity was visualized during squat motion for skill education of squat motion. A subject was expert who had trained on a daily basis in 7 years under instruction of professional athletic trainer and understood important points of squat

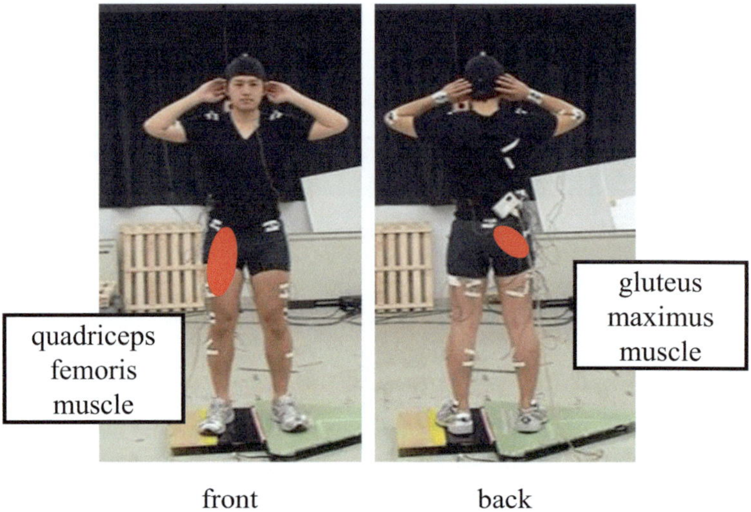

Fig. 4 Target muscle

motion, such as way to use muscles and posture. In this experiment, target muscles of visualization were protagonist (quadriceps femoris (QF) muscle and gluteus maximus (GM) muscle) which are important muscle for squat motion. These muscles are used mainly during squat motion. Figure 4 shows position of target muscles. In this study, non-experts was educated how to use these muscle as physical skill. In the experiment, investigate how to use GM and QF muscles as expert's skill.

3.3 Result

Figure 5 shows sEMG data of expert's trial. Figure 6 shows visualization of muscle activity. In Fig. 6, a whole process of squat motion was regarded as 100 % of rate of motion progress at the horizontal axis. Timing of squatting down the deepest was regard as 50 % of rate of motion progress. Indeed, squatting down phase took 2 s and standing up phase took 1 s. However, for focusing relation body posture and muscle activity change of shoulder vertical position defined as progress of motion. Visualizing muscle activity videos from front and back viewpoints was made for getting easy to observe important muscle activity. Figure 6 indicated that QF and GM were used mainly in standing up phase. QF was used from second half of squatting down to first half of standing up phase and GM was used in second half of standing up phase. These way to use muscle was expert's way. Non-experts need to obtain how to use muscles as this.

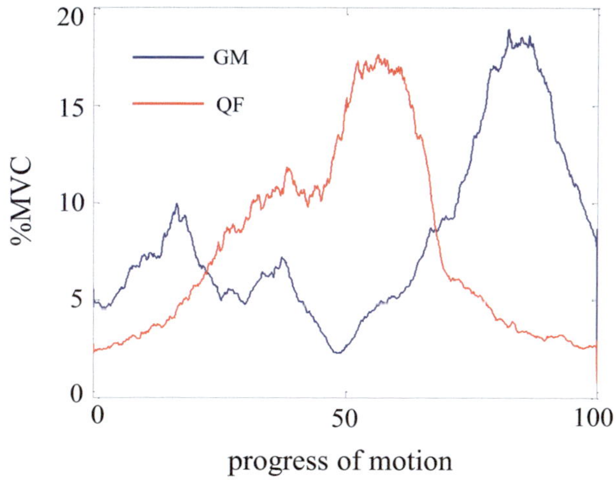

Fig. 5 Muscle activity of expert

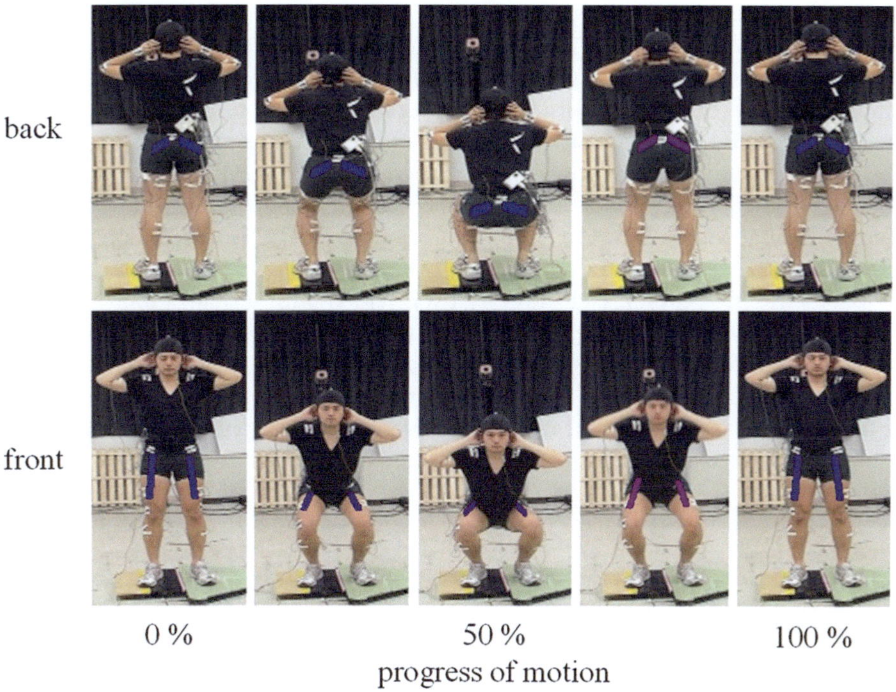

Fig. 6 Visualization of muscle activity

4 Verification of Education

4.1 Participants

We conducted the experiment in order to verify effectiveness of muscle activity visualization in physical skill education of squat motion. Participants were 6 males (age $= 23 \pm 1$) who had not been instructed squat training.

4.2 Method

In this experiment, education was focused on QF activity and GM activity which are important muscle for squat motion. 6 participants were separated to 2 groups (A-group and B-Group) by 3 participants. Three participants (A-1, A-2, A-3) in A-group were educated by using video with visualization of muscle activity. Three participants (B-1, B-2, B-3) in B-group were educated by using video without visualization of muscle activity.

Firstly, expert taught participants in both groups how to perform squat motion with his gesture and words. This is the way to educate usually in education service fields. Expert explained about safety posture that a knee and a waist had few burdens. After explained, participant's posture was safety sufficiently during squat motion. Expert expressed timing and image of using QF and GM with his gesture and words to participants. After teaching, QF's sEMG and GM's sEMG were measured during squat motion. One set consisted of 10 trials. One trial consisted of squatting down in 2 s and standing up in 1 s. Participants performed squat on 1 set.

Secondly, in A-group, education by using video with visualization of muscle activity was conducted. In B-group, education by using video without visualization of muscle activity was conducted. Participants were ordered to watch video of expert's motion. The movie was filmed expert's 8 trials. Participants watched 2 times of the movie on front and back angle. After that, participants performed squat motion on 1 set again and were measured muscle activity.

4.3 Result

Figure 7 shows 6 participant's muscle activities averaged 10 trials of sEMG data. 3 participants on left side belonged to A-group and the others belonged to B-group. The horizontal axis of graph represents progress of squat motion and the vertical axis represents %MVC. In A-1's muscle activity, strength of GM muscle activity increased as a whole and timing of peak got later and strength and timing of using QF muscle did not change. In A-2's muscle activity, strength of GM muscle activity

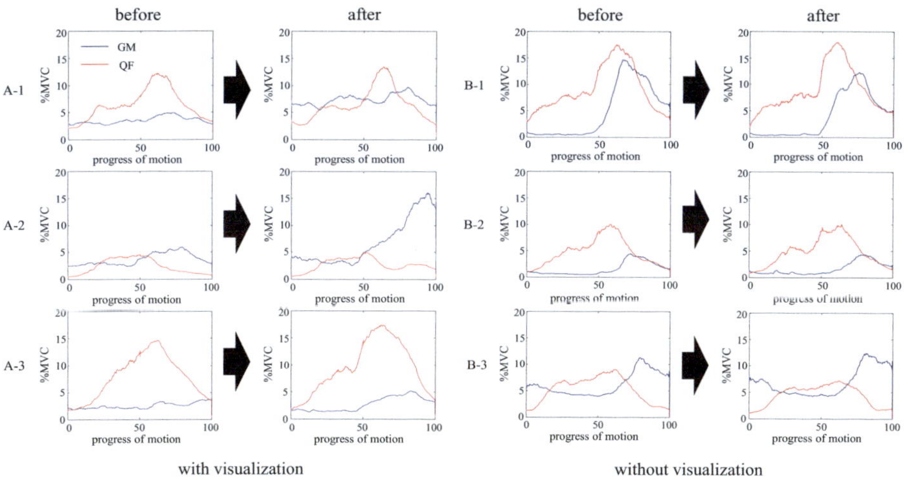

Fig. 7 Muscle activity of non-experts

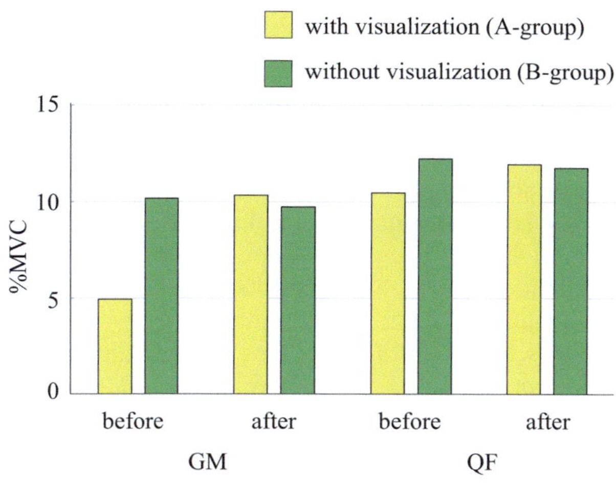

Fig. 8 Change peak values of muscle activity

increased greatly at the peak and timing of peak got later and strength and timing of using QF muscle did not change. In A-3's muscle activity, strength of GM muscle activity increased a little and peak of GM muscle activity appeared and peak of QF muscle increased slightly However, timing of using QF muscle did not change. In B-1's muscle activity, strength of GM muscle activity decreased at the peak of GM muscle activity and strength and timing of using QF muscle did not change. In B-2's muscle activity, strength and timing of using GM and QF muscles did not change. In B-3's muscle activity, strength of GM muscle activity increased slightly at the peak and timing of using QF muscle did not change. Strength and timing of using QF muscle did not change. Peak value of GM and QF muscle activity was analysed. Figure 8 shows change of peak values of muscle activity between before video education and after in education by using video with visualization and without. Value of each bar represents averaged peak values of muscle activity among 3 participants in each group. Peak value of only GM muscle activity of participants in A-group tended to increase notably by educated with visualization system. On the other hand, muscle activity of participants in B-group did not change before video education. A-group participant's results suggested peak of using GM got later than education with expert's gesture and words and closer to expert's timing. However, timing of using QF did not change.

5 Discussion

In the experiment, effectiveness of education by using proposal muscle activity visualization method was verified. Participant's GM muscle activity got stronger owe to proposal method in A-group. Timing of using GM got closer to expert's timing in A-group. On the other hand, muscle activity of participants in B-group did not change very much. It was likely that participants understood relation between progress of squat motion and timing of using muscle by watching visualized muscle activity. In QF muscle activity, change of strength and timing could not be seen on all participants. There was difference in influence to be educated from visualization system by the part of the muscle. It could be that somatic sensation depends on the part of the muscle. In this experiment, participants were indicated that they were aware of two muscle activities. However, if subjects did not understand how to use indicated muscles, it is no mean that subjects know expert's muscle activity. This problem could be solve by visualisation of non-expert's muscle activity with real time feedback. If non-experts can understand their muscle activity, more non-experts learn motion more effectively. In this study, we verified effectiveness to two indicated muscle activity directly. However, in educating more complex motion, it is inefficient to indicate to use more parts of muscle. It is possible that indication to rise one part of muscle activity has a good influence on multiple muscles activity. Investigation physical skill like that is challenges for the future.

6 Conclusion

The purpose of this study is to solve problem about inefficiency and ineffective in squat physical skill education by visualizing muscle activity in physical skill education service fields. We proposed method of visualizing muscle activity and verified effectiveness of proposal method for skill education by using visualization movie of expert's muscle activity. In the experiment, it was suggested proposal method is effective for physical skill education. In the future, we will develop visualization system of muscle activity on real time in order to realize more effective education.

Acknowledgment This work was in part supported by JST RISTEX Service Science, Solutions and Foundation Integrated Research Program.

References

1. Phillips E, Portus M, Davids K, Renshaw I (2012) Performance accuracy and functional variability in elite and developing fast bowlers. J Sci Med Sport 15:182–188
2. Furuya S, Kinoshita H (2008) Biotones: expertise-dependent modulation of muscular and non-muscular torques in multi-joint arm movements during piano keystroke. Neuroscience 156:390–402
3. Tsubouchi Y et al (2010) Biotones: a wearable device for EMG auditory biofeedback. In: Proceedings of the IEEE EMBS, pp 6543–6546
4. Hald RD, Bottjen EJ (1987) Effect of visual feedback on maximal and submaximal isokinetic test measurements of normal quadriceps and hamstrings. J Orthop Sports Phys Ther 9:86–93
5. Lippold OCJ (1952) The relation between integrated action potential in human muscle and its isometric tension. J Physiol 117:492–499

Extraction and Evaluation of Proficiency in Bed Care Motion for Education Service of Nursing Skill

Junki Nakagawa, Qi An, Yuki Ishikawa, Koshiro Yanai, Wen Wen, Hiroshi Yamakawa, Junko Yasuda, Atsushi Yamashita, and Hajime Asama

Abstract In the areas of care nursing, conventional physical skill education which has been conducted face-to-face prevents both experts and beginners from realizing the effects of their education and practice. In order to solve this problem, we aim to develop an effective skill education service of bed care motion with slide sheet which can prevent the outbreak of lumbago. In this study, we calculated the lumbar burden with measured body joint trajectory, foot reaction force, and muscle activities in the motion of experts and non-experts. Based on these analysis results, nursing skill was extracted and evaluated for skill education service.

Keywords Motion analysis • Healthcare • Skill education • Motion measurement

1 Introduction

Aging society leads to increasing a labor burden of nurse and care worker. Especially, in nursing and care facilities, 60 % of nurses and care workers has the occupational low back pain [1]. "No Lifting Policy" is one of the prevention methods of occupational low back pain for nurses and care workers. This policy was suggested by Australian nursing federation in 1996 and prohibit patient transfers with no transfer equipment [2]. Most patient transfers were performed transfer in bed, for example moving up in bed, turning or rolling in bed and moving from lying to sitting in bed [3]. For patient transfer in bed, "slide sheet" has been proposed as an useful tool (Fig. 1). Nurse spread the slide sheet between the bed and the patient, and draw the sheet in order to achieve transfer patients. The slide sheet can lighten burden of both nurse and patients, and is low-cost comparing to electric transfer machines. Thus, slide sheet for patient transfers is worth promoting

J. Nakagawa (✉) • Q. An • Y. Ishikawa • K. Yanai • W. Wen • H. Yamakawa • A. Yamashita • H. Asama
The University of Tokyo, Tokyo 113-8656, Japan
e-mail: nakagawa@robot.t.u-tokyo.ac.jp

J. Yasuda
Japanese No Lifting Association, Kobe, Japan

T. Maeno et al. (eds.), *Serviceology for Designing the Future*,
DOI 10.1007/978-4-431-55861-3_15

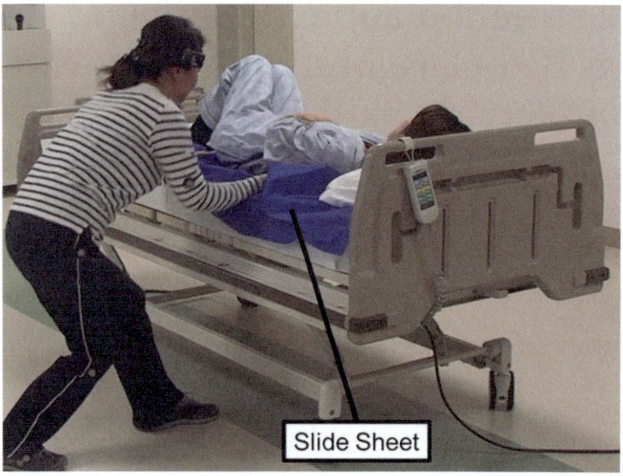

Fig. 1 Bed care motion with slide sheet. In this picture, Nurse is drawing slide sheet between the bed and the patient in order to achieve transfer patients

in many places, such as hospitals, and ordinary homes where health care are necessary.

However, when using the slide sheet, it is important to behave appropriate care movement. Because there is a possibility to increase the burden on not only the nurse but also the patient with wrong movement. In order to disseminate appropriate care motion with slide sheet, skill education of this motion is required.

In the areas of care nursing, the former physical skill education from experts to beginners has been conducted face-to-face. However, face-to-face education is inefficient because experts and beginners have to share their time and space. To deal with this, e-learning service has been widely used in many areas, but still have some unsolved problem in skill education. Specifically, current e-learning service has difficulty to extract skills from movement and valuate skills quantitatively. Thus, it is difficult for both experts and beginners to realize the effects of their education and practice.

In order to solve these problems, we aimed to develop an effective skill education service for e-learning. Figure 2 shows the concept of our suggested skill education service. Our skill education service intends to achieve visualization of the difference between expert's motion and non-expert's motion, expert's motion, and skill for non-expert. Thus, it is necessary to measure expert's motion and non-expert's motion, analyse their motion, and extract skills from expert's motion. In particular, we focused on care motion in the bed (bed care motion) and extracted skills of this motion. This motion particularly promotes the outbreak of lumbago because the motion compels nurses to adjust their posture to the bed heights and lift up the patients, changing patients' body positions and transferring to wheelchair [3]. In the present study, we analysed bed care motion with slide sheet, especially draw sheet motion, which can prevent the outbreak of lumbago.

In this study, skills of care experts of bed care motion with slide sheet were elucidated. First, we inputted measured body joint trajectory and foot reaction force

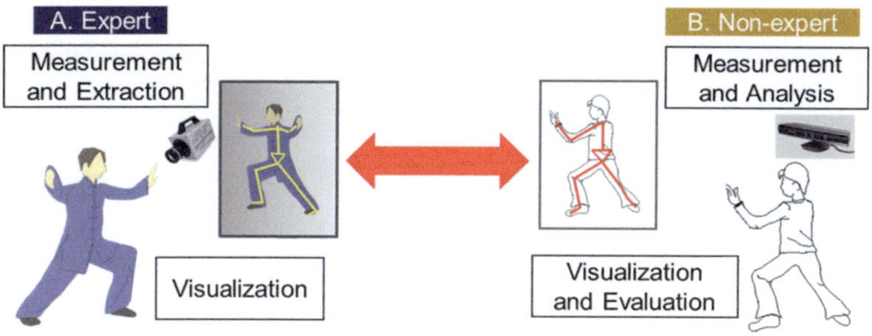

Fig. 2 Concept of skill education service. Our skill education service achieves visualization of the difference between expert's motion and non-expert's motion, expert's motion, and skill for non-expert

data from motion before and after learning into musculoskeletal model, and calculated joint moment. With calculated joint moment and measured muscle activities, we estimated the lumbar burden to evaluate the extracted skills. Then, we analysed measured data based on our extracted skills, to assess skills before and after learning quantitatively.

Previous studies for nursing skill education extracted skills from experts' motion similarly [4]. However they did not focus on an evaluation whether or not extracted skills are appropriate. In contrast, we concluded that extracted skills were appropriate if the lumbar burden decreased after skill learning.

2 Analysis Method

2.1 Skill Extraction

First, we extracted three important skill points of bed care motion with slide sheet with the method suggested by Hashimoto et al. [5]. Then, we compared motion of expert and non-expert with video images, and summarized the differences in Table 1.

In Fig. 3, video images of a specific frame are shown. Left two images shows expert motion, and right two images shows non-expert motion. Comparing the three skill points, the non-expert was found to change trunk angle remarkably, keep a distance between arms and trunk during before and after this motion. The expert was found to flex arm, put close arm and trunk, keep trunk vertically, and utilize weight shift from toward the rear. In contrast, the non-expert used only arm strength rather than weight shift, which is a critical skill for expert.

Skill points of bed care motion with slide sheet were extracted as shown in Table 1. Based on these extracted skill points, measured and calculated data from both expert motion and non-expert motion were compared.

Table 1 Results of skill extraction with interview and video analysis. Three skill points were elucidated, video images of expert and non-expert was compared, and we summarized the differences

Target	Point from interview	Expert skill	Expert motion from video analysis	Non-expert motion from vide analysis
Arm	Arm angle	Keep the 0° position of shoulder flex angle	Flex arm and put arm near to trunk	Keep a distance between arm and trunk
Trunk	Trunk angle	Not change trunk angle during motion	Keep trunk vertically and not bend the body forward	Change trunk angle remarkably
Leg	Weight shift	Promote weight shift with leg motion	Utilize weight shift to move CoM forward	Use arm power instead of weight shift

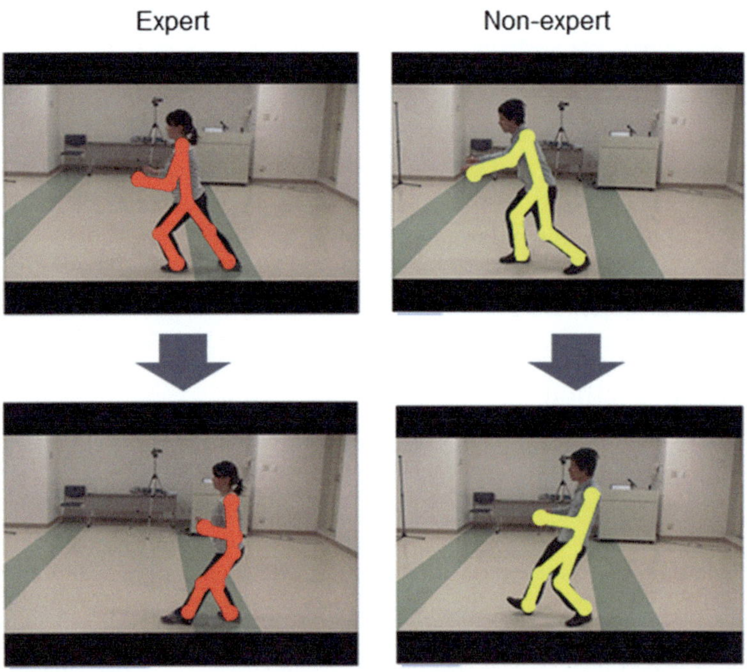

Fig. 3 Different movements of expert and non-expert from video images at a specific frame. *Left* two images are expert motion, and *right* two images are non-expert motion

2.2 Lumbar Burden Estimation

We estimated the lumbar burden with hip flex moment and erector spinae muscle activation. In previous studies, in order to estimate the lumbar burden, hip flex moment and erector spinae muscle activation were used [6, 7].

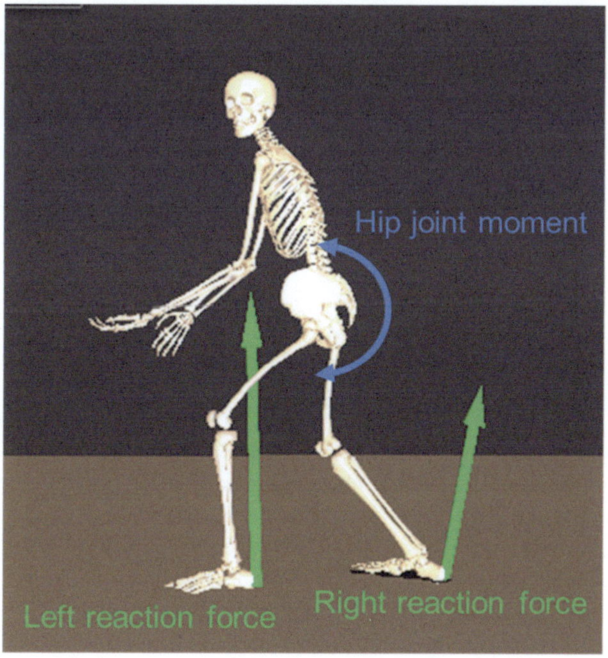

Fig. 4 3-D musculoskeletal model for motion analysis. This model was constructed on SIMM. Measured data was inputted to this model in order to calculate hip joint moment

Hip flex moment was calculated solving the inverse dynamics. Body joint trajectory data and foot reaction force data are the input to the inverse dynamics and the hip flex moment is the output. We used SIMM (MusculoGraphics Corp.), which is a software for calculation of joint moment with inverse dynamics. We constructed a 3-D musculoskeletal model with SIMM, and implement the measured data to the model. Figure 4 shows a 3-D musculoskeletal model which we used in this study. The green arrow from left foot shows reaction force from left foot, the green arrow from right foot shows reaction force from right foot, and the blue arrow shows hip flex moment. In this simulation, hip flex moment indicates bending moment to the lumbar.

Erector spinae muscle activation was measured with surface electromyography (SEMG) sensors attached to the participant's skin. The increase of bending moment and erector spinae muscle activation lead to increase the lumbar burden and develop the lumbago [8, 9].

2.3 Proficiency Evaluation

In order to evaluate a proficiency of motion, measured data were analysed based on three extracted skill points (Fig. 5). Arm angle was compared calculating the standard deviation of elbow and shoulder angle during motion before and after learning. Using these results, the proficiency was investigated focusing on whether

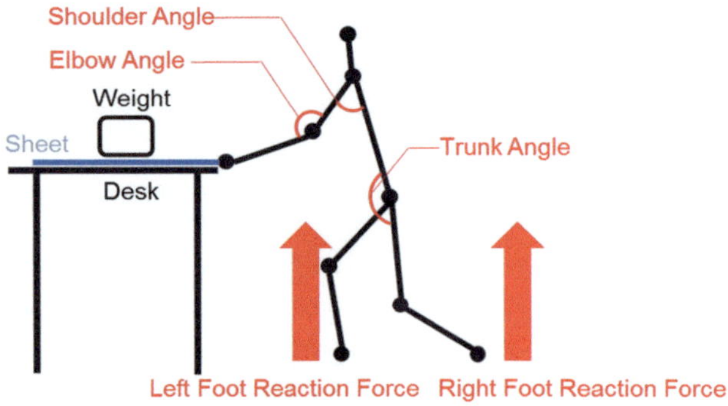

Fig. 5 Quantitative evaluation of three skill point. Arm skill is evaluated with shoulder angle and elbow angle. Trunk skill is evaluated with trunk angle. Leg skill is evaluated with *left* and *right* foot reaction force

motion after learning keep the flex position of elbow and shoulder or not. Likewise, trunk angle was analysed using the standard deviation of trunk angle during motion. The proficiency of trunk angle was investigated focusing on whether motion after learning keep trunk vertically or not. Weight shift was analysed using measured reaction force from both foot. The evaluation of proficiency on weight shift was focused on whether motion after learning utilize weight shift from toward the rear or not.

3 Experimental Setup

3.1 Equipments

In order to investigate the effect on hip flex moment by motion difference. Figure 6 shows our experimental environment. Optimal motion capture system (MAC3D) with eight cameras (HMK-200RT; Motion Analysis Corp.) was used to measure body trajectories in 200 Hz. Based on Helen Hayes marker set, measured body parts were decided. In addition, the reaction force sensor (Nitta Corp.) was used to measure foot reaction force from both foot in 64 Hz. Muscle activation was measured in 1000 Hz from muscle of lumbar, especially erector spinae muscle, in DL-3100 (S&ME Corp.).

3.2 Subject

One healthy male (age: 23, height: 1.73 m, weight: 70 kg) participated in our experiment. This experiment was conducted with approval of the ethics committee of Faculty of Engineering, The University of Tokyo, and informed consent was obtained from the participant.

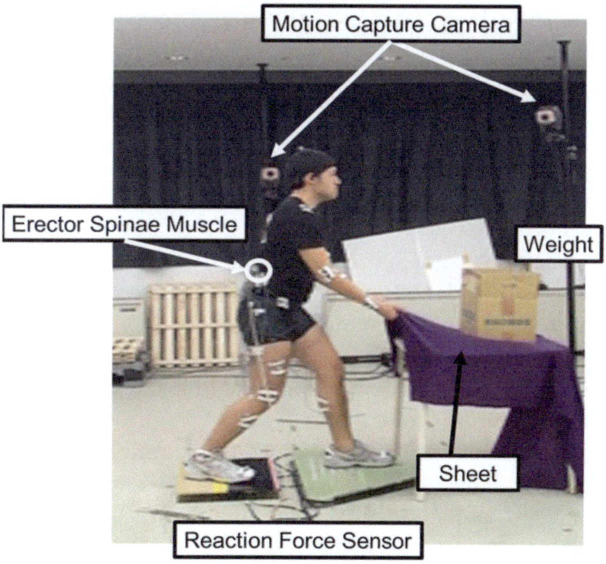

Fig. 6 Experimental environment. There are optimal motion capture system with eight cameras, reaction force sensor, weight and sheet

3.3 Procedure

In this experiment, a desk height was set to 1.0 m as bed and fabric sheet on the desk was used. Additionally, on the sheet, weight (5.0 kg) was put. Improvised environment was constructed using slide sheet and patient on bed. In the first experiment, the participant drew slide sheet with no experience and skill of drawing slide sheet. After the first experiment, the participant was informed extracted skills in Table 1 and watched the video image of expert's draw sheet motion. In the second experiment, the participant drew slide sheet with information of expert's skill.

3.4 Data Processing

In the case of data unavailability of body trajectory, spline interpolation was conducted. Measured body trajectory data was filtered with low-pass filter in 10 Hz. Likewise, foot reaction force data from both foot was filtered with low-pass filter in 25 Hz and resampled from 64 to 200 Hz. Muscle activation data is centred and filtered using 300 Hz low-pass and 10 Hz high-pass filters. In addition, these data were rectified and normalized using maximum values which measured respectively before our experiment.

4 Result & Discussion

4.1 Lumbar Burden

Figure 7a shows the average of hip flex moment with difference between motion after learning and motion before learning. Compared with motion after learning, motion before learning increased hip flex moment. In addition, Fig. 7b shows the average of erector spinae muscle activation during motion before and after learning. Error bars indicate standard deviation during one draw sheet motion. Motion after learning decreased erector spinae muscle activation in comparison with motion before learning. In the previous research, erector spinae muscle is related to the lumbago [9]. Thus, these results indicate that non-expert motion induced the increase of the lumbar burden and is possibility of developing the lumbago.

Definite number which develop the lumbago has been unclear, but it could be concluded that motion before learning is likely to induce the lumbago. This is why a small difference make a big difference in the case of repetitive motion such as nursing motion. In addition, from this result, our extracted skills were appropriate because the lumbar burden decreased after skill learning.

However, in this experiment, three skill points were transferred to the participant at the same time, and the most important skill to decrease the lumbar burden was not clarified. For the future education service, it is necessary to elucidate the most important skill to decrease the lumbar burden and the risk of developing lumbago in order to prioritize teaching skills.

4.2 Proficiency

Based on extracted skill points, proficiency on each skill point was evaluated. Figure 8a–b shows the standard deviation of elbow and shoulder angle during one draw sheet motion respectively. Each angle was calculated with measured joint trajectory data. As shown in Table 1, expert keep the 0° position of shoulder flex angle and put close arm and trunk. Thus, expert fix their flex arm near trunk and

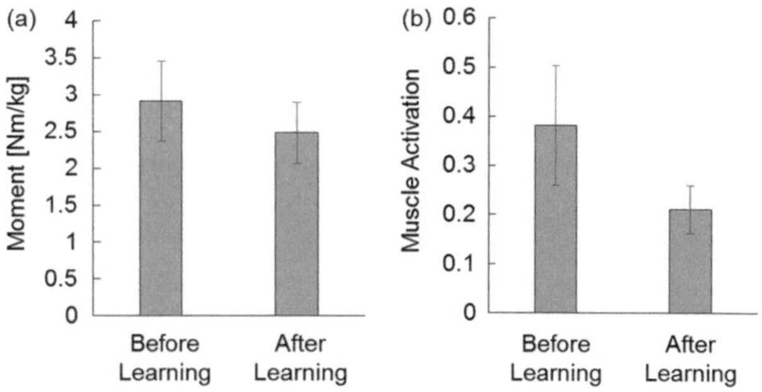

Fig. 7 Lumbar burden between expert and non-expert motion. Compared with expert motion, non-expert motion increased (**a**) the lower part of FIG of moment and (**b**) the lower part of FIG of muscle activation

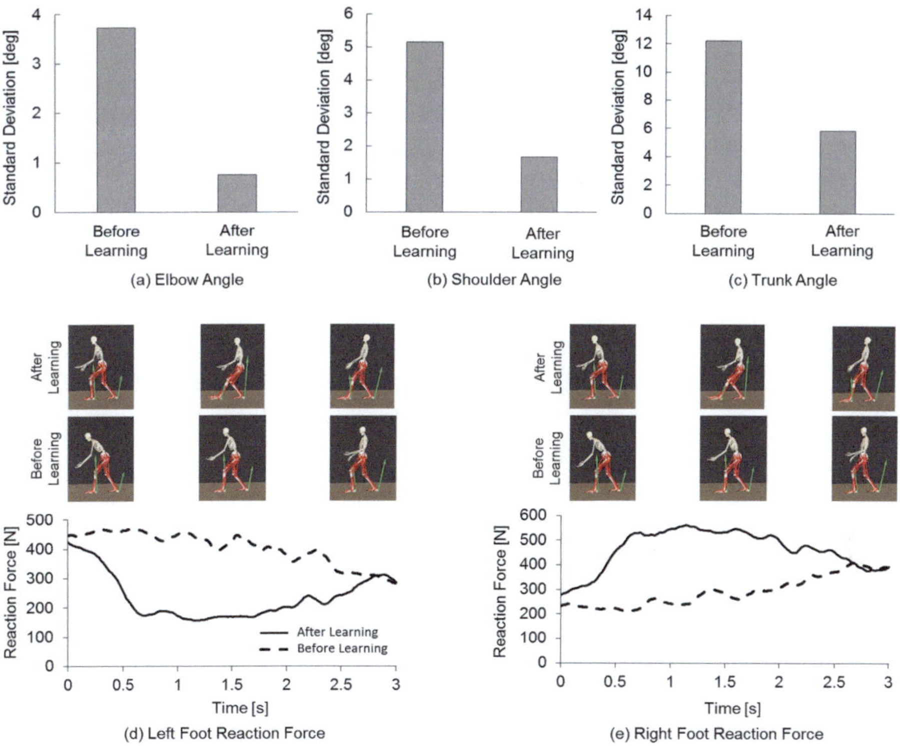

Fig. 8 Measured body data comparison between expert and non-expert motion. (**a–c**) shows the standard deviation of elbow, shoulder and trunk angle during draw sheet motion respectively. (**d–e**) shows reaction force from *left* foot and reaction force from *right* foot during draw sheet motion

decrease variance of elbow, shoulder angle. As shown in Fig. 8a, b, motion after learning decrease variance of elbow, shoulder angle compared to motion before learning. The participant moved and utilized arm largely before learning. Dependence of arm in draw sheet motion is high before learning.

Figure 8c shows the standard deviation of trunk joint angle during draw sheet motion. As with the elbow and shoulder angle, trunk angle during motion after leaning were kept vertically. On the other hand, during motion before learning, trunk angle were changed remarkably and the trunk was bent forward. This results leads to the increase of hip flex moment.

Figure 8d–e show the vertical reaction force from left foot and from right foot during motion respectively. As shown in these figures, motion after learning induced weight shift compared with motion before learning. Therefore it is implied that, in motion after learning, weight shift is utilized without relying on arm movement and hip flex moment is reduced.

In this experiment, motion after learning was able to achieve three skill points of bed care motion which was extracted from interview and video analysis. Before learning, the participant draw sheet using arm power without weight shift, and move upper body largely. On the other hand, the participant after learning draw using weight shift without relying on arm power.

Especially it was defined that trunk movement has important function for posture and balance in the previous research [10]. In a first step of skill education service for non-expert student, it may be important to teach trunk movement of expert.

5 Conclusion

Nursing skill was extracted from interview and video analysis. Three skill points were extracted focusing on arm, trunk and leg. Based on these extracted skills, we analysed draw sheet motion with slide sheet. Our analysis results show that motion after skill learning reduces the lumbar burden compared to motion before skill learning. This is why hip flex moment and erector spinae muscle activation, which is related to lumbago, was reduced in motion after learning. From this result, our extracted skill points were appropriate.

However, if motion after learning were not to achieve the acquisition of skills, we could not conclude this result. In order to confirm that our skill education was conducted in an appropriate way, proficiency on each skill point was evaluated after learning based on these skill points which was extracted from interview and video analysis. Our analysis suggested that motion after learning was able to achieve these three skill points.

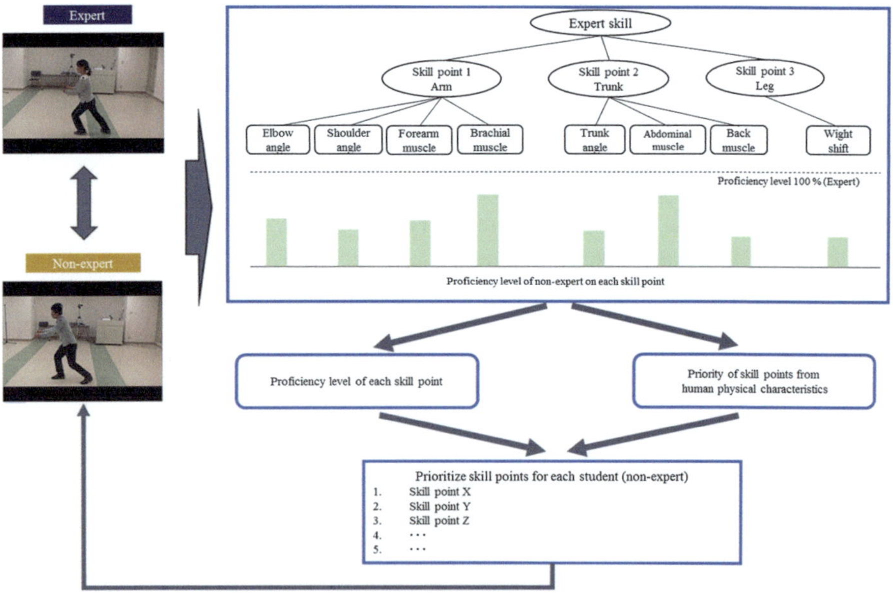

Fig. 9 Our future service concept for skill education. In this service, based on proficiency level of each skill point and priority of skill points from human physical characteristics, skill points prioritize for each non-expert student in order to develop the most effective skill education service

In this study, the experiment was conducted in the improvised condition. Thus, we will conduct future experiment in a condition which is close to a real nursing scene using hospital bed, real slide sheet, and another participant as a patient.

Figure 9 shows our future service concept for skill education. In our future work, based on proficiency level of each skill point and priority of skill points from human physical characteristics, skill points will be prioritized for each non-expert student in order to develop the most effective skill education service. On this account, we will need to investigate the priority of skill points in terms of human physical characteristics.

It is necessary to design the education service of nursing skill for not only nurses but also patients. Therefore we should take the burden reduction of patients into consideration for more practical education service of nursing skill.

Acknowledgement This work was in part supported by JST RISTEX Service Science, Solutions and Foundation Integrated Research Program, the MEXT KAKENHI, Gtant-in-Aid for Scientific Research (B) 24300198.

References

1. Smedley J, Egger P, Cooper C, Coggon D (1995) Manual handling activities and risk of low back pain in nurses. Occup Environ Med 52(3):160–163
2. Australian Nursing Federation (Vic Branch) Policy. http://www.anmfvic.asn.au/multiversions/3555/FileFile/NoLifting.pdf. 18 April 2014
3. Engkvist IL (2006) Evaluation of an intervention comprising a no lifting policy in Australian hospitals. Appl Ergon 37(2):141–148
4. Huang Z, Nagata A, Kanai-Pak M, Maeda J, Kitajima Y, Nakamura M, Aida K, Kuwabara N, Ogata T, Ota J (2014) Automatic evaluation of trainee nurses' patient transfer skills using multiple kinect sensors. IEICE Trans Inf Syst 97(1):107–118
5. Hashimoto H, Yoshida I, Teramoto Y, Tabata H, Han C (2011) Extraction of tacit knowledge as expert engineer's skill based on mixed human sensing. In: Proceedings of the 20th IEEE international symposium on robot and human interactive communication, pp 413–418
6. Cohen SP, Raja SN (2007) Pathogenesis, diagnosis, and treatment of lumbar zygapophysial (facet) joint pain. Anesthesiology 106(3):591–614
7. Ekholm J, Arborelius UP, Nemeth G (1982) The load on the lumbo-sacral joint and trunk muscle activity during lifting. Ergonomics 25(2):145–161
8. Stuart MM, Robert WN (1985) Dynamically and statically determined low back moments during lifting. J Biomech 18(12):877–885
9. Radebold A, Cholewicki J, Panjabi MM, Patel TC (2000) Muscle response pattern to sudden trunk loading in healthy individuals and in patients with chronic low back pain. Spine 25 (8):947–954
10. Krebs DE, Wong D, Jevsevar D, Riley PO, Hodge WA (1992) Trunk kinematics during locomotor activities. Phys Ther 72(7):505–514

Exploratory Analysis of Factors of Patient Satisfaction in HCAHPS Databases

Masumi Okuda, Akira Yasuda, and Shusaku Tsumoto

Abstract Hospital consumer assessment of healthcare providers and systems (HCAHPS) patient survey data were analysed to identify relationships between survey response rate and overall rating of patient satisfaction by survey response size and to identify similarities among variables of patient satisfaction by response size and hospital characteristics. The survey response rates were correlated with patient satisfaction, suggesting that patients choice of ratings with a neutral or good connotation actually means lower ratings. The structure of patient satisfaction differed with hospital size and hospital specialty. Some variables did not change membership in a cluster, but other variables did so in other clusters. To improve overall patient satisfaction, more attention should be paid to the relationships between the factors of patient satisfaction and their changes. Distance scaling may reveal different aspects of patient satisfaction.

Keywords Structure of patient satisfaction • Survey response rate • Clustering • Correspondence analysis • HCAHPS

1 Introduction

Donabedian [1] suggested that evaluation of the quality of medical care is determined by "structure," "process," and "outcome." "Structure" often refers to facilities and staff; "process" refers to medical care and "outcome" refers to heath status such as mortality rate, social restoration and patient satisfaction. A systematic review by William B. found that the concept of patient satisfaction has been developed using various ideas of which one example is the improvement of compliance, given that high-quality clinical outcomes are dependent on compliance; and another being the rise of consumerism [2]. The number of studies of determinants of patient satisfaction based on mathematical models has increased

M. Okuda (✉)
Matsue Red Cross Hospital, Matsue 690-8506, Japan
e-mail: okuda@med.shimane-u.ac.jp

A. Yasuda • S. Tsumoto
Department of Medical Informatics, Shimane University, Izumo, Shimane, Japan

T. Maeno et al. (eds.), *Serviceology for Designing the Future*,
DOI 10.1007/978-4-431-55861-3_16

since 1980s. In an early example, a hypothesis derived from the theory of reasoned action by Fishbein & Ajzen was proposed and tested by Linder-Pelz. In this hypothesis, satisfaction is the sum of products of belief strength and evaluation [3–5]. A recent systematic review of the measurement of satisfaction with healthcare described two major issues: methodology and determinants of satisfaction. Methodology included issues such as survey instruments, survey modes and response rates; determinants of satisfaction comprised two topics. One was patient characteristics such as sociodemographic factors, health status, and patient expectations and the other was health service delivery stimuli such as the activities, attitude, and appearance of human resources and the environment and organizational aspects of care [6]. Regressions and correlations are often calculated to identify determinants, and overall assessment or intentions to recommend/return are often defined as objectives. But focusing on the determinants of satisfaction and not on interrelationships among variables has disadvantages, because the concept of patient satisfaction is hard to define [1] and complex [2, 5, 7, 8], and non-determinants tend remain to be identified. Only a few studies have examined the structure of patient satisfaction [8], and these focused on the relationships between the variables of patient satisfaction based on their ratings.

Another activity in recent years has been to assess the relationships between patient satisfaction, health status as "outcome" and public reporting as "process" in massive datasets, with the arrival of the Internet era and fast-improving information technology in addition to the growing demand for the quality of medical care and patient-centred approach [9]. Hospital performance datasets are now available online for public access. One example is Hospital Compare data provided by the US government. The dataset includes not only standard national patient survey results but also structural measures as "structure" and clinical indicators such as rates of compliance with treatment guidelines by hospital as "process" and hospital performance such as mortality rates as "outcome". The number of hospitals submitting data for Hospital Compare, even though partially, was over 4000 as of 2013 [10]. Survey response rates are also studied, but many studies of survey response rates focus on the patient characteristics of non-respondents and survey mode effects on the response rate [4] and not on the relationship between the survey response rate and patient satisfaction score. The single exception is a study of the impacts of low response rates comparing the first 30 % of respondents against all respondents and finding that late respondents assessments may either improve or worsen the score [11].

To address the issues mentioned above and the massive body of complex data, a new approach such as data mining should be applied. One data-mining activity is searching massive datasets to discover unexpected structures, relationships, patterns, trends, clusters, and outlines in the data. Data mining has been developed in response to rapidly growing computational technology, the accumulation of massive datasets on a daily basis and a need to overcome the limitations of classical methods of multivariate analysis [12].

Medical care as a service has different aspects from customer service in general. Medical care is directly associated with life and death, temporary indisposition must be accepted to regain health or to reach the best health status, and medical information tends to be disproportionately available for medical personnel [13]. The interpretation of patient satisfaction survey data must accordingly be performed cautiously. Data mining has the potential to address this problem.

A study of the US Hospital Compare data will support not only the customers "active decision-making process but also the hospitals" effort to improve the quality of care and management, given that the dataset describes three aspects of quality of care, "structure," "process," and "outcome", as suggested by Donabedian.

2 Objectives

(1) To identify the relationships between overall patient satisfaction ratings and survey response rates.
(2) To identify similarities among variables of patient satisfaction by distance scaling.

3 Methods

3.1 Data Source

Official Hospital Compare data are available on the Medicare.gov Hospital Compare Website provided by the Centers for Medicare & Medicaid Services (CMS) [10], which has been developed over several years [14–16]. Hospital Compare is a tool that provides information about the quality of care to help consumers to make informed health-care decisions. Hospital Compare contains hospital consumer assessment of healthcare providers and systems (HCAHPS) patient survey data. The datasets were retrieved online.

3.2 Data Selection

3.2.1 Hospital Selection

The original data were aggregated by provider ID and the following hospitals were extracted: (1) those submitting completed patient surveys; (2) those whose survey response size was 50 and above, given that data with fewer than 50 responses may be too limited for reliable assessment of hospital performance, according to CMS; (3) those submitting survey response rates and (4) those whose data showed no discrepancies in the data collecting process according to CMS.

3.2.2 Hospital Characteristics

The registry of specialties and care was used: cardiac surgery, general surgery, nursing care, and stroke care.

Hospital types were also used: acute-care hospital (ACH) and critical-access hospital (CAH). The former is a hospital that provides inpatient medical care and other related services for surgery, acute medical conditions, or injuries (usually for a short-term illness or condition) [17]. The latter is a smaller rural community hospital that receives cost-based reimbursement [18].

3.2.3 Patient Survey Results

The HCAHPS scores were collected from July 2012 to June 2013. The survey contained 32 items for measuring patients "perceptions of their hospital experience and was administered to a random sample of adult inpatients between 48 h and 6 weeks after discharge [10]. Six composite questions, two individual questions about environment and two global questions were reported as follows: how often nurses and doctors communicated with patients (nurse communication, doctor communication), how often patients received help quickly from hospital staff (staff responsiveness), how often patients" pain was well controlled (pain management), how often the staff explained medicines before giving them to patients (medication information), how often the hospital environment was kept clean and quiet (cleanliness, quietness), whether patients were given information about what to do after discharge (discharge information), overall rating of the hospital (overall rating), and willingness to recommend the hospital (recommendation). A description of the patient satisfaction survey is shown in Table 1. Questions 1 through 7 required three-choice frequency ratings: "sometimes or never (low)," "usually (medium)," and "always (high)". Question 8 was a closed question. Question 9 was a scale question from 1 to 10: 10 and 9 were labeled "high", 8 and 7 were labeled "medium," and 6 and below were labeled "low". Question 10 was a three-choice rating: no, probably yes, definitely yes. The survey response rates and the number of completed surveys (response size) were also used. The latter was categorized into three sizes; between 50 and 99, between 100 and 299, and 300 and above.

3.3 Analysis

The data were analyzed by response size, which was expected to correspond to hospital bed size, given a prior report that hospital bed size affects patients' overall rating of the hospital in the HCAHPS data bases [19]. R 3.1.0 was used for statistical calculations. R is a free software environment for statistical computing and graphics [20]

Table 1 HCAHPS patient survey

Number	Questions
1	How often did nurses communicate well with patients?
2	How often did doctors communicate well with patients?
3	How often did patients receive help quickly from hospital staff?
4	How often was patients' pain well controlled?
5	How often did staff explain about medicines before giving them to patients?
6	How often were the patients' rooms and bathrooms kept clean?
7	How often was the area around patients' rooms kept quiet at night?
8	Were patients given information about what to do during their recovery at home?
9	How do patients rate the hospital overall?
10	Would patients recommend the hospital to friends and family?

3.3.1 Hospital Characteristics

To compare the proportions of hospital types and specialties by response size, multiple-proportion tests were performed with significant differences determined by a p value <0.05.

3.3.2 Relationship Between Overall Rating and Survey Response Rate

(1) Correlations between the ratio of "high" rating to the overall rating and survey response rate by response size were calculated.
(2) (1) was also calculated with logarithmically converted data.

3.3.3 Structure of Patient Satisfaction by Distance Scaling

The percentages of each rating were cumulated by question with respect to response size. The structure of patient satisfaction was then investigated by correspondence analysis, clustering, and multidimensional scaling (MDS), based on the contingency tables of patient survey questions (Q.1–Q.7, Q.9, and Q.10) and their ratings by response size. Question 8, "discharge information," was excluded from the analysis, owing to its dichotomous nature.

Clustering is a statistical tool for arranging large quantities of multivariate data into natural groups and can be applied to clustering observations, variables or both [21]. Correspondence analysis is an exploratory multivariate technique for simultaneously displaying scores representing the row and column categories of a two-way contingency table [22]. MDS is primarily a data visualization method for identifying "clusters" of points such that points in a given cluster are viewed as "closer" to the other points in that cluster than to points in other clusters [23].

The results were discussed with a study of patient satisfaction by distance scaling in Japan which found that various interpersonal skills of hospital staff were assessed similarly by outpatients, and that for inpatients, environment factors were evaluated dissimilarly to other variables of patient satisfaction [8].

4 Result

Of the 4677 hospitals, 3711 were selected.

4.1 Hospital Characteristics

The proportions of ACHs and CAHs were statistically different by response size ($p < 0.001$). Hospitals with larger response size had larger proportions of ACHs (Table 2).

Hospitals whose response sizes were 300 and above showed larger proportions of hospitals with specialty and care than hospitals whose response size were less than 300 ($p < 0.001$) (Table 3).

4.2 Relationships Between Overall Rating and Survey Response Rates

The average survey response rates by response size were 33.8 % [standard deviation (SD) 10.41] for hospitals with response size between 50 and 99, 33.1 % (SD 10.44) for hospitals with response size between 100 and 299 and 32.8 % (SD 8.21) for hospitals with response size 300 and above.

The proportions of "low" in "overall rating" were negatively correlated (−0.545 to −0.594) with the survey response rate in all response sizes, as were the proportions of "medium" (−0.0399 to −0.518) (Figs. 1, 2 and Table 4), whereas the proportions of "high" showed positive correlations (0.534–0.620) (Fig. 3 and Table 4). Because some scatterplots exhibited possible exponential relationships, such as between survey response rates and the proportions of "low" in all response sizes and "medium" and "high" in response sizes that were 300 and above, logarithmic conversion was investigated. However, only the correlations between the converted proportions of "low" and the original survey response rates showed increased coefficients (−0.573 to −0.615) (Table 4).

Table 2 Hospital type by response size and multiple-proportion test

Response size												Tests of multiple proportions		
50–99 (A) n = 220				100–299 (B) n = 692				≥300 (C) n = 2799						
ACH		CAH		ACH		CAH		ACH		CAH		A vs. B	A vs. C	B vs. C
n	%	n	%	n	%	n	%	n	%	n	%	p	p	p
78	35.5	142	64.5	358	51.7	334	48.3	2709	96.8	90	3.2	***	***	***

Abbreviations: *ACH* acute-care hospital, *CAH* critical-access hospital, *vs* versus
***$p < 0.001$

Table 3 Hospital structure by response size and multiple-proportion test

	Response Size												Tests of multiple proportions		
	50–99 (A) n = 220				100–299 (B) n = 692				≥300 (C) n = 2799				A vs. B	A vs. C	B vs. C
Hospital Structure	Yes	%	Others[a]	%	Yes	%	Others[a]	%	Yes	%	Others[a]	%	p	p	p
Cardiac surgery	–	0	220	100	8	1.2	684	98.8	1005	35.9	1794	64.1		***	***
General surgery	3	1.4	217	98.6	27	3.9	665	96.1	590	21.1	2209	78.9		***	***
Nursing care	5	2.3	215	97.7	71	10.3	621	89.7	1505	53.8	1294	46.2	***	***	***
Stroke care	11	5	209	95	54	7.8	638	92.2	1518	54.2	1281	45.8		***	***

[a]"Others" is the sum of "no", "not available" and blanks
Abbreviation; *vs* versus
***$p < 0.001$ Tests of multiple proportions

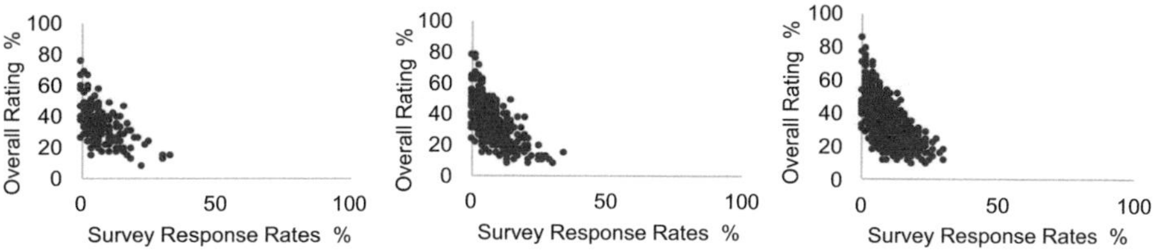

Fig. 1 Scatterplots of survey response rate against the percentage of "low" by response size. From *left*, hospitals with response size between 50 and 99 (n = 220); hospitals with response size between 100 and 299 (n = 692) and hospitals with response size 300 and over (n = 2799)

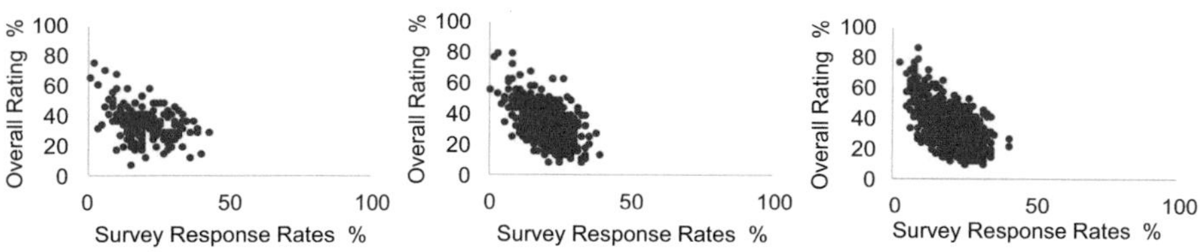

Fig. 2 Scatterplots of survey response rate against the percentage of "medium" by response size. From *left*, hospitals with response size between 50 and 99 (n = 220); hospitals with response size between 100 and 299 (n = 692) and hospitals with response size 300 and over (n = 2799)

Table 4 Correlation between the percentage of each rating in "overall rating" and survey response rate by response size

			Logarithmically converted data		
Rating	Response size	Original data	Overall rating	Survey response rates	Both
Low n = 220	50–99	−0.555	NaN	−0.615	NaN
	100–299	−0.594	NaN	−0.661	NaN
	≥300	−0.545	NaN	−0.573	NaN
Medium n = 692	50–99	−0.399	−0.477	−0.341	−0.386
	100–299	−0.518	−0.533	−0.489	−0.473
	≥300	−0.455	−0.488	−0.423	−0.433
High n = 2799	50–99	0.562	0.548	0.553	0.553
	100–299	0.620	0.610	0.634	0.643
	≥300	0.534	0.521	0.526	0.524

Abbreviation; *NaN* not a number

4.3 *The Structure of Patient Satisfaction*

Hospitals whose response size were between 50 and 99 and between 100 and 299 showed two clusters with the same item combination. "Cleanliness," "nurse communication," and "doctor communication" formed a cluster and the other cluster comprised the rest of the items including "overall rating" (Figs. 4 and 5).

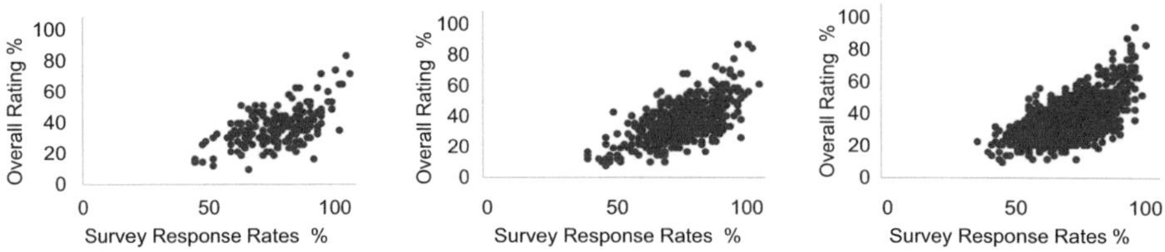

Fig. 3 Scatterplots of survey response rate against percentage of "High" by response size. From *left*, hospitals with response size between 50 and 99 (n = 220); hospitals with response size between 100 and 299 (n = 692) and hospitals with response size 300 and over (n = 2799)

The former cluster showed a larger percentage of "high" rating than the latter. Hospitals whose response size were 300 and above showed four clusters (Fig. 6); the top cluster had only "medication information" with the largest "low" rating; the second cluster had "staff responsiveness" and "quietness" with the largest "medium" rating; the third cluster comprised "nurse communication" and "doctor communication" with the largest "high" rating; and the bottom cluster included "cleanliness," "recommendation," "pain management," and "overall rating," with the second largest "high" rating. In all response sizes. "nurse communication" and "doctor communication" showed the largest "high" rating.

The correspondence analysis reveals relationships among items and their ratings. In all the response sizes, "nurse communication" and "doctor communication" were placed furthest away from "low" and "medium" and closest to "high," meaning that the communication items received the best evaluation. Though "medication information" was placed closest to "low" in all the response sizes due to its largest percentages of "low" ratings, in the response size less than 300, it was included in the cluster with the other low-evaluated items (Figs. 4 and 5). In the response size 300 and above, "quietness" and "staff responsiveness", forming one cluster, were placed closest to "medium". "Cleanliness," "recommendation," "pain management," and "overall rating", forming one cluster were placed closer to "high" (Fig. 6).

Another distance scaling, MDS, indicates that items near the coordinate origin are similar and items on the edge are dissimilar. In all the response sizes, "staff responsiveness," "pain management," and "overall rating" showed similarities, whereas "medication information" showed dissimilarities (Figs. 4, 5, and 6). In the response sizes less than 300, "quietness" and "recommendation" showed slight dissimilarities (Figs. 4 and 5). In the response size 300 and above, the "quietness" and communication items showed dissimilarities (Fig. 6).

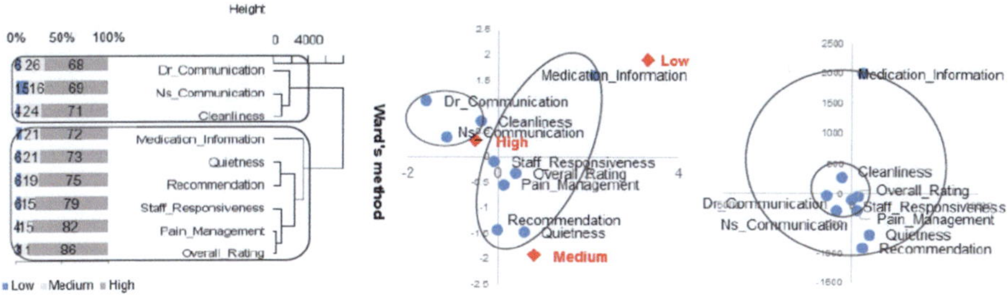

Fig. 4 Distribution graph and distance scaling for hospitals with the response size between 50 and 99 (n = 220). From *left*, clustering results, correpondence analysis, multidimensional scaling (*MDS*). The rectangles in the clustering results indicate clusters, the *circles* in correspondence analysis correspond to clustering results and in MDS, the *inner circle* indicates similarities and the *outer circle* indicates dissimilarities

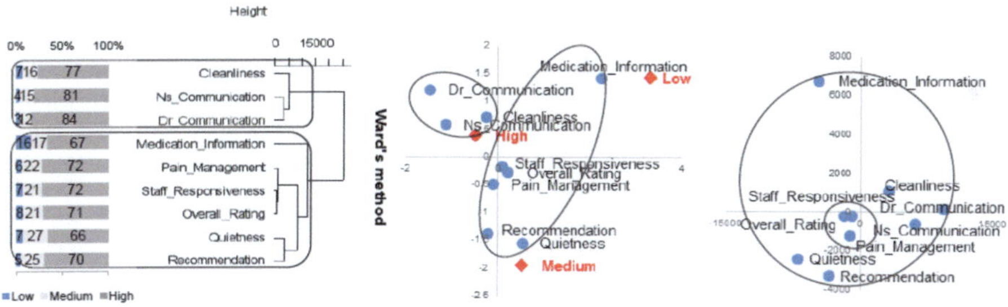

Fig. 5 Distribution graph and distance scaling for hospitals with response size between 100 and 299 (n = 692). From *left*, clustering results, correpondence analysis, multidimensional scaling (*MDS*). The rectangles in the clustering results indicate clusters, the *circles* in correspondence analysis correspond to clustering results and in MDS, the *inner circle* indicates similarities and the *outer circle* indicates dissimilarities

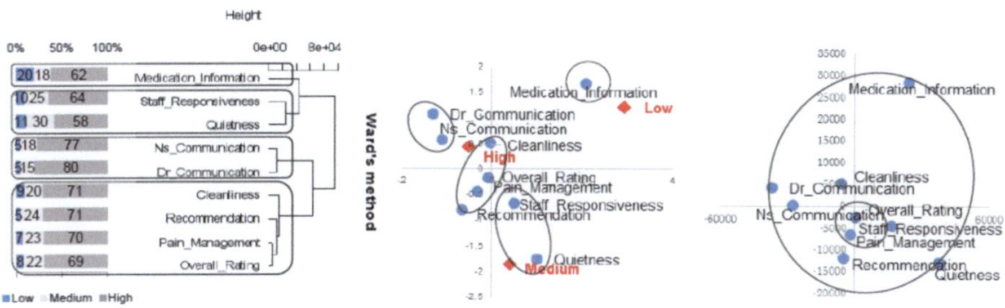

Fig. 6 Distribution graph and distance scaling for hospitals with response size 300 and over (n = 2799). From *left*, clustering results, correpondence analysis, multidimensional scaling (*MDS*). The rectangles in the clustering results indicate clusters, the *circles* in correspondence analysis correspond to clustering results and in MDS, the *inner circle* indicates similarities and the *outer circle* indicates dissimilarities

5 Discussion

5.1 *The Relationship Between "Overall Rating" and Survey Response Rate*

The positive correlations between survey response rate and percentage of "high" rating and the negative exponential correlation between survey response rate and percentage of "low" rating suggest that survey response rates fall drastically as patient satisfaction falls. The negative correlations between the survey response rates and the percentages of both "low" and "medium" ratings suggest that "medium" be considered as "low," without a connotation of being neither neutral nor good. This finding is very suggestive, because "medium" in the questionnaire is actually 7 or 8 on a ten-point scale, which is 4 on a five-point scale; and those ratings are usually combined with "very good" and labeled "good", as a five-point Likert scale yields a distribution resembling a normal distribution [24]. Furthermore, the fact that the "medium" rating was based on the word "usually" also suggests that patients" expect more frequent contact and care delivery from the hospital staff than "usual," which has the neutral connotation of normal or regular [25].

It could thus be surmised that patients tend not to respond to patient satisfaction surveys, especially when dissatisfied and that when patients opt to answer the questionnaire, they tend to give better ratings than their real perceptions by choosing neutral and better-than neutral answers because of the nature of medical care as a service [13]. This interpretation may explain the relatively high level of satisfaction in some quantitative studies [26, 27]. Similar results were reported from a qualitative study in which patients clearly differentiated between "satisfied" and "very satisfied" with healthcare [28]. A quantitative study using decision trees, another data mining method, also suggested that "satisfied" is different from "very satisfied" [5]. But one study found that a substantial assessment score difference between the first 30 % of respondents versus all respondents to whom a reminder mail was sent does not always yield better rating [11].

Survey results with low survey response rates will reveal a better interpretation than the real perception of patients, because non-respondents tend to assess hospital performance differently. However, the possibility that the survey response rates could become one of the patient satisfaction indicators exists. Hospitals whose survey response rates are low should consider the reason.

5.2 *The Structure of Patient Satisfaction by Survey Response Size*

Though studies have revealed that communications with doctors or nurses play an important role in patient assessment of hospital performance [27–31], in the present study, communication items were not grouped with "overall rating" in all the response sizes, an observation that suggests that although communication with nurses and doctors is highly valued, they receive assessments that differ greatly from overall satisfaction. Patients do not differentiate communication with nurses from that with doctors. The communication factor as a part of the structure of patient satisfaction is not influenced by hospital characteristics.

Pain management has also been reported as an important determinant of overall satisfaction [31, 32]. This observation was supported in the present study, with "pain management" showing proximity to "overall rating" in all response sizes. MDS demonstrated that "staff responsiveness" also showed proximity to "pain management" and "overall rating" in all the response sizes. This observation suggests that these three items have a similar basis and describe needs that are common to all patients irrespective of the hospital size. MDS also shows that "medication information" received quite a different and lowest assessment of all. This result suggests that patients need information about medications much more strongly than what the medical staff think. These three aspects of patient satisfaction were not strongly influenced by hospital characteristics.

In response size 300 and above, "medication information", "staff responsiveness" and "quietness" were grouped in one cluster with lower evaluation, though those three were grouped with other items in the other response sizes. Because this response size probably corresponds with large hospitals, given that most of the hospitals with a response size of 300 and above were acute-care hospitals with greater capacity for medical treatment, this finding suggests that the different and lower staff responsiveness evaluation is accounted for by the likelihood that patients receive invasive treatment. The OECD reports that the average lengths of stay in the US is 4.8 days, in contrast to those in other countries, such as 5.6 days in France, 7.0 days in U.K. and 17.5 days in Japan [33]. Many inpatients in the US may still suffer from illnesses or injuries and need help from the hospital staff, accounting for the lower evaluation of quietness, staff responsiveness and especially requirement for providing more information about medications. These are more difficult to provide as part of a patient-centered approach, owing to the characteristics of health care service in acute care.

Patients give higher assessment of communication, which represents the evaluation of approaches by doctors and nurses in the HCAHPS patient survey, but such communication may not fulfill the patients' actual needs. This finding suggests that even if pain is well controlled and patients are, in general, relatively well satisfied with the communication they receive, they need swifter, more individualized attention from medical staff and a quieter environment, and especially medication information. Donabedian suggested that two elements of the performance of medical practitioners are technical performance and interpersonal performance [34] and that they should be provided together in a more patient-centered manner. The structure of patient satisfaction may differ by hospital characteristics.

In Japan, a different structure of patient satisfaction was reported, in which overall rating, communication and environmental variables formed the same cluster despite differences in the lengths of stay [8]. The structure was observed especially among elderly people. This observation could result from the relatively long hospitalization in Japan because, with longer hospital stays, patients may expect more comfort in both their environment and human relations.

5.3 To Improve Quality of Care

Hospitals with low survey response rates should seek the reasons. Improving staff responsiveness and pain management will have a direct effect on overall rating, but more attention should be paid to providing medication information and a quieter environment with more individualized interpersonal skills. Education in attending to personal needs should be more focused.

5.4 Limitations

The response size may not indicate the actual hospital bed size. Non-respondents' assessments were not reflected. Further stratification by hospital characteristics is needed. Other factors of patient satisfaction should be considered to identify the structure of patient satisfaction.

5.5 Future Work

Another data mining activity is to build models and procedures/techniques and assess the predictive accuracy of those models when applied to new data [12]. Constructing models of the structure of patient satisfaction by distance scaling using the dataset of "structure", "process" and "outcome" will contribute to the improvement of quality of care, focusing on patient-centered approach.

6 Conclusions

As patient satisfaction falls, survey response rates fall exponentially. The structure of patient satisfaction differs by hospital size and hospital specialties. Some factors do not change the relationships within a given cluster but with other clusters. To improve overall patient satisfaction, more attention should be paid to the relationships among the factors of patient satisfaction and their changes.

Distance scaling has the potential to identify different aspects of patient satisfaction.

References

1. Donabedian A (1966) Evaluating the quality of medical care. Millbank Mem Fund Q 44: 166–206
2. William B (1994) Patient satisfaction a valid concept? Soc Sci Med 38(4):509–516
3. Linder-Pelz S (1982) Toward a theory of patient satisfaction. Soc Sci Med 16:577–582
4. Linder-Pelz S (1982) Social psychological determinants of patient satisfaction: a test of five hypotheses. Soc Sci Med 16:583–589
5. Fishbein M, Ajzen I (1975) Belief, attitude, intention, and behavior: an introduction to theory and research. Addison-Wesley, Reading. http://people.umass.edu/aizen/f&a1975.html. Accessed 26 Mar 2015
6. Crow R, Gage H, Hampson S, Hart J, Kimber A, Storey T, Thomas H (2006) The measurement of satisfaction with healthcare: implications for practice from a systematic review of the literature. Health Technol Assess 6(32):1–244
7. Sitzia J, Wood N (1997) Patient satisfaction: a review of issues and concepts. Soc Sci Med 45(12):1829–1843
8. Okuda M, Yasuda A, Tsumoto S (2013) The relationship between wait time and outpatient satisfaction, and between admission length and inpatient satisfaction. J Shimane Med Assoc 33(3):8–13
9. Faber M, Bosch M, Wollersheim H, Leatherman S, Grol R (2009) Public reporting in health care: how do consumers use quality-of-care information? A systematic review. Med Care 47(1):1–8
10. https://data.medicare.gov/. Accessed 4 Apr 2014
11. Barkley WM, Furse DH (1996) Changing priorities for improvement: the impact of low response rates in patient satisfaction. J Qual Improv 22(6):427–433
12. Izenman AJ (2008) Modern multivariate statistical techniques: regression, classification, and manifold learning, Springer texts in statistics. Springer, New York, pp 1–16
13. Ibe T, Nakanishi M (2011) Kango keiei keizairon. Japanese Nursing Association Publishing Company, Tokyo
14. Goldstein E, Farquhar M, Crofton C, Darby C, Garfinkel S (2005) Measuring hospital care from the patients' perspective: an overview of the CAHPS hospital survey development process. Health Serv Res 40:1977–1995
15. O'Malley AJ, Zaslavsky AM, Hays RD, Hepner KA, Keller S, Cleary PD (2005) Exploratory factor analyses of the CAHPS Hospital Pilot Survey responses across and within medical, surgical, and obstetric services. Health Serv Res 40(6):2078–2095
16. Giordano LA, Elliott MN, Goldstein E, Lehrman WG, Spencer PA (2010) Development implementation, and public reporting of the HCAHPS Survey. Med Care Res Rev 67:27–37
17. http://publicapps.odh.ohio.gov/facilityinformation/HospitalMeasuresHospitalTypeDefinition.aspx?HospitalTypeName=Acute%20Care%20Hospital
18. http://www.aha.org/. Access date, 24 Apr 2014
19. Jha AK, Orav EJ, Zheng J, Epstein AM (2008) Patients' perception of hospital care in the United States. N Engl J Med 359:1921–1931
20. http://www.r-project.org/. Accessed 4 Feb 2015
21. Izenman AJ (2008) Modern multivariate statistical techniques, cluster analysis, Springer texts in statistics. Springer, New York, pp 407–462
22. Izenman AJ (2008) Modern multivariate statistical techniques, correspondence analysis, Springer texts in statistics. Springer, New York, pp 633–666
23. Izenman AJ (2008) Modern multivariate statistical techniques, multidimensional scaling and distance geometry, Springer texts in statistics. Springer, New York, pp 463–504
24. Likert R (1932) A technique for the measurement of attitudes. Arch Psychol 140:5–55
25. http://www.merriam-webster.com/. Accessed 2 Apr 2015
26. Hall JA, Dornan MC (1990) Patient sociodemographic characteristics as predictors of satisfaction with medical care: a meta-analysis. Soc Sci Med 30:811–818

27. Calnan M (1988) Towards a conceptual framework of lay evaluation of health care. Soc Sci Med 27(9):927–933
28. Collins K, O'Cathain A (2003) The continuum of patient satisfaction from satisfied to very satisfied. Soc Sci Med 57:2465–2470
29. Larrabee JH, Bolden LV (2001) Defining patient-perceived quality of nursing care. J Nurs Care Qual 16(1):34–60
30. O'Connell B, Young J, Twigg D (1999) Patient satisfaction with nursing care: a measurement conundrum. Int J Nurs Pract 5(2):72–77
31. Wagner D, Bear MJ (2009) Patient satisfaction with nursing care: a concept analysis within a nursing framework. J Adv Nurs 65(3):692–701
32. Woodside AG, Frey LL, Daly RT (1989) Linking service quality, customer satisfaction, and behavioural intention. J Health Care Mark 9(4):5–17
33. OECD Health Data (2014) http://www.oecd.org/els/health-systems/health-data.htm
34. Donabedian A (1988) The quality of care, how can it be assessed. JAMA 260(12):23–30

Part IV
Public and Urban Services

One Cycle of Smart Access Vehicle Service Development

Hideyuki Nakashima, Shoji Sano, Keiji Hirata, Yoh Shiraishi, Hitoshi Matsubara, Ryo Kanamori, Hitoshi Koshiba, and Itsuki Noda

Abstract Under JST RISTEX S3FIRE program, we are trying to implement Smart Access Vehicle (SAV) Service in Hakodate. The project adopts the method of service science loop – the repeated cycle of observation, design and implementation. In this paper we report the completion of its first cycle, and discuss how the cycle improved our initial design. We first conducted person trip research in Hakodate. We chose 20 candidates of various age and occupation, and recorded their everyday movements for 4 months. We then analyzed the result and made a person trip model. The model was then fed into our multi-agent simulator for Hakodate public transportation system. We conducted a small field test with five vehicles for 1 week. The most significant achievement is that we confirmed that our design of SAV system works. We succeeded in automatically dispatching five vehicles for 11 h without any significant trouble or human supervision.

Keywords Smart access vehicle • Hakodate • Demand responsive transit

1 Introduction

We initiated the project named "Smart City Hakodate" in 2009 as an envelope project, without attachment to any particular funding. Various activities followed. Future University Hakodate (FUN hereafter) signed mutual agreement on research collaboration with IBM in 2009. An NPO Smart City Hakodate was founded in

---

H. Nakashima (✉) • S. Sano • K. Hirata • Y. Shiraishi • H. Matsubara
Future University Hakodate, Hakodate, Japan
e-mail: h.nakashima@fun.ac.jp

R. Kanamori
Nagoya University, Nagoya, Japan

H. Koshiba
National Institute of Science and Technology Policy, Tokyo, Japan

Future University Hakodate, Hakodate, Japan

I. Noda
National Institute of Advanced Industrial Science and Technology, Tsukuba, Japan

T. Maeno et al. (eds.), *Serviceology for Designing the Future*,
DOI 10.1007/978-4-431-55861-3_17

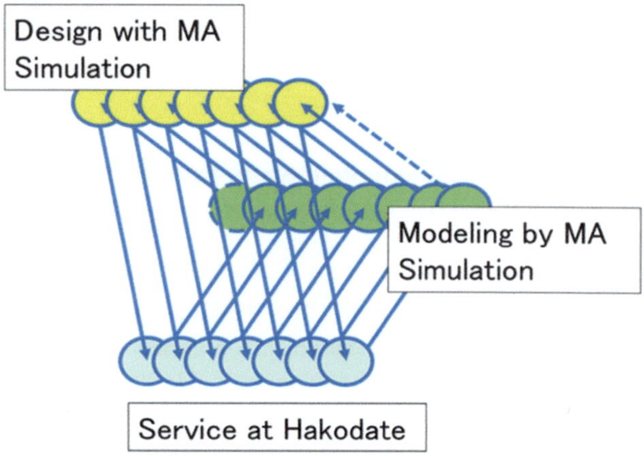

Fig. 1 Design-service loop of SAVS

2010. FUN provided special research funding. And finally we were able to obtain JST RISTEX funding for "IT-enabled Novel Societal Service Design".

As a part of Smart City Hakodate project, we are designing and implementing a new transportation system named Smart Access Vehicle System (SAVS). The key is extensive use of information technology (IT). IT should not be regarded as just a replacement mechanism for traditional mechanical ones. Just as the invention of steam engine changed the whole societal structure, toward "industrial era" in sixteenth century, IT will open up a new "information era".

Proper design is essential for IT to be effectively used in the societal system. We use the implementation methodology induced by Serviceology. We claimed that service provision essentially goes through a design-service loop [1], and our project is designed as such (Fig. 1). Although service, or field test, at Hakodate is the main process, the modeling of person trip in Hakodate and design of the algorithm for the service are also essential to the implementation of SAVS. We plan to repeat the cycle several times. The purpose of this paper is to report its first cycle taking closer look at those each steps.

2 Smart Access Vehicle System

Smart Access Vehicle System (hereafter, we use SAV for each vehicle and SAVS for the whole system/service) is a new public transportation service that unifies bus and taxi services (Fig. 2).

SAVS falls into a category called Demand Responsive Transportation (DRT) system, which is further classified into the following:

1. Detour/free stop

   Fixed route + {detour/stop} on demand
   Pre-scheduling
   Examples: many rural cities

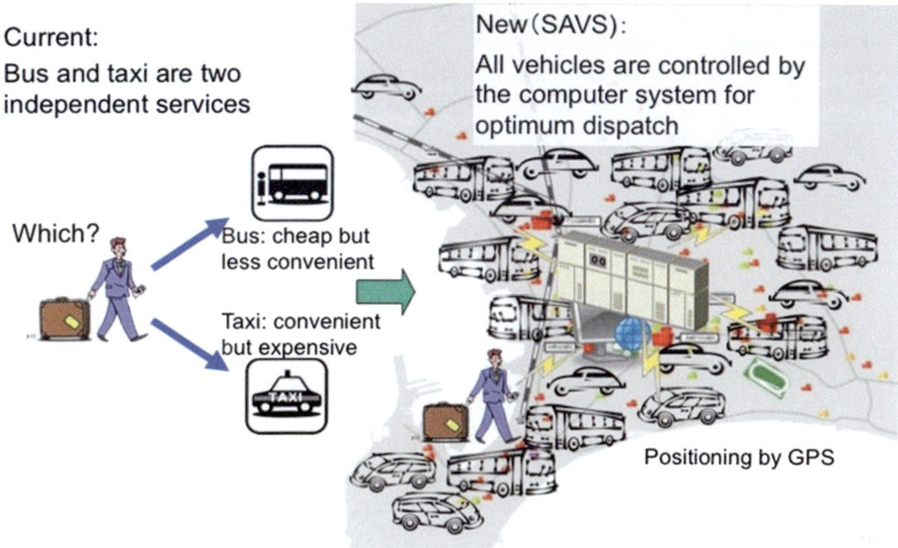

Fig. 2 The basic concept of SAVS

2. Flex-routing

   Fixed stops with on-demand routing
   Pre-scheduling
   Examples: EU project DRT's [2], Soja city, Todai combinicles [3]

3. Full-demand

   3.1 Low demand areas (mainly pre-scheduling)

      3.1.1 Full-demand bus

         Example: Nakamura (Shimanto) city bus

      3.1.2 Shared taxi

         Example: SAVS

   3.2 Urban areas (real-time scheduling)

SAVS is classified as 3.2 above, and is designed to replace current urban local public transportation systems. As far as we know of, SAVS is the only DRT designed for urban areas.

From users' point of view, the process of calling a SAV is very similar to reserving a demand bus:

1. A user contacts the system with the demand (the current location and the destination).

2. The system searches for a best vehicle considering their current position and future route.
3. The system tells the user the pickup point, the estimated time of pickup, and the estimated time to the destination (with a small margin of delay). The user has a choice to either accept or decline the service.

The differences are:

1. SAV's operate in real time (reservation is optional). A user may call a SAV when the actual demand emerges.
2. Many (in the order of 1000 or more) vehicles are involved so that the operation is efficient.

The system knows the locations and routes (destinations of passengers on board) of all vehicles. When a new demand arrives, the system searches for a vehicle that can pick up and deliver the passenger with minimum detour. Even when a vehicle is very close to the request point, it may not be selected if it is heading toward the wrong direction or if a large detour, beyond the limit of promised tie of delivery for already on-board passengers, is required. If the system cannot find any vehicle, it must decline the request. However, we are hoping that denial of service is very rare case such as a large accident or wide area disaster (including heavy storm or snow) as long as sufficient number of vehicles exist in the system.

The central dispatch system runs on MA simulation of the city traffic. If we aim for the best solution, the computation may be too heavy. We use near-optimum solution (see Sect. 5 for more detail).

3 SAVS Project as Serviceology

Implementation of a novel transportation system such as SAVS encounters several practical problems, which supplies several interesting issues for Serviceology.

3.1 U-Shape Transition

The first issue is what we call the U-shape transition of the service (Fig. 3). When we gradually introduce a full-demand bus system mixed with the traditional bus system, the total efficiency drops initially. Actually, after several field test of the full-demand-responsive transportation (DRT) system with a few number of vehicles, it is commonly accepted that the full-demand bus system is inefficient in high-demand areas; Full-DRT is suitable only for low-demand areas [2]. However, using MA simulation, we found that it is not necessarily the case: Full-DRT, operated fully, becomes more and more efficient as the demand increases [4]. Therefore, we have to find a practical tactics to jump over the U-shape valley.

Fig. 3 U-shape Valley

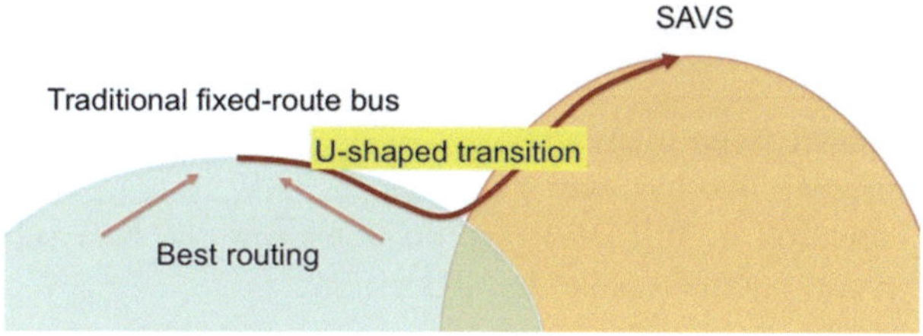

Note: This U-Shape valley is conceptually different from the "death valley" that lies between research and industrial development. The latter is the problem of development risk, and the former is an inherent property of our proposed system. However, we believe "U-shape" problem itself is universal to most of innovational systems, since innovation is a jump.

3.2 Value Co-creation

Service in general should be viewed as value co-creation of the provider and the user [5]. SAVS should be a good example of value co-creation in service. Since SAVS provides the transportation infrastructure for urban life, it should have a large impact on the life style of the users. Therefore, it is expected that the introduction of SAVS changes the trip pattern of people living in the area. People may give up using their private cars within the city. It will push up the importance of SAVS. It may add new value to the public transportation. Therefore, value co-created by SAVS and its users is expected to be quite large and unpredictable.

At the same time that new values are created, new requirements for public transportation may emerge. Design of SAVS should be kept changing as depicted in Fig. 1.

Unfortunately, the above scenario of value co-creation is not realized yet. So far, we could try only small-scale operation tests for short periods – far from enough to change life style of the users.

However, without such expectation of change, taxi companies would not have joined the project. If we assume that the same number of people use public transportation with the same life-pattern, introduction of SAVS just decreases income of taxi companies, because the cost for users should be smaller than the current taxi system. Expectation of increased number of users is mandatory.

In short, value co-creation is not happening yet, but it is already an essential part of the plan.

3.3 User Involvement

Another issue is involvement of users during the design phase (inclusive design). Since no one has ever experienced SAV system, which we regard as one implementation of full-DRT, we do not know the best service parameters. There are several parameters yet to be decided:

- the size of the vehicles (passenger capacity),
- the number of vehicles per area or per population,
- type of stops (predetermined or free),
- prior reservation,
- fare, and
- special services (such as priority delivery).

These parameters are to be decided while the service is carried out.

3.4 Law Issues

The third issue is the law restrictions. Currently in Japan, bus and taxi systems are strictly divided by the law to protect niches of each system. Taxis are not allowed to pickup multiple group of passengers (with only a few exceptions), and buses are not allowed to run free routes but have to load and/or unload passengers only at predetermined bus stops. Vehicles that carry up to 9 people are defined to be taxis and vehicles that carry more than 9 people are defined to be buses. Since our SAV system unifies both of them, it cannot be operated under the current law. We need a special designated area for SAV operation.

4 Person Trip Model

4.1 Acquisition of Person Trip Data

Figure 4 is the result from our previous work [6] to show the superiority of the full-demand bus system over conventional fixed-route fixed-timetable bus systems. Horizontal axis is number of vehicles (increases proportional to population). The vertical axis is average trip time to the destination. As the population – therefore number of vehicles – increases, the average trip time decreases. Conventional bus system, plotted with "x" marks and thicker lines, becomes more efficient as the population glows – buses in large cities are more convenient than buses in rural districts. Full demand bus system is less efficient when the population is small, but quickly become more efficient than fixed route buses as the population grow larger.

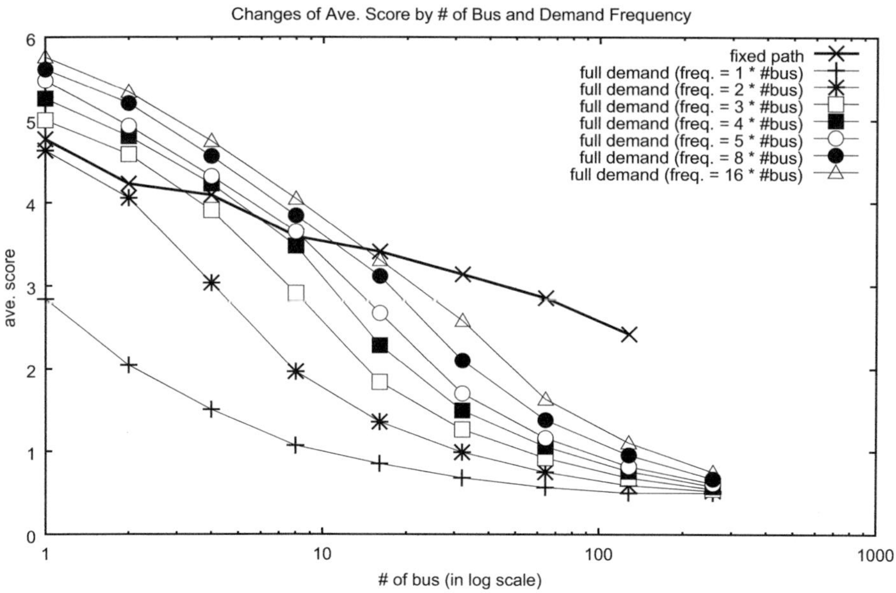

Fig. 4 Efficiency of SAVS *vs.* traditional bus [4]

However, Fig. 4 is a result of MA simulation of an artificial city, and just shows qualitative tendency. We do not know the quantitative population nor the number of vehicles of the cross point. To figure out those quantitative numbers of Hakodate, we plan to do the survey and MA simulation based on the survey.

The first step is to grasp the current status. Of course, we understand that this person trip pattern changes when the transportation system changes. Thus, this survey only supplies the initial value to our simulation.

We used two methods. (A) Direct observation of person trip using smart phones; (B) buying statistical data of mobile phone records.

For (A) direct observation, we developed a smart phone application to record person trip. We asked 20 people of various age and occupation to use the application for 4 months. As the total we gathered 2400 day*person trip records (Fig. 5 left). We also know the transportation mode (bus, taxi, private car, tram, bicycle, or on foot) and trip purpose (shopping, hospital, restaurant, work, school, or other) for each trajectory.

For (B) statistical data, we obtained OD (origin and destination) statistics. Although the mobile phone company has the trace of individual users, we could only get the statistical data for the sake of privacy protection. Figure 5 right, for example, shows the total number of people who left the 2.5 km square area within 2 h of a particular date. Similar data is obtained for entering the area.

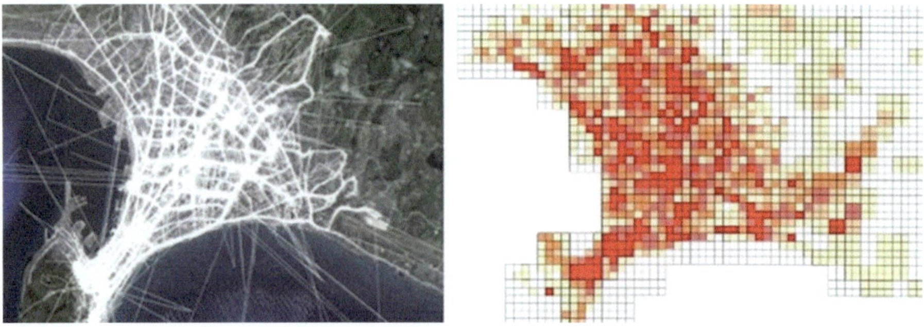

Fig. 5 Person Trip data from GPS – individual (*left*) and from mobile phone – mass (*right*)

4.2 Travel Mode Choice Model

We developed a travel mode choice model to predict travel demands. Using this model, the number of passengers for each transportation means can be estimated when an origin-destination data (Sect. 4.1) are given.

In order to develop a travel mode choice model we introduce the *logit* model, which is one of a discrete choice model based on microeconomic utility maximization. In this model, each citizen is assumed to be a rational individual, and make a decision with attributes of the person and to attributes of the alternatives. The utility for using each travel mode (U_m), where $m = \{$private car, public transportation, walk$\}$, is set as follows:

$$\begin{aligned} U_{car} &= V_{car} + \epsilon = \beta_1 \cdot T_{car} + \beta_2 \cdot C_{car} + \beta_3 D + \epsilon, \\ \epsilon U_{pub} &= V_{pub} + \epsilon = \beta_1 \cdot T_{pub} + \beta_2 \cdot C_{pub} + \beta_3 \cdot D + \beta_4 + \epsilon, \\ U_{wlk} &= V_{wlk} + \epsilon = \beta_1 \cdot T_{wlk} + \beta_2 \cdot C_{car} + \beta_3 \cdot D + \beta_5 + \epsilon. \end{aligned}$$

V_m: the systematic components,
ϵ: the random term whose distribution is independent and identically Gumbel distributed,
T_m: travel time using mode m,
C_m: cost for using mode m,
D: the distance from origin to destination,
β_i: parameters that need to be estimated from real citizen's activity pattern data

Now we can calculate the probability of using the public transport mode (*i.e.*, tram or bus) by the following formula:

$$P_{pub} = exp\left(V_{pub}\right) / \left(exp(V_{car}) + exp\left(V_{pub}\right) + exp(V_{wlk})\right)$$

Next we estimate the parameters in the logit model by a maximum likelihood method. Since the service level of each travel mode should be preset, we prepared the data of time and cost for each travel mode according to Google Map's and Yahoo's information.

The estimated parameters are shown in Table 1. All parameters meet the normal sign condition even though the cost parameter (β_2) does not have a statistical superiority. And the value of time calculated from our model is 42.9 JPY/min, that is an appropriate one compared with the conventional value.

Finally, we calculated the number of passengers of public transportation, using both our travel mode choice model and the total origin-destination trips calculated from the mobile phone data. The estimation results of the number of public transportation passengers are shown in the following two figures: One is the result of the number of daily passengers for a week (Fig. 6), and the other is the hourly average number of passengers in a day (Fig. 7). We can see that the number of

Table 1 Computed parameter values for Hakodate

Parameter	β_1	β_2	β_3	β_4	β_5
Value	−0.057	−0.133	−0.768	1.216	−0.003
t-value	−4.0	−0.3	−1.6	3.8	−0.0

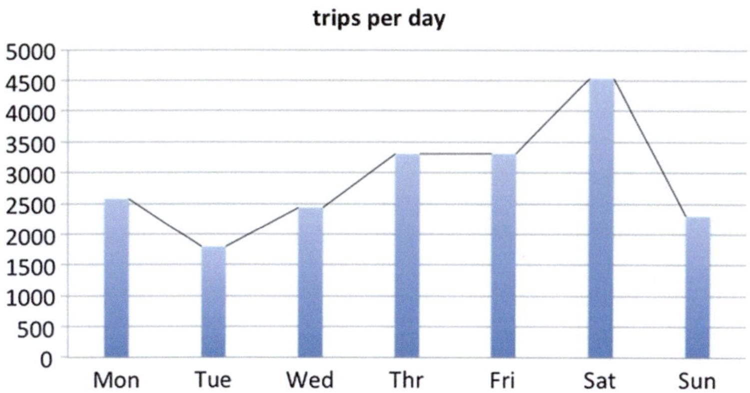

Fig. 6 Estimated trips for a week

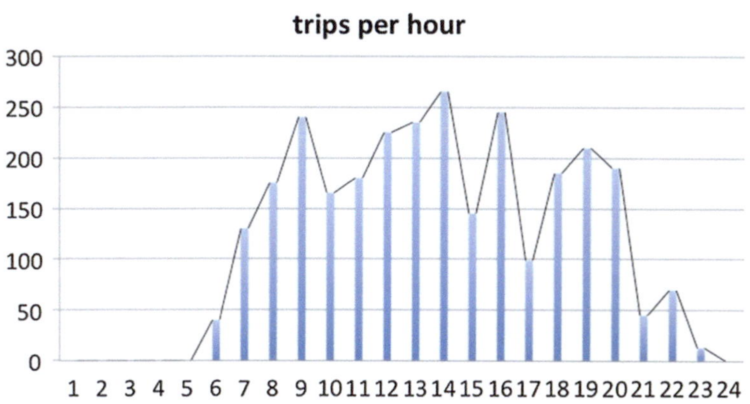

Fig. 7 Estimated trips for a day

passengers on Saturday is the maximum in a week, and the number of passengers in daytime is approximately 200 per hour.

Using those figures, we plan to run a full-scale multiagent simulation of all Hakodate (about 300 K people), and compare the efficiencies of the current traffic system and SAVS. For SAVS, we will also compare the cases (1) only those using public transportation use SAVS, and (2) all transportation, including private cars, are shifted to SAVS. We are hoping to prove that SAVS is actually more efficient than traditional public transportation system using Hakodate's real data.

5 Multiagent Simulation and Vehicle Dispatch Algorithm

5.1 Roles of Multiagent Simulation

There are several roles for multiagent simulation. The first one is to show the efficiency of SAVS in Hakodate (as described in Sect. 4). Another role is to design the detailed algorithm for SAVS operation. There are several unsolved issues for SAVS besides those listed in Sect. 3.3:

- how to reflect learned data from previous operations for dispatch algorithm, and
- where to locate unoccupied cars.

These are future research issues.

5.2 Dispatch Algorithm

The system has to determine vehicle assignment (which vehicle is the best to pick up a new user) and its new route in response to demands in real-time.

In Hakodate, we have about 900 public transportation vehicles (200 buses and 700 taxis) currently. Generally, it is hard to find the optimum routing for a large number of demands, because assignment problem belongs to the class of traveling salesman problems, i.e., NP hard. Instead, we try to find semi-optimum assignment by the *successive best insertion* method [4].

In this method, when the request (1) arrives (Fig. 8), the Dispatch system puts each demand to an auction table in DB, where each vehicle (an agent program representing the vehicle) bets with a cost value to accept the demand. Each vehicle calculates the cost by traffic simulation based on the current holding demands and traffic situations (Fig. 9).

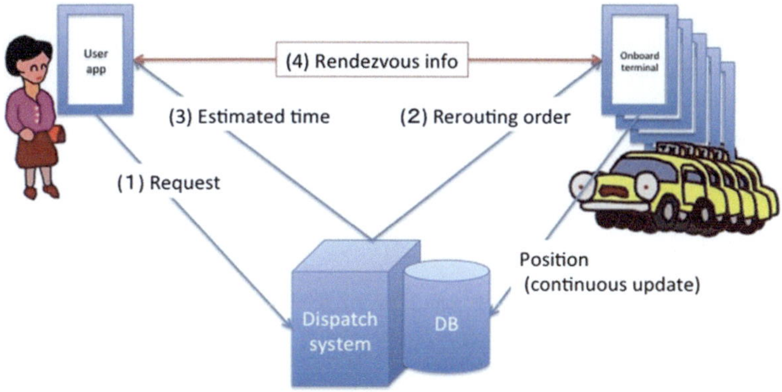

Fig. 8 Communication network of SAVS

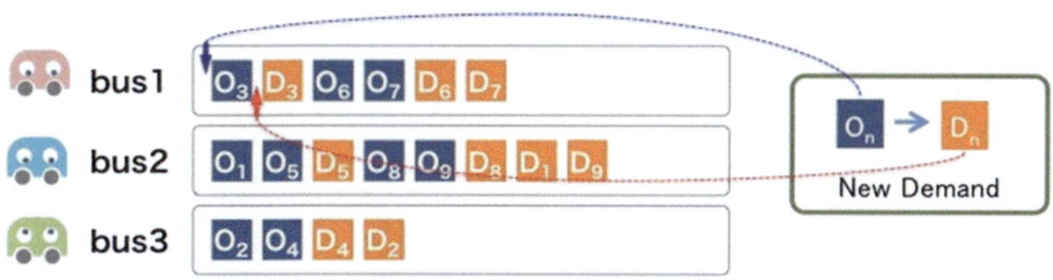

Fig. 9 Successive best insertion method used to determine vehicle assignment

This semi-optimization mechanism is simple and thus flexible enough to introduce several realistic restrictions and novel services in addition to real-time problem solving.

However, although it is a rather simple computation, yet the computational complexity is NP hard. For a large number of vehicles, in the magnitude of 1000 or more, some faster algorithm such as described in [7] is needed.

Once the best vehicle is determined, the new request is relayed to the on-board device of the vehicle (2) and estimated pick-up time is relayed to the user (3). Rendezvous information, such as locations of the vehicle and the passenger, should be displayed for both terminals (4), but this function is not implemented yet.

6 Field Test

We conducted a field test in Hakodate for a week in Oct 2013. The main target is to check system's operability. We hired five taxis and about 20 volunteers (mostly students of Future University Hakodate) to randomly call vehicles. Since we had only small number of vehicles, we could not cover whole Hakodate. We limited the area of operation to inside the designated area shown in Fig. 10.

Fig. 10 Area of Oct. 2013 field test

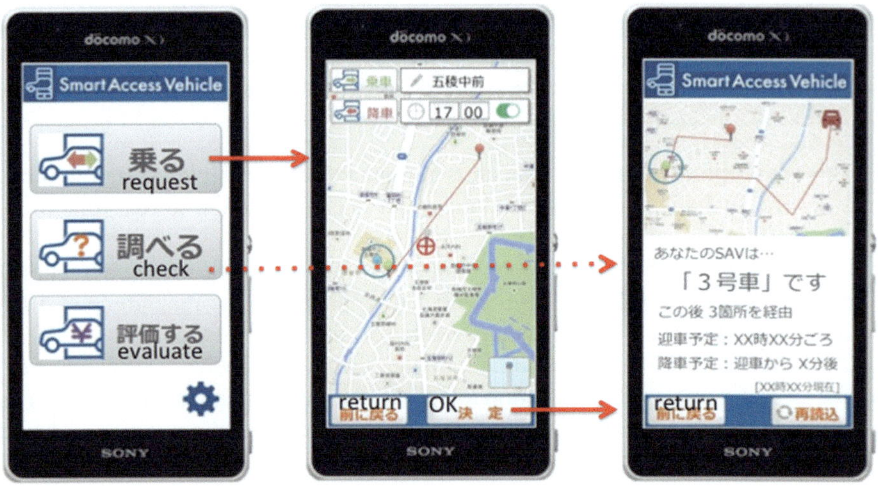

Fig. 11 SAVS application

Each passenger uses a smart phone application to request a SAV with pick-up and delivery locations on the map (with optional description of the locations). Figure 11 shows an image of the application (Pictures shown are improved design after the field test, but the functions are essentially the same).

Figure 12 shows the record of all pick-up (red) and delivery (blue) points. For the last 4 days when all operations were automatic, there were about 680 requests and the average operation time (from request to delivery) was 22 min.

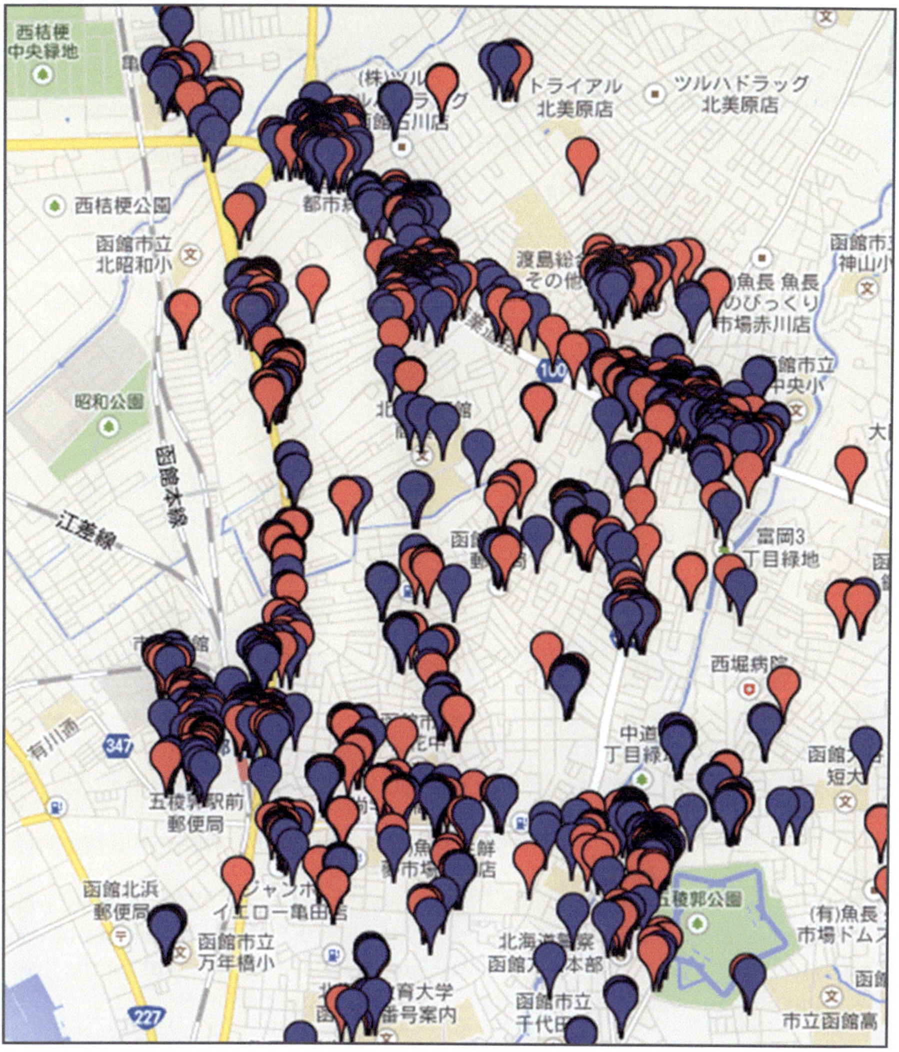

Fig. 12 Record of all pick-up (*red*) and delivery (*blue*) points

The field test proved that SAVS actually operates fully automatically for several days. As far as we are aware of, this is the world first operation of multi-vehicle real-time DRT system.

Figure 13 shows the graph of total number of operations during the field test week. The system was not fully functional for the first several days, and thus could not accept all requests. The system operated without any significant errors for the last 4 days. The taxi company told us that their average operation of one taxy per day is 20. Therefore total of 100 operations with car is the minimum threshold we

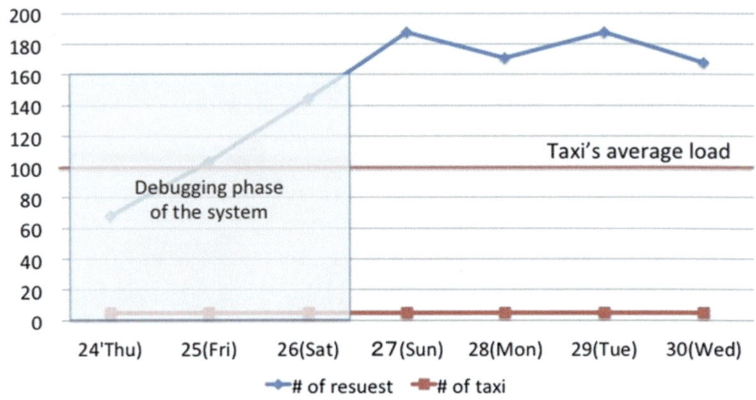

Fig. 13 Number of SAV's and their total operations for each day

Table 2 Operation statistics

Date (Oct 2013)	27	28	29	30
Total # of demands	191	172	178	168
Average total time	22.8	23.0	20.6	18.1
Waiting: request to pickup (min)	13.2	13.5	12.3	9.3
On board: pickup to delivery (min)	9.6	9.4	8.3	8.8
If on foot (min)	32.9	30.9	32.0	30.2

aimed at. We could actually take more than 1.5 times of the threshold. Full automatic operation of 5 cars for 4 days over average taxi's operation load in Hakodate is something we can be proud of, as the world first successful operation of real-time DRT system.

Table 2 shows the statistics of the last 4 days' operations. Waiting time, from request to actual pick up, is rather large. It suggests that the total number of cars (5) was too small compared to the test area. However, we have to wait for the total-city simulation to decide the optimal number of vehicles per area.

7 Further Work and Future Plans

7.1 Virtual Transportation System

SAVS unifies bus and taxi systems. Since SAVS is center-controlled by a computer system, any vehicle in the system can be operated either as a bus or as a taxi. In this sense, SAVS creates a virtual layer for the public transportation system. In the future, trains, ferry boats, and even airplanes may be assimilated, so that transfer from airplane to local buses become seamless not only connection time wise but also payment wise.

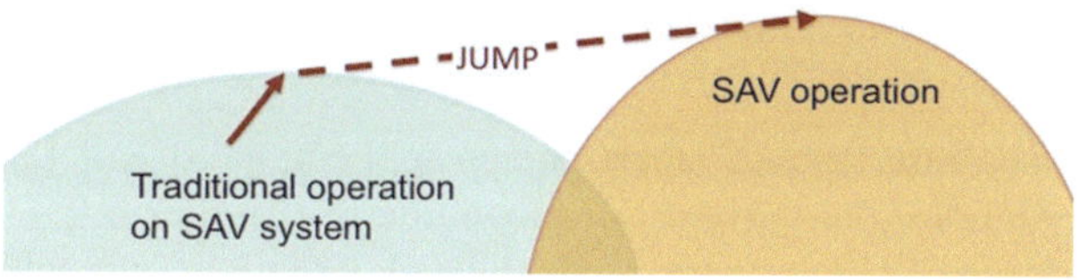

Fig. 14 Jumping over the U-shape valley

This virtualization generates large operational flexibility. Here is a possibility to jump over the U-shape described in Sect. 4.1 (Fig. 14).

Even after all preparation for SAVS is complete, with all on-board devices and central computer system, the traditional operation of buses and taxis can be continued, since SAVS subsumes those operations. And then on particular occasions, such as Hakodate's "no my-car day", the whole vehicles are switched to SAVS mode. It can be switched back to the normal (traditional) operation mode for the next day. By switching to SAVS and back several times, people gradually get accustomed to the new operation. We can also gather data and compare their efficiencies of two modes. Only after all stake holders are convinced with the superiority of SAVS, the whole transportation system can be changed, without going through the valley in between.

7.2 *Disaster Response*

As one of the immediate extension of the concept of virtual transportation, the flexibility suits for emergency transportation in cases of disasters such as large fire and earthquakes.

Road configuration may change due to several roadblocks, or some path may become unusable due to extreme traffic congestion. SAV's can not only compensate for those changes but also function as probes for those blocks.

In case of communication shutdown due to power shortage or physical damage of transmission infrastructure, some alternative means must be prepared. Short-range inter-vehicle communication system or DSRC (Dedicated Short Range Communication) is one of the alternatives.

7.3 *Service Unification*

Unification of services may not be restricted within transportation services. Other services such as restaurant, shopping, library, and health-care can be unified [8]. For example, when you buy some goods using the Internet, you may pick it up on board a SAV.

7.4 User Interface Issues

User Interface of smart-phone application to call SAV must be upgraded. For example, the current UI allows the user either accepts or declines the proposed service. Much more sophisticated negotiation mechanism, such as requesting quicker service with higher fare, should be implemented.

8 Conclusion

The field test proved that SAVS actually operates fully automatically for several days. As far as we are aware of, this is the world first operation of multi-vehicle real-time DRT system.

The most significant result gained by the first cycle of SAVS development is that the system actually runs with only small number of devices. In other words, infrastructure such as GPS and mobile phone communication are sufficient for SAVS operation. User applications on smart-phones and on-board tablet terminals together with a server (Fig. 8) are only necessary additions. In other words, SAVS can be introduced into a city with relatively low cost.

References

1. Nakashima H, Fujii H, Suwa M (2013) Designing methodology for innovative service systems. In: Proceedings of the ICServ2013, pp 187–192
2. Ambrosino J, Nelson MR (eds) (2003) Demand responsive transport servicest Towards the flexible mobility agency. Italian National Agency for New Technologies, Energy and the Environment, ISBN 88-8286-043-4
3. Hiekata K, Yamato H, Tsubouchi K (2008) A demand responsive transit service with emphasizing on better punctuality. The 5th international workshop on intelligent transportation, pp 93–98
4. Noda I, Ohta M, Kumada Y, Shinoda K, Nakashima H (2005) Usability of dial-a-ride systems. In: Proceedings of the AAMAS 2005, pp 1281–1282
5. Vargo SL, Maglio PP, Akaka MA (2008) On value and value co-creation: a service systems and service logic perspective. Eur Manag J 26:145–152
6. Noda I, Ohta M, Shinoda K, Kumada Y, Nakashima H (2003) Evaluation of usability of dial-a-ride systems by social simulation. In: Multi-agent-based simulation III, 4th international workshop, MABS 2003 (LNAI-2927). Springer, pp 167–181
7. Kishimoto A, Sturtevant NR (2008) Optimized algorithms for multi-agent routing. In: Proceedings of the AAMAS 2008, pp 1585–1588
8. Nakashima H, Shiraishi Y, Matsubara H (2011) Design and implementation of a smart access vehicle system as the core of "Smart City Hakodate" (in Japanese). Kanko to Joho, Society for Tourism Informatics, pp 19–28, !SSN 1349-919X

The Value of Community for Resolving Social Isolation

Keiko Aoki, Kenju Akai, and Nariaki Nishino

Abstract This study investigates the value of belonging to community for resolving social isolation problem. To examine this purpose, we employ the choice experiment used to estimate the willingness to pay for services and conduct it in internet survey. The estimation results show that the isolation group more prefers relationship with neighborhoods to that with relative and anonymity than non-isolation group does.

Keywords Social security service • Social isolation • The value of community • Choice experiment • Internet survey

1 Introduction

The OECD biennial report "Society at a Glance 2005" survey social isolation among OECD countries and a few Asian counties as Japan and Korea from 1999 to 2002 [1]. This report shows that Japan is the most social isolation country on the question such that respondents rarely or never spend time with friends, colleagues, or others in social group among counties. Also Japan, Japanese do not become to feel happy when they age [2]. As shown in the results, the problems based on social isolation and unhappiness happen in Japan. For example, the solitary death is the most increasing in old people who are 70–74 years old [3] and the estimated number of people of solitary death is about 10,000 persons per year [4]. Japanese government takes measures for preventing the solitary death of old people. However, the problems based on social isolation do not always happen old people only. Since this problem is influenced by social economics condition as recession and restructuring, related people spread out them with several back ground as out of work, students refusing to go to school as well as single parent. Although Japanese government

K. Aoki (✉)
Research Initiatives and Promotion Organization, Yokohama National University, 79-5 Tokiwadai, Hodogaya-ku, Yokohama 2408501, Japan
e-mail: k-aoki@ynu.ac.jp

K. Akai • N. Nishino
The University of Tokyo, Tokyo, Japan

T. Maeno et al. (eds.), *Serviceology for Designing the Future*,
DOI 10.1007/978-4-431-55861-3_18

tries to take measures also, a variety of back ground and working age make the measures difficult. To have solutions to the social isolation for them, it needs to take detailed measures at the private level as service industries also.

In Japan, there are services sectors as health, welfare and sports, which are about 51.8 trillion JPY in the estimated market size [5]. Since the sectors are fueled by a various demand for service recently, it is important to consider in the idea of service engineering. The service engineering has been proposed and studied so far [6, 7]. To improve efficiency and productivity in services, some characteristics such as intangibility, inseparability, heterogeneity and perishability make them difficult. Regarding an optimization problem, it is very difficult for service providers to seek the optimum solution due to heterogeneity because service receivers have a different sense of worth. Moreover, the service receivers would value service in respective various ways even if the service quality were quite the same. Therefore, service providers cannot directly use conventional engineering approaches.

There is an approach to consider service mechanism for this condition. Ueda et al. [8] show the value creation model is modeled, separating value-creating structures into three classes. Ueda et al. noted that Class I should be a situation where the values for the service provider and receiver can be specified independently and the environment can be determined in advance. Thus, a problem to be addressed in Class I is the search for the optimal solution. Next, it is described that in Class II, the environment changes and thus it makes a situation difficult. The model in Class II is said to be a system that is open to the environment, where the adaptive strategy should be addressed. In Class III, the values for the service provider and the receiver cannot be determined independently and they cannot be separated simultaneously interacting each other; the problem here to be addressed is value co-creation.

There are a few services on social isolation in the service sector. For example, a safety confirmation system is to confirm the condition of elderly people living alone by security service providers. However, the service provider on social isolation mainly depends on public service sector and the market scale is not large yet to enter it. Moreover, people related in social isolation are minority and their value is not clear in the society. Therefore, it needs to investigate the value of people in isolation problem and to search the valid class for it from results of investigating the value.

Since there are three classes in the service above and it is not clear which class is valid for the isolation problem,

The study investigates the value of belonging to community for resolving social isolation problem by using choice experiment [9]. The choice experiment is one of methods evaluated stated preference of individuals for non-market goods as risk and service. The choice experiment is estimated the willingness to pay (WTP) for goods or services. In the choice experiment, respondent are asked to select to belong to community as the decision making, which consists of two alternative and six attributes. The alternative consists of communities A and B. Six attributes consist of relation, communication, time, number of community, anonymity and cost of participant, which mean factor shown the characteristics in each community.

We use the choice experiment in internet survey and focus on people under four million JPY per annual income as respondent in the survey. The reason is that more than half of Japanese business people belong to the class under four million JPY [10]. In the survey, respondents are asked for filling into the questions on social isolation background as between jobs, out of work, single parent before the choice experiment. Then respondents make a decision in total 16 rounds. Although there are studies used the choice experiment [11–13], there few studies evaluated value of the isolation used it.

Regarding studies on the social isolation, we can go back to Aristoteles who said that human is social animal. This is, ancient people understood that it is necessary for living to connect the society. Recently, the results on previous studies found such as the orstracization is equal to bodily pain [14], as the isolation influences Alzheimer's dementia [15] and make people receive economic inequality [16]. Moreover, Takahashi et al. [17] show that Japanese cooperative behavior is supported by public eye and punishment in social psychology and Okada and Riedl [18] show that irrational reciprocity in economic condition influence to make a group. They imply that social isolation condition decrease a social character and a cooperative behavior and promote a social anxiety.

The rest of this article is organized as follows. Section 2 provide the model structure of choice experiment, Sect. 3 explains the experimental design and procedures. Section 4 analyzes the results and, Sect. 5 summarizes the conclusions.

2 Model Structure

2.1 Random Utility Model

The central idea behind choice experiment is random utility theory. The basic assumption embodied in the random utility approach to choice experiment is that decision makers are utility maximizers, which means that the decision makers will choose the alternative that maximizes his/her utility, given a set of alternatives. The utility of an alternative for individual (U) cannot be observed, however, if it could be assumed to consist of a deterministic (observable) component (V) and random error (unobservable) component (ε). Formally, individual q's utility of alternative i can be expressed as:

$$U_{iq} = V_{iq} + \varepsilon_{iq} \tag{1}$$

Hence, the probability that individual q chooses alternative i from a particular set J that comprises j alternative can be written as:

$$\begin{aligned} P_{iq} &= P\left(U_{iq} > U_{jq};\ \forall_j(\neq i) \in J\right) \\ &= P\left(\varepsilon_{jq} > \varepsilon_{iq} + V_{iq} - V_{jq};\ \forall_j(\neq i) \in J\right) \end{aligned} \tag{2}$$

To transform the random utility model into a choice experiment, certain assumptions about joint distribution of the vector of random error components is required. If random error components are assumed to follow the type I extreme value (EV1) distribution and be independently and identically distributed (IID) across alternatives and cases (or observations), conditional logit model is obtained [19]. Then, making further assumption for the deterministic component of utility to be linear and additive in parameters, $V_{iq} = \beta X_{iq}$, in the conditional logit model, the choice probability in Eq. (2) is expressed as:

$$P_{iq} = \frac{\exp\left(\mu\beta X_{iq}\right)}{\sum_{j=1}^{J} \exp\left(\mu\beta X_{jq}\right)} \tag{3}$$

where μ represents a scale parameter that determines the scale of the utility, which is proportional to the inverse of the distribution of the error components. It is typically normalized to 1.0.

2.2 Econometric Models

Based on the above model, the main effect are estimated using CL model as follows:

$$\begin{aligned} V_{iq} = {} & \beta_1 Cost_i + \beta_2 School_i + \beta_3 Neigherhood_i + \beta_4 Internet_i + \beta_5 Time_i \\ & + \beta_6 Community_i + \beta_7 Anonymity_i + e_{iq} \end{aligned} \tag{4}$$

where $Cost_i$ is the cost level of participation from the alternative i. $School_i$and$Neigherhood_i$are dummy variables indicating relations to people in the school or work place and neighborhood from the alternative i, respectively. Both baselines are relation to relative. $Internet_i$ is a dummy variable indicating using internet from the alternative i when people communicate. $Time_i$ is the time level from home to place to participate from the alternative i. $Community_i$ is the number of people structured in community level from the alternative i. $Anonymity_i$ is a dummy variable indicating anonymity condition from the alternative i. β_1, β_2, β_3, β_4, β_5, β_6, and β_7, are parameters to be estimated by independent variables of attributes, respectively.

2.3 *Willingness to Pay*

The willingness to pay (WTP) shows the expected value for attributes based on price parameter. We can directly compare WTP between attributes in models because the estimated price parameter normalizes. There are two kinds of WTP; total WTP and marginal WTP. Since our model does not include "no-purchase option" in choice sets, estimated WTP is marginal WTP. Estimated marginal WTP derived from estimated price parameter as:

$$mWTP = -\frac{\beta_k}{\beta_1}, \; k = 2, \ldots, 7 \tag{5}$$

There are six estimated mWTP because of six β_k existence. Since β_2 is a parameter of cost, mWTP on β_2 means the value from relation to people in the school or work place as compared to relation to relative. Regarding β_3, mWTP means the value from relation to neighborhood as compared to relation to relative. Regarding β_4, mWTP means the value from using internet as compared to meeting someone directly. Regarding β_5, mWTP means the value from time. Regarding β_6, mWTP means the value from the number of people structured in community. Regarding β_7, mWTP means the value from anonymity.

3 Design

3.1 *Sample*

The data sample is gathered by Cross Marketing Inc. and conducted the internet survey in Japan, 2012. Since the study investigate the preference for people isolated from the society, we focus on people with the following condition as shown in Table 1.

3.2 *Design*

Here we show the detailed design of the choice experiment. Each alternative consists of six attributes as relation, communication, time, number of community, anonymity and cost of participant. Table 2 shows the definition and levels in each attribution. Table 3 shows an example of the designated choice sets and three alternative community types, A and B.

Table 1 The structure of isolation group

	Item	Sample
1	People who are between jobs because of out of work	58
2	People are out of work but not between jobs	26
3	People are welfare recipients	42
4	People are applying for welfare	4
5	People who had apply for welfare until 3 years ago	9
6	Dependent children of single female (male) parent	21
7	single female (male) parent with dependent children	38
8	Students refusing to go to school	4
9	People who withdraw from society	54
	Total participants	200

Table 2 Attributes and levels

Attribute	Levels
Cost of participation (JPY)	0, 1000, 2000, 3000
Relation	Relative, People in the school or work place, Neighborhood,
Communication methods	Internet, Real
Time (minutes)	0, 30, 60
Number of people structured in community (person)	5, 15, 25
Anonymity	Yes, No

Table 3 An example of choice set

Alternative	Community A	Community B
Relation	Relative	Neighborhood
Communication methods	Internet	Real
Time	30 (minutes)	0 (minutes)
Number of people structured in community	15 (persons)	25 (persons)
Anonymity	Yes	Yes
Cost of participation	2000 (JPY)	1000 (JPY)
I would choose...		

4 Results

At first, we concern on the hypothesis of equal utility parameters among each sub-sample and the pooled samples as isolation group, non-isolation group and pooled sample both groups. The null hypotheses can be formally stated as $\beta^{pooled} = \beta^{isolation} = \beta^{non-isolation}$. The statistics of the likelihood ratio test suggested by Swait and Louviere (1993) are as LR = −2[−2868.06-(−1926.8-934.08)] = 14.2. The values in the LR test are used the CL estimation results in Model

1 from LIMDEP 9.0 and NLOGIT 4.0. The critical value of the χ^2 distribution is 5.991 at the 5 % significance level on 2 of freedom. Thus, the hypothesis that the vector of common utility parameters is equal across sub-samples can be rejected in all cases (p-value < 0.00). We explain the estimation results in each group as below.

4.1 Analysis of Main Effect

Table 4 shows the estimation results of main effect. Regarding the isolation group, the variables Cost, Neighborhood, Time and Community are significant and negative signs. They imply that people prefer low cost to participate, relation to relative as compared to neighborhood, nearby place and small group. The variable Anonymity is significant and positive sign. This implies that people prefer anonymity condition. Moreover, since the most number of parameter in the model is that on the variable Anonymity and this variable is a significant, the anonymity condition most influence isolation group when they belong to community. On the other hand, the variables Cost, Time and Community in the non-isolation group are significant and negative signs. They imply that people prefer low cost to participate, nearby place and small group. Since the most number of parameter in the model is that on the variable School but this variable is not significant, it is not clear that the relation to people in the school or work place most influence the non-isolation group when they belong to community.

Table 4 Estimation results in Model 1

Variables	Isolation	Non-isolation
Cost	−0.0005***	−0.0006***
	(0.00)	(0.00)
School[a]	0.02	0.16
	(0.07)	(0.1)
Neighborhood,[a]	−0.18***	−0.1
	(0.06)	(0.09)
Internet	0.08	0.01
	(0.06)	(0.08)
Time	−0.009***	−0.01***
	(0.00)	(0.00)
Community	−0.02***	−0.02***
	(0.00)	(0.00)
Anonymity	0.38***	0.12
	(0.05)	(0.07)
Log-likelihood	−1926.8	−934.0
Pseudo-R^2	0.13	0.15
Observation	3200	1600
# of participants	200	100

Notes: Standard errors are in parentheses. *** denotes 1 % significance levels. [a]shows the dummy variables. The baseline is relative

4.2 WTP for the Isolation

To consider both of groups detailedly, we use the mWTP in each group. In the isolation group, mWTPs are calculated by parameter results in significant variables as follows: 360 JPY in relation to neighborhood, 18 JPY per minute in time, 40 JPY per person in the number of people structured in the community and 760 JPY in anonymity. Therefore, the isolation group set high value to relation to relative as compared to that to neighborhood even though they further pay 360 JPY. As same interpretation, the isolation group set high value to decrease time and the number of person structured in community even though they further pay 18 JPY per minute and 40 JPY per person, respectively. Regarding anonymity condition, the isolation group has willingness to pay as 760 JPY when they can be anonymity one. On the other hand, mWTP in the non-isolation group are calculated as follows: 16.6 JPY in time and 33.3 JPY in the community. Therefore, the non-isolation group set high value to decrease time and the number of person structured in community even though they further pay 16.6 JPY per minute and 33.3 JPY per person, respectively.

In summary, the isolation group set higher value on condition structured a shallow relationship than that in non-isolation group when they belong to community.

5 Conclusions

The study investigates the value of belonging to community for resolving social isolation problem. In the analysis used the estimation model, sample are divided into two groups as isolation group and non-isolation group. The isolation group consists of people who are between jobs, out of work, single parent and withdrawing from society. On the other hand, the non-isolation group consists of others. The estimation results show that the isolation group intends to prefer less relation to people who are not family and anonymity condition as compared to the non-isolation group. Regarding interaction results with socioeconomics background, in the isolation group, female intend to have a consciousness on the isolation environment as the number of people structured in the community and anonymity and single and young people intend to be isolated from people excluding relatives. On the other hand, single and young people intend to be isolated from people in the school or work place as compared to relatives.

Since the study found the environment and socioeconomic characteristics isolated, we will consider which class is valid for the isolation problem. Considering the solution of the isolation problem as service sector, it is possible that the interaction class as III has solutions to it. The reasons are that representative individual isolated prefers being closed. Moreover we imply that the community which he belongs is not rich in diversity as a kind of members and how to think. Although the problem in Class III is value co-creation, we consider that important thing is to interact with people excluding relatives also for solution of the isolation problem. For example, it may be good to talk with someone in the not anonymity condition as telephone.

Acknowledgements This study was supported by Zenrousai, Service Science, Solutions and Foundation Integrated Research Program (S3FIRE), Research Institute of Science and Technology for Society (RISTEX), and Japan Science and Technology Agency (JST).

References

1. OECD Social Indicators (2005) Social at a glance 2005. http://www.oecd-ilibrary.org/social-issues-migration-health/society-at-a-glance-2005_soc_glance-2005-en
2. Cabinet Office, Government of Japan (2008) Annual lifestyle white paper. http://www5.cao.go.jp/seikatsu/whitepaper/h20/10_pdf/01_honpen/index.html
3. Hirose K (2012) Risk for soritary death (Kodokushi no risuku to mukiau). Gerontol J 25:11–14 (in Japanese)
4. Ministry of internal Affairs and Communications (2013) http://www.soumu.go.jp/main_content/000217416.pdf
5. Ministry of Economy, Trade and Industry (2006) http://www.meti.go.jp/committee/materials/downloadfiles/g60512b04-2j.pdf
6. Arai T, Shimomura Y (2004) Proposal of service CAD system – a tool for service engineering. CIRP Ann Manuf Technol 53(1):397–400
7. Arai T, Shimomura Y (2005) Service CAD system – evaluation and quantification. CIRP Ann Manuf Technol 54(1):463–466
8. Ueda K, Takenaka T, Váncza J, Monostori L (2009) Value creation and decision-making in sustainable society. CIRP Ann Manuf Technol 58(2):681–700
9. Louviere JJ, Hensher DA, Swait JD (2000) Stated choice methods: analysis and application. Cambridge University Press, Cambridge/New York
10. National Tax Agency (2012) Fact finding survey for salary in private companies. http://www.nta.go.jp/kohyo/tokei/kokuzeicho/minkan2012/pdf/001.pdf
11. Caussade S, Ortúzar JD, Rizzi LI, Hensher DA (2005) Assessing the influence of design dimensions on stated choice experiment estimates. Transp Res B Methodol 39(7):621–640
12. Loureiro ML, Umberger WJ (2007) A choice experiment model for beef: what US consumer responses tell us about relative preferences for food safety, country-of-origin labeling and traceability. Food Policy 32(4):496–514
13. Birol E, Karousakis K, Koundouri P (2006) Using a choice experiment to account for preference heterogeneity in wetland attributes: The case of Cheimaditida wetland in Greece. Ecol Econ 60(1):145–156
14. Eisenberger NI, Lieberman MD, Kipling KD (2003) Does rejection hurt? An fMRI study of social exclusion. Science 302(5643):290–292
15. Wilson RS, Krueger KR, Arnold SE, Schneider JA, Kelly JF, Barnes LL, Tang Y, Bennett DA (2007) Loneliness and risk of Alzheimer disease. Arch Gen Psychiatry 64(2):234
16. Cacioppo JT, Pattrick W (2009) Loneliness : human nature and the need for social connection. W. W. Norton & Company, New York, Reprint edition
17. Takahashi T, Yamagishi T, Hashimoto H (2009) Interdependent self as a form of self-presentation in response to a threat of exclusion from the group. Jpn J Soc Psychol 25 (2):113–120 (in Japanese)
18. Okada A, Riedl A (2005) Inefficiency and social exclusion in a coalition. Games Econ Behav 50:278–311
19. McFadden D (1974) Conditional logit analysis of qualitative choice behavior. In: Zarembka P (ed) Frontiers in econometrics. Academic, New York, pp 105–142

Basic Study of Mobility of Elderly People from the Perspective of Their Emotional Value

Ryoichi Tamura, Yasuyuki Hirai, and Nermin Elokla

Abstract This study investigated the needs of the elderly in terms of sensibility with respect to services relating to mobility. The proportion of elderly individuals in the population has recently been increasing, and most of them are healthy. To date, the welfare of the elderly has been a primary concern, but goods and services that are appealing in terms of sensibility have not been developed. In this study, 24 active elderly individuals were interviewed. Individuals mentioned causes of a positive impression or a negative impression from past experiences of using 7 types of public transportation, including bus, train and airplane. The content was classified based on the 16 items of the Reiss Desire Profile, and relationships with the transportation types, steps and causes of a positive impression or a negative impression were investigated. The results of this study are expected to be helpful for improving the quality of mobility for the elderly.

Keywords Emotional value • Elderly people • Mobility

1 Introduction

The proportion of elderly individuals in the population has recently been increasing, and most of them are healthy. To date, the welfare of the elderly has been of primary concern, but goods and services that appeal to the specific sensibilities of the elderly have not been developed. Here, to provide a starting point for research on this topic, we focused on how mobility is related to quality of life in elderly people.

This study seeks to obtain valuable basic information for considering service design from the perspective of emotional value of the elderly, through investigation

R. Tamura (✉) • Y. Hirai
Kyushu University, 4-9-1, Shiobaru Minami-ku, Fukuoka 815-8540, Japan
e-mail: tamura@design.kyushu-u.ac.jp

N. Elokla
Helwan University, Cairo, Egypt

T. Maeno et al. (eds.), *Serviceology for Designing the Future*,
DOI 10.1007/978-4-431-55861-3_19

and analysis of cases in which elderly people obtained a positive impression or a negative impression after taking public transportation.

The present study is part of our continuous study, 'Study on Investigation of User Satisfaction after Service Implementation about Public Transportation [1]'.

2 Changes in Aging and the Lifestyle of the Elderly

According to research conducted by the Ministry of Internal Affairs and Communications, as of September 2008, the proportion of individuals over 70 years old stood at 15.8 % of the total population, or roughly 1 in 6 people. As shown in Fig. 1, this is predicted to increase to 19.3 % by 2015, 24.6 % by 2025, and 30.8 % by 2045.

To date, elderly people, such as those in need of nursing care, have been considered to be passive members of society due to the stigma surrounding welfare [2]. However, according to comparisons in leisure consumption by age noted in the White Paper on Leisure 2008 by the Japan Productive Center, the elderly have become the leading age group in leisure consumption, a position previously held by the young generation. Consumption related to travel in particular has increased for individuals over 70 [3]. Furthermore, due to reports that seniors with free time and money appear to prefer consuming services instead of products, 47 % of travel agencies have elected to increase their services aimed at individuals in their 60s [4].

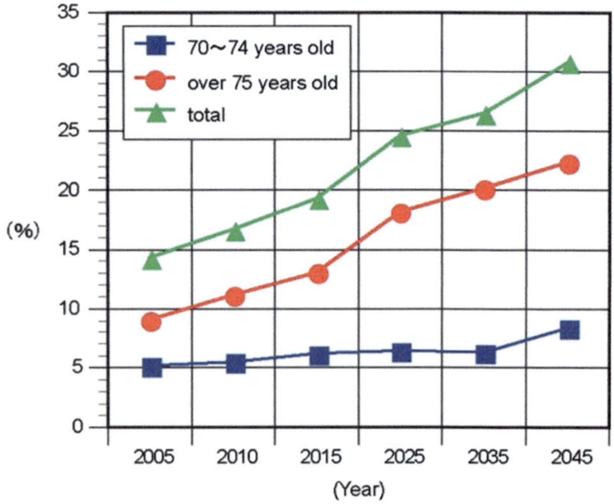

Fig. 1 Changes in aging

Considering this lifestyle of the active elderly, the above reasons underscore the significance of considering public transportation from the viewpoint of their emotional needs.

3 How to Capture Service

3.1 Constituent Elements of Service Processes

Based on our previous study [1], we identify five of the main constituent elements of service processes, namely, (1) the service provider (hereinafter, 'the provider'), (2) the medium through which the service is provided ('the medium'), (3) the user of service ('the user'), (4) the environment in which the service is provided ('the environment'), and (5) the exchanges between the medium through which the service is provided and the user of the service ('exchanges'). The relationship between these constituent elements can be expressed as shown in Fig. 2.

3.2 Steps of Service Processes

Based on our previous study [1], we have decided to take a step to be "a state where there is no change in the three constituent elements of (2) 'medium', (4) 'environment' and (5) 'exchanges', excluding the other two constituent elements, namely, element (1) 'provider' (changes in which are not directly relevant to the user) and element (3) 'user' (which is not relevant when focusing on a particular user). In other words, as shown in Fig. 3, we decided to view a service process as having progressed to the next step when any of following took place: for element (2) 'medium', a change to the 'person or thing' providing the service; for element (4) 'environment', a change to the location where the service is provided; for element (5) 'exchanges', an interruption to a sequence of exchanges.

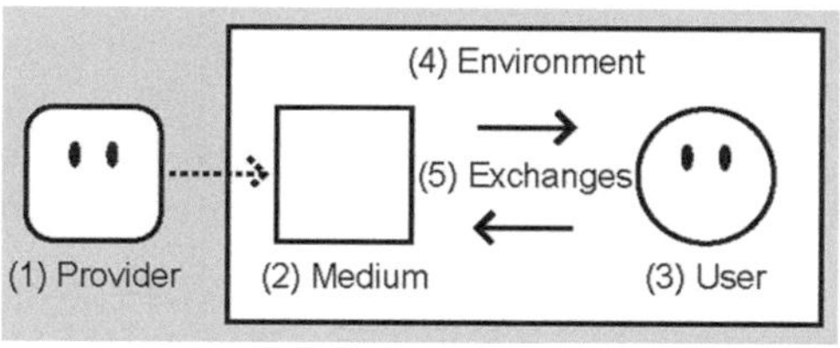

Fig. 2 Relationship between main constituent elements of service processes

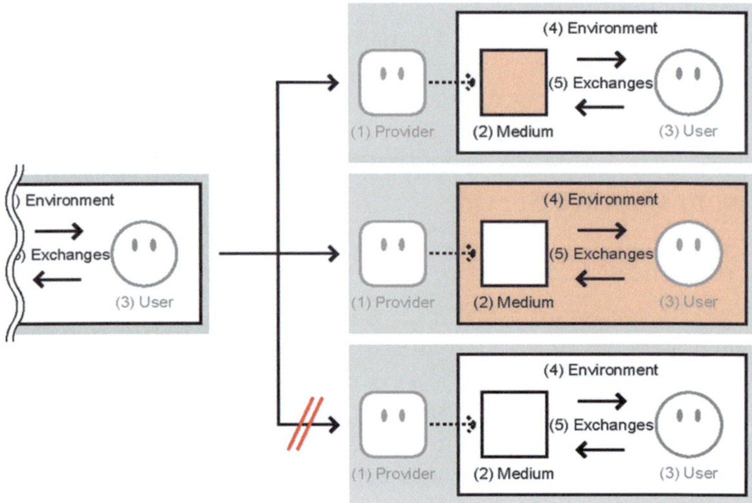

Fig. 3 Three kinds of capturing steps

4 Survey

4.1 Selecting Types of Public Transportation for Investigation

We examined 7 types of public transportation: 3 types of public transportation taken in the course of daily life (taxi, bus, and train), and 4 types taken for longer-distance travel, such as for holidays (express bus, bullet train, airplane, and ship).

4.2 Participants and Interview Survey

A total of 24 active elderly people who belonged to the National Human Resources Center Association [5] and who were over 65 years old, comprising 12 men (average: 74.7 years old, standard deviation: 3.2 years) and 12 women (average: 70.1 years old standard deviation: 4.1 years), were selected randomly.

They were interviewed in January or February 2014. Interviews lasted for approximately 30 min. Questions concerned cases within the past 2 years in which participants were left with a positive or negative impression after utilizing the 7 investigated types of public transportation, and participants were asked to explain the feelings, as well as the steps in the service process, that related to their positive or negative impressions. Participants were asked to respond with their first impression even if they found their feelings somewhat hard to describe in detail.

4.3 *Results*

4.3.1 Types of Public Transportation for Analysis

Interviews yielded cases from 5 people for taxis (3 men and 2 women), 19 people for buses (9 men and 10 women), 13 people for trains (5 men and 8 women), 2 people for express buses (2 women), 3 people for bullet trains (2 men and 1 woman), 7 people for airplanes (4 men and 3 women), and 1 person for ships (1 man) (Table 1).

Because a large variation in the number of responses was observed, we considered only the modes of public transportation for which cases could be collected from over about 30 % of subjects, namely taxis, buses, trains, and airplanes.

4.3.2 Cases for Analysis

The cases that we were able to collect from the 4 types of public transportation selected as objects for investigation included some unsuitable cases, such as those that corresponded to steps caused by other users, who are not included in 'the medium' category. Therefore, these cases were excluded.

As a result, this left the following cases for analysis: a total of 148 objects from 40 people, comprising 5 people for taxis (14 cases), 18 people for buses (92 cases), 10 people for trains (30 cases), and 7 people for airplanes (12 cases) were considered (Table 2).

Table 1 Number of people

	Total	Man	Woman
Taxis	5	3	2
Buses	19	9	10
Trains	13	5	8
Express buses	2	0	2
Bullet trains	3	2	1
Airplanes	7	4	3
Ships	1	1	0

Table 2 Number of people and cases for analysis

	Number of people	Number of cases
Taxis	5	14
Buses	18	92
Trains	10	30
Airplanes	7	12
Total	40	148

5 Cases Where Users Had Positive or Negative Impressions

5.1 Relationship to Desires

5.1.1 Types of Desires

Based on our previous study [1], we considered the responses according to Steven Reiss' profile of 16 basic desires (Table 3).

5.1.2 Analysis

We classified collected cases based on Reiss Profile.

In cases where users had a positive impression, we were able to identify 4 types of desires that were satisfied when using taxies (e.g., 'acceptance'), 6 types when using buses (e.g., 'independence'), 4 types when using trains (e.g., 'saving'), and 3 types when using airplanes (e.g., 'eating') (Table 4).

Each of the 4 types of public transportation satisfied users' desire for 'tranquillity'; 3 types for 'saving' and 'social status'; 2 types for 'acceptance', 'independence', and 'social contact'; and 1 type for 'eating'. Although there was variation in the proportion of responses for each desire, and desires were present in multiple types of public transport, these findings suggest that it is advisable to consider the content of services to satisfy users' desire for 'tranquillity' when designing services for the elderly. The remaining 9 desires—'curiosity', 'family', 'honour', 'idealism', 'order', 'physical activity', 'power', 'romance', and 'vengeance'—were not present in the 4 types of public transportation; however, it is possible that future services could be designed to satisfy some of these desires.

As shown in Table 5, in cases where users had a negative impression, we were able to identify 4 types of desires that were not satisfied when using taxies (e.g., 'order'), 3 types when using buses (e.g., 'tranquillity'), 2 types when using trains (e.g., 'tranquillity'), and 4 types when using airplanes (e.g., 'order').

Desire for 'order' and 'tranquillity' was unsatisfied in all 4 types of public transportation and was reported by a high proportion of participants. This suggests that it will be necessary to reconsider the current content of services that do not adequately satisfy users' desire for 'order' and 'tranquillity'.

Table 3 Stephen Reiss's profile of 16 basic desires

1	Acceptance	2	Curiosity	3	Eating	4	Family
5	Honour	6	Idealism	7	Independence	8	Order
9	Physical activity	10	Power	11	Romance	12	Saving
13	Social contact	14	Social status	15	Tranquillity	16	Vengeance

Table 4 Types of public transportation and satisfied desires related to positive impression

	1	2	3	4	5	6	7	8	9	10	11	12	13	14	15	16
	Acceptance	Curiosity	Eating	Family	Honour	Idealism	Independence	Order	Physical activity	Power	Romance	Saving	Social contact	Social status	Tranquillity	Vengeance
Taxis	3												1	2	1	
	42.9												14.3	28.6	14.3	
Buses	12						19					10	2	2	16	
	19.7						31.1					16.4	3.3	3.3	26.2	
Trains							9					9		1	2	
							42.9					42.9		4.8	9.5	
Airplanes			1									2			1	
			25.0									50.0			25.0	

Upper: number of cases, lower: %

Table 5 Types of public transportation and unsatisfied desires related to negative impression

	1	2	3	4	5	6	7	8	9	10	11	12	13	14	15	16
	Acceptance	Curiosity	Eating	Family	Honour	Idealism	Independence	Order	Physical activity	Power	Romance	Saving	Social contact	Social status	Tranquillity	Vengeance
Taxis								4					1	1	1	
								51.7					14.3	14.3	14.3	
Buses							4	8							19	
							12.9	25.8							61.3	
Trains								1							8	
								11.1							88.9	
Airplanes							1	4						1	2	
							12.5	50.0						12.5	25.0	

Upper: number of cases, lower: %

5.2 *Relationship to Service Processes*

5.2.1 Types of Steps in the Service Process

Based on our previous study [1], we classified 5 steps of the service process for taxies, 4 steps for buses, 6 steps for trains, and 8 steps for airplanes (Table 6).

We focus on the core service, which constitutes only the 'travel' step, as well as steps before and after ('prior steps' and 'subsequent steps', respectively).

5.2.2 Analysis

We considered the relationship between the cases obtained and steps in each type of public transportation.

In cases where users had a positive impression, we were able to identify steps that led to this positive impression: 4 steps for taxies, all 4 steps for buses, 4 steps for trains, and 3 for airplanes (Table 7). In all 4 types of public transportation, the first step, 'embark' for taxies, 'embark' for buses, 'ticket purchase' for trains, and 'reservation/payment' for airplanes, was relevant to positive impression.

Although there were cases that corresponded to nearly all steps for taxies, buses, and trains, positive impression corresponded to only a few steps for airplanes, suggesting that it is necessary to design services to promote positive impression at those steps. On the other hand, for taxies, buses, and trains, we found a high number of cases that corresponded not to the travel step but instead to prior steps and subsequent steps, as well as a large bias in responses, suggesting that it is necessary to improve the contents of the services related to the travel step.

As shown in Table 8, in cases where users had a negative impression, we were able to identify steps that led to this negative impression' 2 steps for taxies, 3 steps for buses, 3 steps for trains, and 4 steps for airplanes.

There was a high ratio of cases where the travel step corresponded to users' negative impression for all 4 types of public transportation. The travel step should therefore be immediately considered when designing the contents of services. For 'trains' and 'airplanes', some steps were related to only negative impression, and never to positive impression. The contents of services of such steps must be reconsidered as well.

5.3 *Relationship to Causes*

5.3.1 Types of Causes

Based on our previous study [1], we divided causes into 4 types: human factors, material factors, environmental factors, and system factors.

Table 6 Steps of the 4 types of public transportation

Taxis	Step1	Step2	Step3	Step4	Step5	
	Embark	Destination	Travel	Payment	Disembark	
Buses	Step1	Step2	Step3	Step4		
	Embark	Travel	Payment	Disembark		
Trains	Step1	Step2	Step3	Step4	Step5	Step6
	Purchase ticket	Ticket inspection	Embark	Travel	Disembark	Ticket inspection
Airplanes	Step1	Step2	Step3	Step4	Step5	Step6
	Reservation → Payment	Check-in	Baggage check-in	Security check	Board	Travel
	Step7	Step8				
	Disembark	Baggage pickup				

Table 7 Types of public transportation and steps related to positive impression

	Step1	Step2	Step3	Step4	Step5	Step6	Step7	Step8
Taxis	2	2	1		2			
	28.6	28.6	14.3		28.6			
Buses	24	11	10	16				
	39.3	18.0	16.4	26.2				
Trains	8	6		1		6		
	38.1	28.6		4.8		28.6		
Airplanes	2					1		1
	50.0					25.0		25.0

Upper: number of cases, lower: %

Table 8 Types of public transportation and steps related to negative impression

	Step1	Step2	Step3	Step4	Step5	Step6	Step7	Step8
Taxis		2	5					
		28.6	71.4					
Buses	6	21		4				
	19.4	67.7		12.9				
Trains			3	3	3			
			33.3	33.3	33.3			
Airplanes		2		1	1			4
		25.0		12.5	12.5			50.0

Upper: number of cases, lower: %

5.3.2 Analysis

We examined the relationship between the collected cases and their causes as they related to 4 types of public transportation.

In cases where users had a positive impression, 1 type of cause related to this impression was found for taxies ('human factors'), all 4 types for buses, 3 types for trains (e.g., 'material factors'), and 3 types for airplanes (e.g., 'material factors') (Table 9).

These causes reflected the characteristics of each type of transportation, but it is also possible that services could be proposed to address non-related factors as well.

As shown in Table 10, in cases where users had a negative impression, 2 types of factors related to this impression were found for taxies (e.g., 'human factors'), 3 types for buses (e.g., 'human factors'), 2 types for trains (e.g., 'material factors'), and 1 type of factor for airplanes (e.g., 'system factors').

Similarly to the case for positive impression the results were taken to reflect the characteristics of each type of transportation. There were some factors that did not correspond to the cases in where users had positive impression, suggesting that the contents of services should be reconsidered accordingly.

Table 9 Types of public transportation and causes of positive impression

	1	2	3	4
	Human factors	Material factors	Environmental factors	System factors
Taxis	7			
	100.0			
Buses	33	5	3	20
	54.1	8.2	4.9	32.8
Trains	2	1		18
	9.5	4.8		85.7
Airplanes		1	1	2
		25.0	25.0	50.0

Upper: number of cases, lower: %

Table 10 Types of public transportation and causes of negative impression

	1	2	3	4
	Human factors	Material factors	Environmental factors	System factors
Taxis	6	1		
	85.7	14.3		
Buses	4	15	12	
	12.9	48.4	38.7	
Trains		3	6	
		33.3	66.7	
Airplanes				8
				100.0

Upper: number of cases, lower: %

6 Conclusion

As a starting point for research on elderly people's quality of life, we defined public transportation to be a mobility-related service related to quality of life, and considered services provided by public transportation from the viewpoint of elderly people's emotional value.

The present study revealed the kinds of desire, the steps of the service processes, and the types of causes related to cases where elderly people had positive or negative impressions after using various types of public transportation. Based on the results, we laid out the general matters that will be important to future service design for elderly people.

In the future, it will be necessary to conduct in-depth analysis with a more limited number of subjects and focusing on individual forms of public transportation, to further refine the result of the present research.

Acknowledgement This work was supported by JSPS KAKENHI (23611023).

References

1. Tamura R, Yoshioka H (2013) Study on investigation of user satisfaction after service implementation about public transportation. In: International conference on serviceology. AIST Tokyo Waterfront, Tokyo
2. Phil-Hwa JO (2000) Study on the quality of life among the elderly-concepts and theoretical framework. Policy Sci 8(1):117–132
3. The healthy elderly people over 70 years old, Nissui, http://www.nissui.co.jp/academy/data/08/. Accessed 15 Oct 2013
4. The healthy elderly people? Driving force of service industry, Nihon Keizai Shimbun, Inc., 2012/11/7. http://www.nikkci.com/article/DGXNASDC06007_W2A101C1EA2000/. Accessed 15 Oct 2013
5. An organization that offers chances to get something to live through working for the elderly people, and contributes to the revitalization of local communities. It is set up a municipal unit as a general rule

Workshop-Based Policy Platform for Public-Private Partnership (WP5): Designing Co-creative Policy-Making Platform for Regional Development of *Nagano*

Toshiyuki Yasui, Takashi Maeno, Seiko Shirasaka, Yoshikazu Tomita, and Kanenori Ishibashi

Abstract Most of local governments are now facing citizens' disillusion to government services. This illusion is mainly coming from their unmatched desires for co-creative policy-making process. Corresponding to the recent paradigm shift of service design from the value exchange model to the value co-creation model, local governments are pressed to develop co-designing policy platform with private and non-profit stakeholders for co-producing better government services. This paper is to propose the model of Workshop-based Policy Platform for Public-Private Partnership (WP5 Model), a co-creative policy-making platform for regional development based upon both system thinking and design thinking. The authors qualitatively and quantitatively validated efficacy of the WP5 Model with two cases of *Nagano* local government workshops in October 2013 and in March 2014, respectively.

Keywords Government service • Regional development • Co-creation • System x design thinking • Workshop

1 Introduction: Toward Co-creative Policy

1.1 Problem

In many industrialized country, governments face frustrated and disillusioned taxpayers because of poor quality of government services delivered to taxpayers [1, 2]. Major source of frustration and disillusion is coming from the lack of participation in policy-making process in governments [3].

T. Yasui (✉) • S. Shirasaka • Y. Tomita • K. Ishibashi
Graduate School of System Design and Management, Keio University, Collaboration Complex, 4-1-1 Hiyoshi, Kohoku-ku, Yokohama City, Kanagawa Prefecture 223-8526, Japan
e-mail: t.yasui@z2.keio.jp

T. Maeno
Graduate School of System Design and Management, Keio University, Yokohama 223-0061, Japan

T. Maeno et al. (eds.), *Serviceology for Designing the Future*,
DOI 10.1007/978-4-431-55861-3_20

Addressing to taxpayers' quest of participation in policy-making process, there has recently emerged several concepts on how government provide services in participatory way. Some concepts drew attentions of the academic circle and government officials such as; the participatory budgeting [4], new public governance [5], and open government [6, 7]. However, taxpayers' quest of participation never ceased particularly in policy-making site of local governements since the previous concepts contain little elements of co-creative design of policy with goverment officials and other local stakeholders [8].

1.2 Purpose and Perspective

This paper is to build a co-creative and co-designing model for collective policy-making at local government level with multi-stakeholders in regions, and to validate qualitatively and quantitatively efficacy of the model to create innovative regional development policy with diversified stakeholders. The Workshop-based Policy Platform for Public Partnership Model (WP5 Model), proposed in this paper, is expected to contrubute to enhance better government service provisions for local area.

The WP5 Model reflects the recent paradigm shift of service theory. The serviceology is observing that major qualified services are now shifting from the model of value exchange to the model of value co-creation [9, 10].

Paralleling to this paradigm shift of service design, the model of government service is also shifting from the new public management (NPM) [11], the exchanging value-based model to more co-creating network-based model [12, 13]. The WP5 Model is proposed as a concrete policy-design tool to materialize this paradigm shift of government service design to the era of co-creation.

The WP5 Model is relevant to be applied for policy-design when a local government plans to create innovative regional development policy with local stakeholders from private and non-profit sectors, because the face-to-face workshop is embedded in the model for architecting *'ba',* namely an organic grounds for better knowledge creation [14] in order to accommodate diversified policy ideas into a innovative policy solution.

2 Conceptual Background: System x Design Thinking

2.1 System x Design Thinking

The WP5 Model is conceptually rooted upon the system x design thinking, a hybrid methodology to formulate an innovative solution for social and business problem [15]. The system x design thinking emerged from project-based learning experiences in the Graduate School of System Design and Management of Keio University from 2008 to 2013 [16, 17].

The system x design thinking inherited two major methodologies for solving a real-world problem by socially implementing a better solution; system thinking [18, 19] and design thinking [20, 21].

2.2 *Innovation Cycles of System x Design Thinking*

The system x design thinking has cycled sequence for reaching to innovative solution. A team of stakeholders iterates to trail like a loop three dimensions until break-through; observation, ideation and prototyping after they start to co-design until they will gain breakthrough innovation (see Fig. 1).

Observation stage is for identifying and sharing a problem to which stakeholders want to address collectively. Ideation stage is for systematic ideation of solutions and their systemic visualization. Prototyping is for empathy and reflection of the early-stage solutions, as well as their verification and validation.

The system x design thinking frequently utilizes the workshop as human-centred platform to stimulate collectively creativity for innovative design. Workshop represents a standard way as group-setting to enhance high-performance work by teams in the field of management study [22]. And work by teams in itself is known to have remarkable impact to increase intellectual performance by inspiring collective intelligence [23]. Thus the system x design thinking embeds a workshop to let team participants pop up collective intelligence.

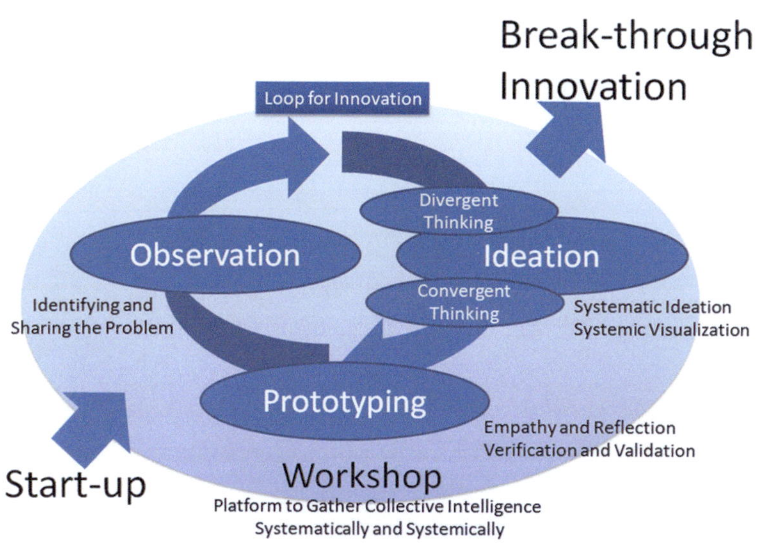

Fig. 1 Cycled sequence of the system x design thinking

3 Architercture: WP5 Model

3.1 *Five Phases of WP5 Model*

The WP5 model is an application for policy-dialogue tool relied on the system x design thinking. It has five phases to co-create an innovative policy of regional development; start-up, policy observation, policy ideation, policy prototyping, and innovative policy formulation (see Fig. 2). The whole phases are implemented in most cases through a workshop designed for creating certain regional policy agenda.

3.1.1 Start-Up

Local government officials working for scattered sectors, regional organizers, NPO officials, corporate workers and retired citizens gather at one place for teaming-up.

They have a single purpose to be there for formulating collectively an innovative regional policy. Participants typically organize themselves into groups of 5–6 persons.

3.1.2 Policy Observation

A facilitator of WP5 Model invites groups of participants to go outdoors to implement fieldwork and interview with town residents.

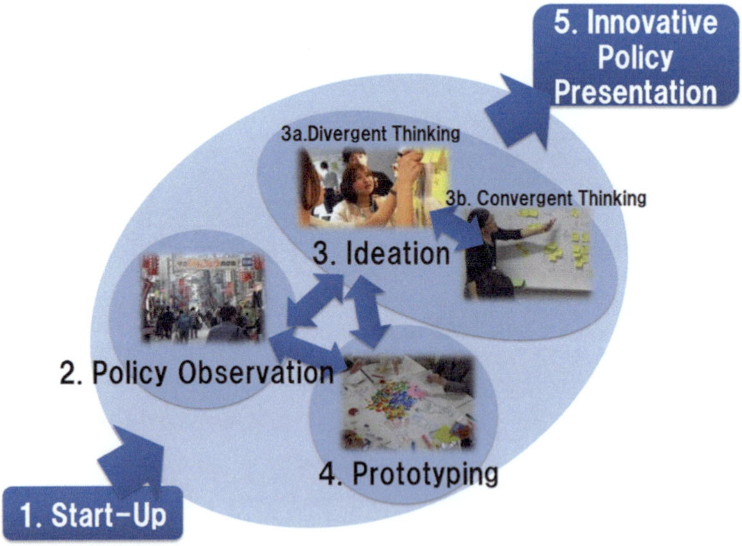

Fig. 2 Five phases of MP5 model

By ethnography and research, participants as teams share problems and identifying gravity and width of the problems which they observed in their fieldwork and interview.

3.1.3 Policy Ideation

The teamed participants organize with facilitator an ideation meeting. This phase is composed of two sub-phases; divergent thinking phase to enlarge solution-space and to produce flexible policy ideas as many as possible; and convergent thinking phase to shift and structure into a solid and truly innovative idea.

3.1.4 Policy Prototyping

Based upon a structured policy idea, teams make rapidly a prototype of policy solution to visualize and share it with local residents. After teams get feedbacks of their policy prototypes, they decide whether they will return to the observation phase again or exit to the presentation phase.

3.1.5 Innovative Policy Presentation

When participants feel the right time to exit the iterative three phases of policy observation, policy ideation and policy prototyping, they come out of workshop and make an innovative policy presentation for decision-maker of that policy. Typical decision-makers include mayor, prefectural governor, Member of Parliament, city manager, and executive officer of local government.

3.2 Toolbox Approach

The iterative three phases of WP5 Model adopt the toolbox approach for tools to be used in those phases. There is no one-by-one fixed tool to specific phase, but rather participants in workshop can select their suitable tool among candidate tools in each phase like the system x design thinking workshop [15].

Table 1 shows the toolbox of representative tools which can be used in each phase of policy observation, policy-ideation (divergent phase and convergent phase respectively), and policy prototyping.

Table 1 Toolbox of WP5 model

Phase	Representative tools (For example)
3.1.2 Policy Observation	Fieldwork, Ethnography, Interview, Intensive Research, Data-Mining
3.1.3 a Policy Ideation: Divergent Thinking	Brainstorming, Planned Cell, Brain Writing, Value Graph, Scenario Graph
3.1.3 b Policy Ideation Convergent Thinking	Affinity Diagram, Structural Shift Ideation, Enabler Framework, Causal Relation Diagram, CVCA, WCA, Pugh Concept Selection
3.1.4 Policy Prototyping	Drawing, Painting, Playback Theatre, Improvised Story Play, Storytelling

Note: This table provides categorization of the most typical case that certain tool is used, but not necessarily means that that tool should only for designed phase in this table. Tools may overlap each other and shift their positions among phases when they are used in different phases from one indicated in this table

4 Social Experiments: *Nagano* Workshop

4.1 Case: WP5 Workshops in Nagano

The authors conducted as social experiments two WP5 Model workshops on October the 22nd, 2013 and on March 8th 2014 respectively in *Nagano* city, a prefectural capital of *Nagano* prefecture (see Fig. 3), to validate empirically efficacy of WP5 Model.

The reason why the authors chose *Nagano* city as the field for social experiments of WP5 Model depended on three crucial conditions that that city is facing. *Nagano* city is thought as one of leading cities in Japan which are experiencing social problems which the entire country of Japan will face severely in the near future.

Firstly, the down-town shopping district of Nagano City is experiencing severe aging and fewer population trends. The population of that district at 2007 decreased to three-fourth of the population of 1980 [24]. On the contrary, senior population of that district is increasing beyond 25 % of the total population.

Secondly, most of local governments in *Nagano* prefecture including *Nagano* city deteriorate fiscal conditions. For example, according to the Ministry of General Affairs of Japanese Government, Nagano City's fiscal strength index is 0.68 (FY2012) [25], which is relatively worse number as prefectural capitals. This number is indicating fiscal fragility of *Nagano* city.

Thirdly, while some social indicators are dauntingly severe, *Nagano* City is one of centres in Japan for innovation movement of government services. In *Nagano* prefecture several local governments such as *Obuse* Town and *Shimojo* village are promoting advanced model of regional re-vitalization. Public-private-non-profit collaboration is quite active in *Nagano* prefecture. For example, some junior level officials of *Nagano* prefectural government organized *Shinshu* Innovation Project (SHIP) to innovate government services of their government toward more co-creative way [26].

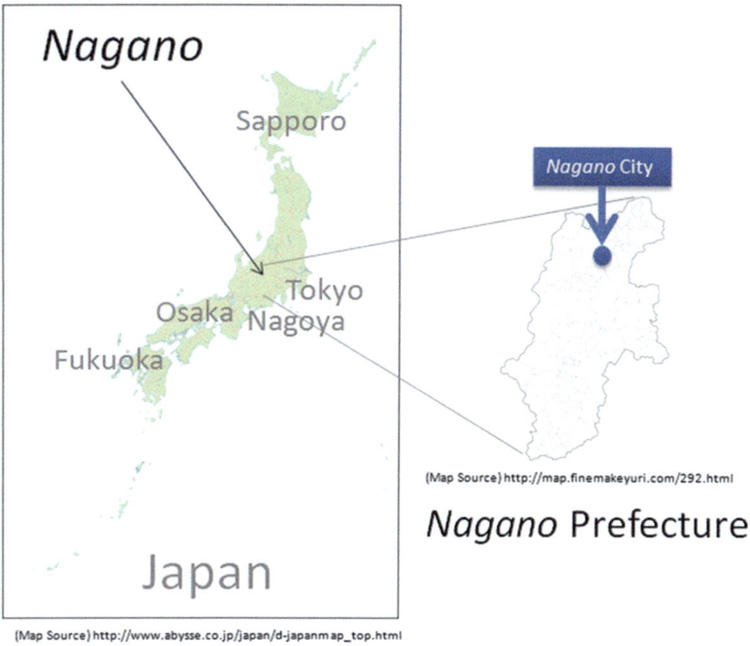

Fig. 3 Geographic location of *Nagano* city

4.2 WP5 Workshops Participants

The topics of two WP5 workshops were the same; the better policy design for regional revitalization of *Nagano*. In the first workshop in October 2013, 35 participants joined in the session and they were randomly teamed up into six workshop sub-groups. In the second workshop, in March 2014, 28 participants joined in the session and they were randomly teamed up into five workshop sub-groups.

Table 2 lists attributions of workshop participants both in October 2013 and in March 2014.

4.3 Tools Used in Two Workshops

Innovation dialogue tools used in two workshops from policy observation phase, to policy ideation phase, then to policy prototyping phase were selected abiding with the toolbox approach referred in 3.2.

Table 3 is the list of tools used in two workshops.

Figures 4 and 5 are the photos of two WP5 Model Workshops held in *Nagano* in October 2013 and in March 2014.

Table 2 Attributions of WP5 model workshop participants in *Nagano*

Workshop #1 (October 22nd, 2013): N = 35	
Gender	Male: 30, Female 5
Age	20s: 7, 30s: 26, 40s: 1, N/A: 1
Vocation	Corporate: 8, Government: 23, NPO and other: 1, N/A: 3
Workshop #2 (March 8th, 2014): N = 28	
Gender	Male: 21, Female: 7
Age	20s: 6, 30s: 12, 40s: 4, 50s: 4, 60s: 2
Vocation	Corporate: 10, Self-Employed 2, Government: 13, NPO and other: 3

Table 3 Innovative dialogue tools used in two WP5 model workshops in *Nagano*

Workshop #1 (October 22nd, 2013): N = 35	
Phase	Used tools
3.1.2 Policy Observation	Fieldwork
3.1.3 a Policy Ideation: Divergent Thinking	Brainstorming
3.1.3 b Policy Ideation Convergent Thinking	Affinity Diagram, Structural Shift Ideation
3.1.4 Policy Prototyping	Improvised Story Play
Workshop #2 (March 8th, 2014): N = 28	
Phase	Used tools
3.1.2 Policy Observation	Fieldwork
3.1.3 a Policy Ideation: Divergent Thinking	Brainstorming
3.1.3 b Policy Ideation Convergent Thinking	Affinity Diagram, Causal Relation Diagram and Leverage Point
3.1.4 Policy Prototyping	Drawing and Doll Play Theatre

Fig. 4 WP5 model workshop (October 22nd, 2013 at *Nagano* city) photo by the authors

Fig. 5 WP5 model workshop (March 8th, 2014 at *Nagano* city) photo by the authors

5 Evaluation: Workshop Results

5.1 Qualitative Evaluation

Eleven of eleven teams participated in two WP5 Model workshops trailed all five phases of the Model and successfully reached to the stage of innovative policy presentation for regional development of *Nagano* city.

This fact qualitatively proved efficacy of WP5 Model for co-creating regional development policy with government and non-government stakeholders.

5.2 Statistical Test

The authors implemented the satisfaction and creativity surveys immediately after two workshops. These surveys are to test quantitatively participants' post-workshop satisfaction and creativity-increase with five-degree scale (5 = very good; 4 = good; 3 = neutral, 2 = poor; 1 = very poor).

This paper interprets the survey results to show efficacy of WP5 Model for co-creating policy for regional development, since better subjective satisfaction and creativity-increase should reflect real outcomes of policy design [27] according to the subjective approach of social policy theory [28].

As for creativity-increase evaluation, these surveys adopt four-point scales of creativity applied from the Torrance Test of Creative Thinking [29]; fluency; flexibility; originality; and elaboration.

Table 4 Independent t-test to the median (M = 3)

Workshop	Average	Standard deviation	Significant probability
#1: October 2013 (N = 35)			
Satisfaction	4.49	.562	.000
Apprehensibility	4.29	.524	.000
Creativity increase			
Fluency	4.14	.692	.000
Flexibility	4.20	.632	.000
Originality	3.74	.741	.000
Elaboration	3.66	.765	.000
#2: March 2014 (N = 28)			
Satisfaction	4.79	.418	.000
Apprehensibility	4.36	.559	.000
Creativity increase			
Fluency	4.39	.567	.000
Flexibility	4.07	.766	.000
Originality	3.68	.772	.000
Elaboration	3.57	.742	.000

Table 4 show statistical tests results of two WP5 Model workshops implemented in *Nagano* in October 2013 and in March 2014, respectively. Satisfaction rate, apprehensibility rate, and creativity-increase of two workshops are statistically significant in satisfying 1 % significant level.

This study also implemented t-test of averaged scores of the above surveys to verify whether particular attributions of workshop participants such as gender and vocation (i.e. government officials or non-officials) would show significant difference in satisfaction, apprehensibility and creativity-increase in using WP5 Model, the proposed co-creative policy-making tool.

In the October 2013 workshop, gender showed difference only in averaged fluency (male 4.28; female 3.50) and this difference was statistically significant at 1 % level ($p = .010$, two-sided). Other than that any attributions resulted at no differences in satisfaction, apprehensibility and creativity-increase at statistically significant level.

In the March 2014 workshop, gender also made difference in averaged fluency (male 4.48; female 4.14), but this difference did not satisfy the 10 % significant level with relatively narrow margin ($p = .183$, two-sided) In contrast, this workshop led the difference in averaged fluency between government officials and non-officials (government officials 4.21; non-officials 4.57). This difference satisfied the 10 % significant level ($p = .096$, two-sided).

The above test results weakly implied that the WP5 Model may cause fluency gap to some gender and vacation (government officials and non-officials) with some degree in their co-creative policy-making works.

Table 5 Pearson's correlation coefficients

Workshop #1: October 2013 (N = 35)				
Factor	Fluency	Flexibility	Originality	Elaboration
Fluency	1	.403*	.475**	.373*
Flexibility	.403*	1	.113	.207
Originality	.475**	.113	1	.411*
Elaboration	.373*	.207	411*.	1
Workshop #2: March 2014 (N = 28)				
Factor	Fluency	Flexibility	Originality	Elaboration
Fluency	1	.530**	.045	.151
Flexibility	.530**	1	.228	.577**
Originality	.045	.228	1	.074
Elaboration	.151	.577**	.074	1

Note: **satisfied at 1 % significant level (two-sided); *satisfied at 5 % significant level (two-sided)

5.3 *Correlation Analysis*

The authors conducted Pearson's correlation analysis to see which factor of creativity is correlated in co-creating public policy for regional development in workshops.

Table 5 shows Pearson's correlation analysis for two workshop results. Fluency and originality are correlated at highly significant level (**) in the October 2013 workshop. In that workshop, fluency and flexibility, fluency and elaboration, and originality and elaboration are significantly correlated (*).

The March 2014 workshop showed somewhat different results from the previous workshop, except that fluency and flexibility are highly correlated. Fluency and flexibility as well as flexibility and elaboration had high significance (**), but none was similar to the previous workshop.

Two workshops thus validated that fluency and flexibility are two highly correlated factors to lead creativity in co-creative policy-making workshop for regional development of *Nagano*.

6 Discussion, Conclusion and Further Research Agenda

6.1 *Discussion*

Evaluation results qualitatively and quantitatively supported efficacy of WP5 Model. Satisfaction and apprehensibility of the Model was very high.

The authors observed creativity-increase of participants through workshops and it was statistically significant. Pearson' correlation analysis for two workshop results made it clear that fluency and flexibility were correlated to some extent.

This result indicates that participants with WP5 Model could go beyond traditional way of thinking in policy design with fluent and flexible manner.

No particular differences among attributions of participants were observed on averages of satisfaction, apprehensibility and four-point scales of creativity-increase, except for gender (male/female: fluency in the October 2013 workshop) and vocation (government official/non-government: fluency in the March 2014 workshop) at weak significance level. This result implies that facilitators of WP5 Model may design carefully their workshop to help some participants' fluent thinking, especially that of female participants and government officials.

6.2 Conclusion

This paper proposed the WP5 Model for co-creative and co-designing platform of regional development policy. The MP5 model has the features as collaborative dialogue tools for policy innovation based upon the system x design thinking.

The MP5 Model is qualitatively and quantitatively validated its efficacy both in method and in outcome. This Model contributed significantly to help participants in extending spontaneous dialogue and in formulating innovative policy for regional development.

The MP5 Model was proved to stimulate participants' creativity-increase for policy formulation. Among creativity factors, fluency and flexibility were observed to have correlation to some extent in two workshops. No particular difference of creativity-increase was observed through workshops in-between attributions of participants (i.e. gender and vacation) except that only fluency made weak difference sometimes in workshops.

6.3 Further Research Agenda

This paper built the WP5 Model and explained its architecture and sequence. For empirical validation, it used as social experiments two policy formulation workshops in *Nagano*. The authors will apply the Model for more cases to show robustness of this Model.

At this stage, the WP5 Model is designed to be the policy dialogue tool for regional development and revitalization. For the next stage of study, the WP5 Model shall be tested for other domains of government services than regional policy.

Acknowledgements The authors of this paper acknowledge generous support extended from officials working for *Nagano* prefectural government, *Nagano* city government, *Nagano* Local Finance Office, *Shinshu* Innovation Project, *Hachijuni* Bank, JA *Nagano* and Development Bank of Japan, when the authors held two workshops in *Nagano* city.

Appendices

Appendix 1: Post-Workshop Survey Sheet (on October 22nd, 2013)

(Original in Japanese, translated to English)

Answer: 5 degrees (5 = very good; 4 = good; 3 = neutral, 2 = poor; 1 = very poor)

Q1 (satisfaction): Were you satisfied with the workshop?

A: 5, 4, 3, 2, 1

Q2 (apprehensibility): Did you understand the contents of workshop?

A: 5, 4, 3, 2, 1

Q3 (Creativity-Increase): By compared with other ideation methods which you experienced in the past, how did you feel with this model particular in these points;

Q3-1 (Fluency): Was it easy to ideate?

A: 5, 4, 3, 2, 1

Q3-2 (Flexibility): Was it flexible to ideate?

A: 5, 4, 3, 2, 1

Q3-3 (Originality): Did you get original idea?

A: 5, 4, 3, 2, 1

Q3-4 (Elaboration): Did you elaborate to new idea?

A: 5, 4, 3, 2, 1

Q4 (Stage of Innovation): On which stage did you sense shift to innovative idea?

Q4-1: A. Purpose B. Function C. Component

Q4-2: Please answer freely what shift to innovative idea you experienced in this workshop.

A. (Free Answer)

Appendix 2: Post-Workshop Survey Sheet (on March 8th, 2014)

(Original in Japanese, translated to English)

Answer: 5 degrees (5 = very good; 4 = good; 3 = neutral, 2 = poor; 1 = very poor)

Q1 (satisfaction): Were you satisfied with the workshop?

A: 5, 4, 3, 2, 1

Q2 (apprehensibility): Did you understand the contents of workshop?

A: 5, 4, 3, 2, 1

Q3 (Creativity-Increase): By compared with other ideation methods which you experienced in the past, how did you feel with this model particular in these points;

Q3-1 (Fluency): Was it easy to ideate?

A: 5, 4, 3, 2, 1

Q3-2 (Flexibility): Was it flexible to ideate?
A: 5, 4, 3, 2, 1
Q3-3 (Originality): Did you get original idea?
A: 5, 4, 3, 2, 1
Q3-4 (Elaboration): Did you elaborate to new idea?
A: 5, 4, 3, 2, 1
Q4 (Stage of Innovation): On which stage did you sense shift to innovative idea?
Q4-1: A. Brainstorming B. Affinity Diagram C. Causal Relations Diagram and Leverage Points D. Prototyping and Story-Telling
Q4-2: Please answer freely what shift to innovative idea you experienced in this workshop.
A. (Free Answer)

References

1. Karmark EC (2007) The end of government…as we know it: making public policy work. Lynne Rienner Publishers, Inc., Boulder
2. Kettl DF (2005) The global public management revolution, 2nd edn. Brookings Institution Press, Washington, DC
3. Kettl DF (2008) The next government of the United States: why our institutions fail us and how to fix them. W.W. Norton and Company, New York
4. Cabannes Y (2004) Participatory budgeting: a significant contribution to participatory democracy. Environ Urban 16(1):27–46
5. Osborne SP (2006) The new public governance? Public Manage Rev 8(3):377–387
6. OECD (2003) Open government: fostering dialogue with civil society. OECD Publishing, Paris
7. Lathrop D, Ruma L (eds) (2010) Open government: collaboration, transparency, and participation in practice. O'Reilly, Sebastopol
8. Yasui T, Shirasaka S, Maeno T (2013) Designing public policy by structural shift ideation through the case of revitalizing decaying local shopping malls'. In: Proceedings of the 7th Asia-Pacific Council on Systems Engineering Conference (APCOSE 2013), Yokohama, Japan, September 8–11, 2013
9. Ueda K, Tanaka T, Fujita K (2008) Toward value co-creation in manufacturing and servicing. CIPR J Manuf Sci Technol 1:53–58
10. Rawaswamy V, Gouillart F (2010) The power of co-creation: built it with them to boost growth, productivity, and profits. Free Press, New York
11. Hood C (1991) A public management for all the seasons? Public Adm 69(Spring):3–19
12. Dunleavy P, Margetts H, Bastow S, Tinker J (2005) New public management is dead: long live digital-era governance. J Public Adm Res Theory 16:467–494
13. Ball A, Osborne SP (eds) (2011) Social accounting and public management: accountability for the common good. Routledge, New York
14. Nonaka I, Konno N (1998) The concept of "Ba": building a foundation for knowledge creation. Calif Manag Rev 40(3, Spring):40–54
15. Maeno T, Yasui T, Shirasaka S, Tomita Y, Ishibashi K, Iwata T, Yagita H (2014) System x design thinking to change the world: Keio SDM style how to make innovation. Nikkei BP, Tokyo (in Japanese)

16. Haruyama S, Kim SK, Beiter KA, Dijkema GPJ, de Weck OC (2012) A new project-based curriculum of design thinking with systems engineering techniques. In: Proceedings, CESUN 2012, Delft University of Technology, Delft, the Netherlands, 18–20 June 2012
17. Maeno T, Yasui T, Shirasaka S, Bosch O (2012) Social system design using models of social and human mind network: CVCA, WCA and Bayesian network modeling. In: Proceedings of the 4th international conference on Applied Human Factors and Ergonomics (AHFE 2012), San Francisco, 21–25 July 2012
18. Senge PM (1990) The fifth discipline: the art & practice of the learning organization. Doubleday, New York
19. Sterman JD (2000) Business dynamics: system thinking and modeling for a complex world. McGraw Hill Higher Education, Boston
20. Simon HA (1967) The sciences of the artificial. MIT Press, Cambridge
21. Brown T (2009) Change by design: how design thinking transforms organizations and inspire innovation. Harper Business, New York
22. Manfred FR, de Vries K (2005) Leadership group coaching in action: the Zen of creating high performance teams. Acad Manage Perspect 19(1):61–76
23. Woolley WA, Chabris CF, Pentland A, Hashimi N, Malone TW (2010) Evidence for a collective intelligence factor in the performance of human groups. Science 330:686–688
24. *Nagano* City Website. http://www.machidukuri-nagano.jp/070221-1-soukai-1-2.pdf. Last Accessed 30 April 2014
25. Ministry of General Affairs Website. http://www.soumu.go.jp/iken/zaisei/H24_chiho.html. Last Accessed 30 April 2014
26. Shinshu Innovation Project (SHIP) Facebook Page. https://www.facebook.com/ship.shinshu Last Accessed 30 April 2014
27. Bardach E (2012) A practical guide for policy analysis: the eightfold path to more effective problem solving, 4th edn. CQ Press, London
28. Veehoven R (2002) Why social policy needs subjective indicators. Soc Indic Res 58 (1–3):33–46
29. Torrance EP (1979) The search for satori and creativity. Creative Education Foundation, New York

System Design of Happy Town Using Four Factors of Happiness

Shiko Kurihara and Takashi Maeno

Abstract We tried the citizen-based town planning using "Four factors of the happiness". A building and public construction-centered design continued for a long time how was the conventional citizen-based town planning. The citizen-based town planning is gradually shifting from concrete to a human being, but still only one factor of "Four factors of the happiness" is used. Therefore it is epoch-making to do the citizen-based town planning using all four factors of the happiness. We collected knowledge from practice works and made the concept of the town of the happiness in a workshop and we crossed this and four factors of the happiness and got an idea. The idea that performs a workshop, and came out is eight. The town which supports a dream of somebody. The town where we give lessons and take lessons from. The town which makes a meal in turn. The town where a variety of people are accepted. The town where private space is followed. The town where common use space is installed in. LRT (next-generation streetcar system). The share village. As a result of having carried out a questionnaire based on prototyping in a completed "happy town", "the ratio of person who replied it saying "it is happy" was 64.2 %, and, thus, a voice that the thing that was effective for the citizen-based town planning using four factors of the happiness became clear and expected agriculture as a result of interview again was asked a lot about the town which just suggested it this time than the town where we lived in. In future work, it should be used for real design for making people happier.

Keywords Four factors of happiness • System design • Happy study • Community development • Workshop

---

S. Kurihara (✉)
Graduate School of System Design and Management, Keio University, Collaboration Complex, 4-1-1 Hiyoshi, Kohoku-ku, Yokohama 223-8526, Japan
e-mail: kuri@z3.keio.jp

T. Maeno
Graduate School of System Design and Management, Keio University, Yokohama 223-0061, Japan

T. Maeno et al. (eds.), *Serviceology for Designing the Future*,
DOI 10.1007/978-4-431-55861-3_21

1 Introduction

1.1 Research Background

Yamasaki said that, "Up to now, the motif of urban development has focused on material things such as the development of land, residences, and roads based on the policies of the local government" [1]. As Yamasaki states, unnecessary public construction has disguised itself under the name of urban development for many years. These types of urban development posses a strong affinity with public construction advanced for economic purposes. As a result, have the residents become happier? Unfortunately, the level of happiness is not improving. Hisashige said, "We are always chasing after economic abundance [2]. That is why community development is unsuccessful. There is an overproduction of various policies and success stories about how 'It will boost the economy in the town' that end up amounting to nothing but empty promises." He points to the reality that economic abundance rarely boosts community development, neither does it improve the level of happiness of residents.

It is disappointing that this is the case, when it states in our country's constitution that we have the basic right to pursue happiness. Furthermore, Japan is a country with the 3rd highest GDP in the world. However, according to OECD's research, Japan's level of happiness ranks in 21st place, out of a survey of 36 industrialized countries [3]. It is clear from that research that economic abundance does not contribute to the level of happiness of residents.

In the first place, urban development is about "How we can create a common and convenient environment for people living together in a particular area that supports their lives and allows them to live like human beings" [4]. People are able pursue happiness in that common environment.

This paper will examine ways of creating a "Happy town". It will also highlight the problems of the current urban development system and use the "Four Factors of the Happiness" in order to create system designs for a happy town.

1.2 The Purpose and Outline of the Study

The purpose of this paper is to demonstrate the efficiency of system designs that use the "Four Factors of the Happiness" in urban development. In chapter "Employee Satisfaction Analysis in Food Service Industry – Resultant of Questionnaire to the Restaurant Staff", we explain the result of factor analysis of happiness referring our article [5]. Then, in chapter "Exploration of Service System and Value Co-creation Mechanism in Islamic Banking in Pakistan", procedure of generation of ideas for happy town is explains. Chapter "The Ordering of Fast Food Using Menu" is for example of real application. Chapter "Evaluation of Taxiing at a Large Airport Considering Customer Satisfaction" shows the results followed by conclusions in chapter "Research of the Social New Transportation Service on Electric Full Flat Floor Bus".

2 Four Factors of the Happiness

2.1 Demand of Happiness Research

There has been increasing numbers of researches on well-being (happiness) since 1990s mainly by psychologists and economists. One of the reasons of the increasing numbers of researches in this field is that people are noticing that economical wealth is not the only way for people to be happy in life in this modern world. Many of the researches are on the relationship between well-being and facts. For example, (1) women are slightly happier than men, (2) healthier people are happier, (3) increase of income makes human happier when the income is less than $75,000, (4) knowing various people makes human happier compared with knowing numbers of people, (5) having purpose in life makes humans happy, (6) believing religion makes people happy, (7) optimists are happy, (8) kind people are happy, (9) gratitude makes human happier, (10) creating beautiful things makes people happier than looking at beautiful things. Researchers on regional difference of happiness are important topics as well. It is interesting that many things are in connection to well-being. However, no research has been done to take an overlook at well-being from a systemic point of view. Hence, in the present study, the result of factor analysis is shown. It is shown that four factors of happiness are found in relation to non-locational goods. Then we show that four factors can be used for community design through workshops.

2.2 Result of Factor Analysis

Goods are classified into positional goods and non-positional goods. Goods are not necessarily physical things. They can be abstract concepts. The positional goods include money, things and fame. One's positional goods can be compared with others' because those position can be measured by its amount. Nettle (2005) has shown that happiness by obtaining those positional goods does not last for a long time. However, happiness given by non-positional goods continues for a long time [6]. The non-positional goods include love, health, autonomy, social embededness, and quality of environment. Kahneman has shown that it is a "focusing illusion". It seems that happiness comes to you if you obtain positional goods. So people wish to get those positional goods. However, when they obtain the positional goods, happiness fades away very quickly somewhat like an illusion. Focus for happiness is not at the adequate place. That is why it is called the focusing illusion.

Table 1 Results of factor analysis by principal factor method

	Factors				
	1	2	3	4	Communality
Competence	**.739**	.231	.192	.068	.641
Meaning in life	**.722**	.356	.290	.112	.745
Mastery	**.696**	.278	.340	.015	.678
Self-actualization	**.676**	.246	.375	.135	.677
Self esteem	**.644**	.381	.323	.033	.666
Autonomy	**.638**	.269	.142	.039	.501
Thought and religion	**.606**	.235	.156	.289	.531
Sense of societal connection	**.602**	.233	.261	.244	.545
Hope	**.587**	.479	.327	.055	.684
Personal growth	**.570**	.488	.143	.120	.599
Kindness	**.539**	.492	.054	.026	.536
Work motivation	**.526**	.330	.069	.199	.431
Clear purpose of life	.493	.321	.324	.115	.465
Preference to make someone happy	.415	**.734**	−.019	−.002	.712
Gratitude	.259	**.729**	.172	.127	.644
Love	.301	**.656**	.315	.054	.624
Humor	.279	**.586**	.200	−.150	.484
Positive relationship with others	.302	**.509**	.372	.314	.587
Savoring	.473	.480	.350	.203	.617
Optimism	.245	.216	**.622**	.136	.512
Self-acceptance	.438	.361	**.556**	.266	.702
Having no anxiety	.227	.007	**.529**	.476	.557
Switching minds	.366	.204	.423	.293	.440
Self-concept clarity	.018	−.021	.114	**.693**	.493
Social comparison	.101	.044	.083	**.538**	.308
Factor loadings	6.214	4.122	2.438	1.604	14.378
Cumulative contribution ratio	24.856	41.344	51.094	57.511	

Hence, we have conducted factor analysis of non-positional goods in relation to happiness. First, we have surveyed researches on well-being that shows the relationship between happiness and various factors. Factors are limited to non-positional goods related to the state and characteristics of mind because those are only goods that people can directly treat. Peace, safety and health are excluded because they are not state or characteristics of mind even though those are important factors for happiness. Then we have made a questionnaire for asking those questions to 1500 Japanese through use of the internet. Titles of those questions are shown at the left hand side of Table 1. Then we have conducted a factor analysis using the software SPSS. Result is shown in the Table 1.

As shown in the table, we obtained four factors. Looking at the table we named four factors as follows:

1. Having purpose and learning new things (focusing into self)
2. Relationships, gratitude and kindness among people (focusing into relationship among others)
3. Positive and optimistic attitude
4. Keeping one's own pace for independent behavior

It means that numbers of factors shown at the left hand side of the Table 1 are united into four fundamental factors that show the happy state and characteristics of people's mind [6–9].

3 Workshop

3.1 Brainstorming

We held workshops in order to design a "Happy Town" system. First, we had a brainstorming session about states of happiness.

3.2 Affinity Diagram

Next, we used an affinity diagram to group the words that emerged from the brainstorming session. The words were grouped into the following headlines: Physiological desire, Desire for safety, Desire for self-realization, Desire for recognition, Desire for love and belonging, Lucky society, Coincidence (not through effort), Cooperative society, Surprises, Mutual recognition society, Mutual acknowledging society, Self-realization, Optimism and Positiveness, Satisfaction, Feelings of ease, Liveliness, Stability.

3.3 Discussion

We held a discussion in order to extract the meta concept from the items grouped above and the result was "A society with different types of happiness".

3.4 Four Factors' Framework (FFF) of Happy System Design

After a serious of workshops we extracted the concept, "A society with different types of happiness". We took that concept and placed it in the "Four Factors' Framework" to extract a concrete idea from the multiplication of a particular

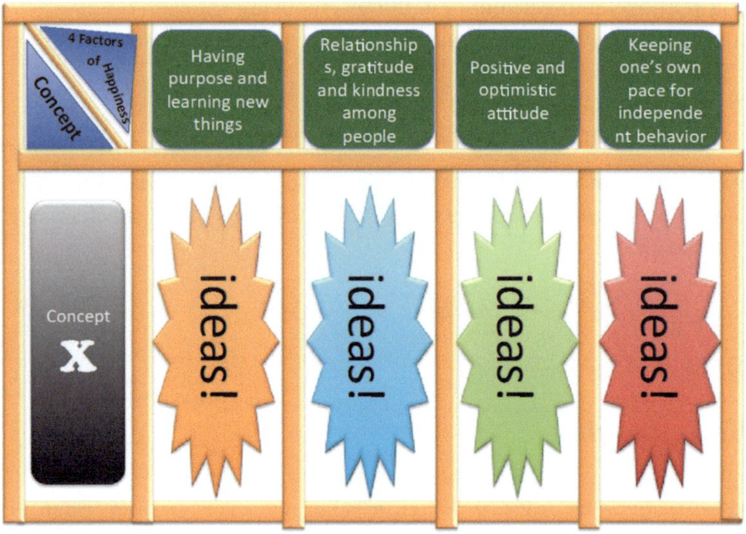

Fig. 1 "FFF" image

keyword [concept x] with the "Four Factors of Happiness" which will now be referred to as "FFF" in order to develop new ideas (see Fig. 1). As a result, we were able to extract ideas ("Ideas of Happiness") that included the "Four Factors of Happiness". Next, we used the Pugh Concept Selection in order to narrow down the ideas extracted from FFF.

3.5 Pugh Concept Selection

We used the Pugh Concept Selection method in order to narrow down the ideas that emerged above.

Pugh Concept Selection is a method of concept selection that ranks the non quantitative values of concepts through comparisons with other alternatives in order to find the most probably option.

The concepts are arranged into a matrix and one of them is selected to be the baseline. The other items are compared in a pairwise fashion against the baseline, using "+" if it is better than the baseline, "−" if it is worse than the baseline, and "S" if it is the same as the baseline. The results are then recorded in the matrix. Our research used the demands of the stakeholder as the baseline and performed comparisons using the Pugh Concept Selection method to choose the best ideas.

Probable ideas that emerged from the Pugh Concept Selection were as follows: "A town that support each other's dreams", "A town that learns from each other", "A town where cooking duties are arranged on a rotation basis", "A town that accepts all types of people", "A town that respects personal space", "A town with a common space".

3.6 Extraction of Criteria

We performed an extraction of criteria for the six ideas that emerged from the Pugh Concept Selection and grouped the results using an affinity diagram. The results were as follows: "Freedom", "Unique Space", "Intercommunication is possible", "Everyone gathers", "Active". We held a discussion about these five items and the idea that emerged was "LRT (Light Rail Transit: The next generation street car system)".

3.7 Value Graph

We used value graphs in order to rank the values of the ideas extracted from Pugh Concept Selection method.

3.8 Affinity Diagrams and Insights

We used affinity diagrams to group the highest values of the concepts extracted from the previous value graph.

The sense of value that emerged as a result of using affinity diagrams to organize the highest values of the value graphs was "I want to have unusual experiences, so that I can know who I really am. However, I do not want to be dragged into any trouble. Having said that, I still have needs and I long for others. The "Insight" gained from that is "The limits of individuals".

3.9 Ideas from Insights

We had a discussion about the "Insight" gained from the previous section, "The limits of individuals", and an idea called "Sharing Village" emerged.

4 Result of Designing Happy Town

4.1 Outline

Here are a list of ideas that have emerged up to this point.

1. Ideas that emerged from FFF: "A town that support each other's dreams", "A town that learns from each other", "A town where cooking duties are arranged on a rotation basis", "A town that accepts all types of people", "A town that respects personal space", "A town with a common space".
2. The idea that emerged from the extraction of criteria was the LRT (Light Rail Transit: The next generation street car system)
3. The idea extracted from the "Insight" was "Sharing Village".

We gathered these ideas in order to create a prototype.

4.2 Prototyping

A prototype is "An early model of a product that is built in order to receive feedback on its requirements in an early stage before the final product is completed" [10]. Prototypes also offer others the opportunity to understand the product's functions and value, and the opportunity to evaluate the product to confirm and gain new insights about it's functions in order to determine whether there is a demand for the product in this world.

We created a prototype with lego Block (see Fig. 2) in order to understand the suitability of a "Happy Town" and conducted interviews with the participants.

4.3 Questionnaire and Observations

We divided the participants into seven groups according to their age group: 10–19 years old, 20–29 years old, 30–39 years old, 40–49 years old, 50–59 years old, 60–69 years old, 70 years old or older. We selected an equal number of men and women, 14 participants in total. We showed the prototype to the participants and explained each of the elements of the idea.

We conducted the questionnaire after briefing the participants about the prototype.

The contents of the questionnaire are as follows. Section 1 was on the front page of the questionnaire. It consisted of boxes that they would check off regarding their sex, marital status, whether they had children or not, the age of the children, and whether or not they lived together. In Sect. 2, we provided them with 7 statements in regards to "Happy Town" and asked them to rate their answers as (1) if they agree,

Fig. 2 The Lego block, which used for prototype

(3) neutral, or (5) if they disagree. Choices (2) and (4) were in the middle of (1) and (3) and (3) and (5) respectively. The 7 statements were as follows: "The people living in this town are happy", "I want to try living in this town", "I am satisfied with the town that I live in", "I want to try having a different lifestyle", "I would live here if I were by myself", "I would live here if I were together with my family or somebody else", "It has a good influence on children".

In the first question of Sect. 3, we asked them, "What kind of town concept would you like to live in?" and offered 4 choices. They were as follows: "A town that pursues functionality", "A town that pursues economic abundance", "A town that pursues happiness", and "Other". They were able to enter their own answer in the "Other" section. Question 2 asked them, "What would make you want to live there?" and provided them with space to write their own answers. Participants who answered (1) and agreed to the previous statement, "I want to live in this town" were excepted from this question.

In question 3, they were given a statement, "If you add (remove) this, then the town will be better" and were asked to fill in their own answers either in the "It is better if you add this" or in the "It is better if you remove this" section. In question 4, they were asked to fill in their own answers in response to the statement, "I liked this part about the town".

At the back of the questionnaire, we asked them questions in regards to the eight ideas. For example, Idea A was "A town that support each other's dreams" (By supporting others, they will in turn receive support. The methods for support are diverse.). We gave them two statements in response to Idea A and asked them to rate their answers as (1) if they agree, (3) neutral, or (5) if they disagree. Choices

(2) and (4) were in the middle of (1) and (3) and (3) and (5) respectively. The statements were as follows: "The people living in this town are happy", "I want to try living in this town". The remaining questions for the other seven ideas were conducted in a similar manner.

5 Conclusions

64.2 % of the participants answered, "This town is happier than the town that I live in". It is clear from the result that urban development systems that include the "Four Factors of Happiness" are effective.

There are still challenges up ahead. Other results include, "I think it will have a good influence on children, but if I think about my self and society then moving becomes difficult". It is necessary to consider solutions for problems that prevent people from changing their current situation.

In addition, we found from our interviews that a big majority of the participants are interested in agriculture, 50 % of the participants suggested including community farming. It will be necessary to consider the possibility of including community farming.

Receiving answers such as "This town looks happy" does not mean that this research is completed. It is ideal if they could physically live in the town and experience happiness. Urban development has focused too much on money for so long and it is time to shift the focus to things that lead to happiness. If not, then the future of Japan is at stake. Many people have realized that Japan's expanding economy has already come to a halt. It is time to change our sense of values and make Japan into a world where children are hopeful about the future.

This research is only an introduction to human happiness and peace.

We hope that the knowledge gained from this research can be used not only to shoulder Japan's future, but also the future of the entire world through real life applications and further research.

References

1. Takeo Yamasaki (2000) An introduction to urban development policies. Japanese Institute of Local Government, Tokyo, P5
2. Tetsunosuke Hisashige (2010) The pitfalls of community development. Kodansha Gendai, Tokyo
3. OECD website, OECD better life index country reports. Retrieved January 9, 2014 from http://www.oecd.org/newsroom/BLI2013-Country-Notes.pdf
4. Akira Tamura (1987) The idea behind urban development. Iwanami, Tokyo
5. Satoshi Honda, Darko Radvic, Takashi Maeno et al (2014) Mn'M Workbook 3, Future urban intensities, measuring the non-measurable, (Maeno: factor analysis of well-being and its application to community design). Tac Co., Ltd, Tokyo

6. Nettle D (2005) Happiness, the science behind your smile. Oxford University Press, Oxford
7. Kahneman D et al (2006) Would you be happier if you were richer? A focusing illusion. Science 312(5782):1908–1910
8. Kosuke Ishii, Olivier de Weck, Shinichiro Haruyama, Takashi Maeno, Sun K. Kim, Whitfield Fowler (2009) Active learning project sequence: capstone experience for multi-disciplinary system design and management education. In: Proceedings of the international conference on engineering design (ICED'09), August 2009, Stanford, CA, USA, pp 57–68
9. Naohiko Kohtake, Takashi Maeno, Hidekazu Nishimura and Yoshiaki Ohkami (2010) Graduate education for multi-disciplinary system design and management, developing leaders of large-scale complex system design and management, synthesiology, English edition, Ibaraki, vol 3, No 2, pp 124–139
10. Project Management Institute (2008) Guide to the project management body of knowledge. Project Management Institute, Newtown Square, P109

Part V
System Design and Management

Evaluation of the Productivity Improvement by Information Presentation in Surveillance Service

Mitsunari Uozumi, Kouichi Yamada, Shuto Murai, Hajime Asama, and Kaoru Takakusaki

Abstract System surveillance is a function required for continuous operation of a system which consists of various apparatus and networks. Many service providers try to raise their productivity. Their methods are to show some information to a operator after alarm occurs. As another method of the rationalization, a surveillance system shows some information to the operator in the waiting time. We have built the simulated environment of surveillance and compared 12 subjects. Operators are working efficiently most in the type which displays the information. It is an effective method of a productivity improvement to display the information on waiting time.

Keywords Surveillance service • Service engineering • Human interface

1 Introduction

System surveillance is a function required for continuous operation of a system which consists of various apparatus and networks. These system components break and deteriorate with age or when tasks that are carried out are beyond its capacity. Surveillance system detects these abnormalities and the surveillance operator performs predetermined disposal for returning to normal (Fig. 1).

The entrepreneur who provides service may maintain a system at low cost. So it is not rare to contract out surveillance to a service provider. The entrepreneur who offers surveillance as a service has the equipment which performs the service

M. Uozumi (✉) • K. Yamada • S. Murai
MITSUBISHI ELECTORIC Corp., 5-1-1 Ofuna, Kamakura 247-8501, Japan
e-mail: uozumi.mitsunari@ab.mitsubishielectric.co.jp

H. Asama
Department of Precision Mechanical Engineering, University of Tokyo, Tokyo, Japan

K. Takakusaki
Department of Physiology, Division of Neural Function, Asahikawa Medical College, Asahikawa, Japan

T. Maeno et al. (eds.), *Serviceology for Designing the Future*,
DOI 10.1007/978-4-431-55861-3_22

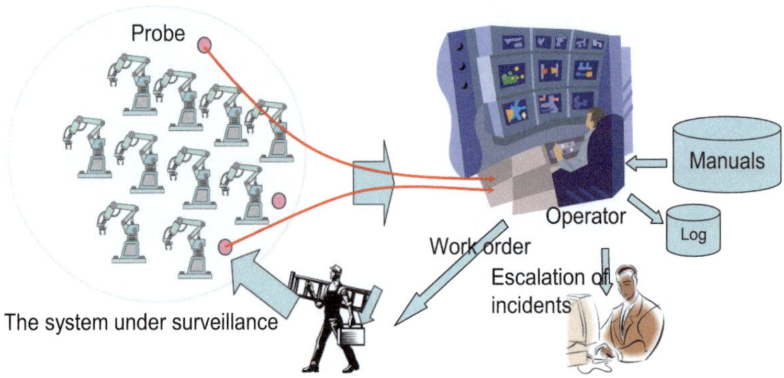

Fig. 1 System surveillance

effectively and trains operators who handle the system and carry out the surveillance 24/7 [1].

The entrepreneur of supervisory service is advancing rationalization of surveillance business. They increase their productivity and offer a competitive charge [2, 3]. Their methods are to show some information to operators after alarm occurs. This information helps the operator to dispose problem [4].

As another methods of the rationalization, it can be considered that a surveillance system shows some information to operators in there waiting time. Tamura et al. have reported that activeness positively affects cognitive performance [5]. Bechara et al. have reported that emotion can affect the decision-making [6].

In the surveillance operation, it can be considered that the productivity can be changed by controlling the state at the operator receives an alarm. Using the information shown, the disposal time of operators are shortened by doing active work [7].

The simulated environment of surveillance has been built and the results of 12 subjects are evaluated.

2 Surveillance Operation Model

An operator's work consists of the following three states.

- The operator waits for the alarm
- The operator recognizes the alarm
- The operator disposes of the alarm

An operator changes these states in order (Fig. 2).

In the waiting state, an operator waits for a change of display in the screen.

In the recognizing state, an operator notices a change of display in the screen and starts to dispose of the alarm.

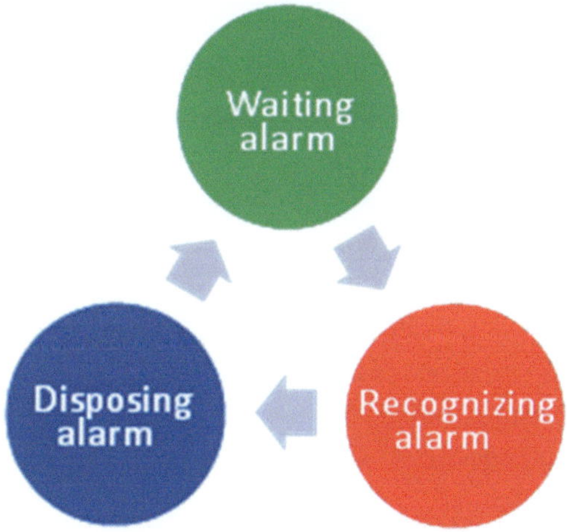

Fig. 2 Surveillance operator model

In the disposing state, an operator deals with the alarm according to the instructions stated in the manual and completes the disposal.

3 Experiments with Simulated Environment

3.1 Simulated Environment

The simulated environment of the surveillance system using a computer has been built.

A subject waits for a change of alarm by viewing a computer screen (Fig. 3).

Alarms are generated randomly and the generating interval average is 25 s. The generated alarm will enable the display of a message on the screen which the subject is looking at. In each test, the subject disposes 20 alarms.

3.2 User Interface

There are three kinds of screens that are shown to the subject on waiting state and are as follows:

Passive case test uses passive waiting screen (A).
Directed case test uses directed waiting screen (B).
Active case test uses active waiting screen (C).

- Passive waiting screen (A):

  No additional information is shown on this screen until an alarm is generated. The generated alarm is displayed on the upper row of the screen. After the

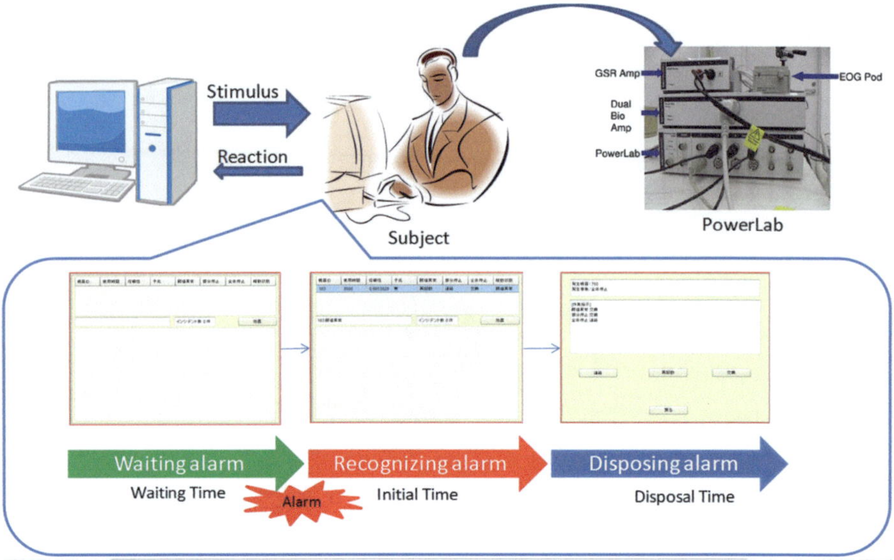

Fig. 3 Simulated environment

display of an alarm is recognized, a subject will click the button on the screen for disposal. Figure 4 shows an example of the screen being displayed before an alarm will be generated.

- Directed waiting screen (B):

  Explanation of the operation method is shown as the information on this screen until an alarm is generated. The generated alarm is displayed on the upper row of the screen. After the display of an alarm is recognized, a subject will click the button on the screen for disposal. Figure 5 shows an example of the screen being displayed before an alarm will be generated.

- Active waiting screen (C):

  A list for surveillance is always shown in the lower part of the screen. While waiting for an alarm, a subject can scroll this list or can sort a list by items, such as apparatus ID or hours of use. After the display of an alarm is recognized, a subject will click the button on the screen for disposal. Figure 6 shows an example of the screen being displayed before an alarm will be generated.

This list contains hours of use, reliability, probability sign, etc. But these items have no correlativity in this simulated environment.

In a real system, the probability of occurrence of an alarm being influenced by these items is little. In this test, the subject disposes 20 alarms and they are not influenced by these items. But the subject is interesting in this list.

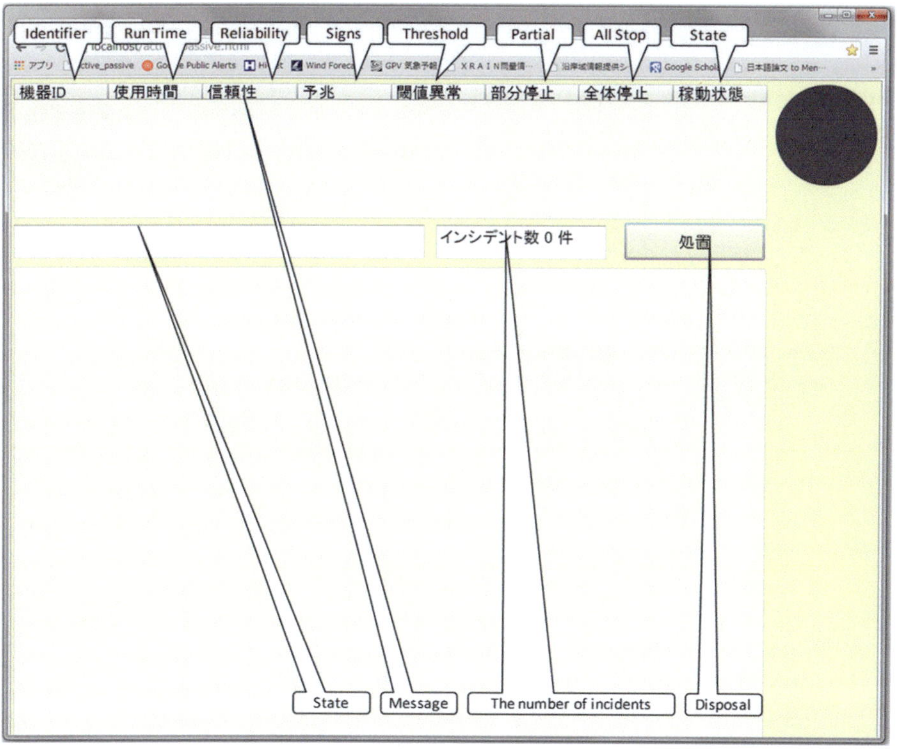

Fig. 4 Passive waiting screen (A)

3.3 Method of Measurement

In all the used screens, the circle is displayed on the upper right (Figs. 4, 5, and 6). The luminosity of this circle is different by changes of a screen.

The difference in the luminosity of this circle is measuring using the phototransistor. A measuring result is shown in Fig. 7. The upper row of Fig. 7 is a measurement value of luminosity. It can check changing with changes of a screen.

The potential difference between the terminals of the microswitch of the mouse which a subject operates is measured. If a subject pushes a mouse, since a microswitch short-circuits, potential difference will be 0 v. The lower row of Fig. 7 is a measurement value of potential difference. The timing which pushed the mouse, and the released timing are recorded.

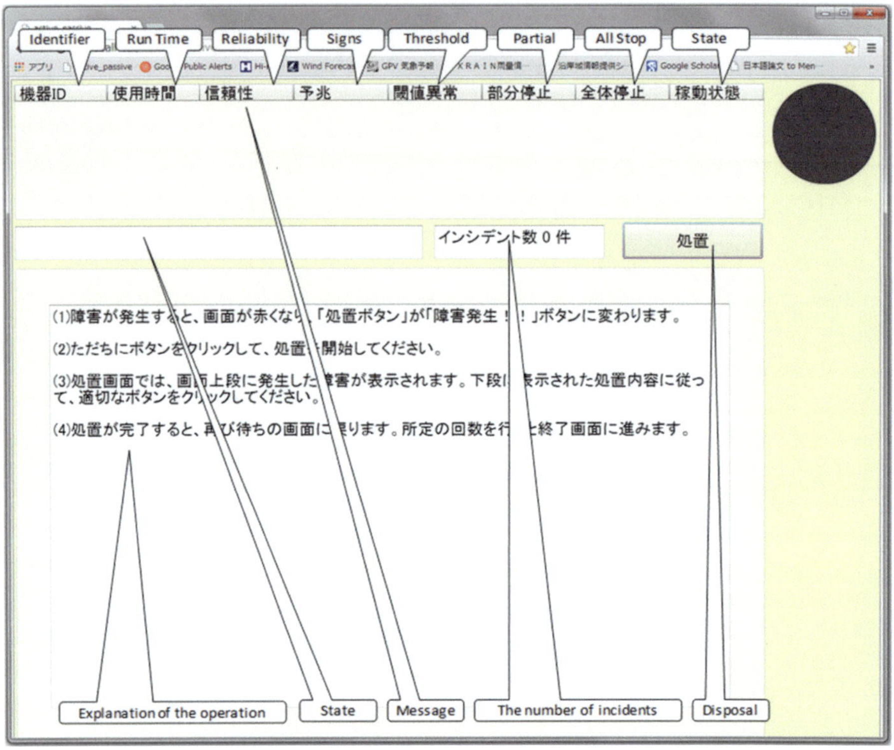

Fig. 5 Directed waiting screen (B)

The time-axis of this record is in agreement with record of the physiology measurement shown in Fig. 3, and a subject's operation and a physiological state can be compared in the unit for 1/1000 s.

As shown in Fig. 7, the period until alarm is generated and the luminosity of a screen changes is defined as working hours (WORK).

The period until processing of alarm is completed and it changes on the screen of the waiting for alarm again is defined as waiting time (WAIT).

Moreover, the back before the test which alarm does not generate starts is defined as IDLE.

After the screen of alarm is displayed, the period until a mouse is pushed is defined as a subject's initial time.

After a disposal screen is displayed, the period until a mouse is pushed is defined as disposal time.

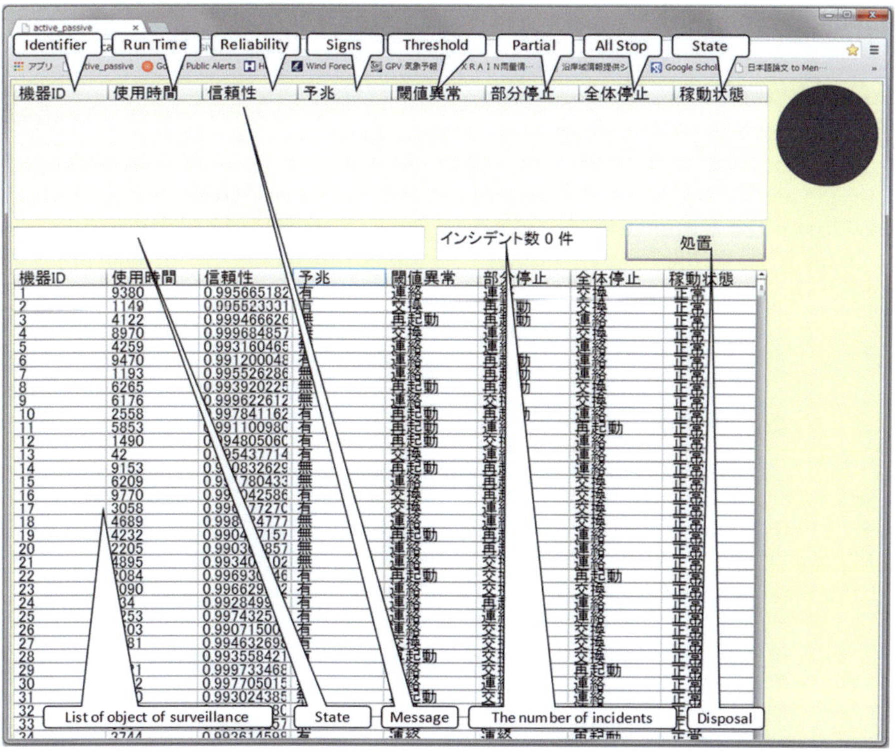

Fig. 6 Active waiting screen (C)

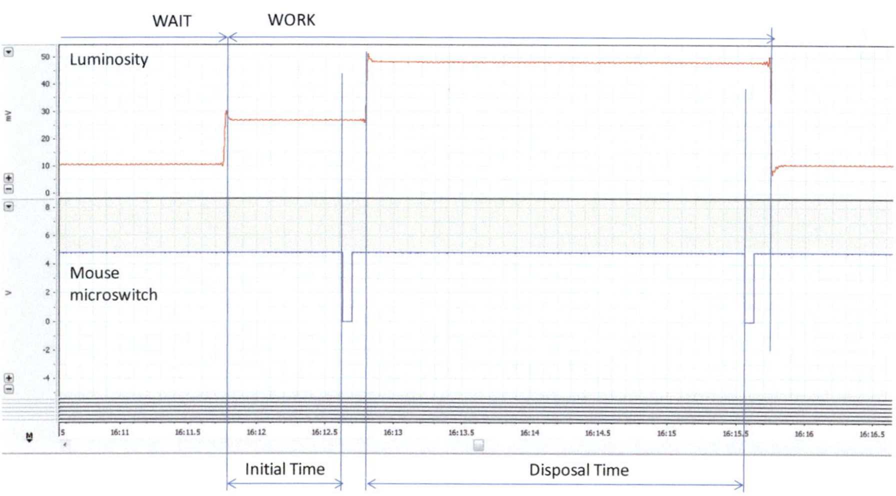

Fig. 7 Example of measurement data

4 Experimental Results and Discussion

4.1 Subjects

In this simulated environment, measurement using 12 subjects was performed. Subjects are 7 employees of corporate laboratory and 5 graduate students. They are 11 men and 1 woman and aged 23–54 years (Average = 32.6, SD = 10.3).

In many surveillance environment, generation structure is broad. This group of subjects is close to this.

4.2 Reaction Time of the Subjects

The total of a measuring result is shown in Fig. 8 as a scatter diagram. A horizontal axis is initial time and a vertical axis is disposal time. The average of a subject's initial shock time and disposal time is computed for every test.

The passive type taste is set to A. The directed type test is set to B. The active type test is set to C.

About disposal time, it can read that it is about the tendency of A > B > C.

The difference in a remarkable tendency cannot be read about disposal time.

Fig. 8 Reaction time of the subjects

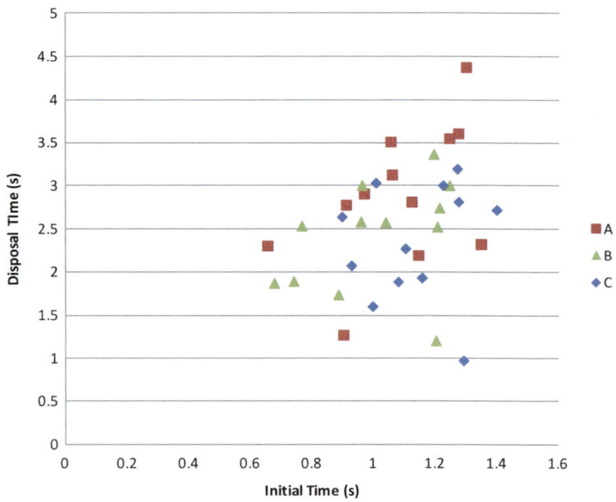

4.3 Mastery of Operation

This measurement was carried out in the following procedures in all the subjects.

- Practice of the disposal of three alarms
- A passive type (A) test
- A directed type (B) test
- An Active type (C) test

The break for several minutes is placed between each.

It is thought that operation is easy and the influence of mastery is small.

However, when evaluating the difference in each type of measuring result, mastery of operation should be examined.

The average of a subject's reaction time is compared about the test of the first half and the second half.

Both were evaluated using the T-test ($p < 0.05$).

In Table 1, the section which had a significant difference in the average value of the first half and the second half is indicated to be 1.

Some subjects' reaction time is different in the first half and the second half of a test.

However, it can be said that the influence by mastery is small as an overall trend.

Table 1 The phase with change of reaction time

	A		B		C	
	Initial	Disposal	Initial	Disposal	Initial	Disposal
#1						1
#2						
#3	1					
#4						
#5		1				
#6				1	1	
#7		1				
#8						
#9	1					1
#10						1
#11					1	1
#12					1	
	2	2	0	1	3	4

4.4 Evaluation to Initial Time

The average of each subject's initial time is shown in Fig. 9. The bar graph shows the average value of the passive type (A), the directed type (B), and the active type (C).

The average value of the passive type (A), the directed type (B), and the active type (C) is estimated by the T-test. The astarisc shows the item a significant difference ($p < 0.05$) is accepted.

All subjects' average initial time of each types is 1.08 s (SD = 0.33), 1.01 s (SD = 0.48), 1.14 s (SD = 0.45).

Significant differences ($p < 0.05$) are accepted between passive type (A) and the directed type (B) and between passive type (B) and the directed type (C) (Table 2). But there are only little differences (from 5.6 to 6.5 %) in these average time.

In initial time, it can be said that there is no big difference between the passive type (A), the directed type (B), and the active type (C).

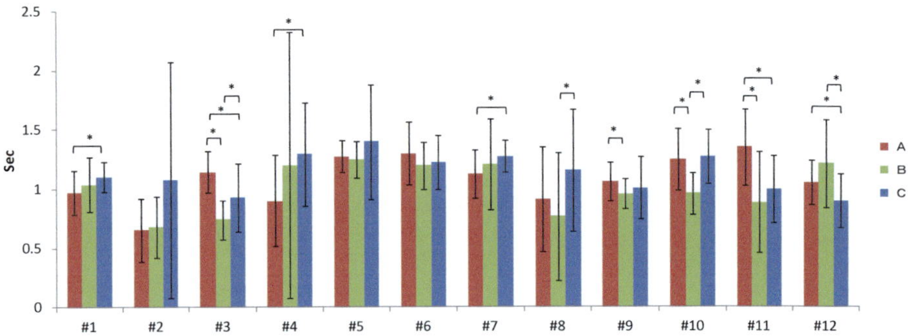

Fig. 9 Initial time

Table 2 Differences between each test

	Initial time		Disposal time	
	Average	SD	Average	SD
A	1.08 s	0.33	2.90 s	0.95
B	1.01 s	0.48	2.42 s	0.78
C	1.14 s	0.45	2.35 s	0.81
	F-TEST	T-TEST	F-TEST	T-TEST
A–B	4.67E-09	**4.65E-02**	1.07E-03	**3.53E-09**
A–C	1.26E-06	1.37E-01	8.23E-03	**2.44E-11**
B–C	1.43E-01	**2.57E-03**	2.50E-01	3.03E-01

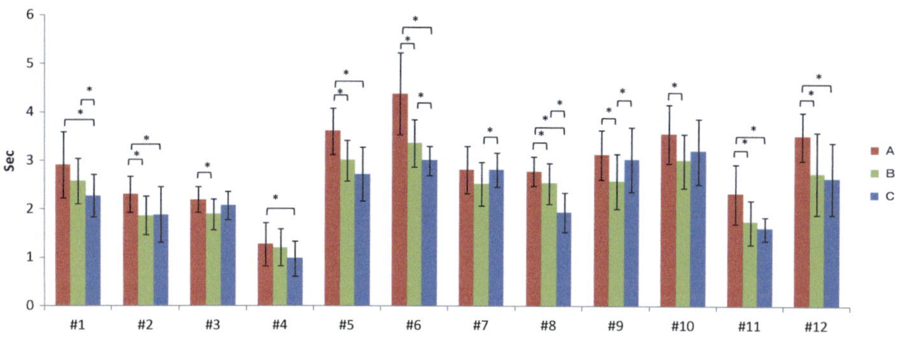

Fig. 10 Disposal time

4.5 *Evaluation to Disposal Time*

The average of each subject's disposal time is similarly shown in Fig. 10. It can be said that the directed type (B) and the active type (C) of disposal time are shorter to the passive type (A) in 11 subjects among 12 subjects. Moreover, in 8 subjects, it can be said that the active type (C) has an effect of a disposal time.

All subjects' average disposal time of each types is 2.90s (SD = 0.95), 2.42 s (SD = 0.78), 2.35 s (SD = 0.81).

In disposal time, Significant differences ($p < 0.05$) are accepted between the passive type (A) and the directed type (B) and between the passive type (A) and the active type (C) (Table 2).

4.6 *Evaluation to Productivity*

The rate of the time savings in the directed type (B) and the active type (C) is computed on the basis of the passive type (A). The scatter diagram which makes the horizontal axis as initial time and the vertical axis as disposal is shown in Fig. 11.

Maximum of 25–35 % of shortening effect is seen initial time. However, the difference is minute on the average.

Average 16.4 % of shortening effect on the directed type (B) and average 19.0 % of shortening effect on the active type (C) are seen disposal time.

Improvement of the productivity by a time savings is accepted.

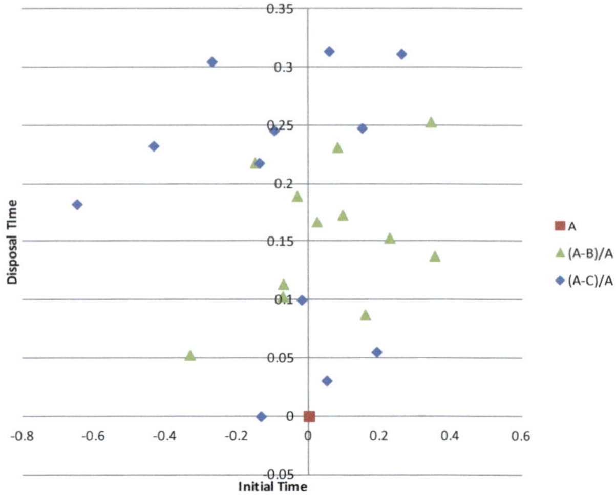

Fig. 11 Time saving effect

5 Conclusion

As methods of the rationalization, it can be considered that a surveillance system shows some information to operators in there waiting time.

In simulated environment, shortening of 16.4–19.0 % of processing time was obtained by displaying information on waiting time to operators.

An effect is acquired also by a simple method like a directed type. Therefore, it is thought that this method is applicable to general work with a worker's waiting time.

The surveillance center was assumed in this paper. However, this method can apply also to the work in a factory or a call center.

References

1. Integrated Control Center Website. http://www.mind.ad.jp/service/icc/
2. Ogata H (2010) Cloud-oriented data center platform. NEC Tech J 5(2):10–15
3. Ito H (2011) Operation efficiency improvement for IT infrastructure thought automation technology. FUJITSU 62:42–46
4. Otsuki M (2009) Study of service operation under unattended ICT system. Unisys Technol Rev 100:53–64
5. Tamura Y et al (2013) Activeness improves cognitive performance in human-machine interaction. J Adv Comput Intell Intell Inform 17(3):425–432
6. Bechara A et al (2005) The somatic marker hypothesis: a neural theory of economic decision. Games Econ Behav 52:336–372
7. Uozumi M et al (2012) Automation of supervisory service and examination of the subject of operation. In: SICE annual conference 2012, pp 903–908

Personalized Information Service Model that Reflects Individual's Will

Yuri Nakagawa, Yuuki Matsuda, and Tetsuro Ogi

Abstract Recently personal data is attractive for the companies to use for marketing and provide a personalized service. However, it is difficult for individuals to access these data, and they don't have any means to benefit from their own records. To solve these problems, we suggested framework for handling personal data to give the power back to individuals. Especially we focus on buying information, to build this framework, we conducted survey research how to handle it. In this paper, to build this framework to give rewards back to individual without any stress, we discussed about a new digital signage system.

Keywords Personal data • Life log • Big data • Buying information • Digital signage

1 Introduction

Recently, with the development of the information communication technologies and the pervasive applications, personal data such as buying information and action history have been recorded throughout the society.

The personal data consists of personal attribute data such as address, birth, title and so on and personal history data such as buying information, action history, food history and so on. These data can be analysed as big data and are attractive for not only the companies to use for the marketing and for providing a new personalized service but also the government. For instance, the Japanese government is offering a 5.3 billion yen subsidy in 2012 [1]. JIPDEC, which is a non-profit organization established with the aim of enhancing the development of economy and community offers many services for personal information protection and information security [2].

However, it takes a long time to build social systems, and individuals have difficulties to access these data, and they usually do not have any means to benefit

Y. Nakagawa (✉) • Y. Matsuda • T. Ogi
Graduate School of System Design and Management, Keio University, Collaboration Complex, 4-1-1 Hiyoshi, Kohoku-ku, Yokohama, Kanagawa 223-8526, Japan
e-mail: yurinkgw@iam.u-tokyo.ac.jp

T. Maeno et al. (eds.), *Serviceology for Designing the Future*,
DOI 10.1007/978-4-431-55861-3_23

from their own records, even though they generate these data by themselves. In this study, we suggested a framework for handling personal data to give the power of personal data back to individuals in balance between individuals and companies. The company can also use it freely across the companies though most current companies hold only limited personal data that is collected by each company, if individual allow using it.

In this paper, we focus on buying information that is got automatically using the Point of Sales (POS) system [3] that was developed in our previous work. In this system, when the customers use the IC card, the POS system sends the buying information to the server so that the customers can use the recorded data. Especially, in this paper, we classify the individual's will for handling buying information based on the questionnaire survey, and propose a template for the self-control of buying information.

The next section shows the situation about personal data and Sect. 3 explains the proposed framework in handling personal information. In Sect. 4, we discuss the classification of individual's will in handling buying information based on the questionnaire survey and propose the templates for controlling it. In addition, in Sect. 5, we propose the idea of new digital signage system that displays personalized data based on the proposed framework.

2 Market Trend and Related Works

In these days, not only company but also the government is engaging in a lively exchange of opinions regarding "Handling about Personal data" [4]. Moreover, here are many studies about personal data, and also each country has national strategies and policies [5]. In this section, we introduce hot market trend and related works and discuss how different from our study.

2.1 World Situation

The World Economic Forum reported "Rethinking Personal Data: Strengthening Trust" in 2012 [6]. They discussed how the potential value of personal data could be unlocked. But they focus on only for the benefit to companies and organizations. Of cause they are care about individual's benefit but not first priority.

In EU, the Commission proposed the EU legal framework on the protection of personal data in 2012 [7]. They said "Everyone has the right to the protection of personal data." This is very strong law and they forced to obey this law to many countries under EU. But they focus on only benefit to individual, not to balance between companies and individual.

In U.K., BIS (Department for Business Innovation & Skills) opened new projects as midata [8] in April, 2011. It is the one of the Consumer Empowerment

Strategy. This project aim that, (1) get more private sector businesses to release personal data to consumers electronically, (2) make sure consumers can access their own data securely, (3) encourage businesses to develop applications (apps) that will help consumers make effective use of their data. Actually, many companies (ex. Google, British Gas, Lloyds TSB and O2.) are joining this project. It is very similar to our framework, but they focus on only how to access to it which "the company hold". They are unmentioned to "Could download their own data" and "Hold it by themselves".

2.2 Market Trend; About the Companies

In business, there are many companies that use personal data in the world. For instance, Blukai, Tesco, Personal inc., Infochimps, Demdex and so on [1, 9, 10]. These companies use personal data to provide many services which pass on the benefit to individual. For instance, in Japan, McDonald and Zennisshoku give some discount ticket and information of basement merchandise to customer by using and analysing customer's personal information. But unfortunately, this benefit is not enough to customer, and also there is no reasonable way to give our personal data to the companies under our control. And also, Sony and Kawasaki city office made the cloud systems that are gathering personal history data of using medicine in Kawasaki, Japan. Sony collects these data into cloud system and provides it to pharmaceutical company as statistical information. Of cause, customer can receive many kinds of service and live much more convenient, but, they do not ask individuals to allow using it. On the other hand, there is much news that the company sells personal data without any individual's permission in Japan. And also, in U.S.A., the Federal Trade Commission penalizes online Company in Sale of Personal Data for protecting consumer.

There is no way to balance the benefits of both individuals and companies.

2.3 Related Works

There are many related works, which focus on personal data.

For instance, there are some developed systems that are able to stock own personal data such as movies and pictures automatically [11–13]. It becomes much easier to collect personal data automatically than before, and also we can enjoy collecting it much more than before.

And also, there are some systems that are able to manage personal own data. PIMA enhanced both human-computer interaction and application integration for PIM on mobile devices [14]. It becomes more perceived usefulness, ease-of-use, and efficiency of PIM on mobile devices. Open PDS by MIT allows individuals to collect, store, and give fine-grained access to their own data all while protecting their privacy [15]. They provide a secured space. Therefore, individual can manage and control the flow of data much safer than before. With these systems such as PIMA and open PDS, individuals are able to use or manage their own data much easier and safer than before.

They become recognizing and discussing consumers' rights to data ownership in personal data and how important individual-oriented information privacy model [16, 17]. However these works are discussing only how to collect such as which device use or how to manage. They do not refer to the need to get reward and be balanced the benefits of both individuals and companies.

We aimed to suggest framework for handling personal data to give the power of personal data back to individuals with rewards on each investment. And also, at the same time, the company can also use it freely and across the companies, if individual allow using it. Especially, in this paper, we classify the individual's will for handling buying information based on the questionnaire survey, and propose a template for the self-control of buying information.

3 Problems and Solutions

In this section, we discuss about three problems that we are able to recognize under the situation of current personal data and suggest the framework to solve it.

3.1 Problems

There are three problems to solve, under this situation:

- We don't figure all of our own personal data out, and also there is no reasonable way to control it.
- There is no reasonable way to give our personal data to the companies under our control and also we are not able to receive a proper reward from companies.
- The companies can hold only limited personal data that is collected by each company.

There are many applications that collect their own personal data on mobile phone and iphone, and we are using it easily. We could get personal data and recognize them much more easily than before. However, we could not have gotten a full picture of them yet. And also, for many years, the companies have used personal data, as customer data to better understand consumer and use for marketing without any permission from each individual. However, individuals have not only difficulties to access, but also do not have any means to benefit from their own records, even though they generate. In fact, according to previous research [3], many people are eager to know and handle these personal data by themselves more.

On the other hand, the companies can hold only limited personal data that is collected by each company. Not only in Japan but also in UN, if they can identify an individual, it's forbidden to share personal data among companies by law. However they want to know whole data for marketing. Therefore, sometime they edit it to avoid identifying an individual and then sell or buy it, although it is offensive to some individuals. In fact, we can see it on the headline news throughout Japan and also, in U.S.A, FTC penalized one of the companies that sold personal data.

Therefore, we aim to suggest framework for handling personal data to give the power of personal data back to individuals [18]. And also, at the same time, the company can also use it freely and across the companies, if individual allow using it. We need a new framework that can be balanced the benefits of both individuals and companies.

3.2 Solutions

To solve these three problems, we suggest framework as Fig. 1 for handling personal data to give the power of personal data back to individuals. And also, at the same time, the company can also use it freely and across the companies, if individual allow using it.

About this framework, individuals can hold their own personal data in Information bank, and they are able to manage and access it whenever they want to use.

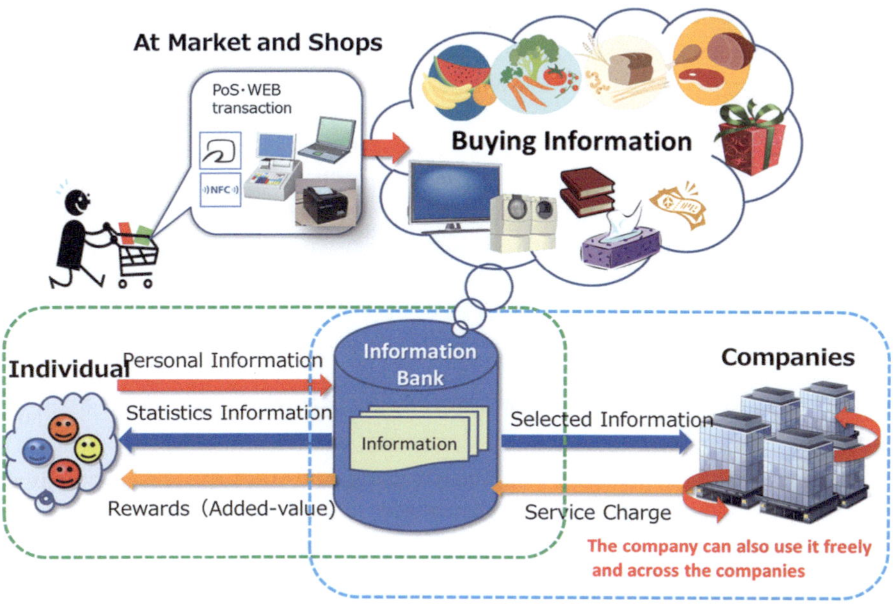

Fig. 1 Overview of proposed framework

Information bank is a secured space and individual can't change their own data by themselves, so personal data can be saved safer. And also, individuals can receive new services, like household account application that can manage how they spend their money and so on. At the same time, the company can also use it freely and across the companies, if individual allow using it.

This is a new framework that can be balanced the benefits of both individuals and companies.

4 How to Handling Personal Data

In this section, we discuss about the structure that is able to handle personal data easily by focusing on buying information.

4.1 Structure

In previous study, to solve the first problems, we studied about system that could get personal data, especially focus on purchase information [3]. We made prototype of household account application getting buying information automatically from Point Of Sales systems by using network, and evaluated its marketability and effectiveness. On household account application, we don't see a row data; we can see well-edited data to help them understand easily. And also, to aim for the realization of this framework and solve the second problems, we studied about system structure that the companies are able to use personal data which individual selected under their control. To build this structure, there are two points: (1) Individual can select which data will they provide and which companies will be able to use it. (2) Individual who put in control their data and allow using it for the companies can receive much more rewards than the other who wouldn't like to participate this framework. With these two points in mind, we discussed how to handle buying information.

4.2 Consideration About Buying Information

Where and when did you buy? What kinds of things did you buy? This information is so-called buying information and we can see these on receipt. On the receipt in Japan, there is much information as below: The name which you bought, the company which made, the place where you bought, the time when you bought, the calorie of the meal which you ate at the restaurant, and so on. These have much personal information.

To be able to select without any resistance and to allow using it for getting more rewards on this framework, we conducted survey research how to handle buying information. Especially wanted to know:

- Is there any difference by individual?
- What kind of information in buying information does individual want to keep a secret?

4.3 The Questionnaire Survey on Handling Buying Information

To investigate the resistance to use it for getting more rewards on this framework, we conducted survey research. The overview of this questionnaire is as below.

We asked open and closed ended questions on the web site and print.131 people, who consist of students and teachers at graduate school of System Design and Management, Keio University, personal friends on Facebook and prints and voluntary bodies and kindergarten mother, answered to us. The rate of this survey is; male 51 %, female 49 %, Married 61 %, Unmarried 39 %, 20's 24 %, 30's 37 %, 40's 18 %, 50's 6 %, 60's 8 %, over 60's 5 %.

In the questionnaire, we asked individuals whether they might give their buying information to the different types of companies under different cases. For instance, under case 1, there is no possibility to identify individuals. Under case 2, there are a few possibilities to identify individuals. Under case 3, there are some possibilities to identify individuals.

About buying information, we grouped it into categories such as food costs, luxury goods costs, entertainment costs, daily necessities, expendable supplies costs, book costs, medical expenses, expense account, educational expenses, the expense that it cost for work, insurance cost, tax, costs of utilities, communications expenses, cut rate, calorie of the food. There are 15 kinds of categories in total.

As for the products in each category, we asked respondents to make judgment whether they allow the buying information that consists of product name, or product name and date information, or product name and place information to be opened to the companies. In this case, the companies were classified into major company like Wal-Mart Stores and Carrefour S.A., small private shop which user knows each other, and advertising company like Omnicom Group and WPP plc.

4.3.1 Result and Discussion

We were able to get results as follow:

There are many kinds of differences between many kinds of cases, which are the results for each expense category, each company type, gender, civil status, and age.

For instance, Fig. 2 shows the differences between married and unmarried people. For instance, more than 47 % of unmarried people feel free to open their Tax information to small private shops. However, there are only about 25 % of married people who are happy to do the same.

Table 1 shows the results of the test of population rate by civil status. In particular, there are many significant differences at the 5 % significance level. In

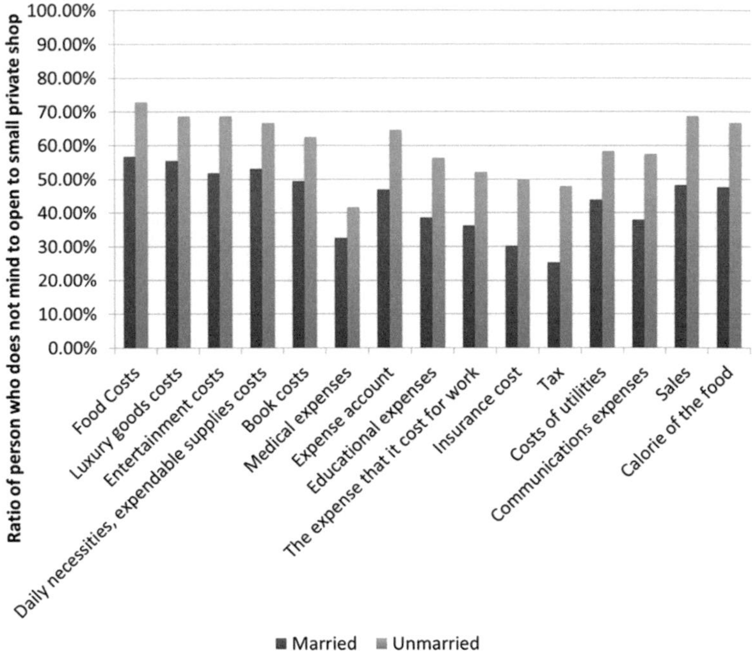

Fig. 2 Differences by civil status

Table 1 Differences by civil status

Category	Ration of person who does not mind to open to small private shop		Data	P value
	Married	Unmarried		
Food costs	44 %	64 %	Trade name + date info	0.027240551
Food costs	42 %	64 %	Trade name + place info	0.013425188
Expense account	7 %	21 %	Trade name + date info	0.035866731
Luxury goods costs	42 %	64 %	Trade name + date info	0.013425188
Entertainment costs	42 %	60 %	Trade name + date info	0.044095502
Educational expenses	30 %	52 %	Trade name + place info	0.018887314
Insurance cost	26 %	45 %	Trade name + date info	0.02402458
Insurance cost	25 %	45 %	Trade name + place info	0.015895168
Tax	22 %	43 %	Trade name + date info	0.012503024
Tax	21 %	41 %	Trade name + place info	0.015180683

Table 2 Reward that answerer want to get (a part)

What do you want to get back as reward?
I do not care about any reward. Even nothing back to me, I can give my personal data
Never open my personal data. Even the companies give a huge amount of rewards
Cash (1000 yen, 2000 yen, 50,000 yen, 10,000 yen and more)
Discount
Coupon or point (In Japan, there are many kinds of point card, and use point as a cash)
Exchange to air miles
Many services (application about house hold accounting and reduce the time and effort spend on counting.)
Attractive/useful information (discount, recommendation, and so on)
Not necessary

many categories we could recognize differences between civil statuses on the 5 % significance level. There are differences between married and unmarried people. This implies that people become more conservative when they get married. (They may be thinking more about risk when they have a family.)

Therefore, there are differences between married and unmarried people.

We also asked open-ended questions. We got results as follow.

Table 2 shows a part of the answers for "What kind of reward do you want to get back, when you open your personal data to the company?"

There are many different kinds of answers. Though some answerers want to get many rewards like cash, services and attractive information, others do not want to any rewards even they open it to the company. It seems that the opinions of recent people have become different compared to previous research [3].

5 Application Idea

In previous section, we could get the results that there are differences among people from the questionnaire. The result shows the differences of among each expense category, gender, age, and the company. And also, there are some different opinions about value of personal information. Some wants to open their personal data, but the other do not. Some wants to manage their own data, but the other do not.

There are differences between many different situations. It means that it is totally different among individual.

Therefore to build this structure to give rewards back to individual without any stress, it is very important to accept each individual's requests.

In this section, based on the result, we discussed about developing a new digital signage system based on this framework for handling personal data.

5.1 Overview of Application System

Digital signage is an electronic display that shows information, advertisement, map and so on. In these days, we are easy to touch it at station, airport, department store, hospital and so on.

Usually digital signage shows information to us one-sidedly, or people have to search to get useful information. To get useful information for them, it takes a little bit long time and sometime it is getting trouble. And also, sometimes we could not find out and miss what they want to get.

Figure 3 shows overview of digital signage system. In our idea, the companies that want to give useful information to individuals are able to give personalized information like order-made service by using their personal data under their control. If individual allowed using own personal data and opening it, it would be much easier. It means that they will be able to get personalized useful information even if the first visit at the place for them. It will help their visit to enjoy. Moreover by using digital signage, people will come to the place and take time there. It helps to the company side to visit to their shops directly. (If we do not use digital signage and use only mobile phone on our service, people will not come to the place and take time).

Thus, people can get personalized information and services from the digital signage system placed in the public space such as information centre, shopping centre, or entrance of shopping street.

5.2 Template for Individual in this System

As based on the result of questionnaire of previous study, Fig. 4 shows around 60 % people do not mind to open their personal data to advertising company except categories of medical expenses and tax. This result means that more than half of people do not mind to open their data to any companies, and also people do not mind to get any advertisement from them. Based on this result, we made a template of that the user can use it for the digital signage system.

In this system, a template shown in Fig. 5 is used, so that the user can select the categories of buying information that is given to the advertising company. On Fig. 5, vertical scale indicates the level on a zero-to-one scale that individual does not mind to open their information to the companies. (Ex: If individual does not mind to open at all, the rate is 1. If do not want to open at all, the rate is 0.) There are 9 kinds of levels on this; Level 0, level 1−, level 1+, level 2−, level 2+, level 3−, level 3+, level 4− and level 4+. Each level is as follow;

On level 0, individual will not open any personal information to anyone, but they want to use this system. If people want to select this level, people can get only information that is not selected personally from restaurant and shops.

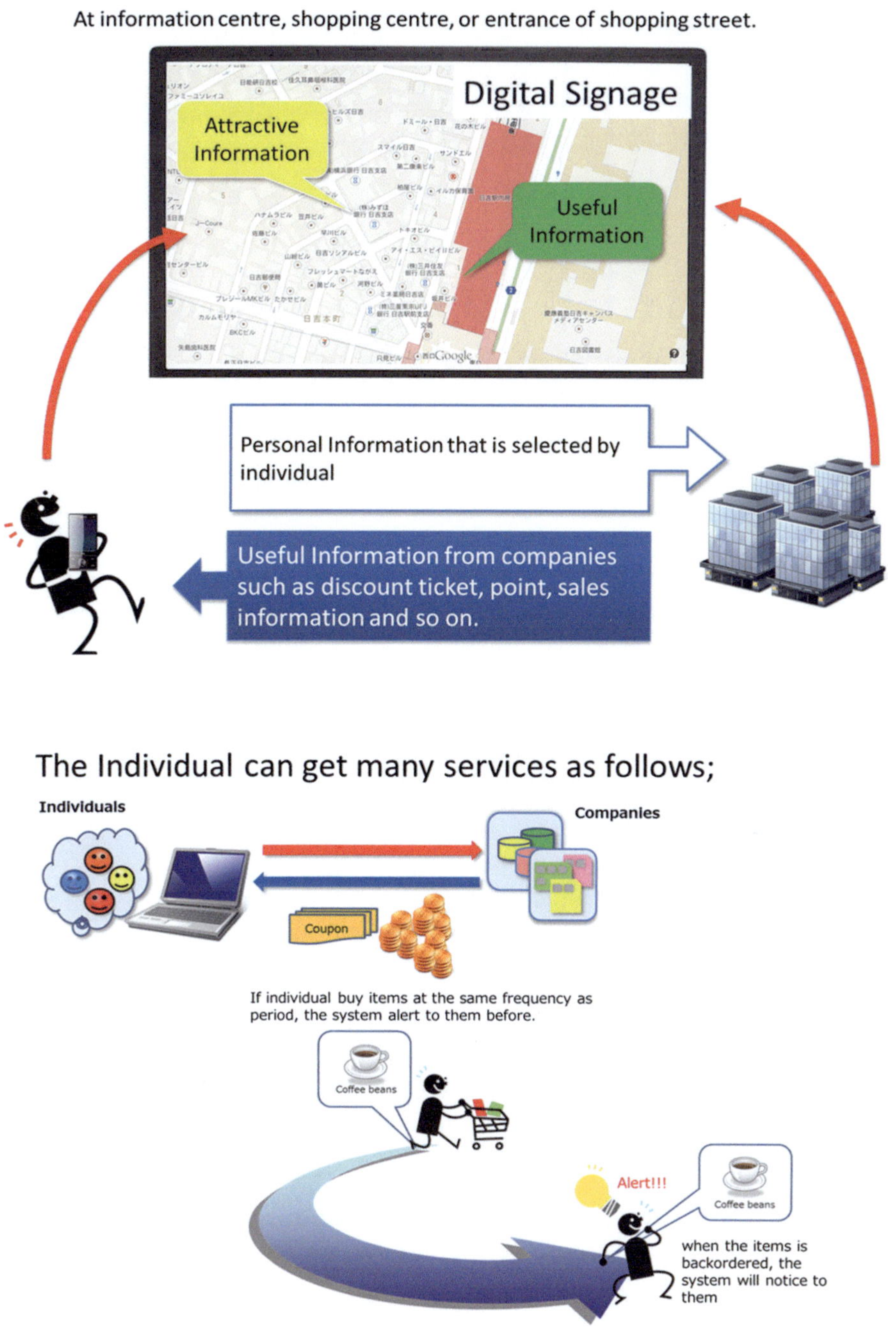

Fig. 3 Overview of digital signage application system

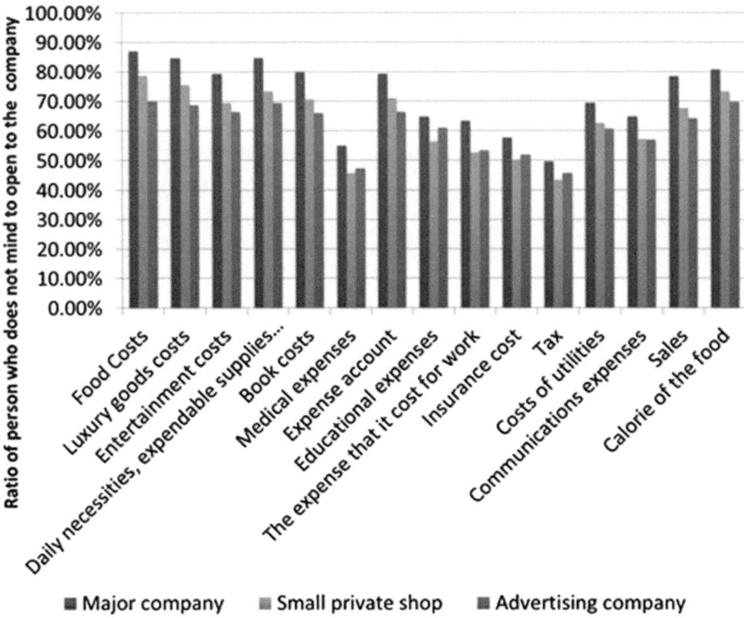

Fig. 4 Ratio of person who does not mind to open to advertising company (Differences by company)

On level 1, the default template is that individual will open only trade name and cost. And also, they will open all categories except medical expenses, insurance costs and tax.

If people want to select this level and they provide their trade name and cost, people can get useful information such as discount ticket, points or coupon from shops and restaurant. For instance, if people bought some pasta, some restaurant, which provide Italian food or have pasta menu, can notice their information and urge them to come to the place. Moreover, if some shops sale it at the same time, they can provide sale information to people directly.

When individual want to select less categories which is different from the default ones, it will be level 1−, and they want to select all categories, it will be level 1+.

On level 2, the default template is that individual will open only trade name, cost and date information. And also, they will open all categories except medical expenses, insurance costs and tax.

If people want to select this level and they provide their trade name, cost and date information, people can get useful information such as timely discount ticket, points or coupon from shops and restaurant. Moreover, the restaurants and shops urge them to come their places. For instance, people bought some coffee beans same frequencies, the shops can alert that "it is the time to buy coffee beans, and we can provide a discount ticket!" The companies can provide much more suitable and personalized information to people than level 1.

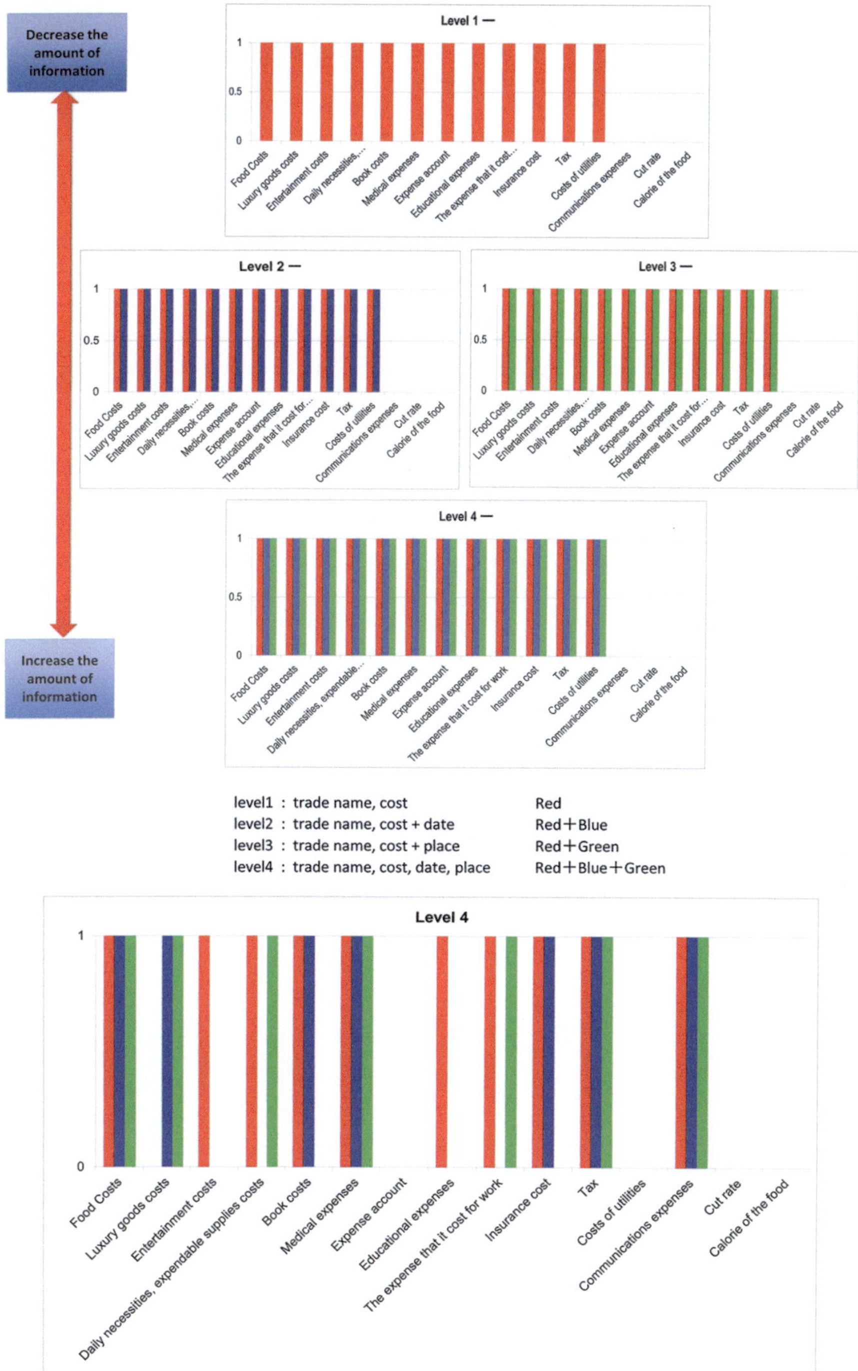

Fig. 5 Sample of template for individual

When individual want to select less categories which is different from the default ones, it will be level 2−, and they want to select all categories, it will be level 2+.

On level 3, the default template is that individual will open only trade name, cost and place information. And also, they will open all categories except medical expenses, insurance costs and tax.

If people want to select this level and they provide their trade name, cost and place information, people can get useful information such as discount ticket, points or coupon from nearest shops and restaurants. The companies can provide much more suitable and personalized information to people than level 1.

When individual want to select less categories which is different from the default ones, it will be level 3−, and they want to select all categories, it will be level 3+.

On level 4, the default template is that individual will open only trade name, cost, date information and place information. And also, they will open all categories except medical expenses, insurance costs and tax.

If people want to select this level and they provide their trade name, cost, place information and date information, people can get all useful and attractive information such as discount ticket, points or coupon from nearest shops and restaurants timely. And also they can get much more discount, points, and coupon than other levels.

On level 4+, if people want to select this level and they provide their trade name more than any other levels. The companies can provide best-personalized information to people.

The default is level 1. If individual want to get much more useful information and services from shops and restaurants, they can choose levels, categories, and personal data such as date or place information. If individual get more information, it means that they should open much more personal data to companies.

In 2020, there will be the Olympic Games in Tokyo. Many tourists will come and see many games, and enjoy staying. If tourist can use this system, they will be able to get personalized useful information even if the first stay in Tokyo for them. It will help their stay to be more joyful in Tokyo.

6 Conclusion

In this paper, we classify the individual's will for handling buying information based on the questionnaire survey, and propose a template for the self-control of buying information.

With using these template, we suggested the new digital signage system based on the result of questionnaire of previous study, to solve the problems that there is no reasonable way to give our personal data to the companies under our control, we are not able to receive a proper reward from companies, and the companies can hold only limited personal data that is collected by each company.

In next study, we will make a prototype of this digital signage system which using a template for the self-control of buying information, and do demonstration experiment at shopping street.

References

1. Ministry of Economy, Trade and Industry (2012) Overseas trends related to personal data. METI, Japan. http://www.meti.go.jp/committee/kenkyukai/shoujo/it_yugo_forum_data_wg2/pdf/001_04_00.pdf
2. Japan Information Processing Development Center (2012) Personal information protection policy. JIPDEC, Japan. http://www.jipdec.or.jp/project/anshinkan/doc/2011/01.pdf
3. Nakagawa Y, Izui T, Isegawa A, Arai K, Sonoda M, Narita M, Ogi T (2012) Proposed framework for collecting and handling personal data. IEICE Transac Inf Syst D J95-D (4):825–833 (in Japanese)
4. Purtova N (2009) Property rights in personal data: learning from the American discourse, Computer law & Security review 25, Netherlands, pp 507–521
5. OECD (2011) National strategies and policies for digital identity management in OECD countries, OECD digital economy papers, No. 177. OECD Publishing, Paris. doi:http://dx.doi.org/10.1787/5kgdzvn5rfs2-en
6. World Economic Forum (2012) Rethinking personal data: strengthening trust, Wednesday 16th of May 2012. Prepared in collaboration with The Boston Consulting Group. http://www3.weforum.org/docs/WEF_IT_RethinkingPersonalData_Report_2012.pdf
7. European Commission (2012) Proposal for a REGULATION OF THE EUROPEAN PARLIAMENT AND OF THE COUNCIL on the protection of individuals with regard to the processing of personal data and on the free movement of such data (General Data Protection Regulation), http://ec.europa.eu/justice/data-protection/document/review2012/com_2012_11_en.pdf
8. Department for Business, Innovation & Skills on GOV.UK. midata, https://www.gov.uk/government/policies/providing-better-information-and-protection-for-consumers
9. JEITA (2013) Information security survey report. JEITA, Japan. http://home.jeita.or.jp/upload_file/20140407115339_i6nm82MoJO.pdf
10. OECD (2012) Exploring data-driven innovation as new source of growth, mapping the policy issues raised by "Big Data". In: OECD Technology Foresight Forum 2012 – Harnessing data as a new source of growth: big data analytics and policies, 22 October, OECD Headquarters, Paris, France
11. Mann S (1998) 'WearCam' (The Wearable Camera): personal imaging systems for long--term use in wearable tetherless computer--mediated reality and personal Photo/Videographic memory prosthesis, at Proceedings of ISWC, Washington, USA, pp 124–131
12. Aizawa K, Hori T, Kawasaki S, Ishikawa T (2004) Capture and efficient retrieval of life log. In: Pervasive 2004 workshop on memory and sharing experiences, pp 15–20, 20 April 2004, Vienna, Austria
13. Kim I-J, Ahn SC, Ko H, Kim H-G (2008) Automatic lifelog media annotation based on heterogeneous sensor fusion, 20–22 Aug 2008 at IEEE international Conference on multisensor fusion and integration for intelligent systems, Seoul, Korea
14. Zhou L, Mohammed AS, Zhang D (2012) Mobile personal information management agent: supporting natural language interface and application integration. Inf Process Manag Int J 48 (1):23–31
15. Montjoye YA, Wang SS, Pentland AS (2012) On the trusted use of large-scale personal data. IEEE Data Eng Bull 35(4):5–8

16. Hoffman DL, Novak TP, Peralta M (1999) Building con trust online: how merchants can win back lost consumer trust in the interests of e-commerce sales. Communications of the ACM 42 (4):80–85
17. Wang T, Srivatsa M, Liu L (2012) Fine-grained access control of personal data, at SACMAT '12 Proceedings of the 17th ACM symposium on access control models and technologies, Newark, NJ, USA, pp 145–156
18. Kanasugi H, Kumagaya J, Thang TD, Shibazaki R (2007) Information bank: development of management platform of personal information, Interaction 2007, 15 March, Japan, (in Japanese)

Ranking Smartphone Apps Based on Users' Behavior Records

Song Luo, Maiko Shigeno, and Wenbo Ma

Abstract This paper discusses the ranking problem of how to assign a linear order to a given collection of smartphone apps by using users' behavior records. The desired linear order is thought of as an observable representation of users' latent aggregate preference of the apps, which balances their conflicting individual preferences as far as possible. The app ranking problem can be interpreted as a stochastic acyclic subgraph problem on a complete digraph with each arc associated with a constant weight obtained from a known probability distribution.

Keywords Usage pattern • Users' preference • Bayesian inference • Stochastic acyclic subgraph problem

1 Background and Motivation

Smartphone usage has grown remarkably in recent years and becomes an increasingly popular choice for mobile communicating and computing. Meanwhile, a huge amount of interface-friendly apps emerge these years and are able to provide diverse functionality and extended capabilities for smartphone users. Accordingly, many practical topics concerning smartphone apps arise and have attracted attention from both academic groups and business organizations.

However, many problems regarding app management and system design remain open. Particularly, the problem of ranking apps into a linear order is probably one of the most interesting and fundamental topics. This is because app ranking is often found to be a useful measurement for evaluating the quality of apps. An app of higher quality often attracts more users and hence will likely gain more benefits, even the app itself is free to use. Moreover, app ranking also plays an indispensable

---

S. Luo • W. Ma (✉)
Graduate School of Systems and Information Engineering, University of Tsukuba, 1-1-1 Tennodai, Tsukuba, Ibaraki 305-8577, Japan
e-mail: maiko@sk.tsukuba.ac.jp

M. Shigeno
Faculty of Engineering, Information, and Systems, University of Tsukuba, 1-1-1 Tennodai, Tsukuba, Ibaraki 305-8577, Japan

T. Maeno et al. (eds.), *Serviceology for Designing the Future*,
DOI 10.1007/978-4-431-55861-3_24

role in system design such as building recommender system for app users and developing consultant system for application developers and so on. Although app stores such as Google play Android app store and iTune Apple app store provide ranking service and update their ranking results periodically, their methods are hidden to the public: both algorithm design and data remain confidential for reasons. Due to the importance of app ranking problem, we are motived to seek new app ranking method that can be applied to obtainable datasets.

This task requires a properly processed dataset and a ranking algorithm, while there is no doubt that the dataset is the first key ingredient. App developers are usually allowed to trace and study users' behaviors for management purpose, in order to improve the quality of their apps as well as their service. Therefore, the data concerning users' behaviors may be properly collected and used for ranking purpose. This naturally leads to the problem of how to assign a linear order to a given collection of apps based on users' behavior records.

2 Data Description

The dataset of our interest is provided by FULLER, Inc., which is obtained during the period from Oct. 2012 to Jun. 2013. From the data, it can be learned that which Android device installed what Android app(s) at what month. Each piece of the data is called an Android user's behavior record. Particularly, a device's sequential states of whether it has a specific app installed over a period of time can be represented as a binary sequence, called a usage record. Notice that it may happen that an Android smartphone user has more than one devices and hence several device IDs. However, since who has which device ID(s) cannot be learned from the data and that each device corresponds to a unique device ID, it can be assumed that each user has only one device and hence can be labeled by his/her device ID (Here device ID means alpha-numeric ID given by randomly in order to identify the same device, but not the production serial number of it). So far, the full dataset contains millions of the users' behavior records, and its size is still increasing due to the arrival of new data. With focus on Android apps, we show in Table 1 a small part of the data.

Next, we introduce some notation that will be consistently used throughout this paper, to describe the structure of data. Let $\mathbf{D}$ and $\mathbf{A}$ be collections of n devices (users) and m applications, respectively. The structure of the time range consisting of l consecutive time intervals can be described as an ordered set $\mathbf{T} = \left(\mathrm{U}_{i=1}^{l}\{T_i\}, \prec_t\right)$, where $T_i \prec_t T_{i+1}$ for $i = 1, \cdots, l-1$.

Let $r(D, A, T)$ be the indicate function of $D \in \mathbf{D}$, $A \in \mathbf{A}$ and $T \in \mathbf{T}$, such that $r(D, A, T) = 1$ if app A can be found installed on device D at T otherwise 0. This allows us to define the usage record of D with respect to A during $\mathbf{T}$. Denote by $r(D, A, \mathbf{T})$ the binary sequence of usage record $(r(D, A, T_i))_{i=1}^{l} \in \{0, 1\}^l$. Moreover, without loss of generality, $\mathbf{A}$ is assumed to only contain the active apps that are installed by at least one user during $\mathbf{T}$. That is, for any $A \in \mathbf{A}$ it holds that

Table 1 Example of users' behavior records

Device ID	App	Dec	Jan	Feb	Mar	Apr
5048...a94d	com.google.android.apps.plus	1	1	1	1	1
5048...a94d	com.twitter.android	1	1	0	0	0
5048...a96e	com.facebook.katana	1	1	1	1	0
5048...a96e	jp.mixi	1	1	1	1	1
5048...a96e	com.co mm	1	1	0	0	0
15048...a96e	jp.ameba.candy	0	1	1	1	1
5048...a96e	com.twitter.android	0	0	1	1	1

$\sum_{D \in \mathbf{D}} \sum_{T \in \mathbf{T}} r(D, A, T) \geq 1$. Similarly, **D** is assumed to only contain the users who have ever installed at least one app of **A** during **T**. That is, for any $D \in D$ it holds that $\sum_{A \in \mathbf{A}} \sum_{T \in \mathbf{T}} r(D, A, T) \geq 1$.

3 Related Work and New Challenge

Fujimoto et al. [1] used the same dataset for extracting noticeably "good" apps from a large amount of apps that are not only free to install, but also can be uninstalled and reinstalled repeatedly free from any restriction. In their proceeding paper, score functions were constructed to measure the goodness of apps, and the apps with noticeably high score values were considered "good" and then picked out.

Although the extracted apps could further be ranked into a linear order according to their score values, it may be still interesting to know the rank obtained by pairwise comparisons of the apps. This rank is expected to be able to describe users' aggregate preference of the apps.

If users are thought of as voters and the apps as candidates to be ranked, then the app ranking problem seems closely related to political voting problems, and naturally mathematical voting method is expected to apply and work. However, the app ranking problem at hand is essentially different from any class of the voting problems as far as we know. First, in a typical election each voter is asked to fill a ballot, which can be used to explicitly express a voter's individual preference of the candidates. However, such explicit ballots cannot be directed learned from the dataset of users' behavior records. Thus, new perspectives have to be introduced in order to express not only users' individual preferences but also users' aggregate preference as well. Second, in a typical election, voters are only allowed to vote once, and their historical opinions are not taken into consideration. In this problem, however, from the data we can observe that users' behaviors may change from time to time, which probably results from their changes in preference during their usage of the apps. Thus, the way of expressing users' individual preferences is expected to be capable of reasonably reflecting their individual preferences over a period of time. These new features of the app ranking problem raise new challenges.

4 Users' Indidividual Preference

This section introduces a deterministic model of users' individual preferences of the applications to be ranked. The main idea is to give an ordinal structure to the set of usage patterns, which can be used to reflect users' satisfaction level during their usage of the apps. And, the relation of a user's preference between two apps that she has ever installed within a given period of time can be determined by comparing her satisfaction level of the two apps.

4.1 Usage Pattern

Fujimoto et al. [1] introduced the concept of usage pattern of apps, in order to classify users into a reasonable number of types based on their usage records.

When users' usage records have a sufficient length (here, $l \geq 4$ is assumed, they become quite informative and expressive. However, the number of all possible usage records is also exponentially large, that is, 2^l. This makes it less appropriate to classify users if users' usage records are directly used, especially when l is large.

To overcome this difficulty, the concept of usage pattern was introduced, which can be used to reduce the number of user types. The basic idea is to classify users' usage records into a small number of classes, each of which is called a usage pattern. The basic observation is that, the 1's or 0's near the end of a user's usage record can be given a more detailed meaning, capable of summarizing her historical behavior records. For example, given $r(D, A, T) = (0, 0, 0, 0, 1, 1)$, the 1 at the fifth coordinate can be interpreted as a state of that A is newly installed by D with respect to $\mathbf{T}$ and the following 1 means a state of that A is continuously installed.

This example also casts a light on the basic idea for classifying users based on their usage records. Formally, given $r(D, A, T)$ and some integer h $(1 \leq h \leq l-3)$, for each $l-h+1 \leq i \leq l$, $r(D, A, T_i)$ will be associated with a state p_i taken from the symbol set $\{N, R, C, I, U\}$, such that

$$p_i = \begin{cases} N, & \text{if } r(D, A, T_i) = 1 \text{ and } \Sigma_{j=1}^{i-1} r(D, A, T_j) = 0 \\ R, & \text{if } r(D, A, T_i) = 1,\ r(D, A, T_{i-1}) = 0 \text{ and } \Sigma_{j=1}^{i-2} r(D, A, T_j) \geq 1 \\ C, & \text{if } r(D, A, T_i) = 1 \text{ and } r(D, A, T_{i-1}) = 1 \\ I, & \text{if } r(D, A, T_i) = 0 \text{ and } \Sigma_{j=1}^{i-1} r(D, A, T_j) = 0 \\ U, & \text{if } r(D, A, T_i) = 0 \text{ and } \Sigma_{j=1}^{i-1} r(D, A, T_j) \geq 1 \end{cases}$$

The resulting symbol sequence $p_{l-h+1}p_{l-h+2}\cdots p_h$ is called the h length usage pattern of $r(D, A, \mathbf{T})$. Intuitively speaking, N represents the state of new installation with respect to $\mathbf{T}$. R re-installation, C continuous installation, U uninstallation and I ignorance (the state of that A is unknown, or has been already known but not yet installed) with respect to $\mathbf{T}$. Therefore, any possible usage record $r(D, A, \mathbf{T}) \in$

Table 2 Classifications of usage records with $h = 2$

1	2	...	$l-2$	$l-1$	l	Usage pattern
All zeros				0	0	*II*
Any			1	1	0	*CU*
All zeros			0	1	0	*NU*
At least a one			0	1	0	*RU*
At least a one				0	0	*UU*
At least a one				0	1	*UR*
All zeros				0	1	*IN*
At least a one			0	1	1	*RC*
All zeros				1	1	*NC*
Any				1	1	*CC*

$\{0, 1\}^l$ can be classified by its usage pattern of a prescribed length h, as long as $l \geq 4$ and $1 \leq h \leq l - 3$. Denote by $\mathbf{C}(h)$ the set of all usage patterns of length h. Obviously, $\mathbf{C}(1) = \{N, R, C, I, U\}$. $\mathbf{C}(2)$ is shown in Table 2.

It is not hard to see that $\mathbf{C}(h)$ can be constructed from $\mathbf{C}(h-1)$ and hence recursively from $\mathbf{C}(1)$. For the sake of convenience, let $\mathbf{C}_1 = \mathbf{C}(1)\backslash\{I\}$ and $\mathbf{C}_2 = \mathbf{C}(2)\backslash\{II\}$. This classification method is sensitive to both users' historical behavior records and to their latest records with respect to a time range. Thus, users' changes in preference during a period of time can be summarized and represented. Moreover, with an appropriate value of h, the total number of usage patterns $5 \cdot 2^{h-1}$ can be much smaller than that of all possible usage records, 2^l. Hence, the value of h can be regarded as the compression degree of users' usage records.

4.2 *Linear Ordering of Usage Pattern*

To model a user' individual preference of apps with his/her usage records, one possible way is to introduce an ordinal structure on $\{0, 1\}^l$, by which each user's individual preference of the apps can be obtained from his/her usage records of the apps. Recall that usage pattern is a compact expression of usage record. Thus, it is more practical to introduce an ordinal structure to $\mathbf{C}(h)$. In this paper, we discuss the ordinal structure of $\mathbf{C}(2)$.

Let $\prec_2$ be the order relation on $\mathbf{C}(2)$. At the vey beginning, however, it can be noticed that what order *II* should be assigned under $\prec_2$ has to be discussed. Without having ever installed A within T, user D may hardly make any meaningful comparison between A and those apps that he/she has ever been installed within T. Therefore, it may be reasonable to remove usage pattern *II* from our consideration. With this presumption, we give two possible ordinal structures restricted to $\mathbf{C}_2$.

The first construction of the linear order of $\mathbf{C}_2$ is given as follows:

$$UU \prec_2 NU \prec_2 RU \prec_2 CU \prec_2 IN \prec_2 UR \prec_2 NC \prec_2 RC \prec_2 CC.$$

This construction is due to a lexicographic order $\prec_L$ on the set of all possible usage records (i.e.,$\{0, 1\}^l$), that is, for any two distinct vectors $r,\ r' \in \{0, 1\}^l$, we say $r \prec_L r'$ if there exists a l_0, where $0 \leq l_0 \leq l$, such that $(r)_j = (r')_j$ for $j = l_0 + 1, \ldots, l$, $(r)_{l_0} = 0$ and $(r')_{l_0} = 1$. Then, recalling that usage pattern $\boldsymbol{p} \in \mathbf{C}_2$ can be viewed as a subset of $\{0, 1\}^l$, for any two distinct usage patterns $\boldsymbol{p},\ \boldsymbol{p}' \in \mathbf{C}_2$ we say $\boldsymbol{p} \prec_2 \boldsymbol{p}'$ if and only if for any $r \in \boldsymbol{p}$ and $r' \in \boldsymbol{p}$ we have $r \prec_L r'$. In other words, this construction can be viewed as a lexicographic order on $\mathbf{C}_2$, which is obtained by using a linear order on $\mathbf{C}_1$, that is,

$$U \prec_1 N \prec_1 R \prec_1 C.$$

And, for any two distinct usage patterns $p_2p_1,\ p_2'p_1' \in \mathbf{C}_2$, where $p_1, p_2, p_1', p_2' \in \mathbf{C}_1$, we say $p_2p_1 \prec_2 p_2'p_1'$ if and only if $p_1 \prec_1 p_1'$, or $p_1 = p_1'$ and $p_2 \prec_1 p_2'$.

The second construction of the linear order of $\mathbf{C}_2$ is given as follows:

$$CU \prec_2 RU \prec_2 NU \prec_2 UU \prec_2 CC \prec_2 RC \prec_2 NC \prec_2 UR \prec_2 IN.$$

In this construction, usage pattern is thought of as the observable reflection of the change of satisfaction level in the last two consecutive time intervals. Notice that, we specify that the usage patterns implying a transition from uninstallation (or ignorance) to installation will be associated with higher ranks than any other patterns, and that those patterns implying a transition from installation to uninstallation (or ignorance) will be associated with lower ranks. Moreover, if a usage pattern represents fewer installation states in a near past period, then it will be associated with a higher rank.

We call the first construction and the second one the lexicographic order of $\mathbf{C}_2$ and the transition order, respectively. There may be other meaningful ordinal structures we can define on $\mathbf{C}_2$, but the lexicographic order and the transition order seem not only natural in intuition but also imply a general idea of constructing ordinal orders on $\mathbf{C}_2$. However, we will leave the problem of how to construct meaningful ordinal structures on $\mathbf{C}_k$ for $k \geq 1$ as future work.

4.3 A Model of Users' Individual Preference

This part models a user's individual preference of two apps with the ordinal structure of $\mathbf{C}_2$. If the ordinal structures of $\mathbf{C}_h$ for other values of h can be obtained, then our method can be generalize to model a user's individual preference of two apps with the ordinal structure of $\mathbf{C}_h$.

Let $\varphi_2 : \{0, 1\}^l \rightarrow \mathbf{C}_2$ be the classification function of usage record, that is, mapping a usage record of length l to its usage pattern of length h. Denote by $\prec_{D,\mathbf{T}}$ (without causing confusion, $\prec_D$) the relation of user D's individual preference of two apps within $\mathbf{T}$ and by $\sim_{D,\mathbf{T}}$ (without causing confusion, $\sim_D$) the relation of user D's indifference between two apps within $\mathbf{T}$. We say that $A_i \prec_D A_j \Leftrightarrow \varphi_2(r(D, A_i, \mathbf{T})) \prec_2 \varphi_2(r(D, A_j, T))$, and that $A_i \tilde{}_D A_j \Leftrightarrow \varphi_2(r(D, A_i, \mathbf{T})) = \varphi_2(r(D, A_j, T))$, where $\varphi_2(r(D, A_i, \mathbf{T})) \in \mathbf{C}_2$. and $\varphi_2(r(D, A_j, \mathbf{T})) \in \mathbf{C}_2$. Intuitively speaking, there exists a relation of a user's individual preference between two apps within $\mathbf{T}$ only if the user has, at least once, ever installed both apps within $\mathbf{T}$ (but, needless to have to install both in the same time interval). Let $\mathbf{D}(A_i, A_j) = \{D \in \mathbf{D} : \varphi_2(r(D, A_i, \mathbf{T})), \varphi_2(r(D, A_j, \mathbf{T})) \in \mathbf{C}_2\}$ the collection of active users who have ever installed both apps A_i and A_j.

5 Users' Aggregate Prefernce

Dote by $\prec$ the relation of users' aggregate preference between two apps. $A_i \prec A_j$ means A_j is preferred to A_i by $\mathbf{D}$. Our goal is to find a linear order $A_{\sigma(1)} \prec A_{\sigma(2)} \prec \cdots \prec A_{\sigma(m)}$, for some permutation σ on the set $\{1, 2, \ldots, m\}$, which is consistent with users' behavior records as far as possible. In this section, we first model users' aggregate preference between two apps from a Bayesian view, and then interpret the app ranking problem as a stochastic acyclic subgraph problem.

5.1 *Divergence of Users' Individual Preferences*

In the previous section, usage patterns are viewed as original numbers, which leads to a deterministic model of the relation of users' individual preferences between two apps. However, when we proceed to discuss the relation of users' aggregate preference of the two apps, uncertainties arise due to the divergence of users' individual preferences. Therefore, a compromise between users' pros and cons has to be introduced. This raises the problem of how to decide the relation of users' aggregate preference between two apps that could meet users' divergent individual preferences as far as possible.

In our setting, making such a decision seems more involving, due to the existence of the neutral users and the ignorant users. In other words, we have to deal with users' pros and cons in the context of permitting users' indifference and ignorance. For convenience, define

Table 3 Divergences of users' individual preferences

Scenario 1	$a_{ij} = 70$	$b_{ij} = 10$	$c_{ij} = 10$
Scenario 2	$a_{ij} = 2$	$b_{ij} = 0$	$c_{ij} = 0$
Scenario 3	$a_{ij} = 0$	$b_{ij} = 0$	$c_{ij} = 0$
Scenario 4	$a_{ij} = 21$	$b_{ij} = 50$	$c_{ij} = 20$

$$a_{ij} = \left|\left\{D \in \mathbf{D} \ : A_i \prec_D A_j\right\}\right|$$
$$b_{ij} = \left|\left\{D \in \mathbf{D} \ : A_i \,\tilde{}_D\, A_j\right\}\right| \text{and}$$
$$c_{ij} = \left|\left\{D \in \mathbf{D} \ : A_i \succ_D A_j\right\}\right|.$$

We begin with considering several artificial scenarios. Imagine that, A_i and A_j are compared by $\left|\mathbf{D}(A_i, A_j)\right| = 100$ users. What conclusions could be drawn, if the following scenarios shown in Table 3 happen?

In Scenario 1, since 70 out of the 90 users prefer A_i to A_j, A_i is more likely to be preferred to A_j. It also allows a degree of possibility that A_i could be preferred to A_j. In Scenario 2, however, we would have few confidence in concluding so with only two observations, even though they both support that A_j is preferred to A_i. Scenario 3 looks extreme but is still possible to happen, for example, if both apps are relatively new. In this case, we can hardly say anything since no evidence is available to support any conclusion. In scenario 4, it is more comfortable to conclude that A_i is almost as good/bad as A_j, although the number of the users who prefer A_j to A_i is slightly larger. From these scenarios, we can observe that: (1) the percentage $\frac{a_{ij}}{a_{ij}+b_{ij}+c_{ij}}$ is not always practical to reflect the relation of users' aggregate preference; (2) the number of the ignorant users will influence the degree of our certainty to draw a conclusion; (3) $A_i \prec A_j$ and $A_j \prec A_i$ are not mutually exclusive events due to the appearance of the neutral users.

5.2 Comparison of Two Apps

We assume that there is an underlying but unknown probability p_{ij}that the relation of users' aggregate preference is $A_i \prec A_j$. The relations of users' individual preference of A_i and A_j are thought of as a random sample of $|\mathbf{D}(A_i, A_j)|$ independent Bernoulli trials with probability of success p_{ij}. Formally, for each $D \in \mathbf{D}(A_i, A_j)$, let R_{ij}^D be a Bernoulli trial with respect to A_i and A_j such that

$$R_{ij}^D = \begin{cases} 1, \ \textit{if} \ A_i \prec_D A_j \\ 0, \ \textit{otherwise} \end{cases}$$

and $\Pr\left(R_{ij}^D = 1\right) = p_{ij}$, where 1 means "success" and 0 means "failure". If it is further assumed that A_i and A_j are independently compared by the users of

$\mathbf{D}(A_i, A_j)$ then the likelihood function of s successes and f failures given $p_{ij} = x$ can be given in the form

$$l(s, f|p_{ij} = x) = x^s(1-x)^f .$$

Then, our primary purpose is to estimate p_{ij} with the observation of s and f. However, when the value of $s+f$ is very small, the naive estimate of p_{ij} by the percentage $\frac{s}{s+f}$ could be misleading. Here we take Scenario 2 for an example. In this case, based only on two observations, if one claims that $\widehat{p}_{ij} = 1$, then it is nevertheless a bad estimate. This is because we usually begin our estimation with an amount of prior belief concerning prior expectations and uncertainties. For example, before estimating p_{ij} we may have thought of that one app is quite unlikely to be completely preferred to another one, unless sufficiently convincing evidence has already been provided before our estimation. To overcome this difficulty, we adopt Bayesian approach to deal with users' conflicting individual preferences. For an overview of Bayesian statistics, we refer to the books Gelman et al. [3] and Kruschke [8].

To begin with, p_{ij} is assumed to distribute according to a beta distribution. Suppose that a continuous random variable X has a beta distribution with parameter α and β, where $\alpha > 0$ and $\beta > 0$. Then, the probability density function has the following form

$$f(x|\alpha, \beta) = \frac{x^{\alpha-1}(1-x)^{\beta-1}}{B(\alpha, \beta)}, \ 0 \leq x \leq 1,$$

where $B(\alpha, \beta) = \frac{\Gamma(\alpha)\Gamma(\beta)}{\Gamma(\alpha+\beta)}$ is the beta function and where $\Gamma(\alpha)$ is the gamma function and $\Gamma(\alpha) = \int_0^\infty x^{\alpha-1}e^{-x}dx$. The mean and variance of the beta random variable X are $\mu = \frac{\alpha}{\beta}$ and $\sigma^2 = \frac{\alpha\beta}{(\alpha+\beta)^2(\alpha+\beta+1)}$, respectively.

The beta distribution is often viewed as a probability distribution of probabilities, and plays a crucially important role in many Bayesian Statistical analyses. Here we refer to the book Gupta and Nadarajah [7] for an overview. Since the beta distribution can take an amazingly great variety of forms, it can be used for summarizing our prior belief of $p_{ij.}$ More concretely, the values of α and β can be thought of as α successes and β failures that have been imaginarily observed, and thus it allows us to express our prior expectations and uncertainties with properties specified by the values of α and β.

Moreover, it is also very convenient to update our expectations and uncertainties after a collection of observations is obtained, due to the fact that beta distribution is a conjugate prior with respect to the likelihood of Bernoulli trials. Assume that the prior probability distribution of p_{ij} is *beta*$(\alpha_{ij}, \beta_{ij})$. After a_{ij} successes and $b_{ij} + c_{ij}$ failures are observed, the posterior probability distribution of p_{ij} is given by Bayes' theorem (in the form of density probability function),

$$
\begin{aligned}
& f\left(p_{ij} = x \middle| a_{ij},\, b_{ij} + c_{ij}\right) \\
& = \frac{f\left(x \middle| \alpha_{ij},\, \beta_{ij}\right) l\left(a_{ij},\, b_{ij} + c_{ij} \middle| p_{ij} = x\right)}{\int f\left(x \middle| \alpha_{ij},\, \beta_{ij}\right) l\left(a_{ij},\, b_{ij} + c_{ij} \middle| p_{ij} = x\right) dx} \\
& = \frac{\dfrac{x^{\alpha_{ij}-1}(1-x)^{\beta_{ij}-1}}{B\left(\alpha_{ij},\, \beta_{ij}\right)} x^{a_{ij}}(1-x)^{b_{ij}+c_{ij}}}{\displaystyle\int_0^1 \dfrac{y^{\alpha_{ij}-1}(1-y)^{\beta_{ij}-1}}{B\left(\alpha_{ij},\, \beta_{ij}\right)} y^{a_{ij}}(1-y)^{b_{ij}+c_{ij}} dy} \\
& = \frac{x^{\alpha_{ij}+a_{ij}-1}(1-x)^{\beta_{ij}+b_{ij}+c_{ij}-1}}{B\left(\alpha_{ij}+a_{ij},\, \beta_{ij}+b_{ij}+c_{ij}\right)},
\end{aligned}
$$

which is another beta distribution. It implies the posterior expectations and uncertainties of p_{ij}. In this way, each relation of users' aggregate preference can be associated with a beta distribution, representing our posterior belief of the underlying probability that the relation will happen.

If a relation between $A_i \prec A_j$ and $A_j \prec A_i$ has to be decided, then their associated beta distributions may be useful for comparison. For example, if both distributions are highly concentrated at their mean values with "distinguishably" non-overlapping support, then one may conclude that the one with smaller mean value is unlikely to be preferred to the other. The measure of mean-variance ratio MVR, written as $\frac{\mu}{\sigma}$, captures this idea, and provides a criterion for dealing with the case where the non-overlapping support is less "distinguishable". For example, if *beta*(1, 1) is used as the priors of the relations $A_i \prec A_j$ and $A_j \prec A_i$, then the *MVR*s of $A_i \prec A_j$ and $A_j \prec A_i$ are

$$
\sqrt{\frac{\left(3 + a_{ij} + b_{ij} + c_{ij}\right)\left(1 + a_{ij}\right)}{1 + b_{ij} + c_{ij}}}
$$

and

$$
\sqrt{\frac{\left(3 + a_{ij} + b_{ij} + c_{ij}\right)\left(1 + c_{ij}\right)}{1 + a_{ij} + b_{ij}}}
$$

respectively. Notice that, if $a_{ij} > c_{ij}$ is assumed, then $\frac{1+a_{ij}}{1+b_{ij}+c_{ij}} > \frac{1+c_{ij}}{1+a_{ij}+b_{ij}}$ suggests that $A_i \prec A_j$ be chosen since it is more likely to happen. This choice coincides with the majority rule if the neutral users are removed from consideration.

5.3 Stochastic Acyclic Subgraph Problem

Before discussing how one may rank the apps into a linear order based on pairwise comparisons of the apps, we first introduce some necessary notation concerning digraphs. A strict digraph $G(V, E)$ consists of a finite nonempty set V of nodes, and a set E of arcs that are ordered pairs of different elements of V without any loop or parallel arc. If (u, v) is an arc, then (u, v) is said to go from u to v. We also say u dominates v.

If $G(V, E)$ is a digraph, then every digraph $G^{'}(V^{'}, E^{'})$ with $V' \subseteq V$ and $E' \subseteq E$ is called a subdigraph of $G(V, E)$. A nonempty set of arcs $P = \{(v_1, v_2), \cdots, (v_{k-2,}\ v_{k-1}), (v_{k-1},\ v_k)\}$ in $G(V, E)$ such that $v_i \neq v_j$ for $i \neq j$ is called a $(v_1, v_k)-$ dipath of length $k - 1$, and is denoted by $P(v_1, v_k)$. For a (v_1, v_k) $-$ dipath $P(v_1, v_k)$ and $(v_k,\ v_1) \in E$ then $C = P(v_1, v_k) \cup \{(v_k,\ v_1)\}$ is called a $k-$ dicycle. A digraph $G(V, E)$ which contains no dicycle is called acyclic. A standard instance of (weighted) acyclic subgraph problem can be described as follows. Given a digraph $G(V, E)$ with constant arc weight w_{ij} for every $(v_i,\ v_j) \in E$ we look for an acyclic subdigraph $G^{'}(V^{'}, E^{'})$ of $G(V, E)$ such that $w(G) = \sum_{(v_i,\ v_j) \in E'} w_{ij}$ is maximized.

With the model of users' aggregate preference of two apps, app ranking problem can be interpreted as a stochastic acyclic subgraph problem on a complete digraph with each arc associated with a weight distributed according to a beta distribution. To see this, let $G(V, E)$ be a complete digraph in which $V = \mathbf{A}$, that is, node v_i represents app A_i, and arc $(v_i,\ v_j) \in E$ represents the relation $A_j \prec A_i$ for all $i \neq j$. Each arc $(v_i,\ v_j) \in E$ is associated with a beta distribution $beta(\alpha_{ji} + a_{ji},\ \beta_{ji} + b_{ji} + c_{ji})$, representing our posterior belief of that$A_j \prec A_i$ will hold, where α_{ji} and β_{ji} specify our prior belief. Based on the information of the beta distributions associated to the arcs, we are to find an acyclic subgraph $G^{'}(V^{'}, E^{'})$ and hence a linear order of $\mathbf{A}$ that meets our posterior belief of users' aggregate preference as far as possible. This problem is stochastic, since each arc is associated with a random variable of beta distribution; it is combinatorial, since for all $i \neq j$ either $(v_i,\ v_j) \in E'$ or $(v_j,\ v_i) \in E'$ will be chosen, but not both.

Deterministic reformulation method is often used to obtain workable solution of stochastic combinatorial problem. In this problem, one possible approach is to provide a reasonable criterion of "optimality" for choosing the optimum acyclic subgraph. Recall that MVR is a useful criterion for comparing two apps, reflecting our posterior degree of "expectation per unit of uncertainty". Therefore, MVR may be used as a measure for choosing the linear order of the apps based on pairwise comparisons. To do this, each arc$(v_i,\ v_j) \in E$ is associated with weight w_{ij}, which is the MVR of the beta distribution associated to the relation$A_j \prec A_i$, that is,

$$w_{ij} = \sqrt{\frac{\alpha_{ji} + a_{ji}}{\beta_{ji} + b_{ji} + c_{ji}} \left(\alpha_{ji} + \beta_{ji} + a_{ji} + b_{ji} + c_{ji} + 1\right)}\ .$$

Then, the stochastic acyclic subgraph problem is converted to a deterministic acyclic subgraph problem. The latter has been well studied, and has many important applications. The study of acyclic subgraph problem can be found in, for example, Garey and Johnson [2], Grotschel et al. [4], [5] and [6], and Pedings et al. [9].

6 Experimental Result

This section shows experimental results of our approach. We used the data concerning Android smartphone users' behavior records during the period from Jan. 2013 to Jun. 2013. In particular, with focus on the social category and the free game category, we respectively show the ranking results of fifteen social apps and fifteen game apps, which are obtained by using the method proposed by Fujimoto et al. [1]. And, this rank result is called Score Ranking (SR), and each app will be associated with a rank number under SR. Moreover, the rank results obtained by the lexicographic order of $\mathbf{C}_2$ and by the transition order of $\mathbf{C}_2$ are called Lexicographic Ranking (LR) and Transition Ranking (TR), respectively.

In the experiments, uninformative priors were used. That is, $\alpha_{ij} = \beta_{ij} = 1$ for all distinct $i, j = 1, \ldots, 15$. Then, with the data, the MVRs between two apps can be efficiently obtained, since most of the Android users only use a few of the apps of the same category. Finally, to obtain LR and TR of the selected apps, we solved the acyclic subgraph problem on a weighted complete digraph of 15 nodes by using the commercial software FICO Xpress. In the experiments, the exact solutions can be obtained within 0.1 second on PC. Table 4 shows a comparison of SR, LR and TR of the selected social apps.

Table 5 shows a comparison of SR, LR and TR of the selected free game apps.

Unlike the score ranking method in which the basic idea is to create an absolute standard of goodness, our new ranking method is based on pairwise comparisons of the apps from the perspective of users' experience and behavioral response during their usage of the apps. It seems likely that the two methods have different performance. Moreover, we can also see that using different ordinal structures on $\mathbf{C}_2$ could lead to different ranking results, even with the same ranking algorithm. However, at the time being, we have few insights into the problem of how to choose an ordinal structure to model the structure of users' individual preferences of free smartphone apps. This will be left for future research.

Table 4 Ranking results of top fifteen social apps selected with SR method

App	SR	LR	TR
Twitter	1	3	2
Facebook	2	2	4
Mixi	3	4	8
Gree	4	10	12
Ameba	5	7	9
Comm	6	14	14
Bump	7	8	11
Mbga	8	13	10
Instagram	9	12	1
PersonalSpace	10	6	3
Sockets Live	11	1	7
Twicca	12	9	5
MixiSH	13	11	13
2chmate	14	5	6
Saitosan	15	15	15

Table 5 Ranking results of top fifteen free game apps selected with SR method

App	SR	LR	TR
SJLGHB2	1	5	13
soccer2_ja	2	2	6
SJFISHING	3	12	12
bowling	4	6	15
colopl.baseball	5	3	2
derbyimpact	6	1	4
ppt2013	7	14	10
punchhero.glo	8	15	9
airhockey	9	10	8
virgil.basketball	10	8	11
dodgeball	11	7	1
gmo.sumasaka	12	4	7
mapp.tennis	13	11	5
PinballGame	14	9	14
tukikuma.keruhito	15	13	3

7 Conclusion

This paper proposed a new method to rank smartphone apps based on users' behavior records. To obtain a ranking result of the apps of our interest, the basic idea is to first infer users' individual preferences between two apps. And then, based on the pairwise comparisons of the apps, we continue to merge users' individual preferences of the apps into their aggregate preferences. Although our method could be efficient in implementation (even when hundreds of smartphone apps are selected as the candidates to be ranked), the ranking algorithm requires a reasonable choice of an ordinal structure on the set of usage patterns, which so far lacks theoretical work and thorough simulating investigation.

References

1. Fujimoto M, Luo S, Sari MP, Shigeno M, Zago M (2014) Usage pattern analysis for Android applications: classification and measurement. In: The 2th domestic conference on serviceology, Hakodate, Hokkaidou, Japan, 28–29 April
2. Garey MR, Johnson DS (1979) Computers and intractability: a guide to the theory of NP-completeness. Freeman, San Francisco
3. Gelman A, Carlin JB, Stern HS, Dunson DB, Vehtari A, Rubin DB (2013) Bayesian data analysis, Thirdth edn. Chapman and Hall/CRC, Boca Raton
4. Grötschel M, Jünger M, Reinelt G (1984) A cutting plane algorithm for the linear ordering problem. Oper Res 32:1195–1220
5. Grötschel M, Jünger M, Reinelt G (1985) Facets of the linear ordering polytope. Math Program 33:43–60
6. Grötschel M, Jünger M, Reinelt G (1985) On the acyclic subgraph polytope. Math Program 33:28–42
7. Gupta AK, Nadarajah S (2004) Handbook of beta distribution and its applications. CSC Press, New York
8. Kruschke JK (2010) Doing Bayesian data analysis: a tutorial introduction with R. Academic, Burlington
9. Pedings KE, Langville AN, Yamamoto Y (2012) A minimum violations ranking method. Optim Eng 13:349–370

Contribution of ICT Monitoring System in Agricultural Water Management and Environmental Conservation

Koshi Yoshida, Kenji Tanaka, Ryunosuke Hariya, Issaku Azechi, Toshiaki Iida, Shigeya Maeda, and Hisao Kuroda

Abstract Recently, ICT based monitoring system is popularized in some agricultural areas. In such area, farmers can obtain their paddy conditions by using smart phone without going to the paddy. In this study, the effect of ICT based monitoring system on water and labor saving, and nutrient loads reduction was evaluated. Then, cost/benefit of monitoring system installation was compared. Finally, how to promote ICT systems in agricultural engineering field was discussed based on the characteristic of agricultural water management system.

Keywords Agricultural water use • Auto irrigation system • Cost/benefit • Inbanuma land improvement district

1 Introduction

Irrigation is essential for food production especially in paddy rice, and much water has been withdrawn from water sources compare to other water use sector [1]. In Japan, water delivery system for agriculture was highly-developed and managed by LID (Land Improvement District) [2]. Figure 1 shows the relation of provider and receiver of agricultural water delivery services. Provider is LID who is in charge of not only the water transfer but also the operation and maintenance of irrigation facilities. And service receivers are farmers who pay water fee to LID and can get water in any time they want. The 1st priority of irrigation is of course water supply to avoid the water stress, however, saving a labor input for agricultural land management such as weed control also the important factor. Therefore, farmers

K. Yoshida (✉) • I. Azechi • S. Maeda • H. Kuroda
Regional and Environmental Science, Ibaraki University, Ami-machi, Japan
e-mail: koshi.yoshida.agri@vc.ibaraki.ac.jp

K. Tanaka • R. Hariya
Agricultural and Environmental Engineering, Tokyo University of Agriculture and Technology, Fuchu-shi, Japan

T. Iida
Biological and Environmental Engineering, The University of Tokyo, Bunkyo-ku, Japan

T. Maeno et al. (eds.), *Serviceology for Designing the Future*,
DOI 10.1007/978-4-431-55861-3_25

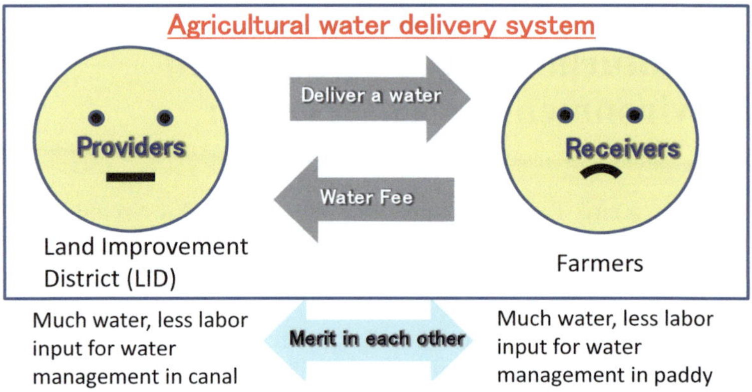

Fig. 1 A relation of service provider and receiver in agricultural water delivery system

Fig. 2 A relation of amount of water use and labor input in agricultural water management

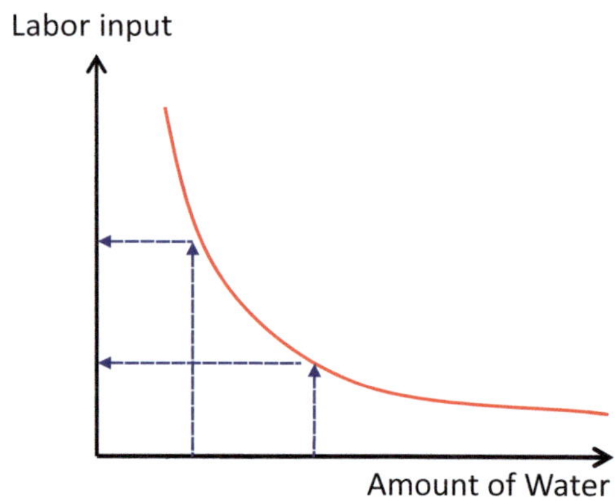

want to input much water for their paddy to save a labor cost [3]. The relation between water use amount and farmer labor input was shown in Fig. 2. Normally, most of paddy farmers are part-time farmers who have another job in weekday, therefore they go to their paddy in the morning to open the water bulb and they go to paddy again to close it in evening after they finish the work. However, if paddy water can be controlled automatically, they don't need to go their paddy field in every day. For example, in case of the paddy which have pipeline water delivery system, farmer just leave the water bulb open and water is charged when the pumping system is working and water stop when pump is stopped.

On the other hand, in the provider side also, much water leads less labor input for the system operation. When water shortage occurred in some area, farmers complain to LID stuffs, therefore it is convenient for LID to take some margin of water to control the system in safety. That is one reason why irrigation water use has not decreased so much compared to the reduction of paddy cultivation area. However, much irrigation water leads much energy consumption for pumping, as a result, water fee also increase. In addition, increase of drainage water cause a much nutrient loading to downstream water environment, because nitrogen load can be calculated multiplying drainage water amount (m^3) and nitrogen concentration (mg/l) [4, 5]. Historically, water delivery system in paddy agriculture has been supply-oriented system. However recent innovation of field monitoring or automatic regulation system will change it to demand-oriented system. In this study, the automatic irrigation system was evaluated to improve the water management system in paddy without the additional labor input. Kashima district, which was managed by Inbanuma land improvement district, was selected for the test site. Water amount and nitrogen concentration were monitored at Iino pumping station and paddy drainage pipe. Meteorological data were also observed at Iino pumping station for calculation of evapotranspiration. From the monitoring data, water balance and nitrogen balance was estimated. Then, by using the water management model, the effect of automatic irrigation system was evaluated in the view point of water and energy saving and nitrogen load mitigation.

And then, install cost of monitoring system and benefit of environmental conservation was evaluated. Recently, ICT based monitoring system is popularized in some agricultural areas. In such area, farmers can obtain their paddy conditions by using smart phone without going to the paddy. In this study, how to promote ICT systems in agricultural engineering field was discussed based on the characteristic of agricultural water management system.

2 Study Area

For the test site, Kashima district was selected which was managed by Inbanuma land improvement district and having 46.1 ha benefit area (Fig. 3). Figure 4 shows the monitoring paddy fields having 6.48 ha. All of these paddy fields were cultivated by same farmer group. Irrigation water was pumped up from Inbanuma lake, and water from paddy was drained to Inbanuma again flowing through Kashima river. In this paddy fields, open type drainage canal was not existed and subsurface drainage system was installed in all paddies so that all drained water, which was percolated through paddy soil layer and overflowed from paddy outlet, gathers to drain pipe. Inbanuma lake is closed water area and eutrophic lake [6, 7]. Therefore, drainage water from paddy fields was one of the pollution sources. Although drainage water from paddy have relatively low nutrient concentration compare to that from upland, drainage water amount from paddy is much more. Therefore, nutrient load from paddy fields could not be negligible for the conservation of lake water environment.

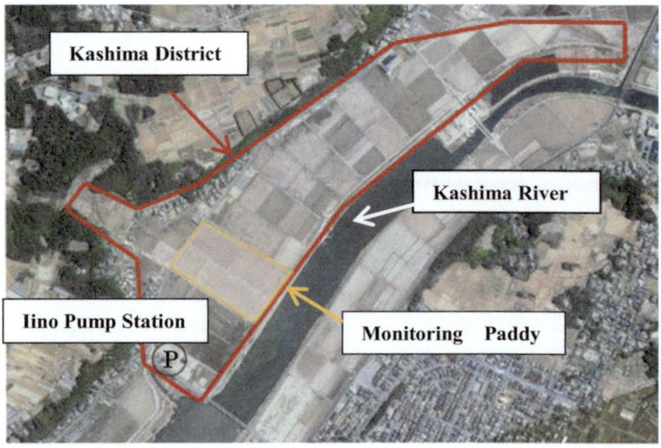

Fig. 3 Outline of study area

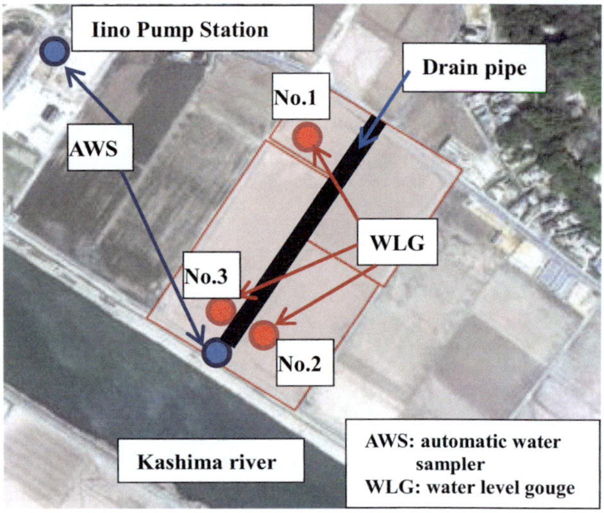

Fig. 4 Monitoring paddy fields

3 Field Measurement

To evaluate the water balance such as how much water was irrigated, evaporated and drained, paddy water level was monitored in hourly basis at 3 paddies by using HOBO U20 water level loggers (Onset Co. Ltd.). Water quality at Iino pumping station and paddy drainage pipe were measured in daily basis (at noon) such as soil sediment (SS), total nitrogen (TN), nitrate nitrogen (NO_3-N), ammonium nitrogen (NH_4-N), chemical oxygen demand (COD), total phosphorous (TP), dissolved oxygen (DO) and PH. Meteorological data also measured by using Field Router system at Iino pumping station for calculating evapotranspiration. The Field Router is Quasi real-time monitoring system that consists of a CPU (Web server), AD converter, DA converter, Ethernet controller, high-intensity LED lighting, and

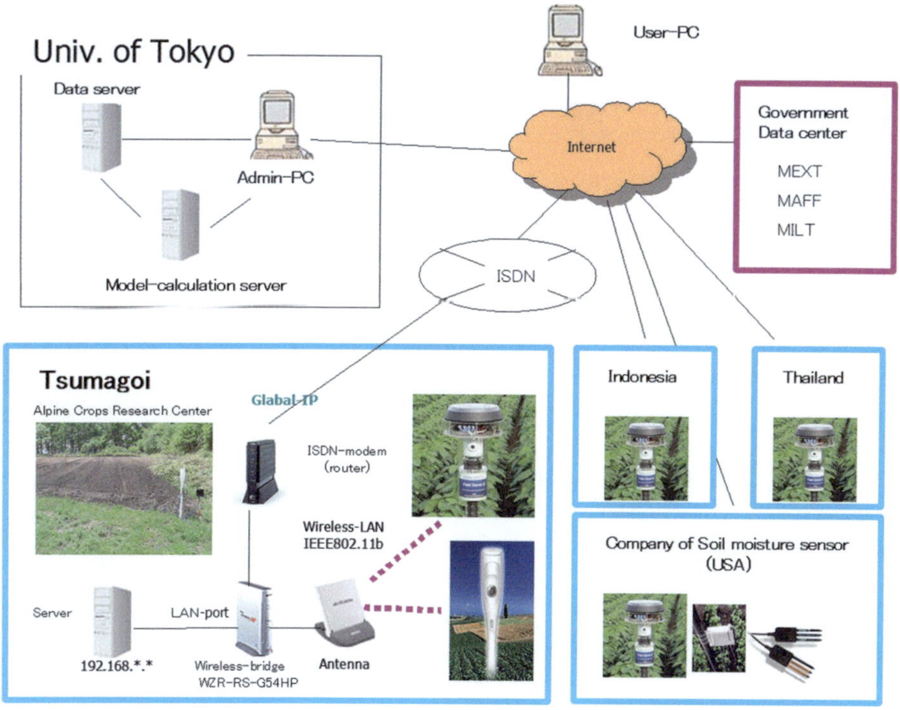

Fig. 5 Field router monitoring system

sensors such as air temperature, relative humidity, solar radiation, soil moisture, soil temperature, wind direction, wind speed and precipitation sensors (Fig. 5). The Field Routers are interconnected by a mobile internet (GSM/3G). Web cameras can be connected to the Field Routers, and high-resolution pictures of fields are transferred through mobile internet networks and stored on Web servers. Therefore farmer can obtain their paddy condition through their smart phone. Rice variety was Momiroman which was not for human food but for livestock feeding. Observation period was from 2012/6/4 to 10/31. A questionnaire to the farmer also conducted to grasp the fertilizer input or land management practices.

4 Water Balance

From the monitored water level, irrigated water and drained water amount was calculated by using paddy water balance equation as follows.

$$h^{t+\Delta t} = h^t + (I^t + R^t - Q^t - P^t - ET^t)\Delta t \quad (1)$$

where h: paddy water level(m), I: irrigation rate(m/h), R: rainfall intensity(m/h), Q: surface drainage (m/h), P: percolation rate(m/h), ET: evapotranspiration rate (m/h), t: time, Δt: time step(=3600 s). Surface drainage and percolation rate were calculated as a function of paddy water level as following equation.

$$Q = C \times B_w \times (\mathrm{h}^t - H_w)^{3/2} \times \Delta t / A \quad (2)$$

$$P = T \times h^t \quad (3)$$

where C: parameter, B_w: a width of drainage weir(m), H_w: a height of drainage weir (m), A: area of paddy field(m^2), T: percolation parameter(=0.005).

Evapotranspiration ET can be calculated from Penman- Monteith equation by using observed meteorological data. Rainfall, paddy water depth also measured so that only irrigation rate was unknown. Therefore, by the model fitting of observed water level and calculated one, irrigation rate can be estimated. Figure 6 shows fitting results of observed and calculated water level. The calculated result shows good agreement with observed data. Figure 7 shows estimated irrigation rate and other parameters related to water balance in paddy field from 6/4 to 10/31. Iino

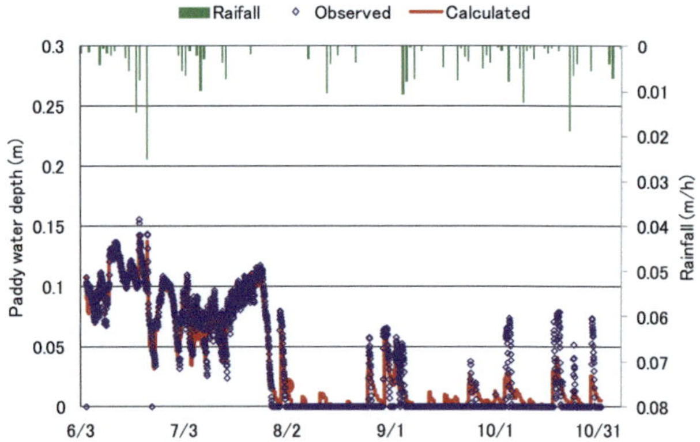

Fig. 6 Calculated and observed paddy water depth

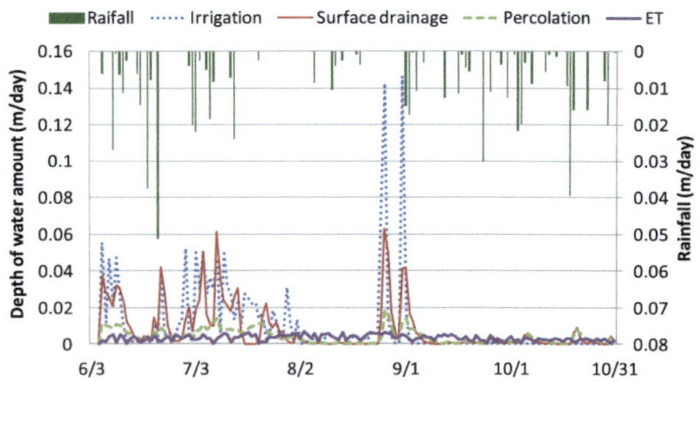

Fig. 7 Estimated irrigation rate and other parameters

Table 1 Water balance at paddy fields (2012)

	Parameters	Amount (mm/period)	Rate (%)
Input	Rainfall	628	29.6
	Irrigation	1495	70.4
	(Sum)	(2123)	(100)
Output	Surface drainage	1168	55.0
	Percolation	602	28.4
	Evapotranspiration	353	16.6
	(Sum)	(2123)	(100)

pumping station was operated until 9/1, therefore irrigation rate was zero after 9/1. Table 1 shows water balance during observation period. At the monitoring paddy fields, 1495 mm water was pumped up for irrigation and 55 % of supplied water was released by surface drainage from paddy outlet in current situation. Therefore there was still large possibility to save water by introducing elaborated water management system to reduce surface drainage water [8].

5 Nitrogen Balance

Figure 8 shows TN concentration of Iino pumping station for irrigation intake and paddy drainage pipe during observation period. TN concentration of Iino station was around 3 mg/l and almost stable except the October. Nitrogen load from paddy fields can be estimated by multiplying nitrogen concentration and drainage water amount. Table 2 shows nitrogen balance. In the conventional paddy field, 90–120 kg/ha nitrogen was recommended in food rice cultivation. However too much nitrogen was fertilized in this monitoring paddy compared with conventional one. Because, in feed rice cultivation, farmers do not need to pay attention to a taste of rice, therefore much amount of fertilizer was input to increase the rice biomass. In addition, farmer can get manure in free from livestock sector by exchanging the rice straw. Estimated nitrogen load from paddy was 36 kg/ha, it was summation of surface drainage 23 kg/ha and percolation 13 kg/ha. In Inbanuma watershed, there are 7370 ha paddy fields so that about 265 ton nitrogen was flow into the Inbanuma lake from all paddy fields. The pollution load through surface runoff can be reduced by controlling the surface drainage water volume [8, 9]. Such water management leads water and irrigation cost saving, and pollution load to downstream water environment also can be reduced.

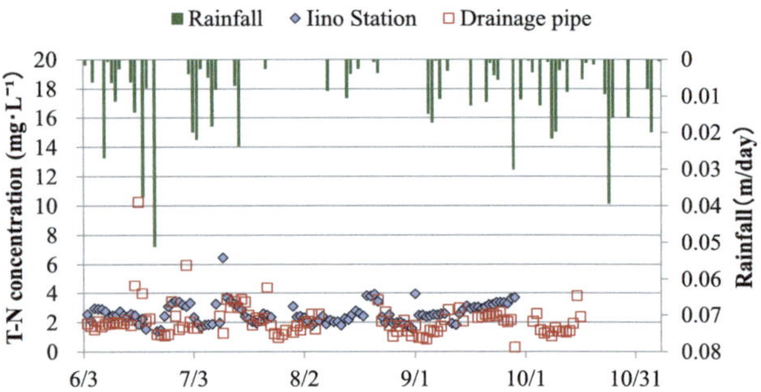

Fig. 8 TN concentration during observation period

Table 2 Nitrogen balance at paddy fields (2012)

	Parameters	Amount (kg/ha)	Rate (%)
Input	Manure	45	15.3
	Chemical	204	69.4
	Irrigation	37	12.6
	Rainfall	8	2.7
	(Sum)	(294)	(100)
Output	Harvest	60	20.4
	Surface drainage	23	7.8
	Percolation	13	4.4
	Denitrification and storage in soil	198	67.3
	(Sum)	(294)	(100)

6 Effect of Automatic Irrigation System

For saving water and farmer's labor input, automatic irrigation system was developed and introduced in some irrigation district. Figure 9 shows schematized description of float auto irrigator. Farmers just set the target water level and auto irrigator supply water until float level increase at target depth. In this study, numerical simulation was conducted to evaluate the effect of this auto irrigation system on water saving and nitrogen load reduction. For the calculation, following assumptions were employed.

1) target water level is 5 cm and it is constant during growing stage
2) auto irrigator performs perfectly. Just reduced water by percolation and evapotranspiration was supplied so that surface drainage does not occur.
3) TN concentration of drainage water is not changed and same as current situation case.

Figure 10 shows nitrogen load change from paddy before and after installing auto irrigator. The water saving effect was about 894 mm (60 %) and energy consumption for pumping also can be saved. As a result, farmer can save water fee which farmer pay to land improvement district for the maintenance and operation of water delivery facilities. Nitrogen load reduction was 21 kg/ha (58 %), and

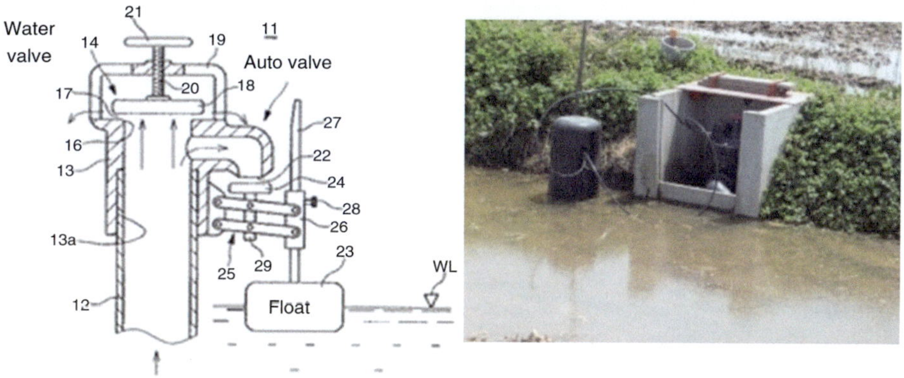

Fig. 9 schematized description of float auto irrigator

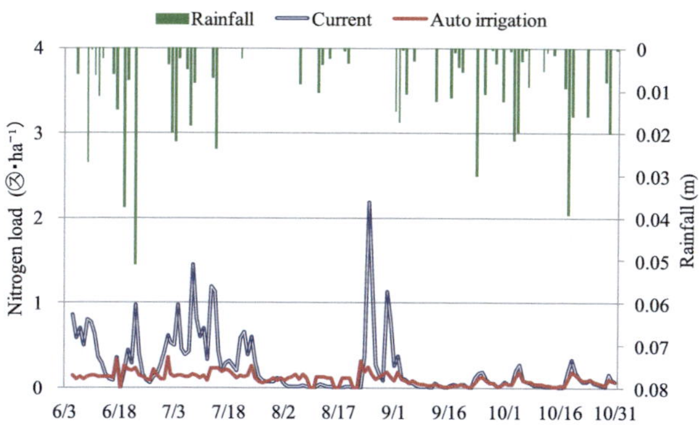

Fig. 10 Nitrogen load change from paddy before and after installing auto irrigator

about 154 ton from all paddies in Inbanuma watershed. Especially, nitrogen load reduction of first growing stage (June, July) was 17 kg/ha (46 %), therefore application of auto irrigation system was relatively effective at that period.

7 Discussion

From the analysis of water and nitrogen balance, auto irrigator has a good performance to save irrigation water, and farmer can save labor input by using it. In addition, 968 kg nitrogen load can be reduced in Kashima district (having 46.1 ha area). This value is equivalent with the nitrogen load of 496 peoples during same 150 days (nitrogen load per 1 person is about 13 g/day). For the construction of water treatment plant in rural area, following equation is available for cost calculation.

$$Y = 11,186 \cdot X^{0.414} + 8.74 \cdot X + 11,027 \quad (4)$$

where Y: plant construction cost (in unit of 1000 Japanese YEN), X: number of treatment people. In case of the treatment plant for 496 peoples, construction cost is estimated about 160 million YEN (≒1.6 million USD). This amount of money can be assumed the benefit of auto irrigator installation in this district. In fact, running cost and maintenance fee also needed to operate the water treatment plant.

On the other hand, a cost of auto irrigator is 60,000 YEN/unit (≒600 USD/unit). Standard size of paddy in Japan is 0.3 ha so that 154 units of auto irrigators are needed to install the all paddies in Kashima district. Total installation cost was 9.24 million YEN (≒92,400 USD). Therefore, ration of benefit/cost from auto irrigator can be estimated as 17.3.

8 Conclusion

In this study, field monitoring was conducted to grasp the current condition of water and nitrogen balance in the paddy field. Large amount of irrigation water and nitrogen fertilizer were inputted in monitoring paddy of Kashima district, because feed rice was cultivated in this paddy. And high input of water and nitrogen leads much nitrogen load to downstream water environment. Recently, feed rice paddy area is increasing under the governmental regulation policy. Therefore, water pollution in closed type lake may become worse if feed rice paddy spread widely. Auto irrigation system is effective to save irrigation water input and pumping energy consumption and to conserve the water environment without more farmer's labor input for water management [10, 11].

Installation cost of auto irrigators was less than 10 % compare with the benefit from nitrogen load reduction effect. However even after farmers install auto irrigator to their paddies, they will go to see and check paddies almost in every day. Because they afraid of their paddies without to check by their eyes. In such case, auto irrigator is not useful and not effective for them. Therefore, auto irrigator should be combined with web-camera system. In this study, Field Router monitoring system was installed to monitoring paddy, therefore famer can check their paddy condition including the picture of paddy through the Internet. However, the cost of Field Router system is still expensive (4000USD). Important point is that if farmer don't need to go to paddy and they can monitor their paddy in their house, for example by internet or using smart phone, auto irrigator is very useful and effective.

In current situation, the distance between famer and their paddy is still close. The meaning of "distance" here is not only physical distance but also mental distance. In the near future, paddy farming structure will be changed and one farmer have to cultivate more than 10 ha. In such case, more new service are needed in the agricultural field also (Fig. 11). Recently, ICT technology in agricultural filed is developed and spread rapidly so that farmer will be able to manage their paddy at their house in near future. In addition, to motivate the farmer to save water, this kind of data should

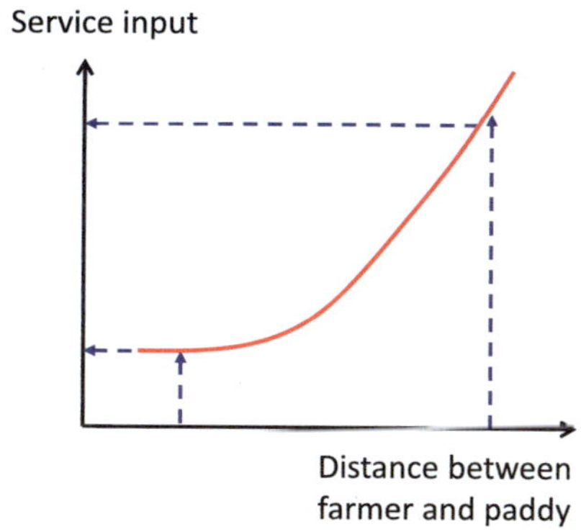

Fig. 11 A relation of service input and distance between farmer and paddy

be informed to them and their merit of saving water fee and effect of water saving on environmental conservation also should be explained. And, of course, such farmers attempt should be informed in public for promoting their effort to the other residents.

Acknowledgements This research was supported by RISTEX Service Science, Solutions and Foundation Integrated Research Program "Quantitative Evaluation and Demand-oriented Provision of Irrigation Service" and the Japan Society for the Promotion of Science (JSPS) Grant-in-Aid for Scientific Research (KAKENHI) no. 24780228.

References

1. Cantrell RP (2004) Challenges and opportunities for rice-based farming in the International Year of Rice and beyond. Editorial Paddy Water Environ 2:1–4
2. Iida T, Kimura M, Yoshida K, Kubo N, Yokoi T (2013) Investigation of irrigation activities by farmers on paddy field plots seeking irrigation service needs. In: Proceedings of the 1st international conference on serviceology (Tokyo), pp 286–290
3. Mizutani M (1999) Advanced paddy field engineering. Shinzansya Publisher Co., Ltd., Tokyo
4. Clemmens AJ, Allen RG, Burt CM (2008) Technical concepts related to conservation of irrigation and rainwater in agricultural systems. Water Resour Res 44:16
5. Sadler EJ, Evans RG, Stone KC, Camp CR (2005) Opportunities for conservation with precision irrigation. J Soil Water Conserv 60(6):371–379
6. Nakamura S, Motohashi K, Kubota H, Masuoka Y, Hayashi K, Yuasa T, Shoji T (2008) A new approach in water quality improvement of Lake Inba-numa using submerged plants. Adv Hydro-Sci Eng 8:645–646
7. Hayashi K, Yuasa T, Nakamura S, Mysiake K, Ono F, Ogura H, Uehara H (2009) Integrated watershed management for restoration of Lake Inba-numa and its watershed. In: 13th World Lake conference, Wuhan, China, pp S17–5
8. Hanson BR, Ayars JE (2002) Strategies for reducing subsurface drainage in irrigated agriculture through improved irrigation. Irrig Drain Syst 16(4):261–277
9. Rice RC, Hunsaker DJ, Adamsen FJ, Clemmens AJ (2001) Irrigation and nitrate movement evaluation in conventional and alternate furrow irrigated cotton. Trans ASAE 44(3):555–568
10. Kim CS, Schaible GD, Daberkow SG (2000) An efficient cost-sharing program to reduce nonpoint-source contamination: theory and an application to groundwater contamination. Environ Geol 39(6):649–659
11. Schaible GD (2000) Economic and conservation tradeoffs of regulatory vs. incentive-based water policy in the Pacific Northwest. Int J Water Resour Dev 16(2):221–238

A Questionnaire Assessment of the Contributing Factors to Empathy

Miki Nishio and Takashi Maeno

Abstract The present study addressed the contributing factors to empathy and the magnitude of influence among the selected factors. The main factors of interest were age, sex, personality, experience, and a desire to be empathic. Results showed that each factor influenced reports of empathy, and factors were ranked according to their strength of influence. A desire to be empathic was most related to reports of empathy. The present findings help extend our understanding of the key factors that predict empathic behaviors.

Keywords Empathy • Emotive value • Contributing factors

1 Introduction

The ability to be empathic, or share in others' feelings, is a unique human trait [1, 2]. Empathy is an important construct as it tends to predict prosocial behavior [3–6]. Additionally, empathy can be useful in other endeavors, such as consumer behavior and marketing. For instance, consumers have several choices at their disposal given the advancement in information technologies [7]. Several products and services are chosen based on economic merit, functionality, and reasonable prices. On the other hand, the value of a good/service can be determined beyond its economic merit. Consumers' interest in spiritual richness rather than material richness was reflected by the success of "Drink 1, Give 10 Campaign" launched in Japan [8]. This was a collaborative project developed by the Kirin Beverage Company, a beverage manufacturer and member of the Japan Committee for UNICEF. The project donated a portion of their sales to well-digging projects in

M. Nishio (✉)
The System Design and Management Research Institute, the Graduate School of System Design and Management, Collaboration Complex, 4-1-1, Hiyoshi, Kohoku-ku, Yokohama, Kanagawa, Japan
e-mail: miki-nishio44@a2.keio.jp

T. Maeno
Graduate School of System Design and Management, Keio University, Yokohama 223-0061, Japan

T. Maeno et al. (eds.), *Serviceology for Designing the Future*,
DOI 10.1007/978-4-431-55861-3_26

Africa. In "Marketing 3.0," Kotler stated, "consumers will be treated as human beings who are active, anxious, and creative. Then they will request more participation in value creation and demand their deepest anxieties and desires—not traditional needs and wants—be identified and fulfilled" [9, 10]. In this scenario, consumers might feel a stronger desire to use these products due to the prosocial tenets the company touts. We considered that consumers feel the experience accompanied by feeling as value.

Another consumer behavior study revealed that the experience of social values attracts attention. Schmitt described that the customer experience has an effect acquired through the use experience and the mental and sensuous value of satisfaction throughout the lifecycle of the goods or services [11, 12].

Emotion is also a central component of consumer behavior research. Cohen and Areni advocated a model in which consumer behavior is formed while a cognitive and emotive route conduct an interaction. They indicate that both intellectual evaluation and emotive evaluation specify actual action [13]. Thompson et al. reported that emotive motive exceeds rational motive [14]. These findings suggest that addressing affective outcomes is necessary to manipulate consumer behavior.

It is possible that anthropocentric values placed on a good/service can emerge through emotive experience. Emotive experience is a state of being "emotively moved" when a paradigm shift or conversion has occurred (such as when an individual is affected by the events of a movie or a beautiful scene). Most individuals experience strong feelings in certain contexts. Previous studies have reported that emotive experiences can affect motivation [15] and creates feelings of happiness and the effect is continued [16, 17]. These effects are useful in the context of consumer behavior or consumptive experience. Therefore, the eventual goal of the present study was to assess ways of addressing value in certain emotive experiences.

To achieve this goal, we need to understand better the contributing factors to an emotive experience of interest. Here, we chose to focus on empathy. In a previous study from our lab, we found that empathy had a role special in positive emotive experiences [18]. We clarified specific elements and relationships relevant to an emotive experience. An emotive experience consists of both emotive events and emotive causes. The emotive event consists of a subject and an object. Moreover, each subject and object is classified into one of four groups: the self, others, things, and circumstances. Emotive causes are categorized into eleven types. From here, we built an emotive experience model considering a precondition what an individual is an object of an emotive experience only when the individual is stimulated by the emotive event. Thus, though the emotive individual is the self, there is the case which the self is not directly concerned with the emotive event. In this case, empathy is the essential factor for developing the emotive experience. Empathy is a condition that elicits an emotive experience within the individual when he/she is not directly intervening in the emotive event. Thus, we need to understand empathy for instances of emotive experience when the self is not directly concerned. If empathy can be designed intentionally, an emotive event in which the self is not directly concerned can emerge. This allows for the expansion of forms of emotive experience.

Studies on empathy originated in the 1970s. Most studies assessing empathy can be categorized into one of three groups: the element of empathy, structural model of empathy, and factors affecting empathy.

Elements of empathy are classified into two main groups: cognitive and emotive elements [19, 20]. These two elements were further divided by Davis into four lower-order levels. These include empathic concern, personal distress, perspective taking, and fantasy. Empathic concern and personal distress are cognitive elements while perspective taking and fantasy are emotive elements [21]. Empathic concern is defined as, "the feeling for others thinking such as sympathy to an unfortunate and concern person." Personal distress is defined as, "self feels pain and displeasure reacts to others' pain." Perspective taking is defined as adopting another person's viewpoint. Finally, fantasy is defined as a tendency to be imaginative and align the self toward fictitious situations [22]. Apart from this four-level distinction, Mahrabian and Epstein divided empathy into three low-order levels [23]. These included emotive warmth, emotive coolness, and a tendency to be easily influenced emotively. However, the research literature is more aligned with Davis' four-level model.

Davis' theory is a well-known model of empathy [2]. The theory argues that empathy is derived from relationships between three elements. These include antecedent conditions, a process, and a result. Antecedent conditions are divided into an individual factor and a situation factor. Results are divided into individual and interpersonal results. Antecedent conditions influence the process and interpersonal and individual results. For instance, the process influences an individual result. In turn, the individual result influences the interpersonal result.

Contributing factors of empathy have been investigated for each element. As for empathic concern, investigations have assessed four contributing factors. The first contributing factor is intelligence, and it has been reported that correlations between intelligence and empathy are quite low [22]. The second contributing factor is family. However, specific relationships between empathy and family are unclear since correlations between empathy and family have been shown to be both positive and negative [24]. The third contributing factor is sex; it has been reported that women tend to be more empathic than men. The fourth contributing factor is age; it has been reported that older individuals have higher rates of empathy [25, 26]. As for personal distress, investigations have assessed two contributing factors. In terms of sex, it has been reported that women have higher correlations with empathy than men. Relationships between ages revealed that empathy was associated with only certain age groups (high school students compared to junior high students) [25, 26]. As for perspective taking, investigations assessed three contributing factors. Regarding the relationship with sex, women had a higher correlation with empathy than men. Regarding the relationship with age, the correlation with empathy was reportedly high [25, 26]. Regarding the relationship with family, there was a negative correlation [26]. Regarding fantasy, investigations have assessed three contributing factors. The relationship with sex has found women have a higher correlation with empathy than man. In terms of age, the correlation with empathy had been reported to be high in the high school students

[25, 26]. Regarding the relationship with family, there was a positive correlation [24].

Relationships between elements have been clarified through several studies. However, factors influencing low-order level factors have been investigated individually. The degree of influence for each factor has not been clarified. We considered it important to understand better the relationships among contributing factors of empathy. Understanding these factors can help determine consequences of promoting or removing factors with a high degree of influence on empathy. Therefore, alternative occurrences of empathy might be useful for an emotive experience design.

The purpose of the present study was to clarify the influence of multiple contributing factors related to the occurrence of empathy.

This can be used to look at a specific person's empathy by taking into consideration the contributing factors and how much they influence empathy. The present study provides an important contribution to understanding special functions of empathy, including how empathy attracts attention and motivates prosocial behavior. Therefore, this result can use organizational management and the promotion of pro-social behavior.

For the current study, contributing factors of empathy, and the degree of each factor's influence, were investigated. We addressed two research questions. Five contributing factors assumed to affect empathy were selected from previous study. We used an Internet questionnaire for data collection. We confirmed the influential degree to be influence of empathy on each contributing factor via the statistical work.

2 Investigation

The contributing factor of empathy was selected and two Research Questions were investigated.

2.1 Research Items

In Davis' theory, antecedent conditions are divided into an individual factor and situation factor. The purpose of this paper is for an emotive experience design to investigate the degree of influence of factors on empathy. Individuals frequently engage in emotive experiences. Thus, it is necessary to consider individual factors when assessing emotive experiences. Previous studies have examined age, sex, and certain personality factors [24–26]. The current study added experience and empathic desires in determining which factors were most likely to influence empathic behaviors/experience. Because we thought that the age was having more experiences and height of empathy is due to being conscious of the empathy

2.2 Research Question

The two main research questions are as follows. The first was, "Has the selected contributing factor affected empathy?" The validity of this investigation can be confirmed by comparing the results of the present study from factor results based on previous studies. The second question was, "do the degrees of incidence of each contributing factor differ?" In other words, how much does each factor actually influence empathy?

2.3 Approach

2.3.1 Implementation

The present investigation was conducted using an Internet questionnaire. We distributed the measure with the help of a questionnaire company. The investigation period was March14–16, 2013.

2.3.2 Participants

Participants were 1638 Japanese adults. Sex distribution was relatively equal. Ages ranged from 20 to 60 years old. Participants lived in regions throughout Japan and worked a variety of jobs. However, the most common job was "housewife," and most participants lived in the Tokyo area. The detail is shown in a Table 1.

2.3.3 Questionnaire Information

1) Age
2) Sex
3) Personality traits (Big Five)
4) Marital status
5) Experience with childcare
6) Experience with frustration
7) Desire to be empathic
8) Conscious of easily being empathic
9) Desire to evoke emotions in others

Item (1) was imputed as a whole number. Items (2), (4) and (6) were reported with "yes" or "no" responses. With regard to (3), respondents' personality traits were determined via 30 questions addressing the Big Five [27] traits on a 5-point rating scale from "very much applies" to "does not apply at all." Desire to be empathic also used a similar rating scale (however, on a 9-point scale).

Table 1 Participant information

Sex/age	N	%
Man/20–24	91	5.6
Man/25–29	91	5.6
Man/30–34	91	5.6
Man/35–39	91	5.6
Man/40–44	91	5.6
Man/45–49	91	5.6
Man/50–54	91	5.6
Man/55–59	91	5.6
Man/60–	91	5.6
Woman/20–24	91	5.6
Woman/25–29	91	5.6
Woman/30–34	91	5.6
Woman/35–39	91	5.6
Woman/40–44	91	5.6
Woman/45–49	91	5.6
Woman/50–54	91	5.6
Woman/55–59	91	5.6
Woman/60–	91	5.6
Total	1638	100.0
Area	N	%
Hokkaido	80	4.9
Tohoku	97	5.9
Kanto	660	40.3
Chubu	242	14.8
Kinki	309	18.9
Chugoku	70	4.3
Shikoku	41	2.5
Kyushu	139	8.5
Total	1638	100.0
Job	N	%
Government employee	48	2.9
Business manager/board member	30	1.8
Businessman (back office)	237	14.5
Businessman (engineer)	177	10.8
Businessman (other)	154	9.4
Independent business	104	6.3
Freelance	29	1.8
Homemaker	345	21.1
Part-time job	219	13.4
Student	106	6.5
Other	80	4.9
Unemployed	109	6.7
Total	1638	100.0

2.4 *Procedure*

The results of the investigation were compared with self-valuation regarding the ease of empathy in relation to the nine steps. The procedure involved statistical work. The influence of each factor on empathy was confirmed.

3 Result

Results addressing our two research questions (RQ) are as follows.

3.1 *RQ 1: Has the Selected Contributing Factor Affected Empathy?*

3.1.1 Age

An ANOVA, analysis of variance, a statistical method in which the variation in a set of observations is divided into distinct components, was conducted to assess the effect of age on empathy. Results showed no differences regarding the effect of age on empathy ($F(8, 1629) = 1.84$, ns). However, comparisons between specific age groups revealed significant differences in empathy as shown in Fig. 1.

3.1.2 Sex

A t-test was used to examine sex differences in empathy. This test revealed a significant sex difference ($t(1636) = 7.24$, $p < 0.01$), whereby women tended to report being more empathic than men as shown Fig. 2.

3.1.3 Sex and Age

As shown in a Fig. 3, examining the combined influence of age and sex on empathy revealed that women tended to report being more empathic than men up until about age 50.

3.1.4 Personality

We assessed the role of the Big Five personality factors on empathy in two steps. First, we assessed each individual's most prominent Big Five trait. We then

Fig. 1 Relationship between age and empathy (***p < 0.001, **p < 0.01, *p < 0.05)

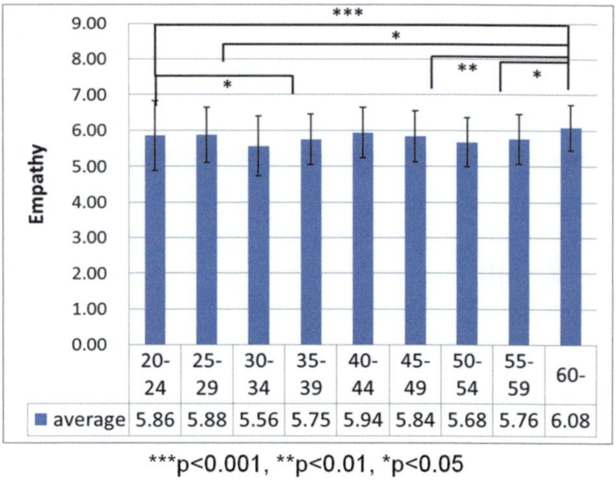

Fig. 2 Sex differences in empathy (***p < 0.001, **p < 0.01, *p < 0.05)

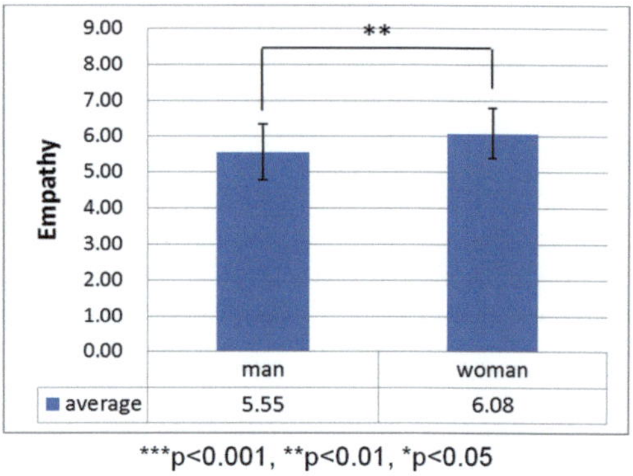

examined reports of empathy across individuals within the Big Five categories with an ANOVA. Intentional desire to be empathic differed across the Big Five traits ($F\ (29, 1608) = 1.92$, $p < 0.01$). A significant difference was shown for empathy between personality classifications. The detail is shown in a Fig. 4.

3.1.5 Marital Status

The influence of marital status on empathy was assessed with a between-groups t-test that included individuals who had or had not ever been married. Figure 5 showed that individuals who had been married at some point reported higher empathy ($t(1636) = 2.90$, $p < 0.01$).

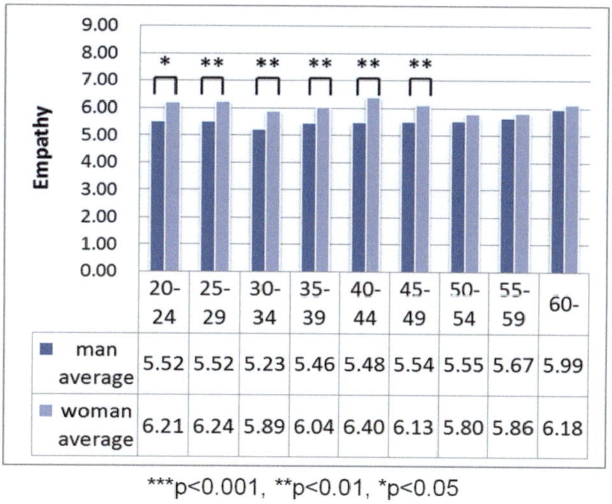

Fig. 3 Relationship between sex and age on empathy (***p < 0.001, **p < 0.01, *p < 0.05)

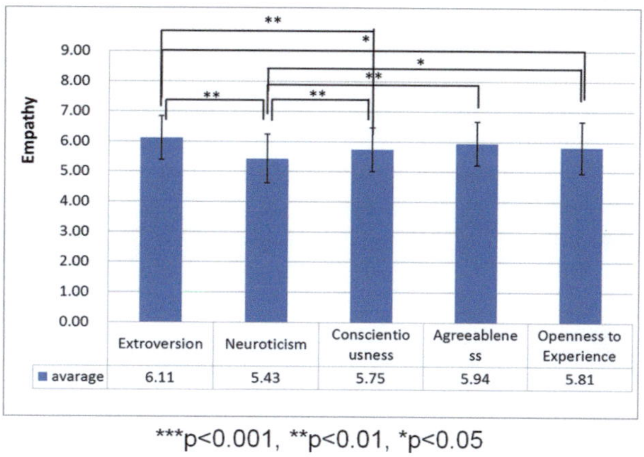

Fig. 4 Relationship between empathy and Big Five personality traits (***p < 0.001, **p < 0.01, *p < 0.05)

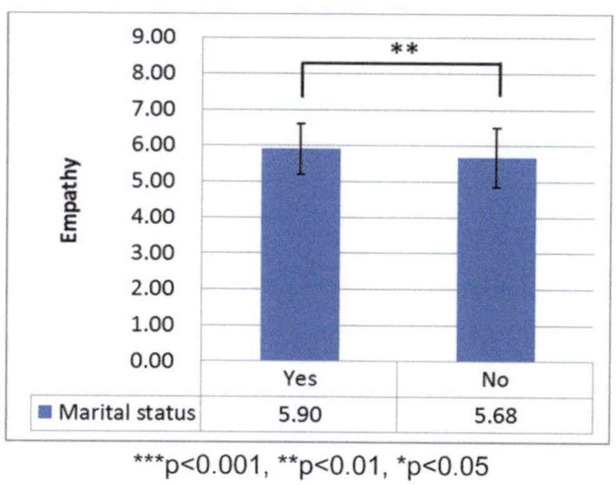

Fig. 5 Relationship between marital status and empathy (***p < 0.001, **p < 0.01, *p < 0.05)

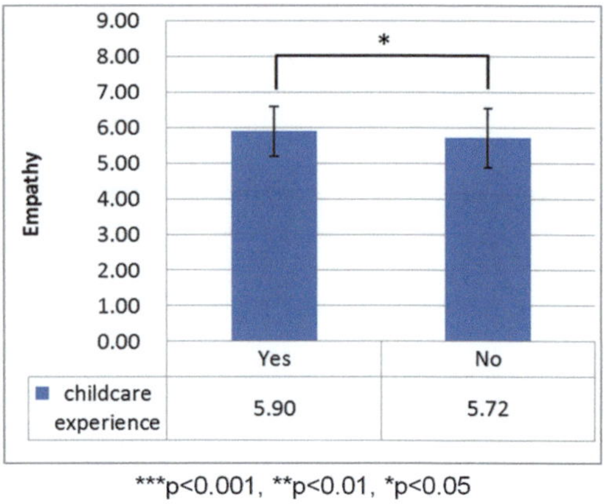

Fig. 6 Relationship between childcare experience and empathy (***$p < 0.001$, **$p < 0.01$, *$p < 0.05$)

3.1.6 Childcare Experience

The influence of childcare experience on empathy was analyzed by another between-groups t-test. Figure 6 showed higher levels of empathy among those who had childcare experience ($t(1636) = 2.49$, $p < 0.05$).

3.1.7 Experience with Frustration

The influence of frustration on empathy was analyzed by another between-groups t-test. Figure 7 showed that experiences of being frustrated did not affect empathy ($t(1636) = 0.72$, ns).

3.1.8 Desires Concerning Empathy

The influence of a desire to be empathic and reports of empathy was assessed with a between-groups t-test. As shown in a Figs. 8, 9, and 10, those who desired to be empathic had higher reports of empathy for all desire items: desire to be empathic ($t(1636) = 9.70$, $p < 0.001$), conscious of being easily empathic ($t(1636) = 11.56$, $p < 0.001$), and desire to affect other people emotively ($t(1636) = 12.36$, $p < 0.001$).

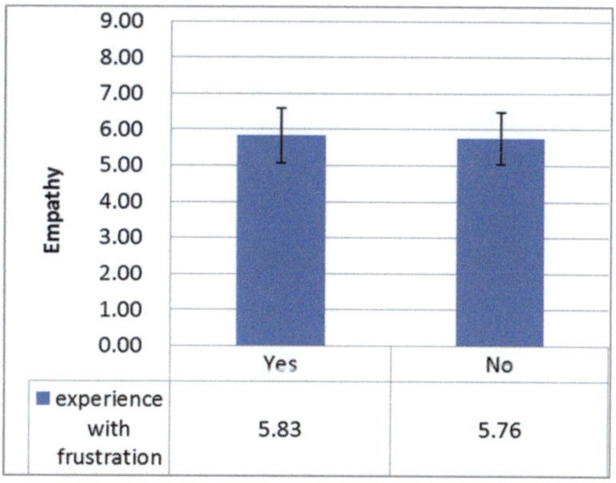

Fig. 7 Relationship between empathy and experience with frustration

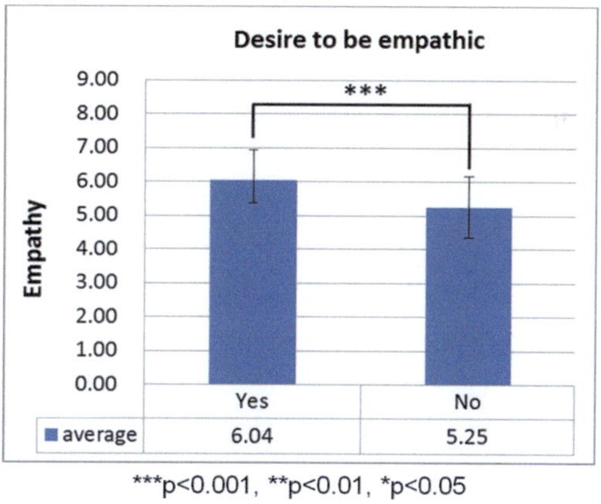

Fig. 8 Relation between empathy and desire to be empathic (***$p < 0.001$, **$p < 0.01$, *$p < 0.05$)

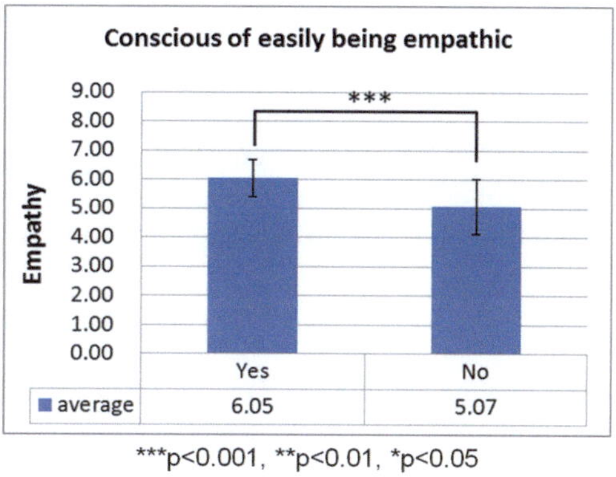

Fig. 9 Relation between empathy and conscious of being easily empathic (***$p < 0.001$, **$p < 0.01$, *$p < 0.05$)

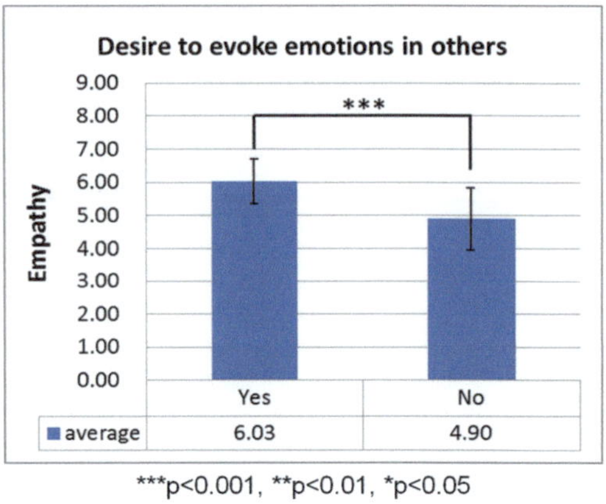

Fig. 10 Relation between empathy and desire to affect other people emotively (***$p < 0.001$, **$p < 0.01$, *$p < 0.05$)

Table 2 Results of the multiple linear regression analysis

Explanatory variable	β	Standard β	t	p value
Desire to evoke emotions in others	.195	.213	8.247	.000
Conscious of easily being empathic	.145	.144	5.591	.000
Openness to experience	.043	.110	4.226	.000
Desire to be empathic	.122	.136	5.616	.000
Sex	.357	.118	5.647	.000
Agreeableness	.039	.102	4.083	.000
Openness to experience	.024	.063	2.530	.011

3.2 *RQ 2: Do the Degrees of Incidence of Each Contributing Factor Differ??*

We confirmed a variety of contributing factors that influenced empathy, as well as one that was not so influential. Next, we examined which factors were most influential and the magnitude of that influence.

We conducted a multiple regression analysis to rank which factors w(**e**)**re** the most influential. The following equation was used:

$$y = a_1x_1 + a_2x_2 + a_3x_3 \cdot \cdots \cdot a_nx_n + a_0 \tag{1}$$

Since the measurement scale for each item differed, each factor was standardized for comparison.

Results are shown in Table 2. This 7-factor model accounted for 31 % of the variance in reported empathy. In assessing the most influential factors, results showed that the desire to be empathic was most influential, followed by personality factors and sex.

4 Discussion

The purpose of the present study was to clarify the influence of multiple contributing factors related to the occurrence of empathy. For the current study, contributing factors of empathy, and the degree of each factor's influence, were investigated.

In terms of sex, results were similar to past studies [26, 27] showing that women report more empathy or empathic behaviors compared to men.

With regard to age, overall, we did not observe age effects. However, when we examined differences between specific age groups, significant results emerged. Previous studies reported that there is an element that is related to age and an element that is not so related with the element of empathy [26, 27]. We confirmed that empathy had a significant difference by comparing each generation. We considered that there are differences between the generations when it comes to the elements of empathy that are of interest to them.

We also observed similar findings as past studies concerning the relationship between personality and empathy [25]. The validity of the result was demonstrated because previous studies reported the same results.

Regarding the influence of family variables on empathy, previous studies have reported conflicting relationships [25]. This was also the case in the current study.

No previous study had examined the role of frustration on empathy. We considered that the frustration experience did not influence empathy because there is nothing concerning remembrance of experiences in the four elements of empathy.

Furthermore, no previous study had examined how the desire to be empathic affects empathy. Not surprisingly, we observed that a desire to be empathic was influential on reports of empathy.

A multiple linear regression analysis showed that this desire was most influential on empathy. Specifically, a desire to affect others emotively was most predictive of empathy. We also observed that a connection between prosocial behavior and a desire to affect others emotively was a strong influence on empathy. One potential reason for why empathic desire had an influence on empathy is that empathic desire motivates empathic behavior. This analysis also revealed that specific personality characteristics also influenced empathy. We argue that specific personality predispositions strengthen empathic concerns, which lead to empathic behaviors. Since such predispositions include curiosity with others and receptiveness toward external stimuli, empathic behaviors would likely result. Finally, in terms of sex, it is likely that women are more socially responsive, which strengthens women's empathic concern and empathic behaviors. Since a woman generally has high social responsiveness, there are the receptiveness and attentiveness to a stimulus.

Overall, the current study investigated the influence of various contributing factors on empathy. Results showed that some factors were influential while others were not. Moreover, we revealed a rank-order of factors predicting empathy. The present findings are useful for effectively designing models of empathy that include the myriad possible factors related to promoting prosocial and adaptive behaviors.

5 Conclusion

The current study assessed various contributing factors to empathy, as well as the magnitude of influences for each factor. We observed that the following four factors significantly contributed to empathy: sex, personality, experience, and desire. Although age did not influence empathy globally, age differences were observed when assessing specific age groups. We also observed that an individual's frustration experience did not influence empathy. Regarding the magnitude of influence for each factor, empathic desire was the most predictive of empathy, followed by personality factors and sex. We considered that it was effective to understand such rank when guiding empathy of the others.

References

1. Plutchik R (1990) Evolutionary bases of empathy. In: Empathy and its development. Cambridge University Press, Cambridge, p 38
2. Davis MH (1994) Empathy: a social psychological approach. Westview Press, Boulder
3. Batson CD, Powell AA (2003) Altruism and prosocial behaviour. In: Handbook of psychology. Wiley, Hoboken
4. Eisenberg N, Fabes RA, Murphy B, Karbon M, Maszk P, Smith M, Suh K (1994) The relations of emotiveity and regulation to dispositional and situational empathy-related responding. J Personal Soc Psychol 66(4):776.
5. Eisenberg N, Miller PA (1987) The relation of empathy to prosocial and related behaviors. Psychol Bull 101(1):91
6. Eisenberg N, Mussen PH (eds) (1989) The roots of prosocial behavior in children. Cambridge University Press, Cambridge/New York
7. Ministry of Economy, Trade and Industry (2013) White paper on international economy and trade
8. Volvic Drink 1 (2013) Give 10 Campaign, Volvic. http://www.volvic.co.jp/csr/1lfor10l/index.html
9. Kotler P (2000) Marketing management: the millennium edition. International edition. Prentice Hall, London
10. Kotler P, Lee N (2008) Corporate social responsibility: doing the most good for your company and your cause. Wiley, Hoboken
11. Schmitt B (1999) Experiential marketing. J Mark Manag 15(1–3):53–67
12. Schmitt BH (2010) Customer experience management: a revolutionary approach to connecting with your customers. Wiley, New York
13. Cohen JB, Areni CS (1991) Affect and consumer behavior. Handb Consum Behav 4(7):188–240
14. Thompson CJ, Locander WB, Pollio HR (1989) Putting consumer experience back into consumer research: the philosophy and method of existential-phenomenology. J Consum Res 16(2):133

15. Tokaji A (2001) The mechanism of impression arouse. Cogn Stud Bull Jpn Cogn Sci Soc 8(4): 360–368
16. Maslow AH (1959) Cognition of being in the peak experiences. J Genet Psychol 94(1):43–66
17. Maslow AH (1961) Peak experiences as acute identity experiences. Am J Psychoanal 21(2): 254–262
18. Nishio M, Makino Y, Shirasaka S, Maeno T (2014) System structure of "being emotively moved" by analyzing emotive causes. In: Business systems laboratory second international symposium, January 2014, Roma, Italy
19. Feshbach ND (1975) Empathy in children: some theoretical and empirical considerations. Couns Psychol 5(2):25–30
20. Hoffman ML (1984) Interaction of affect and cognition in empathy. In: Emotions, cognition, and behavior. Cambridge Universtiy Press, Cambridge/New York, pp 103–131
21. Davis MH (1980) A multidimensional approach to individual differences in empathy. American Psychological Association, Washington, DC
22. Davis MH (1983) Measuring individual differences in empathy: evidence for a multi-dimensional approach. J Pers Soc Psychol 44(1):113
23. Mehrabian A, Epstein N (1972) A measure of emotive empathy1. J Pers 40(4):525–543
24. Eisenberg N, Fabes RA, Schaller M, Miller P, Carlo G, Poulin R, Shell R (1991) Personality and socialization correlates of vicarious emotive responding. J Personal Soc Psychol 61(3):459
25. Wise PS, Cramer SH (1988) Correlates of empathy and cognitive style in early adolescence. Psychol Rep 63(1):179–192
26. Davis MH, Franzoi SL (1991) Stability and change in adolescent self-consciousness and empathy. J Res Pers 25(1):70–87
27. Gough HG, Heilbrun AB (1983) The adjective check list manual. Consulting Psychologists Press, Palo Alto

Part VI
Regional Development and Policymaking

Evaluation of Countermeasures for Low Birthrate and Aging of the Population in a Suburban New Town

Yoshiki Ito, Tomomi Nonaka, and Masaru Nakano

Abstract Tama City is engaged in two measures to counter the declining birthrate and aging of its population: an incentive scheme that encourages companies to invite young people and an action plan to support the development of the next generations. But, the exact process to estimate policy effects and identify measures that lead to good results is unknown. This paper articulates the causes of Tama City's declining birthrate and aging population, and reveals measures derived from a cohort-component method that are suitable for measuring the future impact of policies intended to reduce these trends.

Keywords Urban service • Public service • Sustainable development • Policy analysis • Social system design

1 Introduction

In recent years, in a number of advanced nations, the population has aged and fewer babies have been born due to falling birthrates and longer lifespans. While the working age population can increase in urban areas, populations can shrink dramatically in rural areas [1]. This is a global trend observed in urban areas such as New York, Munich, Beijing, and Tokyo. Aging populations have significant impacts on the economy and society as a whole and thus cannot be ignored in the development of a sustainable urban infrastructure.

Sustainability of urban infrastructures has been evaluated with economic, environmental, and social axes [2–6]. Issues such as energy saving, renewable energy, and efficiency of urban development have also been studied [7–14]. However, there have been few studies on aging issues [15].

Y. Ito • M. Nakano (✉)
Graduate School of System Design and Management, Keio University, 4-1-1 Hiyoshi, Kohoku-ku, Yokohama, Kanagawa 223-8526, Japan
e-mail: nonaka@ise.aoyama.ac.jp

T. Nonaka
Dept. of Industrial and Systems Engineering, Aoyama Gakuin University, 5-10-1 Fuchinobe, Chuo-ku, Sagamihara-shi, Kanagawa 252-5258, Japan

T. Maeno et al. (eds.), *Serviceology for Designing the Future*,
DOI 10.1007/978-4-431-55861-3_27

Conditions of cities vary widely with respect to factors such as development process, historical background, and geographical characteristics. The specific background of a city should be considered for the sustainability of its urban infrastructure. Thus, conducting urban studies to determine common and individual themes among different urban areas, through case studies of typical cities and advanced cities having new problems, is important.

Japan has the most serious aging population problems. Suburban new towns are planned areas of agglomerated dwellings with no neighboring central cities [16]. Tama New Town in Tama City is the largest suburban new town in Japan and has the highest increase in the old-age index (65 years or older) in Japan. In this article, a model is developed for evaluating sustainability of the suburban new town.

In Tama City, an incentive scheme to attract enterprises was established in 2011 for the promotion of business relocation and expansion of job opportunities [17]. Expanded job opportunities is expected to reduce the number of young people moving out and encourage more young people to move in. An action plan to support families raising children, "Action plan for Measures to Support the Development of the Next Generation," was also established in FY2011 to encourage families raising children to remain in Tama City and increase the birthrate [18]. For both measures, however, neither their effects on the current situation nor their impact on future population composition have been clarified quantitatively.

Various studies have been conducted in Japan and overseas regarding the impact of countermeasures to motivate women to bear children [19]. Few studies, however, reveal the impact of those measures on the future population composition. Though studies for population estimates in specific geographic areas have been made in Japan and overseas, many simply estimate future population compositions [20, 21]. They do not consider the effects of the current countermeasures to mitigate the problem. Future population composition is estimated through studies of the collective housing area, suggesting that measures to counter the declining birthrate and aging population are needed [22].

This study analyzes Tama City's population estimates from the past to clarify the causes of population aging. Next, a population estimate model is developed to assess the incentive scheme for attracting enterprises and the action plan to support families raising children.

Section 2 analyzes population estimate models used in this study. Section 3 describes preconditions to the study in detail. In Sect. 4, causes of the problems in Tama City are clarified through scenario analysis. In Sect. 5, population in the future are estimated in consideration of countermeasures for the problems of. Section 6 gives the conclusion of this study and future issues.

2 Population Estimate Model

2.1 A Cohort-Component Method

The purpose of this study is to clearly articulate the causes of Tama City's declining birthrate and aging population, and to evaluate the impacts of Tama City's ongoing measures for handling these issues on future population composition. The former is clarified through scenario analysis of population estimates between 1970 (when the initial residents of Tama New Town moved into the area) and 2010 (the year before the countermeasures came into effect). The latter is evaluated through population estimates between 2010 and 2035. For population estimates, a model based on a cohort-component method [23], with demographic equations and census figures, is used.

Using survival rates and move-in and move-out rates by gender and age in the base year, five-year-later populations by gender and age are estimated. Ten-year-later populations are derived by entering the five-year-later populations. The population estimate model is described in detail in subsequent sections.

2.2 Calculation Processes of the Cohort-Component Method

Mathematical formulae of the proposed model are given as follows:

$$M_{0,t+5} = (p_{0,t}^M + m_{0,t}^M)B_t^M \tag{1}$$

$$M_{85,t+5} = (p_{85,t}^M + m_{85,t}^M)(M_{80,t} + M_{85,t}) \tag{2}$$

$$M_{x,t+5} = (p_{x,t}^M + m_{x,t}^M)M_{x-5,t}, 5 \leq x \leq 80 \tag{3}$$

$$F_{0,t+5} = (p_{0,t}^F + m_{0,t}^F)B_t^F \tag{4}$$

$$F_{85,t+5} = (p_{85,t}^F + m_{85,t}^F)(F_{80,t} + F_{85,t}) \tag{5}$$

$$F_{x,t+5} = (p_{x,t}^F + m_{x,t}^F)F_{x-5,t}, 5 \leq x \leq 80 \tag{6}$$

$$B_t = \sum_{x=15}^{45} (F_{x,t} + F_{x,t+5})b_{x,t} \times \frac{5}{2} \tag{7}$$

$$B_t^M = \alpha \times B_t \tag{8}$$

$$B_t^F = \beta \times B_t \tag{9}$$

$$m_{x,t}^M = m_{x,t,in}^M - m_{x,t,out}^M \tag{10}$$

$$m_{x,t}^F = m_{x,t,in}^F - m_{x,t,out}^F \tag{11}$$

$M_{x,t}$, $F_{x,t}$: Male and female populations in the age segments $(x\ to\ x+4)$ in year t $x = 0, 5, \ldots, 80$; $M_{85,t}$ and $F_{85,t}$ indicate male and female populations of people aged 85 years or older

$b_{x,t}$: Birthrates by age of females in the age segments $(x\ to\ x+4)$ between years t and $t+5$

B_t: Birthrates between years t and $t+5$

B_t^M, B_t^F: Number of male and female babies born between years t and $t+5$

α, β: Ratio of male and female births

The sex ratio at birth in recent years has been 106 male births per 100 female births. In this study, $\alpha = 106/206$ and $\beta = 100/206$ are used [24].

$p_{x,t}^M$, $p_{x,t}^F$: Rates of male and female populations in the age segments $(x-5\ \mathrm{to} x-1)$ in year t who survive in the age segments $(x \mathrm{to} x+4)$ in year $t+5$

$p_{0,t}^M$, $p_{0,t}^F$: Rates of male and female babies born between years t and $t+5$ who survive in the age segments (0–4) in year $t+5$

$p_{85,t}^M$, $p_{85,t}^F$: Rates of males and females aged 85 years or older in year t who survive in the age segments (85 years or older) in year $t+5$

$m_{x,t}^M$, $m_{x,t}^F$: Net migration rates of male and female populations in the age segment $(x-5\ \mathrm{to} x-1)$ who fall into $(x\ to\ x+4)$ in year $t+5$

$m_{x,t,in}^M$, $m_{x,t,in}^F$: Move-in rates of male and female populations in the age segment $(x-5\ \mathrm{to} x-1)$ who fall into $(x \mathrm{to} x+4)$ in year $t+5$

$m_{x,t,out}^M$, $m_{x,t,out}^F$: Move-out rates of male and female populations in the age segment $(x-5\ \mathrm{to} x-1)$ who fall into $(x\ to\ x+4)$ in year $t+5$

$m_{0,t}^M$, $m_{0,t}^F$: Migration rates of male and female babies born between years t and $t+5$ who fall into the age segment (0 to 4) in year $t+5$

$m_{85,t}^M$, $m_{85,t}^F$: Migration rates of male and females aged 80 years or older in year t who fall into the age segment (85 years or older) in year $t+5$

2.3 Reflection of Effects of Measures in the Cohort-Component Method

This section describes the method to analyze the effects of countermeasures for declining birthrate and aging population in the calculation processes of the model. The following two measures have been implemented by Tama City: an incentive scheme to attract enterprises and an action plan to support families raising children.

The incentive scheme to attract enterprises is assumed to reduce the move-out rates. Reasons for move-out are categorized into eight items, such as housing, environmental issues, commuting to workplace, and living together with or apart from parents or children [18]. The survey results are categorized by gender and age. The implementation of the incentive scheme is assumed to encourage people to have their workplace near their homes. Those who cited convenience of commute as a reason for move-out are considered to all remain in Tama City.

Calculation processes of the cohort-component method, which reflect the effects of the countermeasures, are formulated in Eqs. (12) and (13). Equations (12) and (13) were obtained by adding the effects of the action plan to Eqs. (10) and (11).

$$m_{x,t}^{M} = \left\{m_{x,t,in}^{M} \times \left(1 - R_{x,work,in}^{M}\right)\right\} - \left\{m_{x,t,out}^{M} \times \left(1 - R_{x,work,out}^{M}\right)\right\} \tag{12}$$

$$m_{x,t}^{F} = \left\{m_{x,t,in}^{F} \times \left(1 - R_{x,work,in}^{F}\right)\right\} - \left\{m_{x,t,out}^{F} \times \left(1 - R_{x,work,out}^{F}\right)\right\} \tag{13}$$

$R_{x,work,in}^{M}, R_{x,work,in}^{F}, R_{x,work,out}^{M}, R_{x,work,out}^{F}$: Percentage of male and female people who cite commute as the reason for move-in and move-out to the total move-out population in the age segment $(x-5 \text{ to} x-1)$ who fall into $(x \text{ to } x+4)$ in year $t+5$

The Action Plan for Measures to Support the Development of the Next Generation is assumed to affect both move-out rates and birthrates. Similar to the study of the incentive scheme, Tama City's survey data of reasons for move-out are used to include the move-out rates in the cohort-component method. Those who cited childrearing circumstances as the reason for move-out are assumed to all remain in Tama City due to the action plan. Calculation processes of the cohort-component method, which reflect the effects of the measures, are formulated in Eqs. (14) and (15):

$$m_{x,t}^{M} = \left\{m_{x,t,in}^{M} \times \left(1 - R_{x,child,in}^{M}\right)\right\} - \left\{m_{x,t,out}^{M} \times \left(1 - R_{x,child,out}^{M}\right)\right\} \tag{14}$$

$$m_{x,t}^{F} = \left\{m_{x,t,in}^{F} \times \left(1 - R_{x,child,in}^{F}\right)\right\} - \left\{m_{x,t,out}^{F} \times \left(1 - R_{x,child,out}^{F}\right)\right\} \tag{15}$$

$R_{x,child,in}^{M}, R_{x,child,in}^{F}, R_{x,child,out}^{M}, R_{x,child,out}^{F}$,: Percentage of male and female people who indicate childrearing circumstances as the reason for move-in and move-out to the total move-out population in the age segment $(\mathrm{x}-5 \text{ to } \mathrm{x}-1)$ who fall into $(\mathrm{x} \text{ to } \mathrm{x}+4)$ in year $\mathrm{t}+5$

Next, the effects of the action plan on birthrates are formulated. In the action plan, nurseries are built in respective city blocks and every applicant is able to enter the nursery as he or she wishes. As shown in previous studies [19], implementation of childrearing support measures to facilitate the availability of nurseries increases the childbearing intention rate by 8.8 %; this rate is used in this study. The measures are considered to have a similar impact on childbearing intention in the age segment (15–49) in this study, although this assumption was not covered in previous studies. In this context, calculation processes of the cohort-component method, which reflect the effects of the countermeasures, are formulated in Eq. (16):

$$B_t = \sum_{x=15}^{45} (F_{x,t} + F_{x,t+5}) \{b_{x,t} \times (1 + W)\} \times \frac{5}{2} \tag{16}$$

B_t: Birthrates between years t and $t + 5$
$b_{x,t}$: Birthrates by age of females in the age segments $(x\ to\ x + 4)$ between years t and $t + 5$
$F_{x,t}$: Population of females in the age segments $(x\ to\ x + 4)$ in year t
W: Rate of increase in childbearing intention by implementation of the Action plan for Measures to Support the Development of the Next Generation

As described above, W is assumed to be 0.088 in this study. Equation (16) is obtained by adding the effects of the action plan to Eq. (3).

2.4 Evaluation Index

In this study, the results of population estimates by the cohort-component method are evaluated by old-age indexes. Old-age indexes are percentages of people aged 65 years or older in Japan or specific areas. Lower old-age indexes are considered better. Old-age indexes are formulated in Eq. (17):

$$I_{old\ index} = \frac{\sum_{x=65}^{\infty} P_x}{P_{all}} \tag{17}$$

$I_{old\ index}$: Old-age index
P_{all}: Total population in Japan or a specific area
P_x: Population of x–year old people in Japan or a specific area

3 Data and Preconditions

3.1 Standard Population

The population of Tama City as per the national census is used for the standard population [25]. Census data of 1970 are used for the scenario analysis to analyze causes for the declining birthrate and population aging, and that of 2010 is used for the impact assessment of countermeasures for declining birthrate and population aging.

3.2 Assumed Values for the Cohort-Component Method

Four assumed values are used in the cohort-component method: future survival rates by gender and age, future net migration rates by gender and age, future birthrates of females by age, and the male and female ratio at birth. Known data are used as assumed values for the scenario analysis of population estimates between 1970 and 2010. For the analysis, census data [25, 26] are used.

3.3 Impacts of the Measures for the Declining Birthrate and Aging Population on the Move-Out Rate

As described in the preceding section, reasons for move-out shown by Tama City's survey are used to analyze impacts of the countermeasures for the declining birthrate and aging population on the move-out rate. Populations between 2010 and 2035 are estimated, assuming that Tama City's survey results in 2010 are applicable until 2030.

3.4 Impacts of the Measures for the Declining Birthrate and Aging Population on the Birthrate

As described in the preceding section, findings of previous studies are used to analyze impacts of countermeasures for declining birthrate and aging population on the birthrate. Populations between 2010 and 2035 are estimated, assuming that the findings of the aforementioned previous studies are applicable between 2010 and 2035.

4 Factor Analysis for the Declining Birthrate and Aging Population

This section describes simulation results of factor analysis for the declining birthrate and aging population.

4.1 Scenario Building

Table 1 shows scenarios used to analyze factors that have affected declining birthrate and aging population. Populations are estimated by changing a factor

Table 1 Factor analysis scenarios for the declining birthrates and aging population

Data scenario	Base population	Birthrate by age	Migration (Move-in + Move-out)	Survival rate
Reference: "Reference data"	Tama City data	Tama City data	Tama City data	Tama City data
Initial: "Dispersal of the initial residents" scenario	Half of the initial residents move in between 1970 and 1974 and the rest between 1975 and 1979	Tama City data	Tama City data	Tama City data
Birthrate: "High birthrate" scenario	Tama City data	Japan data	Tama City data	Tama City data
Migration: "Migration rate of working population" scenario	Tama City data	Tama City data	Japan data	Tama City data

relevant to population change and keeping the other factors fixed. Relevant factors to population change are base population (for this study, population in 1970), birthrates by age $b_{x,t}$, migration rates (move-in + move-out) $m^M_{x,t,in}$, $m^F_{x,t,in}$, $m^M_{x,t,out}$, $m^F_{x,t,out}$ and survival rates $p^M_{x,t}$, $p^F_{x,t}$. The Tama City data are obtained from [24] and the Japan Data are collected from [26].

Tama City's data are used for all factors in the reference scenario (Reference). In the "dispersal of the initial residents" scenario (Initial Residence), populations are estimated assuming that half of the initial residents moved in between 1970 and 1974, and the rest moved in between 1975 and 1979.

Tama City is assumed to be suffered from highly declining birthrate and aging population because the city's birthrate is lower than that of Japan as a whole. In the "high birthrate" scenario (Birthrate), populations are estimated using birthrates by age in Japan while Tama City's data are used for other factors. Tama City's move-in migration rate of working population was lower than that of Japan. The fact is assumed to cause highly declining birthrate and aging population in Tama City. In the "migration rate of working population" scenario (Migration), the city's move-in migration rate of the working population in the age segment (15–64) is assumed to be that in Japan as a whole, while Tama City's data are used for other factors.

4.2 Simulation Results

The population compositions for 2010 are simulated by using the proposed model for the abovementioned six scenarios. Figure 1 shows changes in the old-age index calculated using different population compositions.

In the Reference scenario, the old-age index in 2010 is 20.9 %. It is 15.4 % in the Initial scenario, 21.6 % in the Birthrate scenario, and 14.3 % in the Migration scenario. There are small differences in the old-age index for 2010 between the

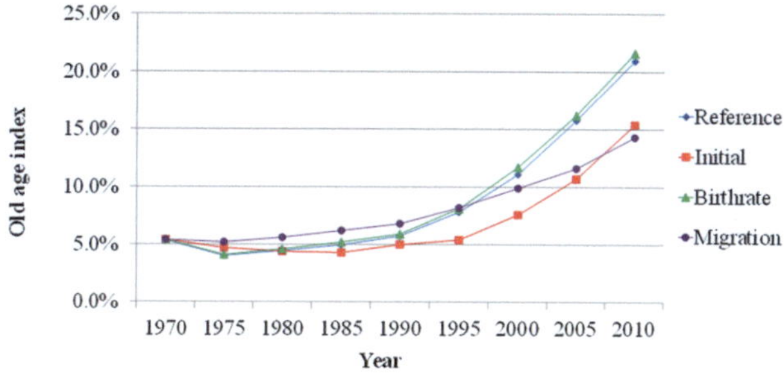

Fig. 1 Changes in old-age index between 1970 and 2010 for each scenario

Birthrate and the Reference Scenarios. In fact, little difference in birthrates can be observed between Tama City and Japan. Old-age indexes in 2010 are lower in Initial and Migration scenarios than that in the Reference scenario. Therefore, Tama City's declining birthrate and aging population problem is considered to become worse due to concentrated initial move-in and low migration rate of working population. Regarding the old-age index, the largest deviation from the Reference scenario is seen in Migration scenario, followed by Initial scenario.

5 Impact Assessment of Measures for the Declining Birthrate and Aging Population

5.1 Scenario Building

This section describes the simulation results of the impact assessment of countermeasures for the declining birthrate and aging population. Table 2 shows four scenarios for the impact assessment of the Incentive Scheme for Attraction of Enterprises and the Action Plan to Support the Development of the Next Generation. For the four scenarios, one of the two measures is implemented, both measures are implemented, or neither of them is implemented, and populations between 2010 and 2035 are estimated. As for the 'Attraction of Enterprises', $R^M_{x,work,out}$, $R^F_{x,work,out}$ are set to zero for the 'Not considered' and given from [24] for the 'Condsidered'. As for the 'Next Generation', $W = 0.88$ and $R^M_{x,child,out}$, $R^F_{x,child,out}$ are set to zero for the 'Not considered' and given from [24] for the 'Condsidered'. Effects of countermeasures for factors having an impact on population have already been described above. The influences for moving effects represented by $R^M_{x,work,in}$, $R^F_{x,work,in}$ $R^M_{x,child,in}$ and $R^F_{x,child,in}$ are not considered in this paper due to lack of data for scenario assumptions.

Table 2 Scenarios for impact assessment of measures for the declining birthrates and aging population

Measures scenario	Incentive scheme for attraction of enterprises	Action plan to support the development of the next generation
Reference: "No measures " scenario	Not considered	Not considered
Attraction: "Attraction of enterprises" scenario	Considered	Not considered
NextGen: "Development of the next generation" scenario	Not considered	Considered
Both: "Attraction of enterprises and support for the next generation" scenario	Considered	Considered

5.2 Simulation Results

Population compositions for the four scenarios in 2035 are simulated using the cohort-component method. Figure 2 shows changes in the old-age index calculated from different population compositions.

The old-age index is 36.9 % for 2035 in the Reference scenario, 28.9 % for the Attraction scenario, 34.4 % for the NextGen scenario, and 26.8 % for the Both scenario. The results of the impact assessment show that over the next 25 years, the old-age index will decrease by 8.0 % with the Attraction scenario, 2.5 % with the Next Gen scenario, and 10.1 % with the Both scenario.

The incentive scheme for attraction of enterprises has more impact than the Action Plan to Support the Development of the Next Generation. In 25 years, the former is expected to have about three times higher impact than the latter. In view of financial restrictions, focusing more on the incentive scheme for attraction of enterprises appears to be more promising.

Two background factors exist for these findings:

1. In Tama City, the percentage of those who cite "workplace environment" as a reason for move-out (move-out to have their workplace near their home) is much higher than that for the "childrearing environment" (move-out for better childrearing environment).
2. Tama City's Action Plan to Support the Development of the Next Generation (fulfillment of requirements for nurseries) has little impact on birthrates. Financial assistance – for example, for monthly tuition of nursery and childcare programs – has more impact on birthrates.

Fig. 2 Changes in old-age index between 2010 and 2035 for each scenario

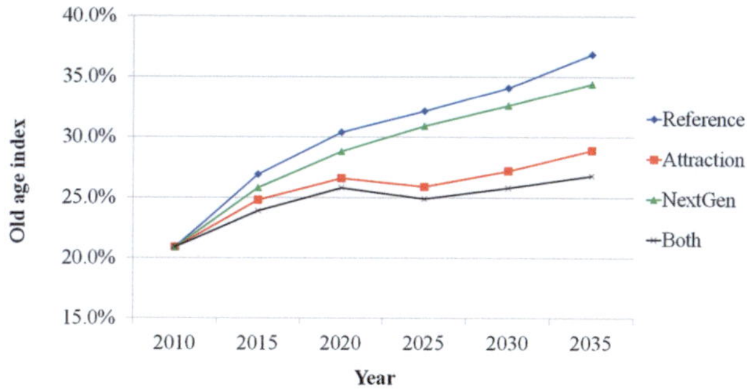

6 Conclusions

In this paper, through case studies of Tama City, which is experiencing the most serious situation in Japan in terms of aging population, a population estimate model was developed using the cohort-component method, and scenario analyses based on survey results and findings of previous studies were performed to clarify primary factors affecting declining birthrate and aging population and to clarify the expected impact of Tama City's countermeasures in the future.

(1) The primary factors for the aging population in Tama City are

1. The migration rate (Move-in – Move-out) of working population (age 15–65) in Tama City is higher than that of Japan as a whole, with excess move-out.
2. A number of residents have moved in for only a short period.

(2) The incentive scheme for attraction of enterprises is predicted to be more effective in decreasing the old-age index compared with the Action Plan to Support the Development of the Next Generation in Tama City.

According to research findings, the following two points are suggested for the housing policy for new towns and incentive scheme for attraction of enterprises in Tama City:

(1) For new town planning in not only Japan but other areas, a large number of initial residents, as compared with the original population of the area where the new town is planned, should ideally not be allowed to move in. System design is required to determine the percentage of the initial move-in to the original population and dispersal of residents.
(2) In Tama City, focusing on the incentive scheme for attraction of enterprises is more efficient than focusing on the action plan to support the development of the next generation to counter the declining birthrate and aging population.

To restrain the declining birthrate and aging population, providing support programs for families raising children and incentive schemes to attract new enterprises is necessary, in consideration of childrearing circumstances and characteristics of local industries (i.e., whether it is an industrial district where people have their workplace near their home or a suburb where people commute to nearby cities). With a population estimate model similar to the one built in this study, evaluation of the effects of multiple measures quantitatively that considers of the characteristics of respective areas is made possible.

The authors recommend these statements to indicate that planning a new town and maintaining it is a fluid process, where the driving factors may need to be reassessed in time or considered during planning.

Acknowledgement This work was supported in part by a Grant-in-Aid from the Association for Sustainable Building Systems and JSPS KAKENHI Grant-in-Aid for Young Scientists (B) Number 26730159. We thank the members of the theme "Development of simulation models for long life housing and sustainable urban infrastructure" under the grant.

References

1. Takayama N, Saito O (2006) Shoshika no keizaibunseki [Economic analysis of declining birth rate] –translated by authors (in Japanese). Toyokeizai Inc., Tokyo
2. Carter K, Fortune C (2007) Sustainable development policy perceptions and practice in the UK Social Housing Sector. Construct Manag Econ 25(4):399–408
3. Howes R, Robinson H (2005) Infrastructure for the built environment: global procurement strategies. Elsevier, Oxford
4. Hugentobler M, Brandle-Stroh M (1997) Sustainable urban development: a conceptual framework and its application. Urban Technol 4(2):85–99
5. Richard CH, Bowen PA (1997) Sustainable construction: principle and a framework for attainment. Construct Manag Econ 15(3):223–239
6. Singh RK, Murty HR, Gupta SK, Dikshit AK (2012) An overview of sustainability assessment methodologies. Ecol Indic 15(1):281–299
7. Adinyira E, Oteng-Seifah S, Adjei-Kumi T (2007) A review of urban sustainability assessment methodologies. International conference on whole life urban sustainability and its assessment. Glasgow Caledonian University, Glasgow
8. Hal AV, Dulski B, Postel AM (2010) Reduction of CO2 emissions in houses of historic and visual importance. Sustainability 2(2):443–460
9. McGranahan G, Satterthwaite D (2003) Urban centers: an assessment of sustainability. Ann Rev Environ Resour 28:243–274
10. Muller MO, Stampfli A, Dold U, Hammer T (2011) Energy Autarky: a conceptual framework for sustainable regional development. Energy Policy 39(10):5800–5810
11. Newton PW (2012) Liveable and sustainable? Socio-technical challenge for twenty-first-century cities. Urban Technol 19(1):81–102
12. Poveda CA, Lipsett MG (2011) A review of sustainability assessment and sustainability/environmental rating systems and credit weighting tools. Sustain Dev 4(6):36–55
13. Stablo J, Winkel CR (2012) The integration of energy conservation into the political goal of renewable energy self-sufficiency – a German case study based on a longitudinal reconstruction. Sustainability 4(5):888–916
14. Zavrl MS, Zeren MT (2010) Sustainability of urban infrastructures. Sustainability 2 (9):2950–2964
15. Han J, Fontanos P, Fukushi K, Herath S, Heeren N, Naso V, Cecchi C, Edwards P, Takeuchi K (2012) Innovation for sustainability: toward a sustainable urban future in industrialized cities. Sustain Sci 7(1):91–100

16. Daido Y (1996) Nihongata no nyutaun [Japanese-Style New Town] (in Japanese). Keizai to Boeki [Economy and Trade] –translated by authors. 90:86–92
17. Tama City (2011a) Kigyou yuuchishourei seido no goshoukai [Introduction of the welcome enterprises system] (in Japanese) –translated by authors. http://www.city.tama.lg.jp/16962/4289/004282.html. Accessed 26 Apr 2014
18. Tama City (2011b) Tamashi kosodate, kosodachi, kodomo pulan <kaiteiban>, 2011 [Tama City raising child, child growing, children plan <Revision>, 2011] (in Japanese), –translated by authors. http://www.city.tama.lg.jp/plan/947/10875/index.html. Accessed 26 Apr 2014.
19. Matsuda S (2009) Jisedai ikusei shiensaku ni yottsute shuttusanikou ha takamaruka -vinettuto chousa niyoruseisakukouka no kentou [Is measures to support raising next-generation children effective? – estimates of political efficacy by using vignette survey] –translated by authors (in Japanese). Life Des Rep 1(2):16–23
20. Arai T (2011) Population projections for Onomichi City: 2011 (in Japanese). J Econ Manag Inf Sci Onomichi Univ Keizai Joho Ronshu 11(1):93–103
21. Fujii T (1999) Forecasting population of Osaka City by cohort-component method (in Japanese). Q J Econ Stud 22(2):25–37
22. Katayama N, Kaido K (2007) Characteristics and prediction of demographic structure in Suburban residential areas: case study of fifteen residential estates in Kani City, Gifu Prefecture (in Japanese). Urban Sci Stud 12:53–62
23. Yamaguchi K (1990) Jinkou suikei nyumon [Introduction to population estimates] (in Japanese) –translated by authors. Kokonshoin, Tokyo, pp 1–221
24. Tama City (2011c) Tamashi yoron chousa [Public opinion research of Tama City] –translated by authors (in Japanese). http://www.city.tama.lg.jp/dbps_data/_material_/common/siminsoudan/34yorontyousagaiyouban.pdf. Accessed 26 Apr 2014
25. Ministry of Internal Affairs and Communications (2012) e-Stat: portal site of official statistics of Japan. https://www.e-stat.go.jp/SG1/estat/eStatTopPortalE. do (Accessed 26 Apr 2014)
26. National Institute of Population and Social Security Research (2006) Population projection for Japan: 2006–2055 (in Japanese). http://www.ipss.go.jp/syoushika/tohkei/suikei07/P_HP_H1812_A/index2.html. Accessed 26 Apr 2014

Analysis of Value Co-creation Between Farmers and Land Improvement District in Japan Through Irrigation Service Improvement by Good Water Quality

Tasuku Kato, Kana Nakano, and Toshiaki Iida

Abstract In Japan, agricultural sector including rice production has several problems that are complex of farmers aging, small scale farming, and high cost management. To improve this situation, Japanese government pushed to increase efficiency to change high profit production or large scale agriculture. In this research, objectives is analyzed that farmer's acceptability of new innovative irrigation water distribution technology to improve rice quality as high profitable production. As a methodology, water quality control management was assumed in a good rice quality control scenario for Agent Based Model (ABM). In water quality management for rice production, former research results showed that when nitrogen water quality was higher after flowering in rice growth stage, nitrogen concentration and protein contents ratio was increasing, then, taste of rice was decreasing. In study area as an example of application, Imbanuma basin is chose as high nitrogen concentration area. As results, the modified ABM was applied that farmer's preference would be no change in prior the cost, and if new irrigation would be accepted the profit would be higher than a threshold, though interview results for farmers were more complicated because of cost, profit, and risk management.

Keywords Rice quality • Water distribution • Agent based model

1 Introduction

Nowadays, Japanese agricultural sector, especially in rice production, has several problems that come from complex situations of farmers aging, small scale farming, and high cost management. To improve this situation, Japanese government pushed

T. Kato (✉) • K. Nakano
Tokyo University of Agriculture and Technology, 3-5-8, Saiwaicho, Fuchushi, Tokyo 185-8503, Japan
e-mail: taskkato@cc.tuat.ac.jp

T. Iida
Graduate School of Agricultural and Life Sciences, The University of Tokyo, Tokyo, Japan

T. Maeno et al. (eds.), *Serviceology for Designing the Future*,
DOI 10.1007/978-4-431-55861-3_28

to increase efficiency to change high profit production or large scale agriculture to increase efficiency.

Rice production is required a lot of water consumption, it is estimated around 1200 mm for whole cropping period, and over half amount of water is supplied by artificial irrigation [1]. Amount of water consumption is equal to sum of rainfall and irrigation, and irrigation amount is near to rainfall. Irrigation water is supplied by water resources in watershed, e.g., ground water, reservoir, or river water with headwork. And irrigation supply is needed a water distribution system and hydraulic facilities.

In irrigation distribution, efficient water management is an important role, because new development of water resources is quite limited effect as high cost. Innovative management is a desirable way to improve water resource distribution under current facilities and organization, and create new additional value in agricultural production.

In current water resources distribution organization is Land Improvement District (LID). LID is elected by farmers, and it manages water distribution system from headwork to agricultural plot [2]. Farmers involved in LID are charged cost for water distribution system management, and LID operates and manages the irrigation system and facilities. As a service aspect, stability and reliability for water distribution are most important service for farmers. Also, under drought condition, LID try to save water in system operation and to mediate between stakeholders to share water resources and draught risk. Besides, in flood season, LID cooperates with river management organization and mitigates flood damage.

However, LID has not committed to improve rice productivity and farmers economic condition. Because, other sales and market organization "Nokyo" covered for farming technology and supporting farmers economics. LID is focused to supply management of water resources. To improve service, water management and production sales should be combined, and water resources distribution should be controlled by demand side in farmers.

Former research to concern with water distribution service regarding with stakeholders agreement was conducted at SUBAK system, Bali, Indonesia, by Lansing et al. [3, 4]. A traditional water distribution organization "SUBAK" and its management way were analysed by Agent Based Model (ABM). In this ABM, traditional water distribution rule is well adapted pest control and optimized rice production.

In water environmental aspect, irrigation water is contained nutrients load that comes from excess fertilizer or livestock waste especially in agricultural area. Eventually, irrigation water quality is high nutrients concentration in nitrogen and phosphorous [5, 6].

High nitrogen water quality is influenced to nitrogen contain ratio in rice crop, and it occurs to increase protein ratio. High protein ration in rice is degraded rice taste. So, irrigation water quality is desirable to lower nitrogen water quality.

In the other aspect of nitrogen water quality, paddy field drainage is reducing nitrogen by denitrification that is microbiological activities. If, repeatedly, drainage water is reused as irrigation, rice quality will be improved, and it emerges new additional value for rice crops. As this service would be provided by LID, farmers can receive a new service, and their economic success is reflected by LID performance. The problem is possibilities of new service installation and acceptance for farmers.

In this research, it is aimed to improve water supply service of LID and to create new water demand for farmers' profit increase through increasing rice quality.

Then, it is analyzed that farmer's acceptability of new innovative irrigation water distribution technology to improve rice quality as high profitable production.

As a research framework, case study research in Imbanuma area is conducted where rice quality is not high value because irrigation water quality is degraded. Irrigation water quality management is assumed to emerge a good rice quality as a scenario of LID management. For scenario evaluation, ABM is developed and applied for service distribution. To develop ABM, existing rules regarding irrigation management is clearly identified by interview with farmers, and officer of LID is interviewed for evaluation of ABM model.

2 Materials and Method

2.1 *Study Area*

2.1.1 General Information and Landscape

As a study area, Imbanuma watershed was chosen, because enough water resources for agricultural sector could be provided by river discharge in the watershed, though water quality was degraded by urbanization and excess fertilizer for vegetable fields. Major rice variety is profitable one 'Koshihikari', but price of average is 223 yen/kg is lower than national average 236–252 yen/kg (Chiba prefecture, http://www.pref.chiba.lg.jp/index.html). This area is located near to Tokyo and good access to a huge market, then, selling price would be increased if potential demand for rice quality had been clearly found.

Imbanuma lake was reclaimed for paddy fields development in 1969, and now it was separated two small lakes (Fig. 1). Kashima River is the largest river in the Imbanuma watershed (541 km^2), and its watershed covered 156 km^2 (Imbanuma water quality conservation authority http://www.insuikyo.jp/). Observation and target paddy plots were settled in upper Kashima river and paddy block.

2.1.2 Landscape and Agriculture in Imbanuma

In 2011, land use in Imbanuma basin is composed of forest 19 %, vegetable fields 22 %, paddy fields 15 %, and the rest is urban area. Urban area and vegetable fields are spread in upland, and paddy fields are in lower area along to riversides. Urban

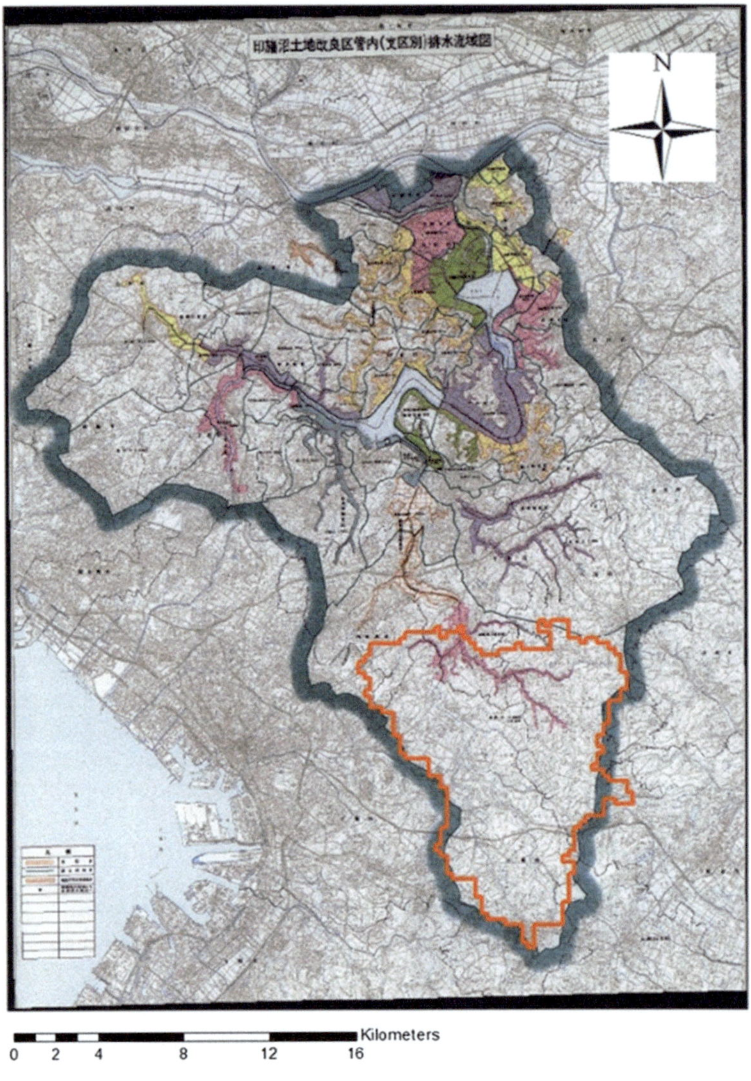

Fig. 1 Study area map (Imbanuma LID web site, http://www.inbanuma-lid.jp/04/prof.htm)

area is still spreading because of expansion of Tokyo Metropolitan area (Chiba prefecture, http://www.pref.chiba.lg.jp/index.html).

Originally, this area was used agricultural production, especially rice production, and paddy fields covers 11 % to a whole watershed. Agricultural stats in total related county shows 49,000 tons of rice production and it shares 14 % for this prefecture (Food stat in MAFF http://www.maff.go.jp/j/tokei/kouhyou/syokuryo_nenkan/index.html).

2.1.3 Water Quality in Imbanuma Watershed

Imbanuma lake is eutrophic lake because of urbanization and excess fertilizer [7].

Imbanuma lake (center of west lake) total nitrogen concentration is 2.4 mg/L averaged 1988–2011. Kashima river water quality is 4.6 mg/L averaged 2004–2011. River water is highly polluted by urbanization and excess fertilizer. A sewage system was installed but nitrogen and phosphorous concentrations are not improved, because excess fertilizer is infiltrated and leached to ground water, then discharged to the basin rivers. As the sewage system was settled for purification of urbanized waste water, it was not effective to reduce nitrogen pollution from agricultural area.

Seasonally, nitrogen water quality is higher in winter season and lower in summer season, and it comes from dilution effect of irrigation water and denitrification in paddy soils.

2.2 *Imbanuma LID*

Imbanuma LID is an organization for operation and management of irrigation water supply service. Number of participatory farmers are 8640 in 2009, and covered paddy fields are 6417 ha. This is occupied 11.9 % for whole Imbanuma watershed. Facilities are mainly pumps for irrigation and drainage channels, and the number is 384. Other headworks and dams are 7 and 2, respectively. Water resources are taken from both river water and ground water [8].

Upper Kashima river paddy fields is a branch of Imbanuma LID. It is located 10 km far from Imba lake. In this branch, the number of farmers joined LID is 804, covered area is 348 ha. Landscape is hilly and small valley, then, it is difficult to enlarge paddy scale. Pump facilities are existing 40; 21 for river water pump and 19 for ground water [9].

Farmers are difficult to find their successors, and labor population is decreasing tendency. This situation will be influenced to LID management. The successor problem is widely known as agricultural structure issue, and one reason is low income coming from small scale and low additional value of production.

2.3 *Monitoring*

Water quality of paddy fields drainage and hydrological data in Kahima river are investigated from 2012 to 2013, Sampling was conducted daily by auto sampler (ISCO 6412 series, USA). Hydrological data was observed hourly water depth by pressure logger (Hobo U20, USA), and bi-weekly, discharge volume was investigated to estimate rating curve (water table-discharge curve).

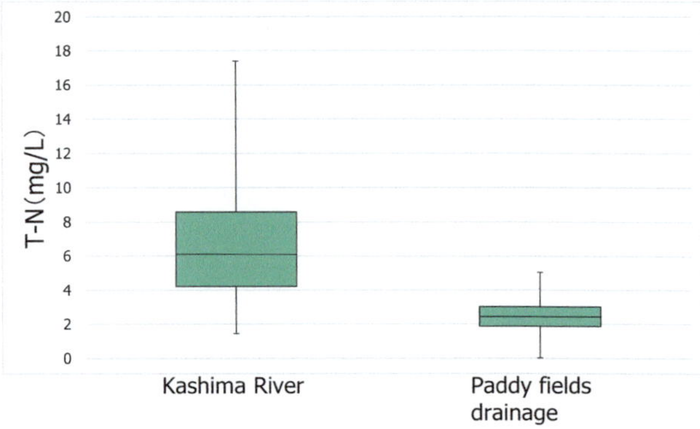

Fig. 2 Summary of observation data in Kashima River and paddy fields drainage

Regarding the sampling water, total nitrogen (T-N) concentration was analysed by spectrophoto meter (Shimadu UV 1800, Japan) followed procedure JISK103.

Results of water quality T-N median are 6.1 mg/L and 2.4 mg/L in Kashima River and paddy fields drainage channel, respectively (Fig. 2).

This figure shows there is nitrogen purification function in paddy fields because ponded situation in irrigation period is optimized microbiological activities for nitrification and denitrification. Then, T-N concentration in river water is diluted by paddy fields drainage.

2.4 *Water Quality and Quality of Rice*

2.4.1 Increase Quality of Rice by Irrigation

As results of investigation, paddy fields drainage water is considered as alternative water resources in better water quality. As new innovative irrigation, repeatedly irrigation scheme is proposed (Fig. 3).

In Fig. 3, left is current situation that irrigation water is intake from river of high T-N concentration, and right is proposed irrigation method that irrigation is intake from drainage channel of low T-N concentration.

As high Nitrogen concentration is occurred degraded rice taste quality, this proposed irrigation method would increase taste quality and additional value for rice production.

2.4.2 Water Quality Purification

Former researches revealed nitrogen removal or purification in paddy fields [10–12]. In those researches, it is shown that paddy drainage dilutes polluted river water. In watershed discharge mechanism aspect, water purification is mainly caused by denitrification in paddy soils. In this area, ordinary vegetable fields are located in

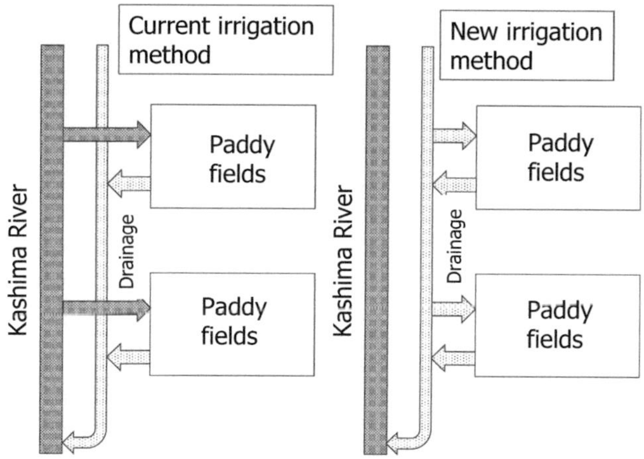

Fig. 3 Schematic image for current and proposed irrigation method

hills or upper place in the watershed, and drainage from vegetable fields contains nitrogen load coming from excess fertilizer. Irrigation water for lower paddy fields is influenced by drainage from vegetable fields. Dilution effect is emerged from high background water quality and denitrification at paddy fields along to riverside.

2.4.3 Nitrogen Contents Ratio and Rice Quality

In general, it is shown that rice taste quality is influenced with protein contents ratio in rice [13]. Protein contents ratio is related with application amount of nitrogen fertilizer. In nitrogen fertilizer application, the timing of application is important factor [14]. In rice growing stage, if nitrogen fertilizer is applied excess amount after flowering stage, protein contents ratio is increased. High protein contents degraded taste quality [15]. This relationships is already formulated below.

$$FQ = 4.79 - 0.567PR \tag{1}$$

$$PR = 7.45N + 5.87 \tag{2}$$

$$N = 0.026x + 0.012 \tag{3}$$

where, FQ is taste quality, PR is protein contents ratio (%), N is nitrogen contents in a pinnacle (mg), x is nitrogen absorption (g/m^2) for nitrogen application by fertilizer or irrigation water.

If nitrogen concentration were high in irrigation water after flowering stage, it occurs high protein contents ration in rice grain. Then, irrigation water quality control is a way to increase additional value for rice production. If it is available to supply low nitrogen concentration water in after flowering stage, it reflects to increase rice quality and selling price. And it is available low nitrogen concentration water inside paddy drainage channel.

Paddy drainage is temporally discharged by surface and ground water flow, and it is caught in drainage channel. In facility aspect, pump facilities easily change the inlet from river to drainage channel. Then, change intake water resources between river water and drainage channel is a feasible technology and initial cost is not high.

If lower nitrogen concentration water from paddy drainage were available to be supplied repeatedly to irrigation, rice quality would be increasing. But, interview result to farmers show that reaction of farmers to new irrigation way was more complicated because of cost, profit, and risk management.

It is not clear to evaluate this new irrigation management way in practical installation to Imbanuma LID. But virtually, this new irrigation management way are evaluated by ABM.

2.5 Agent Based Model

Agent Based Model (ABM) is a simulation model well applied in the computational research fields for environment, economics, and artificial intelligence. This model is basically application of game theory. It analyze how a decision of individual player is influenced whole system performance.

To develop ABM, Lansing and Kremer [3] model was considered. This model is simulated SUBAK irrigation system in Bali, Indonesia. Water distribution in SUBAK system is promoted by two reasons; a irrigation schedule and a pest control. Synchronized irrigation schedule results in conflict of water resources between SUBAKs, and diversity of plant schedule is increasing pests. Then some SUBAKs are simultaneously synchronized plant schedule, and the other are diverted by several groups. In optimized results show that current SUBAK system is quite reasonable distribution system [3, 4].

In this research, each paddy farmer is treated as a different agent, and each decision of agent is followed several rules.

2.5.1 Extract Decision Rules from Farmers Interview

An interview with farmers was conducted to extract decision rules for irrigation management. The interview was sampled two farmers (one family 67 and 40 years old) for mainly rice farming. Interview topics and items are current condition, farming schedule, concerning to irrigation water quality, service from LID, and impression of new irrigation management. Based on these view points, decision rule on irrigation management was extracted (Table 1).

Table 1 Decision rules from interview results

Field	IF	THEN
Rice production system	Water shortage in puddling season	Domestic small pump use
	Slightly broken channels	Cover the recovery cost by themselves
	Large broken channels	Ask LID
	Flood situation	Recovered by themselves
	Difficulties to sustain rice production by aging or economical condition	Ask some farmers to maintain their holding paddy fields
Water quality	Not required new cost and no contamination by sewage drainage	New water resources and new irrigation way is preferred
	River water quality degraded rice taste quality	Individually, holding and developing new water resources from ground water
Crop production	Selling price is reflected by crop production management	Try to increase additional value
	Profit comes from additional value is larger than risk of reducing harvest	Try to new methods, i.e. eco-farming or organic rice.
	Equality between farmers and efficiency are conserved	New plant methods are preferred
	Income larger than cost	Sustain their farming
Scale of management	Relationship between collaborators is good	change from individual farming to group/enterprise farm
	Concentration of time for using agricultural machine is avoided	Sharing agricultural machine is preferred

2.5.2 Framework of ABM

This model shows that in case of new water quality irrigation way were installed virtually in Upper Kashima.

LID branch, other farmers decision would be simulated. Netlogo (http://ccl.northwestern.edu/netlogo/) is applied to model development environment. In this simulation, two scenarios were evaluated. Three pumps and six paddy fields (as same as six agents) are settled to evaluate decision rules. In initial condition, current irrigation and new water quality irrigation way are distributed randomly, then, as time processing, each agent decide to choose irrigation way between two options.

Scenario 1: Cost weighted scenario

Decision to apply new water quality irrigation way is assumed cost comparison between other farmers. From the interview results, it is found that farmers prefer to avoid cost increase.

This scenario is assumed a situation that there is not enough water quality information delivered to farmers. This means profits from rice quality increase is in vain for no reliability to LID and other farmers.

Scenario2: Income weighted scenario

Decision to apply new water quality irrigation way is assumed income comparison between other farmers. In this scenario, water quality information is regularly delivered by LID. It also assumed that probability for increasing income is randomized function in initial condition.

$$\text{Probability} = 0.6 + (\text{random}\ ())\ /\ 1000 \tag{4}$$

where random() is function of generate random number 0–400. For farmers are compared an expected value of rice production income to adjacent farmers. If other farmers success to new water quality irrigation way, a farmer decides to install new water quality irrigation way. And probability is improved to time processing. As evaluation function, cost and income are assumed.

3 Results

3.1 Cost Weighted Scenario

In this scenario, all results are converged to current irrigation way regardless each initial condition. This situation is an extreme example but if farmers would like to sustain their small scale farming to hold their land, situation is not changed. Strong incentives are required to farmers toward change of agricultural sector structure.

3.2 Income Weighted Scenario

In this simulation, results are converged two patterns. It depends on initial condition, if majority of agents are taking new water quality irrigation way, most results are converged new water quality irrigation way. In contrast, in case that current irrigation way were settled as majority in initial condition, most results converged to current irrigation way.

This result means that expected value as reliability coming from good quality water resources is important factor for agent choice. Therefore, to change new irrigation way, LID should increase reliability of water quality demand and expected value for the income that is emerged from combined plant and water management.

4 Discussion

4.1 Model Result Evaluation by Imbanuma LID

Regarding model results, interview with an administrator of LID was conducted. Interviewed topics are evaluation of model results and future possibilities of development on new irrigation service.

Interview results show the limitation of LID to improve water distribution service. Because a main purpose of LID is delivery of water resources by irrigation in reliable and sustainable method. Individual farmer should focus on the increase of income or profit, and LID is not in service to plant production technology. Basically, water is delivered by farmers demand, but water delivery is dominated by supply side 'LID' management in current condition. In this interview process, an administrator brought about an idea that if LID developed new delivery service in future, water distribution and plant production technology should be packaged. It seems that a step fo change LID and farmers, and ABM model will be implemented based on this idea for further development.

Also administrator mentioned public value in paddy fields water management. Actually, this aspect is important for ecosystem service in local area though it is not reflect in this model analysis. This value should be clearly presented for further model development.

4.2 Scenario Evaluation by ABM

Scenario analysis shows possibilities on application for new irrigation service. However, model assumption is too simplified context and behavior of agents. The model development should be included market price, competitive producers not only domestic but international, history and tendency of farmers decision.

It would be profitable new water quality irrigation for farmers, and this service would be valuable for LID. Because, LID function were limited like current condition, LID has not influenced to agricultural sector and productivity. It means that decrease of agricultural production is weakened LID organization and structure. If LID had a function to promote advanced farmers and a selective water delivery function, LID acquires more independent and sustainable organization. In future, more study cooperated with LID will be required to evaluate co creation of service.

5 Conclusion

In this research, new water quality irrigation service was proposed. Toward installation of this service, ABM analyses was conducted. The results shows that strong incentive or profit for farmers are required for installation of service and reliability of irrigation water quality is important factor in new service. This service will be provided by LID, and LID administrator considered to unite water delivery function and rice production technology. There is a possibility to emerge new service between LID and farmers.

Further research will be required to develop ABM to consider context and agent performance. Also, public value in regional ecosystem should be included between LID and residential people.

Contribution to service science, a mechanism of new service distribution/expansion process was analyzed as an example. A emergence process of new service is coming from dynamic process between both demand and supply sides, and at same time, it is important that how will new service be distributed and survived in the external environment. ABM is good supporting tool to analyze distribution or survival process on new service. This research methodology should be developed for comparison of various service competition process.

This research is supported by 'Development and evaluation of demand oriented irrigation service' in S3FIRE Research Institute for Science and Technology for Society (RISTEX) in Japan Science Technology agency (JST).

References

1. Editorial Committee of Advanced Paddy Field Engineering, JSIDRE (1999) Advanced paddy field engineering. Shinzan-sha Sci &Tech, Tokyo, pp 39–50
2. JSIDRE (2010) Handbook for irrigation and drainage engineering. 7th edn. JSIDRE, Tokyo, pp 707–710, in Japanese
3. Lansing SJ, Miller JH (2003) Cooperation in Balinese rice farming. Santa Fe Institute, Santa Fe
4. Lansing JS, Kremer JN (1993) Emergent properties of Balinese water temples. Am Anthropol 95(1):97–114
5. Kato T, Kuroda H, Nakasone H (2009) Runoff characteristics of nutrients from an agricultural watershed with intensive livestock production. J Hydrol 368(1–4):79–87
6. Nakasone H, Kuroda H, Kato T, Tabuchi T (2003) Nitrogen removal from water containing high nitrate nitrogen in a paddy field (wetland). Water Sci Technol 48(10):209–216
7. Water Conservation Council in Imbanuma Watershed (2012) Imbanuma watershed improvement, Conservation planning of water balance and cycle in imbanuma. http://inba-numa.com/data/materials/kenzenkadownload/ in Japanese
8. Nakaamura Y, Mashino M, Takahashi O, Watanabe T (2000) Irrigation system and management of water environment by land improvement district in the Inbanuma basin. J JSIDRE 68 (11):1159–1164 in Japanese
9. Imbanuma LID (2007) The map of irrigation and drainage facilities in Imbanuma Land improvement district. Retrieved February 3, 2016 from http://www.inbanuma-lid.jp/04/prof.htm

10. Matsuno Y, Nakamura K, Masumoto T, Matsui H, Kato T, Sato Y (2006) Prospects for multifunctionality of paddy rice cultivation in Japan and other countries in monsoon Asia. Paddy Water Environ 4(4):189–197
11. Kato T, Kuroda H, Nakasone H, Kiri H (2007) Evaluation of nutrient removal in a constructed irrigation pond. Paddy Water Environ 5(3):189–199
12. Kuroda H, Kato T, Nakasone H (2006) The nitrate nitrogen pollution and the nitrogen removal used by paddy field in agricultural area. J Water Environ Technol 3(2):165–168
13. Ishima T, Taihra H, Taira H, Mikoshiba K, Yoshikawa S (1974) Effect of nitrogenous fertilizer application and protein content in milled rice on olganoleptic quality of cooked rice. Rep Nat Food Res Inst 29:9–15
14. Toriyama K, Iimura K, Sekiya S (1987) Continuous flow water culture method with limited supply 0L nitrogen and quantitative control of nitrogen absorption pattern of rice. J JSSSPN 58 (4):452–459
15. Matsuda H, Fujii H, Shibata Y, Kominami C, Hasegawa S, Oobuchi K, Ando H (1997) Effect of number of grains per amount of N at heading and amount of N in grain on the protein content of milled rice. J JSSSPN 68(5):501–507

Analysis of Multi-language Knowledge Communication Service in Intercultural Agricultural Support

Masayuki Otani, Kaori Kita, Donghui Lin, and Toru Ishida

Abstract This paper proposes the design process of knowledge communication with constructing the common knowledge and evaluates it by improving the knowledge communication between Japanese experts and Vietnamese children in agricultural support project in Vietnam as a case study. From the analysis of experiment in 2nd and 3rd seasons, we have revealed the following implications: (i) the proposed design process is able to improve the knowledge communication environments; and (ii) that communication in each community is important in knowledge communication to encourage users to accept new knowledge and experts to cooperate with other experts.

Keywords Multi-language knowledge communication • Common knowledge • YMC-Viet project • Field experiment

1 Introduction

With the expansion of the globalization, the multi-language knowledge communication has been much attention on. The knowledge communication is defined as the activities where the expertise knowledge is transferred from knowledge experts to knowledge users [6]. This paper defines the multi-language knowledge communication as the knowledge communication among people who speak different languages. In the multi-language knowledge communication, the transferred knowledge is sometimes different from original knowledge because the knowledge is transformed by the language or cultural differences [1]. This problem prevents the activities of several international organizations such as NPOs or NGOs.

To solve this problem and realize the multi-language knowledge communication environment which can be applied to real fields, we started the international agricultural support project called YMC-Viet (Youth Mediated Communication-Vietnam) with NPO Pangaea from 2011 as a field of multi-language knowledge communication. To support the communication between Japanese agricultural

M. Otani (✉) • K. Kita • D. Lin • T. Ishida
Department of Social Informatics, Kyoto University, Kyoto, Japan
e-mail: m-otani@i.kyoto-u.ac.jp

T. Maeno et al. (eds.), *Serviceology for Designing the Future*,
DOI 10.1007/978-4-431-55861-3_29

experts and Vietnamese farmers in this project, we proposed service design process which realize knowledge communication environment by improving the service model iteratively [10]. We constructed the communication environment by using the proposed service design method which includes the several web services registered in Language Grid [9] that we have developed to break the language barrier, and evaluated the effectiveness of the designed service model through the YMC-Viet project.

However, this design process is not able to be applied to knowledge communication because it mainly focuses on multi-language communication. Sometimes knowledge users cannot understand unintelligible expertise which is translated correctly. In another case, knowledge experts may transmit expertise knowledge which is useless for the children and the farmers. To transmit knowledge correctly, the design process which is able to improve knowledge communication.

This paper aims to propose and to evaluate the service design process which is able to be applied to knowledge communication by improving the proposed service design process. Concretely, we construct the framework that enables the experts and users to exchange background information which is necessary in knowledge communication and evaluate its effectiveness through the experiments in 2nd and 3rd seasons in YMC-Viet project.

This paper is organized as follows. In the next section, this paper explain what knowledge communication is. We propose the design process for knowledge communication in Sect. 3. Then we introduces the agricultural support project called YMC-Viet as a field of multi-language knowledge communication in Sect. 4. We conduct the interview survey for the participants of the experiments in second season and analyze it, and try to improve the knowledge communication in Sect. 5. In Sect. 6, this paper discuss about value co-creation of the knowledge communication in YMC-Viet project. Finally, our conclusions are given in Sect. 7.

2 Knowledge Communication

Epplers define the knowledge communication as the (deliberate) activity of interactively conveying and co-constructing insights, assessments, experiences, or skills through verbal and non-verbal means [8]. They also mentioned that it requires to transmit not only the information (*e.g.,* facts, figures, events, situations, developments, etc.) or emotions (*e.g.,* fears, hopes, reservations, commitment) but also the context, background, and basic assumptions each other. Since both of the experts and users has snippets of such information, they have to iteratively communicate each other and reconstruct their mental model to make "common knowledge". This concept is introduced by David Lewis. When members of a group know a fact, they all know that they all know the fact, they all know that they all know that they all know the fact, and so on endlessly, this fact is common knowledge in the group [5]. Similarly, the concept of common ground [4] means that speaker and hearer think that they have the same background. It is revealed that communication or collaboration are failed when the common ground is not constructed correctly.

To construct common knowledge, it is necessary to transmit the context information which include their background or consideration, *i.e.,* what they know, what they believe, or why they want to know. Epplers reported the following problems in knowledge communication [8]: (1) experts use unintelligible terminology or transmit useless expertise due to their lack of understandings on the users' situation or skill; and (2) users are not able to understand the expertise knowledge due to their lack of basic knowledge concerned with the expertise knowledge. To solve these problem, knowledge users should send their current situation, what they know, and what they want to experts, and knowledge experts should send the corresponding knowledge that includes the background information such as elementary knowledge, or why these knowledge are needed.

This paper regards the insights, assessments, experiences, or skills as knowledge, grasping the knowledge communication defined previous paragraph as Fig. 1. In this figure, experts who provide knowledge and users have knowledge

Fig. 1 Knowledge communication process

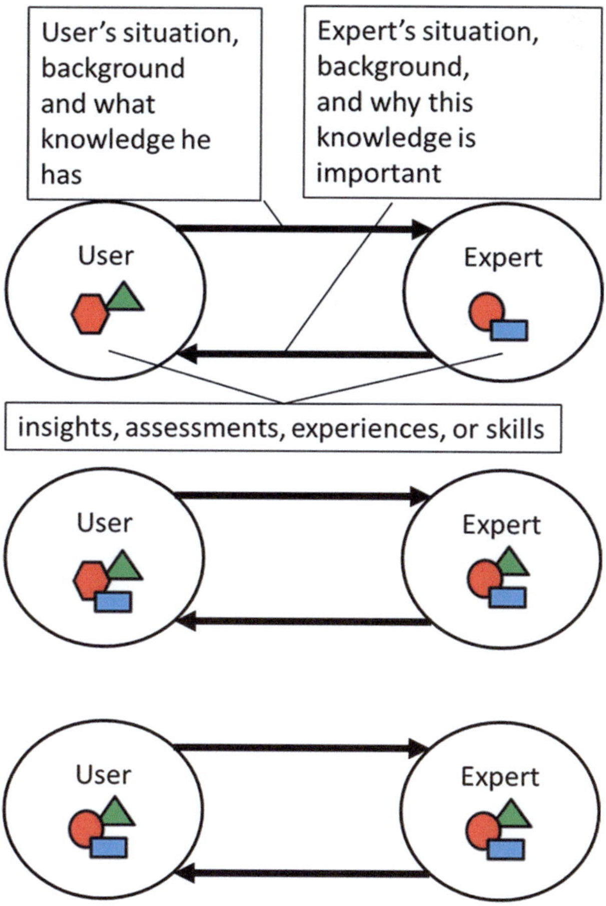

represented by circle, square, triangle, and hexagon. The hexagon means the users' knowledge that is old and conflicted with the experts' knowledge represented by circle. The triangle means the users' situation. The upper side of this figure shows that experts do not understand the users' situation. Both of experts and users communicate and exchange the information, emotion, context, background, and its elementary knowledge iteratively until the common knowledge are constructed.

3 Design Process of Knowledge Communication

Although knowledge communication is applied several fields (*e.g.*, international cooperation, education, intercultural exchange, or so on), the form of communication is different in each field. Since it is difficult to make a model which is able to be applied every field, the design method of communication environment is important.

To overcome this problem, Lin proposed the user-cantered service design process that improves the multi-language knowledge communication iteratively [10]. This design method constructed with following processes: (1) analysing requirements through field study, (2) construct communication model, (3) experiment in field, and (4) evaluate service by analysing the result of experiment. The characteristics of this method is to make a model by analysing participatory simulation in process (2). This paper propose the design process which enables experts and users to exchange the background or context each other by extending Lin's design process. Concretely, Fig. 2 shows the proposed design process.

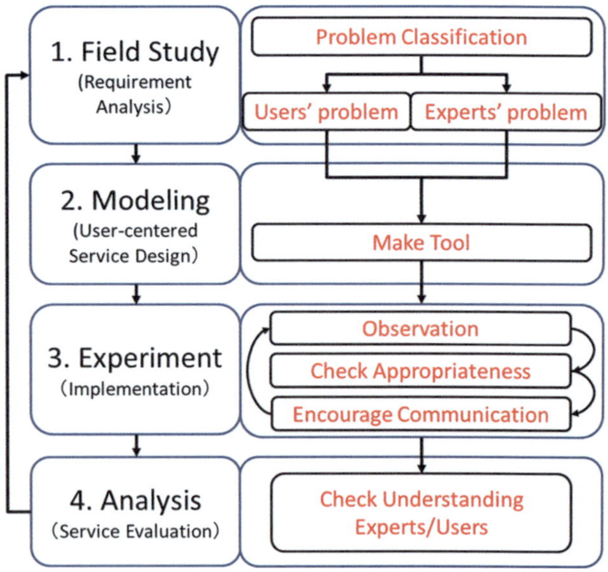

Fig. 2 Design process of knowledge communication

In this process, modelling and experiment are mainly different from original design process. Concretely, the proposed method iteratively improves the communication model in the process of experiment while the Lin's model iteratively improves the communication model in the process of modelling. The detail explanation is described as follows.

(1) Field Study

To realize the knowledge communication environment, it is required to construct tools or communication model that support exchanging the information of each context through verbal and non-verbal communication. To construct the common knowledge, it is necessary to develop the framework that exchange not only expertise knowledge itself but also its background. From this reason, this process reveals the following requirements: (i) the problems or requirements when the users request knowledge; and (ii) the problems or requirements when the experts convey the corresponded knowledge, and enumerates its solution.

(2) Modelling

In Lin's design method, modelling process investigates the quality and problems of the communication model through the participatory simulation iteratively. However, the participatory simulation mainly focus on the language communication and it is difficult to apply on the knowledge communication because the knowledge communication requires not only the verbal communication but also non-verbal communication like cards or pictures. Therefore, we develop the tools or frameworks based on the problems and requirements listed in the process of field study without any participatory simulation.

(3) Experiment

Although the iterative improvement is important to develop the knowledge communication environment, its evaluation requires long time if the experiment requires long span. Since it is not practical in field, we implement the surveillance that checks whether the communication between experts and users are correct or not in real time. When one does not understand the intention of another, the surveillance ask them to resend the improved knowledge to improve or coordinate the communication environment. If there are major version up, the improvement will be reflected from next experiment.

(4) Analysis

In this process, we analyse whether the experts transmit the correct knowledge by understanding the situation of users, the skills, and also we analyse whether the users understand the expertise knowledge and accepts it. This investigation requires interview survey to measure the users and experts understandings correctly.

4 Case Study: Intercultural Agriculture Support

4.1 YMC-Viet Project

We started the intercultural agricultural support project called YMC-Viet (Youth Mediated Communication – Vietnam [11]) with NPO Pangaea from 2011 as a field of the knowledge communication. Since the educated children have high literacy than their parents in Vietnam, the Japanese experts send the expertise agricultural knowledge to the Vietnamese farmers via their children by using ICT as shown in Fig. 3. This project is conducted in cooperation with Ministry of Agriculture and Rural Development of Vietnam (MARD), Vietnam National University (VNU), Tokyo University, Mie University, and other research institute. In this project, we already conducted three seasons of the field experiment as shown in Table 1. We mainly conduct experiments at two fields in Vinh Long province in Vietnam, *i.e.*, Thien My Commune, Tra On District (1st–3rd seasons), Dong Thanh Commune, Binh Minh District (3rd and 4th seasons).

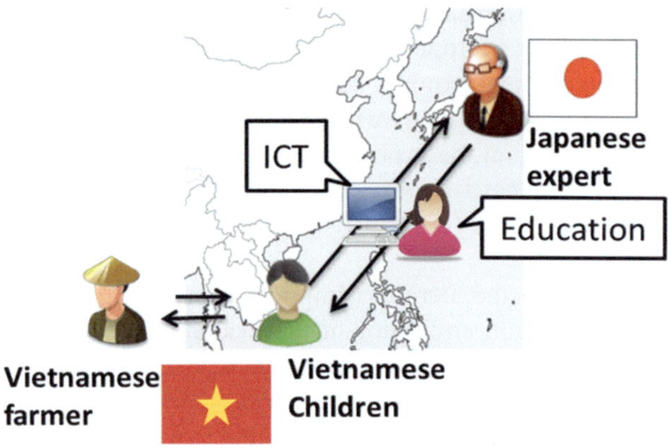

Fig. 3 Youth mediated communication model proposed by NPO Pangaea [11]

Table 1 Experiment schedule of YMC-Viet

	Area	Term	Participants
1st season	Tien My Commune, Tra On District, Vinh Long Province, Vietnam	Feb. 2011–Mar. 2011	29 families (30 youths)
2nd season		Oct. 2012–Feb. 2013	15 families (15 youths)
3rd season		Sep. 2013–Jan. 2014	15 families (15 youths)
4th season	Dong Thanh Commune, Binh Minh District, Vinh Long Province, Vietnam	Oct. 2013–Feb. 2014	15 families (15 youths)
		Mar. 2014–Jun. 2014	15 families (15 youths)

4.2 Knowledge Communication Between Experts and Children

To exchange agricultural knowledge and field data with the Japanese experts, the Vietnamese children communicate with the Japanese experts by an online multi-language communication tool with personal computers (PCs), YMC system (online Q&A BBS) once a week as shown in Fig. 4. The PCs which children use are set in the community centre in Thien My and Dong Thanh commune before experiments. Vietnamese children take pictures of their rice field and post agricultural questions related to the rice via internet. Japanese agricultural experts reply answers including agricultural knowledge in Japanese with some recipe cards where the related knowledge is described as shown in Fig. 5. The transferred knowledge is translated into Vietnamese and feed backed to the children. Children also make a memo by

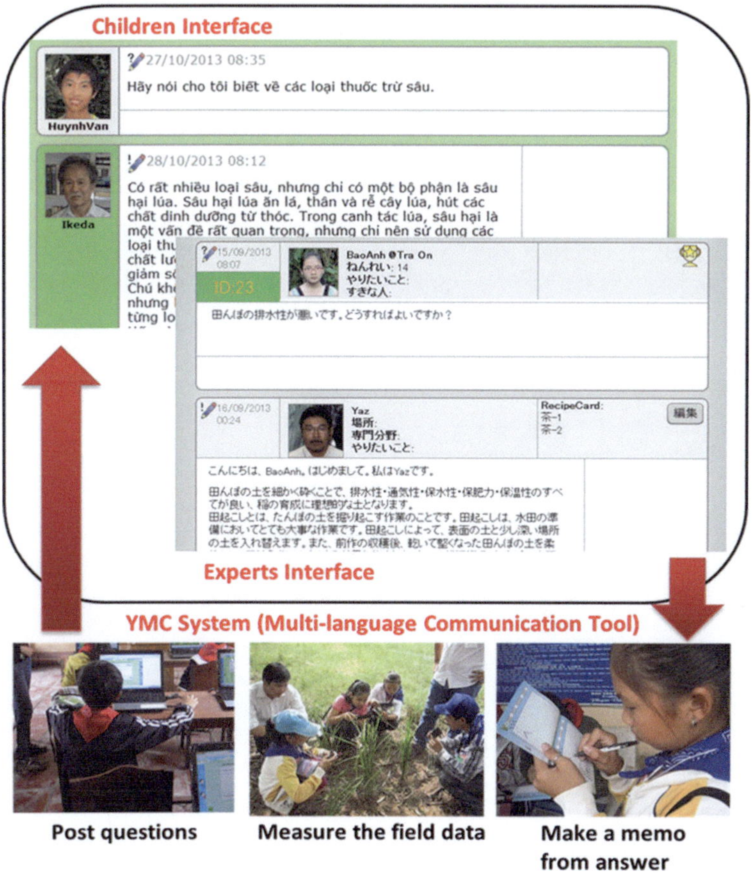

Fig. 4 Communication between experts and children via YMC system [11]

Fig. 5 An example of the recipe cards which includes the agricultural knowledge [11]

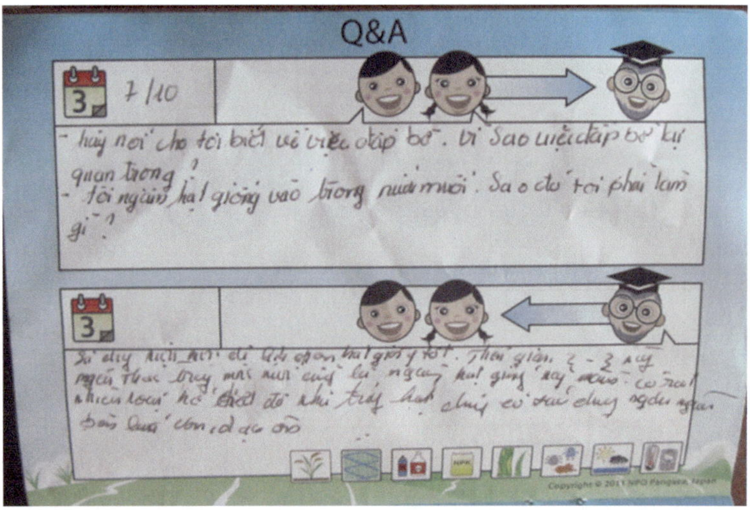

Fig. 6 An example of the memos made by a child on YMC passport [11]

writing it on YMC passport as shown in Fig. 6 when they see the feed backed answer. Children tell the agricultural knowledge to their parents (farmer) by using YMC passport or recipe cards after they return to home.

Table 2 Classification of support tools in knowledge communication

Categories	Tools	Context
Support experts	YMC passport	Experts check whether children understand the expertise or not
	Photo of field	Experts understand the situation of children and farmers
Support users	Mobile phone	Users are able to receive the emergent information sent by experts
	Recipe card	Users are able to refer the detail background of expertise knowledge whenever they want

Table 2 shows that these tools which support knowledge communication are classified into two categories, *i.e.*, support tools for experts or children.

This paper evaluates the effectiveness of proposed design process for the knowledge communication by applying it on the knowledge communication between Japanese experts and Vietnamese children by improving the communication model from experiments in second to third season.

5 Analysis of Knowledge Communication

As described in chapter "A Survey of Business Models in Japanese Restaurant and Retail Industries", some factors prevent value co-creation in knowledge communication. To reveal how these factors affect knowledge communication and cause the problems, we conduct the interview survey and analyze the knowledge communication model according to the following steps. At first, we conduct the interview survey that focuses on the questions and answers (Q&As) which is exchanged between the Japanese experts and the Vietnamese farmers/children who participated in the experiment of 2nd season. Next, we apply the improved communication model to the experiment of 3rd season. Finally, we also conduct the interview survey after the experiment of 3rd season. The detail explanation is described as follows.

5.1 *Interview Survey*

To analyze the interview survey on the factors that prevent the knowledge communication, *i.e.*, the remote communication and the evaluation of knowledge communication, we prepare the interview items as shown in Table 3 by applying the concept of awareness which has been attention on in the domain of computer-mediated collaboration [6, 7]. The aims of these items are to reveal (i) whether the Japanese expert understand the background of the question from the Vietnamese farmer/child, and (ii) whether the Vietnamese farmer/child understand the core

Table 3 Interview questionnaire

	Japanese expert	Vietnamese farmer/child
For question from Vietnamese farmer/child	Which is the reason why do you want to ask this question, A or B?	Which do you think is the reason why the youth want to ask this question, A or B?
	A: Some problem occurred in your paddy	A: Some problem occurred in his/her paddy
	B: There is no problem but I want to know this knowledge	B: There is no problem but he/she want to know this knowledge
For answer from Japanese expert	Which do you think is the most important sentence in this answer?	Which is the most important sentence in this answer?

Table 4 Interview settings

	Japanese expert	Vietnamese farmer/child
Interview period	July, 2013–August, 2013	August 27, 2013–August 30, 2013
Interviewees	4	15 families
Location	Tsukuba	Thien My Commune, Tra On District, Vinh Long Province in Vietnam
	Tokyo	
	Kyoto	
Interview time	2 h/expert (Total 8 h)	30 min/family (Total 8 h)
Interview sheets	380 pages (97 Q&As)	380 pages (97 Q&As)

knowledge of the answer from the Japanese expert. For the questions, we ask the Japanese expert what does he/she imagine why the farmer/child posted the question, then we check with the farmer/child. For the answers, we ask the Vietnamese farmer/child and the Japanese expert who the sentences of the question are most important knowledge, then we compare these important sentences. We employ the semi-structured interview that try to bring out correct information by asking the detail information on the questionnaire items if the answers for items are unclear. Table 4 shows the interview settings. Figure 7 shows an example of interview sheet which we show for Vietnamese farmer/children. Since we investigate 97 Q&As, the number of interview sheets become about 380 pages, which requires 8 h interview time.

5.2 *Problem of Knowledge Communication Protocol*

Figure 8 shows that Q&As are classified by the intersections of the important sentences between the Japanese expert and the Vietnamese farmer/child. We define

【Ngày】27/10/2013

【Câu hỏi】Hãy chỉ cho tôi về sự lựa chọn trọng lượng riêng Có gì khác nhau giữa việc dùng nước muối và amoni sunphát?

【Trả lời】

Nếu tỷ trọng đã được điều chỉnh đến 1.13, cả hai đều như nhau. Bạn có thể chọn một trong hai loại để sử dụng.

Để chọn giống lúa tốt, sử dụng nước muối. Giống lúa tốt chìm trong nước muối trong khi những giống lép sẽ nổi lên

Hạt giống to, chắc là hạt giống tốt.

Hạt lúa tốt sẽ khỏe mạnh, phát triển tốt sau khi nảy mầm.

Không nên bảo quản lúa giống trong một thời gian dài, nó khó nảy mầm đặc biệt là sau hai năm. Kiểm tra bằng cách sử dụng nước muối, và chỉ sử dụng những hạt giống mà chìm trong nước muối loãng để gieo trồng.

Sau khi chọn hạt giống bằng nước muối, để tránh việc nảy mầm bị tổn hại, nhất thiết phải rửa lại bằng nước.

【Thẻ thông tin】Màu xanh lá cây-1

Fig. 7 An example of the interview sheet (for Vietnamese farmer/child)

the degree of intersections as follows: (A) the Vietnamese youth understands the knowledge from Japanese experts completely; (B) the Vietnamese youth understands the knowledge from Japanese experts completely and also absorb other information written on the answer; (C) the Vietnamese youth understands a part of knowledge from the Japanese expert; (D) the Vietnamese youth understand a part of the knowledge from the Japanese expert and also absorb other information; and (E) the Vietnamese youth does not understand the knowledge from the Japanese expert at all. From this result, we revealed that the knowledge communication is failed in 25 % of Q&As which is exchanged between Japanese experts and Vietnamese farmers/children.

Furthermore, the Japanese experts reported that they did not use almost agricultural information which is sent from the field. These problems are caused by the knowledge communication protocol which is applied to the experiment in 2nd season. Figure 9 shows that protocol and its problems. In this protocol, there is no feedback in the question from Vietnamese farmer/child to Japanese expert, and vice versa. Therefore, it is not evaluated whether the Vietnamese farmer/child understands the received knowledge, and also the Japanese expert cannot receive the

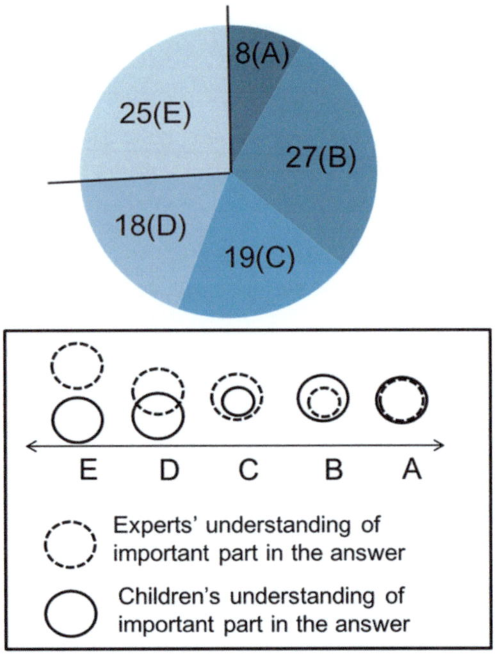

Fig. 8 Intersections of the important point between the Japanese experts and the Vietnamese farmer/children

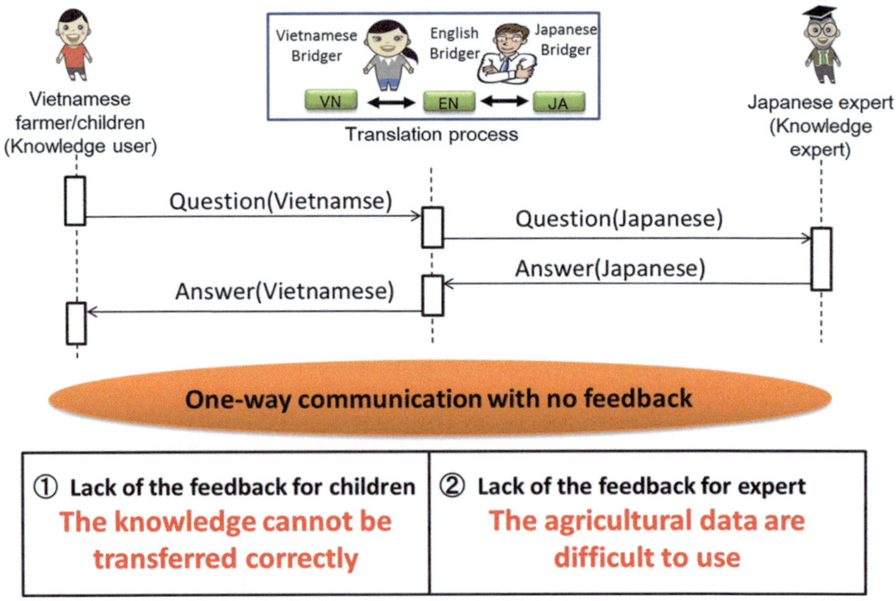

Fig. 9 Problems in knowledge communication protocol

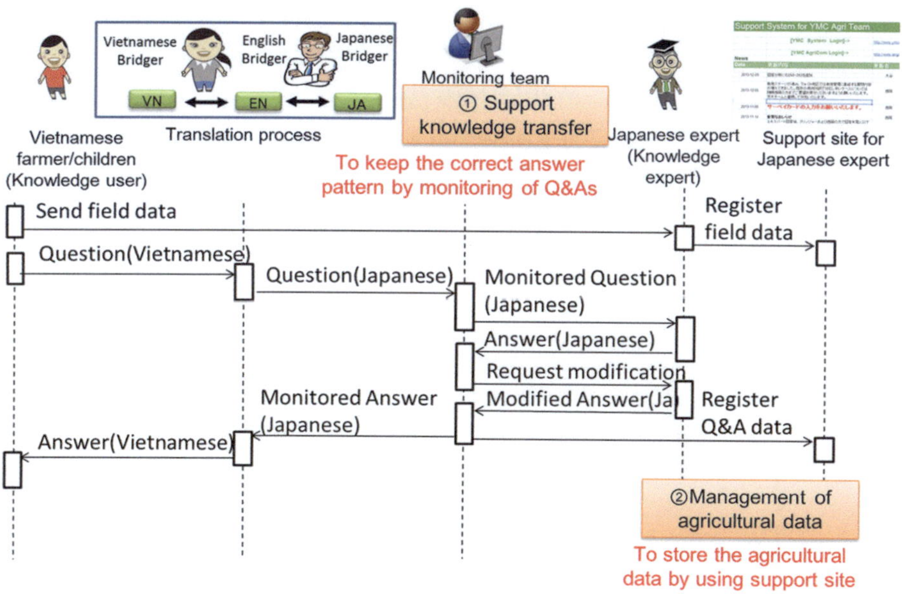

Fig. 10 Improvement of knowledge communication protocol

information whether the transferred knowledge is appropriate for the Vietnamese farmer/child.

To solve these problems, we improve the communication protocol as shown in Fig. 10. The details of the improvement are described as following sections.

5.2.1 Feedback for Vietnamese Children

Although the frequent communication between the Japanese experts and the Vietnamese farmers/children will help the Vietnamese farmers/children understand the answers from the Japanese experts, it is difficult to communicate again and again between remote locations. For this problem, we proposed the communication monitoring team that keeps the communication quality by monitoring the communication between the Japanese expert and the Vietnamese farmer/child. The monitor team knows the answer patterns extracted by the Q&As of the past experiments. This team asks the Japanese experts to modify the answers to keep the correct patterns. The monitoring team is composed from the members of Kyoto University. They monitor the answers on YMC system every Monday.

We analyze the patterns of the knowledge communication by the concept of semantic network [2, 3] which is widely used as knowledge representation. We extract the characteristics of the succeeded pattern by making the graph of related knowledge in the answer. Figure 11 shows the example of the communication pattern. This figure indicates the case where the background information is

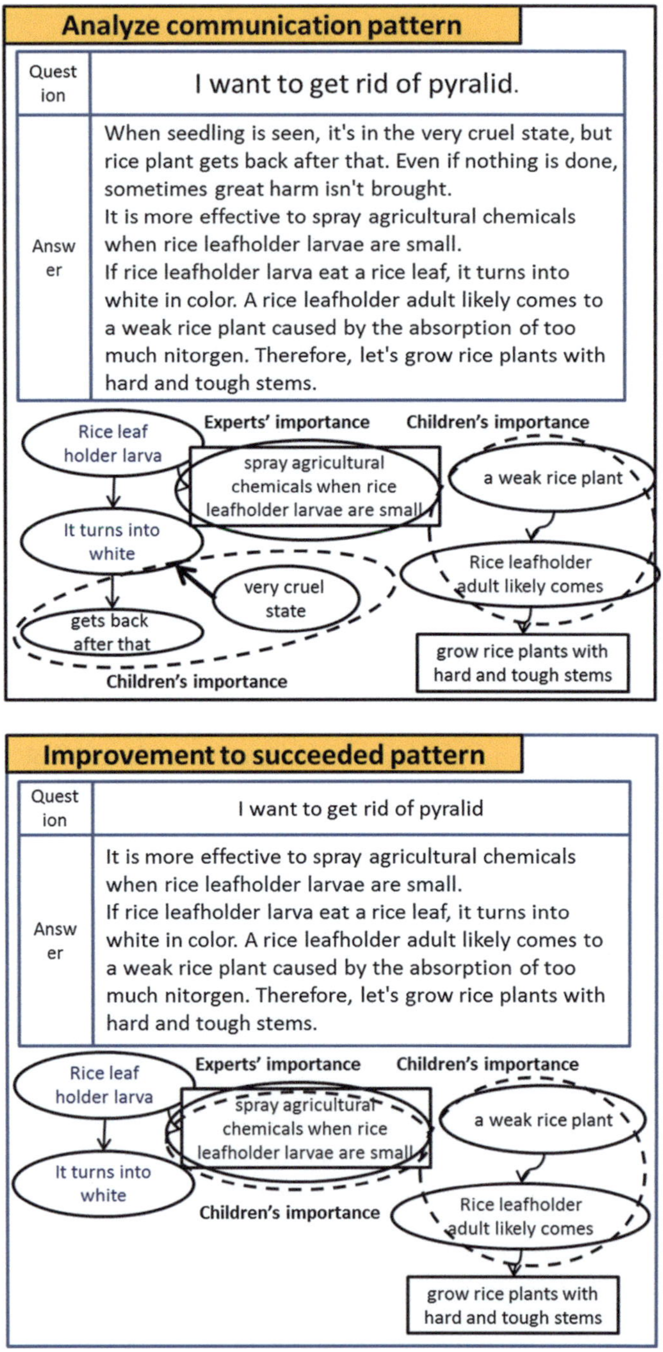

Fig. 11 Analysis of knowledge communication pattern

described in the question as "When should I spread an agricultural chemical?" The farmer/child thinks that the sentences circled by dotted lines are the most important. The Japanese expert thinks that the sentences circled by solid lines are the most important. Although Q&As located in the upper and lower sides of the figure are the same questions, the answers are different. In the upper side of the Q&A, the understandings of the Japanese expert and the Vietnamese farmer/child are different because the important sentences of them are not overlapped. On the other hand, the understandings of the Japanese expert and the Vietnamese farmer/child are partially agreed because the important sentences of them are overlapped. The difference between the upper and lower sides of Q&As is whether the Q&A is included other knowledge except for the knowledge which is the important part of the Japanese expert. In this figure, the farmer/child considered that "rice leaf will recover" is important part.

5.2.2 Feedback for Japanese Experts

Although we ask the Japanese experts how many agriculture data, *e.g.*, temperature, length of rice leaf, or leaf color, did they use, some of them said that they did not use almost data because they do not know where the data are stored. Someone also said that there is no data. This problem is caused by the distribution of the gathered data. Q&As are stored in YMC system. The Japanese experts have the yield data. The pictures of YMC passports are stored in Dropbox. Therefore, we constructed the web site called "Support System for YMC Agri Team" to support the Japanese experts as shown in Fig. 12. We gather the agricultural data, i.e., the yield data, the agricultural data provided by the organizations of Vietnam as MARD or DARD, the pictures taken by the Vietnamese children, the field information that are input by the children on YMC system (weather, rice leaf color, amount of water, or etc.), Q&A data, and also the monitoring data.

5.3 *Improvement of Knowledge Communication*

To evaluate the improved protocol, we applied the improved protocol for the experiment in 3rd season, and conducted the interview survey after the experiment. Although we are analyzing the interview result, we show the current progress of the analysis in Fig. 13. This figure shows the comparison of the success and failure patterns of knowledge communication between the experiments in 2nd and 3rd season. In this figure, the rate of success pattern reaches 90 % in 3rd season. From this result, we can expect that the understanding of the knowledge will be improved.

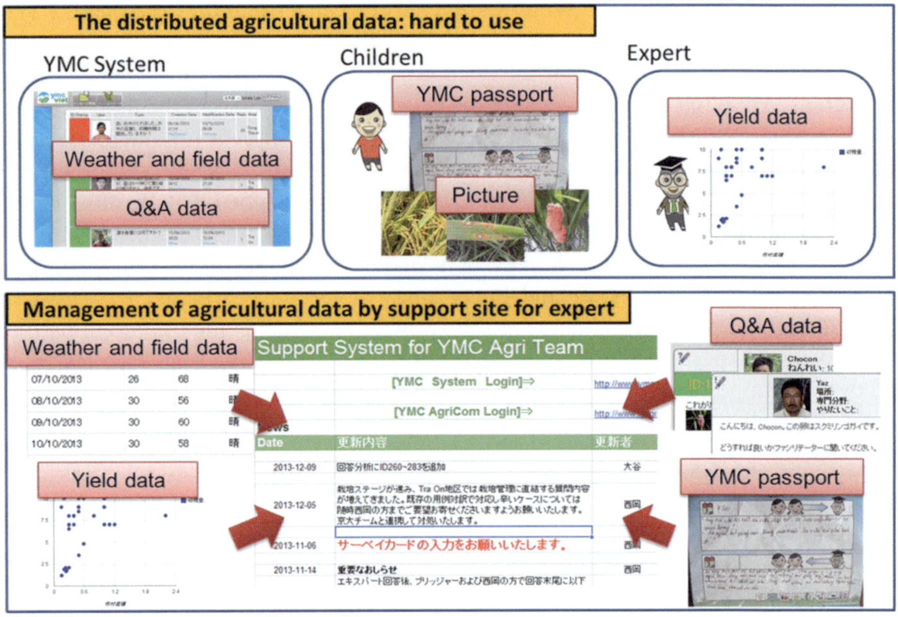

Fig. 12 Support system for YMC agri team

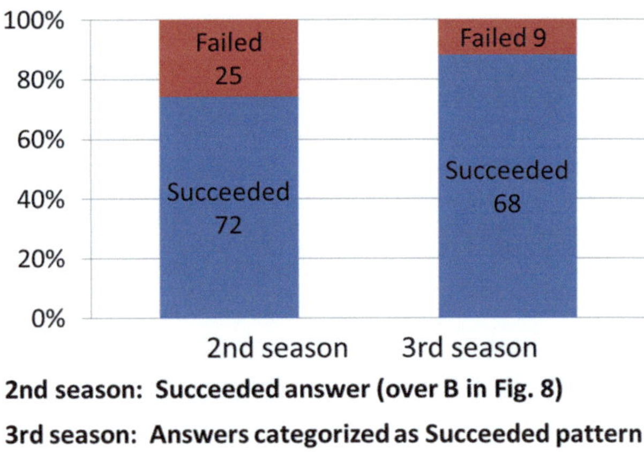

Fig. 13 Improvement of knowledge communication

6 Difficulties in Knowledge Communication

This paper revealed that the proposed design process improves knowledge communication in field. However, the following problems are also appeared.

- Problem of Communication Infrastructure

Although the monitoring team enables knowledge experts to receive feedback indirectly, this feedback does not support knowledge users. Therefore, current communication model should be improved by implementing a framework which enables knowledge users to receive feedback from experts.

- Problem of Organization

When we ask some Vietnamese farmers "whether is this expertise knowledge useful or not?", they answered "Although we understood it but we do not use because we believe our way is correct". This is "the problem caused by organizational context". Several researchers indicated this problem. This problem is common in the persons with a lot of experience as elderly. Although it is quite difficult to solve this problem, it may be able to be solved by communicating among neighbour farmers because some farmers understood the effectiveness of the expertise and accepted it.

- Problem of Expertise

We found other problem that Eppler did not discussed, where the opinions of experts are sometimes conflicted each other. Although the researchers assumed that all experts have the same knowledge and will cooperate each other, there are individual variation of the expertise knowledge and are different answers to Vietnamese children. Therefore, the proposed design process should be extended by implementing a new process which encourage the communication among experts.

7 Conclusion

This paper proposed the design process which improves knowledge communication in fields, and evaluated its effectiveness through the agriculture support project in Vietnam as a case study. From the interview survey, we found the following problems: (1) the Vietnamese farmers/children cannot understand 25 % of Q&As because they cannot have enough communication with the Japanese experts due to the factors of remote location; and (2) the Japanese experts cannot use almost agricultural data of the field because there are no management system of the data. Thus, this paper revealed that the proposed design process is able to improve knowledge communication environments. This paper also found an implication that communication in each community is important in knowledge communication to encourage users to accept new knowledge and experts to cooperate with other experts. The proposed design process may need more processes which realize such communication in community.

For future challenges, the following issues should be tackled in the near future: (a) the interview survey and its analysis should be conducted for the Japanese experts who are participated in the experiment in 3rd season to reveal whether the proposed service design process is able to improve the knowledge communication environment; (b) the interview results should be analysed in more detail to reveal the relation between the questions and answers; and (c) the mathematical model of the multi-language knowledge communication also should be extracted through the interview survey.

Acknowledgment We would like to thanks NPO Pangaea for promoting YMC-Viet project and supporting us. This research is partly supported by Service Science, Solution and Foundation Integrated Research Program from JST RISTEX and a Grant-in-Aid for Scientific Research (S) (24220002, 2012–2016) from Japan Society for the Promotion of Science (JSPS).

References

1. Aiken M, Martin J, Shirani A, Singleton T (1994) A group decision support system for multicultural and multi-language communication. Decis Support Syst 12(2):93–96
2. Collins AM, Quillian MR (1969) Retrieval time from semantic memory. J Verbal Learn Verbal Behav 8(2):240–247. doi:10.1016/S0022-5371(69)80069-1
3. Collins AM, Ross Quillian M (1970) Does category size affect categorization time? J Verbal Learn Verbal Behav 9(4):432–438. doi:10.1016/S0022-5371(70)80084-6
4. Clark HH, Wilks-Gibbs D (1986) Re: ferring as a collaborative process. Cognition 22:1–39
5. David L (1969) Convention: a philosophical study. Blackburn, Oxford. ISBN 0-631-23257-5
6. Dourish P, Bly S (1992a) Portholes: supporting awareness in a distributed work group. In: Proceedings of the SIGCHI conference on human factors in computing systems, the Association for Computing Machinery. Monterey, California. pp 541–547
7. Dourish P, Bellotti V (1992b) Awareness and Coordination in Shared Workspaces. In Proceedings of the ACM Conference on Computer-Supported Cooperative Work (CSCW'92), the Association for Computing Machinery, Toronto, Canada. June. pp 107–114
8. Eppler M (2007) Knowledge communication problems between experts and decision makers: an overview and classification. Electron J Knowl Manag 5(3):291–300
9. Ishida T (ed) (2011) The language grid: service-oriented collective intelligence for language resource interoperability. Springer
10. Lin D, Ishida T (2014) User-centered service design for multi-language knowledge communication. In: Mochimaru M (ed) Serviceology for services. Springer, pp 309–317
11. Mori Y (2009) Youth mediated communication model: new challenge to bring youths for better world. Phase I agriculture. Asia-Pacific Advanced Network: APAN2009, e-Culture Session, Kuala Lumpur, Malaysia

A Value Co-Creation Model for Multi-Language Knowledge Communication

Donghui Lin, Toru Ishida, and Masayuki Otani

Abstract This paper aims at investigating the formation process of value co-creation in the real fields by using a case study of multi-language knowledge communication. We use the YMC (Youth Mediated Communication)-Viet project, an agricultural support project for Vietnamese farmers by Japanese experts through children of the farmers, where the end-to-end value co-creation is between the Japanese experts and Vietnamese farmers. On one hand, Japanese experts provide agriculture knowledge to Vietnamese farmers for solving problems in the paddy or increasing their knowledge. On the other hand, Vietnamese farmers send field data to Japanese experts for providing more accurate knowledge and collecting data for further analysis. However, we have to consider several problems due to the complexity of the real fields. First, translation between different languages should be worked out to support the knowledge transfer. Second, since there are different stakeholders in the project, the various incentives from the stakeholders should be considered. Iterative service design approach is applied to address the complexity of such real field problems, where the formation process of value co-creation can be realized by several stages: proposing individual services, modeling value co-creation for individual services, and designing the structure of different value co-creation layers for the overall service. As a result, we propose a hierarchical value co-creation model for multi-language knowledge communication, including language communication, knowledge communication, and organizational communication. Moreover, we investigate the issues and possible solutions for each layer in the model.

Keywords Service design • Value co-creation • Multi-language knowledge communication

---

D. Lin (✉) • T. Ishida • M. Otani
Department of Social Informatics, Kyoto University, Yoshida-Honmachi, Sakyo-Ku 606-8501, Kyoto, Japan
e-mail: lindh@i.kyoto-u.ac.jp

T. Maeno et al. (eds.), *Serviceology for Designing the Future*,
DOI 10.1007/978-4-431-55861-3_30

1 Introduction

Service-dominant logic is previously proposed with the concept of shifting from value creation to value co-creation, where service is regarded as the fundamental basis of exchange and value-in-use (or value-in-context) is important [1]. In the research area of service design and service science, value co-creation has been much discussed in recent years, which is a form of marketing strategy for realizing mutual value between firms and customers. However, there are several important issues on value co-creation when we focus on the service design in the real fields, where multiple stakeholders co-exist and unexpected happenings continuously occur [2]. First, services, service providers and service receivers are difficult to be defined at the early stages of service design. With the change of incentives of stakeholders in the real fields, roles of service providers and receivers may change constantly and new services always appear to be designed. Moreover, services and value co-creation among them might be dependent with each other.

In this paper, we attempt to investigate the above issues of value co-creation by using a case study of multi-language knowledge communication. Knowledge communication is important for transferring knowledge within an organization or among organizations. In previous research, several issues have been studied, such as problems of knowledge communication between experts and decision makers, the clarity of knowledge communication, and so on [3, 4]. In recent years, knowledge transfer becomes more and more important among people from different nations with the rapid expansion of globalization [5]. In such circumstances, information and communication technology (ICT) is expected to play an essential role. However, the major problem of supporting multi-language knowledge communication using ICT is the variety of requirements considering different types of communication fields, languages and stakeholders. Therefore, it is important to propose a general value co-creation model for designing multi-language knowledge communication.

To provide a model for value co-creation in multi-language knowledge communication, it is important to work together with the real fields. We use the YMC (Youth Mediated Communication)-Viet project, which is an agricultural support project for Vietnamese famers by Japanese experts through children of the farmers [6]. In this project, Japanese experts provide agriculture knowledge to Vietnamese farmers for solving problems in the paddy or increasing their knowledge, while Vietnamese farmers provide field data to Japanese experts for collecting data for further analysis. However, there are several problems to realize the above knowledge communication. First, translation between different languages should be worked out to support the knowledge transfer. Second, since there are different stakeholders, various incentives from the stakeholders should be considered to provide effective knowledge transfer. Therefore, we try to investigate the issues of service design in the real fields and show the process of formation of value co-creation.

Due to the complexity of such real field problems, iterative service design approach are always applied. Based on the iterative service design approach, the formation process of value co-creation can be realized by several stages from proposing individual services to designing the structure of different value co-creation layers for the overall service. We propose a hierarchical value co-creation model for multi-language knowledge communication, including layers of language communication, knowledge communication, and organizational communication. Language communication deals with the value co-creation for realization and improvement of high-quality language translation. Knowledge communication aims at realizing and improving agricultural knowledge transfer. The purpose of value co-creation in organizational communication is to realize sustainable services by the maintenance of various incentives of different organizations.

The rest of this paper is organized as follows. Section 2 introduces multi-language knowledge communication by using the example of YMC-Viet project. The value co-creation issues are discussed in Sect. 3. Section 4 describes the design of value co-creation model for multi-language knowledge communication. Section 5 is the conclusion of this research.

2 The Field: YMC-Viet Project

We focus on service design in the real fields and use a case study of multi-language knowledge communication which aims at providing agriculture support in rural areas in Vietnam by Japanese experts. The YMC-Viet project was conducted in cooperation with Ministry of Agriculture and Rural Development of Vietnam (MARD) as part of a model project to support developing countries with ICT. Since the literacy rate of farmers is low, youths (children) with high literacy acted as mediators between Japanese experts and Vietnamese farmers, which is called Youth-Mediated Communication (YMC) [6].

Several organizations in Japan and Vietnam are involved in this project including NPO Pangaea, Kyoto University, University of Tokyo, Mie University, Vietnam MARD, DARD of Vinh Long Province, Vietnam National University (VNU), Ho Chi Minh City and so on. Different organizations collaborate with each other while focusing on their own specific roles such as multi-language communication service design, agriculture knowledge support, administrative support, server maintenance, and so on.

To support the communication protocol in this project, a multi-language tool called YMC system is implemented, where Vietnamese youths can ask questions while Japanese experts can answer questions. This project has been conducted for four seasons (from the year 2011 to 2014) in Tra On District and Binh Minh District of Vinh Long Province, Vietnam with 15–30 participants each time. Figure 1 shows the activities of Vietnamese youths in this project. More detailed information about this project can be found in our previous work [7–9]. In this paper, we deal with the

Fig. 1 Participation of Vietnamese youths in the YMC-Viet project: measuring the size and color of rice leaf (*left*), using the YMC system for asking questions to Japanese experts (*right*)

issue of designing value co-creation for multi-language knowledge communication service based on our experiences in the YMC-Viet project.

3 Value Co-Creation of Service Design in the Real Fields

Vargo et al. propose the service-dominant logic with several foundational premises, which regard service as the fundamental basis of exchange and move from value-in-exchange to value-in-use or value-in-context [1, 10, 11]. In other words, traditional value creation process in marketing will not only be conducted by firms but also continued by customers through use, which is known as the value co-creation process [1]. The purpose of value has also changed from increasing wealth for the firm to increasing adaptability, survivability, and system wellbeing through service of others [10].

Ueda et al. propose the classification of value creation and summarize into three types of value creation models based on the relations among service providers, service receivers, services, and the environments: providing value model (Class I Value), adaptive value model (Class II Value), and co-creation value model (Class III Value), where the co-creation value model is regarded as a promising concept to integrate values of industries and those of consumers, and is described to be necessary when the objective of service receiver is uncertain and therefore the service provider and receiver cannot be separable [12, 13].

Since the field is a complex area with continuous unexpected happenings where multiple stakeholders co-exist with different motivations [2], there are several issues of designing value co-creation model. First, it is always difficult to define services, identify service providers and service receivers in the early stages of service design. Second, different types of services might be required to solve the real problem in the fields which are dependent with each other. In such cases, the same stakeholder might be involved in multiple services, either as a service

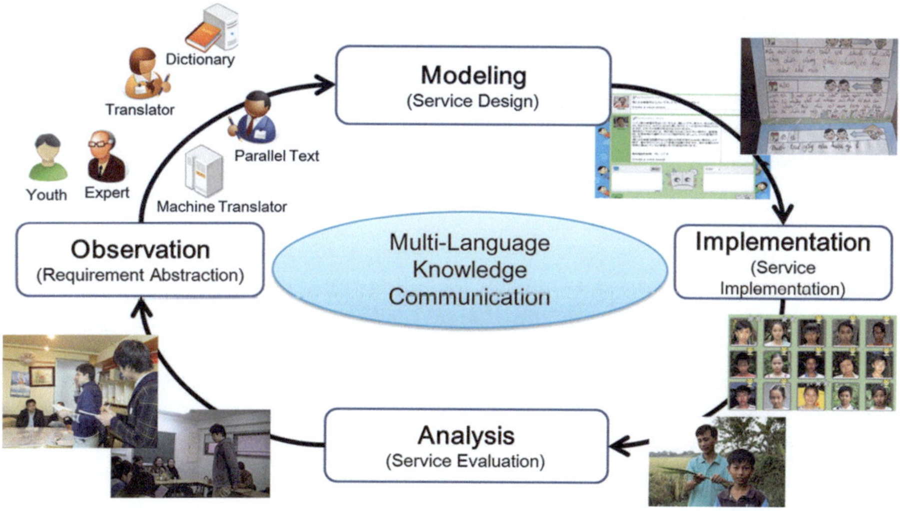

Fig. 2 Iterative service design process for multi-language knowledge communication

provider or a service receiver. Moreover, sustainable services should be designed to deal with the dynamic changes in the field.

Therefore, it is difficult to create a value co-creation model at the very early stages of the field activities. The formation of a value co-creation model is an iterative process. During such a process, definition of services, service providers and service receivers become more and more clear, and changes become predictable in a certain degree. Figure 2 shows a design process for the multi-language knowledge communication in the YMC-Viet project [8, 9].

In other words, value co-creation model is gradually converged during the iterative process in the real fields with multiple stakeholders and complex service requirements, which may include the following stages.

(1) Stage 1: values are proposed by the involved stakeholders. In this stage, individual services are proposed which are linked with potential service providers and service receivers.
(2) Stage 2: values are clearly described including the definition of different services, definition of service providers and service receivers for each service, mutual effects of service providers and service receivers, and the measurement criteria.
(3) Stage 3: the structure of the value co-creation for the overall service is formed. In this stage, the relationships of different services are established. The service for the filed can be designed and realized based on the formed structure of the value co-creation in this stage.

4 Case Study

In the YMC-Viet project, we need to consider the end-to-end value co-creation in the multi-language knowledge communication. However, as has been described in the previous section, it is difficult to design a value co-creation model for the real fields at the early stages. The formation of value co-creation is a gradual process based on the iterative service design.

In the initial stage, several individual services were proposed based on field investigation by different stakeholders including translation service (by the translation process designer), and knowledge service (by the Japanese agriculture experts). Then, the value co-creation model for each individual service is gradually created in the next stage based on iterative service design. It was not until the third season of the experiment could we define the whole value co-creation model for the multi-language knowledge communication, which includes three layers as shown in Fig. 3: language communication, knowledge communication and organizational communication.

In the proposed hierarchical value co-creation model, language communication exists as a channel for knowledge communication, which provides the availability of knowledge in both languages. Knowledge communication deals with the end-to-end value co-creation between Japanese experts and Vietnamese farmers/youths. Organizational communication is indispensable for maintaining the sustainability of the multi-language knowledge communication.

Table 1 shows the proposed three layers in multi-language knowledge communication for the YMC-Viet project. Since we focus on describing the value co-creation of the multi-language knowledge communication, we do not introduce the evaluation results for each layer in details in this paper.

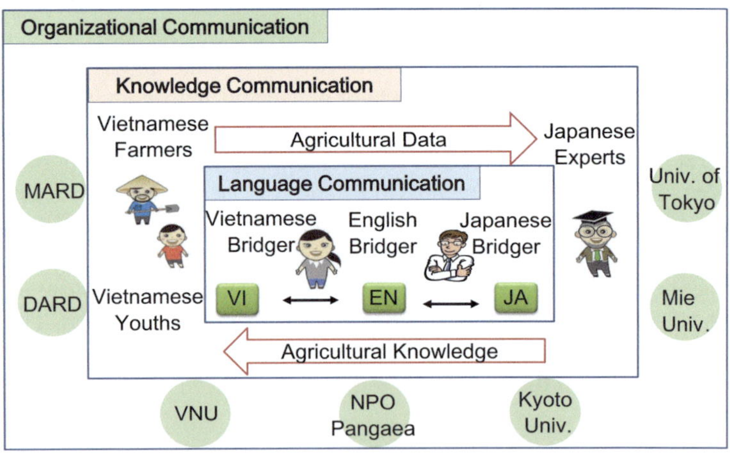

Fig. 3 Hierarchical value co-creation model for multi-language knowledge communication with three layers

Table 1 Description of three layers in multi-language knowledge communication

Value co-creation layer	Creator of value	Purpose of value	Measurement of value
Language communication	Translation process designer, English Bridger, Japanese Bridger, Vietnamese Bridger	Realization and improvement of high-quality language translation between text senders and text receivers	Language translation quality (accuracy), cost
Knowledge communication	Japanese agriculture experts, Vietnamese youths/farmers	Realization and improvement of agricultural knowledge transfer between knowledge senders and knowledge receivers	Effectiveness of knowledge transfer (short term), influence to rice productivity (long term)
Organizational communication	Involved organizations (agriculture, ICT, education, local community, etc.)	Realization of sustainable services for agriculture support in rural areas by the maintenance of various incentives of different organizations	Adaptability, sustainability of the overall service system

4.1 Language Communication

The layer of language communication aims at realizing and improving high-quality language translation between text senders and text receivers. We design the language communication by modeling it as a QoS-aware service composition and optimization problem, which is an important issue in the research area of service-oriented computing [14]. Criteria for measuring the value of language communication like translation quality and cost are defined as QoS attributes.

To realize the language communication, we have been working on technologies and methodologies to support intercultural collaboration through service-based infrastructures since 2006. We have developed the Language Grid,[1] which is a service-oriented language infrastructure that allows users to deploy and combine various language resources as language services [15, 16]. The Language Grid has been operated by Department of Social Informatics of Kyoto University since December 2007; it provides necessary language resources for the language communication in the YMC-Viet project, including machine translators among English, Japanese and Vietnamese, agriculture dictionaries, morphological analyzers, and parallel texts for agriculture.

Although the Language Grid provides many language services, it is difficult to ensure the quality of Japanese-Vietnamese translation by combining language

[1] The Language Grid: http://langrid.org

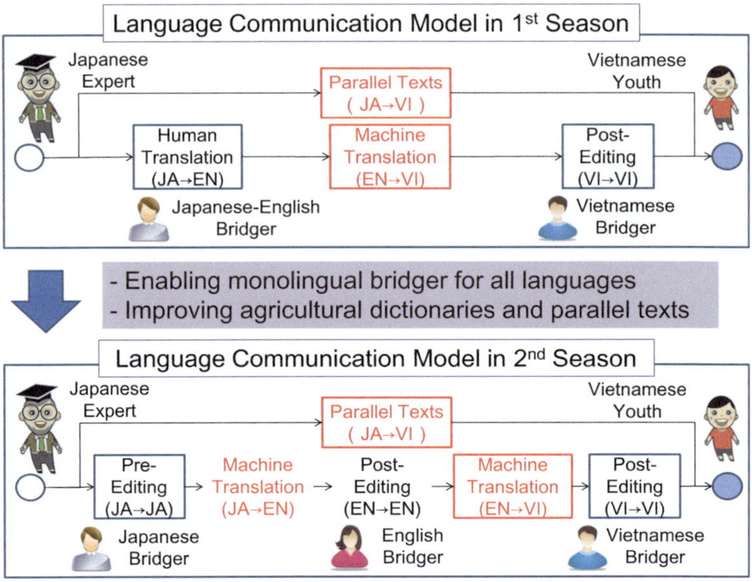

Fig. 4 Service refinement of language communication

services (e.g., machine translators, dictionaries, and parallel texts) for supporting communication between Japanese experts and Vietnamese youths. Therefore, it is necessary to consider combining human activities and language services for language communication. Human activities include pre-editing, post-editing, and human translation, which are named as Bridger in the YMC-Viet Project. Since the quality of machine translators and human activities is difficult to estimate, we use a user-centered service design approach for testing the language communication models, considering the balance of translation quality and translation cost.

During the design process, the language communication model can be improved gradually based on evaluation and refinement of service composition. Therefore, value co-creation of language communication is realized by Bridger of different languages. Figure 4 shows the improvement of the language communication model during the experiments. The refinement process of service composition and evaluation result of the language communication improvement for different seasons of experiments are reported in our previous work [8, 9].

4.2 *Knowledge Communication*

Language communication is a channel for knowledge transfer, and therefore it is supposed to be transparent to knowledge communication. The layer of knowledge communication aims at realizing and improving agricultural knowledge transfer between the service sender (Japanese experts) and the service receiver (Vietnamese

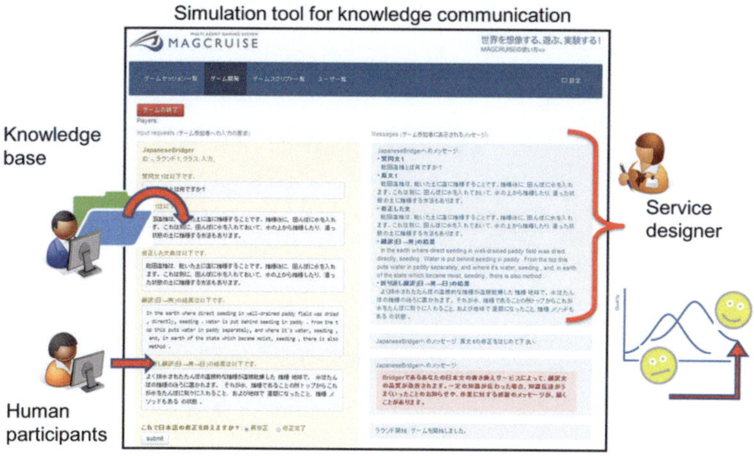

Fig. 5 Participatory service simulation for knowledge communication with human participants

Table 2 Examples of two types of knowledge

Types	Examples
Textbook knowledge	Question from the youths:
	Are there any diseases caused by bacteria?
	Answer from the experts:
	Bacteria are fungi. Various types of bacteria cause a disease on rice plants. Rice blast and rice blight are such diseases.
Situation-dependent knowledge	Question from the youths:
	A rice root doesn't come out. Is it because the water depth of paddy field is too deep?
	Answer from the experts:
	Doesn't root come out from planted seed? Deep flooding could be its cause. If possible, please send me its picture. There may be some different problems.

youths/farmers). We try to design effective protocols for the knowledge communication, which are simulated and evaluated using a participatory design approach with human participation before conducting the experiments in the real fields as shown in Fig. 5.

In the YMC-Viet project, there are two types of knowledge that is transferred between Japanese experts and Vietnamese youths. One type is *textbook knowledge*, which deals with some general common knowledge for agriculture. The other type is *situation-dependent knowledge*, which deals with the real problems occurred in the rice paddy and might affect behaviors of rice cultivation. Table 2 shows examples of the two types of knowledge.

We collect all the log data of knowledge communication between Vietnamese youths and Japanese experts in the YMC systems for experiments from the year 2011 to 2014 in Vinh Long Province, conduct interviews with Vietnamese youths,

(a) (b)

Fig. 6 Interview for knowledge communication in Vietnam: (**a**) Interview with farmers, (**b**) Interview with youths

farmers, and Japanese experts (see Fig. 6) to analyze the knowledge communication. From the analysis of season 2012, we found that 74 % of the knowledge communication deals with textbook knowledge while 26 % of that deals with situation-dependent knowledge.

For different types of knowledge, we use different approaches for measuring the value of knowledge communication. For textbook knowledge, it is necessary to evaluate the effectiveness of knowledge transfer between Japanese experts and Vietnamese youths/farmers in the short term by tests or interviews. On the other hand, we need to evaluate how the situation-dependent knowledge communication affects the rice productivity by conducting agricultural investigations in the long term.

4.3 *Organizational Communication*

The role of organizations is important for realizing language communication and knowledge communication. Due to the complexity of the real fields, it is always difficult to design the overall service system since there are various types of stakeholders. Therefore, service design in the real fields should focus on how to provide a sustainable mechanism for designing services rather than how to design perfect services.

More than seven organizations are involved in the YMC-Viet project as we described in Sect. 2 for different purposes. Different organizations have their own different incentives in the project. For example, different organizations from Japan are involved in this project in different ways: NPO Pangaea organizes and manages the whole project for youth education; University of Tokyo and Mie University participates for agricultural knowledge support and field data analysis; Kyoto University focuses on designing and providing the multi-language communication services. Therefore, it is necessary to consider how to match different incentives of different organization to improve and optimize the communication design. We try to observe and analyze the formation of the community in the YMC-Viet Project, and deal with the mechanism design of incentive matching of organizations for maintaining the sustainability of the service design of multi-language knowledge communication.

4.4 Discussion

The proposed value co-creation model in this paper is not only useful in the YMC-Viet project; it deals with the general research issues in multi-language knowledge communication of similar situations in the real fields. However, there are several topics that we should consider in our future work.

First, although we designed three layers of value co-creation and defined the purpose of value, value creators and measurement of value separately, the layers of language communication, knowledge communication, and organizational communication are not totally independent. For example, language communication is supposed to be only focused on the value co-creation of text messages but it changes the content of knowledge in an unexpected way when translation is conducted from one language to the other language, which is invisible for the knowledge communication layer. Such interdependencies between value co-creation layers are expected to be considered in our future study.

Second, the value co-creation model proposed for multi-language knowledge communication in this paper can be regarded as a hierarchical model. However, other types of value co-creation network may be possible if the complexity of the fields differs. Therefore, it is necessary to conduct a series of systematic case studies to discover various patterns of the value co-creation in the real fields.

Moreover, although we reported the experiences of designing a value co-creation model for multi-language knowledge communication in the YMC-Viet project, we need to abstract the process to propose a methodology for forming value co-creation models of filed-oriented service design in the future.

5 Conclusion

This paper deals with how to design value co-creation model for services in the real fields. We use a case study of multi-language knowledge communication in the YMC-Viet project which aims at providing agriculture support in rural areas of Vietnam. We aimed to propose a value co-creation model for multi-language knowledge communication where three issues are to be solved: how to overcome the language barriers between different nations, how to ensure the effectiveness of knowledge transfer, and how to map various incentives of different stakeholders. The formation process of value co-creation can be realized by several stages based on the iterative service design approach: proposing individual services, modeling value co-creation for individual services, and designing the structure of different value co-creation layers for the overall service.

As a result, we proposed a hierarchical value co-creation model for multi-language knowledge communication to address the above issues, which includes three layers: language communication, knowledge communication and organizational communication. Language communication deals with the value co-creation for realization and improvement of high-quality language translation. Knowledge communication aims at realizing and improving agricultural knowledge transfer between knowledge senders and knowledge receivers. The purpose of value co-creation in organizational communication is to realize sustainable services for agriculture support in rural areas by the maintenance of various incentives of different organizations. Based on the proposed model, we conducted four seasons of experiment of the YMC-Viet project in Vinh Long province, Vietnam from the year 2011 to 2014.

Currently, we are mainly focusing on the layers of language communication and knowledge communication. In the future, we will also study how to design the mechanisms for incentive coordination among different organization to cover the whole model. It is also important to analyze how the three layers affect each other in the process of end-to-end value co-creation. Since the above problems do not only exist in the YMC-Viet project, we will deal with how to provide a general methodology for designing value co-creation model for similar situations in the real fields.

Acknowledgements We would like to express our gratitude to NPO Pangaea for managing the YMC-Viet field experiments and all participants of the experiments in Tra On District and Binh Minh District of Vinh Long Province, Vietnam. This research was partially supported by Service Science, Solutions and Foundation Integrated Research Program from JST RISTEX, and a Grant-in-Aid for Scientific Research (S) (24220002) from Japan Society for the Promotion of Science.

References

1. Vargo SL, Lusch RF (2004) Evolving to a new dominant logic for marketing. J Mark 68 (1):1–17
2. Ishida T (ed) (2012) Field informatics. Springer, Berlin/Heidelberg. ISBN 978-3-642-29005-3
3. Eppler M (2007) Knowledge communication problems between experts and decision makers: an overview and classification. Electron J Know Manage 5(3):291–300
4. Bischof N, Eppler M (2011) Caring for clarity in knowledge communication. J Univers Comput Sci 17(10):1455–1473
5. Aiken M, Martin J, Shirani A, Singleton T (1994) A group decision support system for multicultural and multilingual communication. Decis Support Syst 12(2):93–96

6. Mori Y et al (2012) Youth mediated communication (YMC) – agricultural technology transfer to illiterate farmers through their children. World conference on computers in agriculture, Taipei, 2012
7. Kita K, Takasaki T, Lin D, Nakajima Y, Ishida T (2012) Case study on analyzing multilanguage knowledge communication. In: 2012 international conference on culture and computing (culture and computing 2012), Hangzhou
8. Lin D, Ishida T (2013) Participatory service design based on user-centered QoS. In: 2013 IEEE/WIC/ACM international conference on web intelligence (WI 2013), Atlanta, pp 465–473
9. Lin D, Ishida T (2014) User-centered service design for multi-language knowledge communication. In: Mochimaru M et al (eds) Serviceology for services. Springer, Tokyo, pp 309–317
10. Vargo SL, Maglio PP, Akaka MA (2008) On value and value co-creation: a service systems and service logic perspective. Eur Manage J 26(3):145–152
11. Vargo SL, Lusch RF (2008) Service-dominant logic: continuing the evolution. J Acad Mark Sci 36(1):1–10
12. Ueda K, Takenaka T, Váncza J, Monostori L (2009) Value creation and decision-making in sustainable society. CIRP Ann Manuf Technol 58(2):681–700
13. Ueda K, Takenaka T, Fujita K (2008) Toward value co-creation in manufacturing and servicing. CIRP J Manuf Sci Technol 1(1):53–58
14. Zeng L, Benatallah B, Ngu A, Dumas M, Kalagnanam J, Chang H (2004) QoS-aware middleware for web services composition. IEEE Trans Softw Eng 30(5):311–327
15. Ishida T (2006) Language grid: an infrastructure for intercultural collaboration. In: IEEE/IPSJ Symposium on Applications and the Internet (SAINT-06), Phoenix, pp 96–100
16. Ishida T (ed) (2011) The language grid: service-oriented collective intelligence for language resource interoperability. Springer, Berlin

Part VII
Service Innovation and Design

Field-Oriented Service Design: A Multiagent Approach

Toru Ishida, Donghui Lin, Masayuki Otani, Shigeo Matsubara, Yohei Murakami, Reiko Hishiyama, Yuu Nakajima, Toshiyuki Takasaki, and Yumiko Mori

Abstract Service has been considered as value co-creation through the cooperation of service providers and customers. This paper, however, focuses on service design in problem fields where complex issues exist among various stakeholders, where identifying service providers and customers is not a simple process. In other words, we focus on a very early stage of service design with huge ambiguities; we call it *field-oriented service design*. A typical case is introducing new services in developing countries. The main issue here is to create new services compatible with existing services through action research that considers a wide variety of regional, national and global stakeholders. It is often difficult to identify the influence of/to the services to be designed due to the differences in culture, language and business customs. As a result, unexpected interdependencies among services together with stakeholders are often revealed during the process of action research. To resolve this ambiguity in the design process, we propose a multiagent approach that couples role playing games with participatory simulations; it is based on our experiences in agricultural support projects in Southeast Asia.

Keywords Service design • Field informatics • Multiagent system • Role playing game • Participatory simulation • Gaming simulation

T. Ishida • D. Lin (✉) • M. Otani • S. Matsubara
Department of Social Informatics, Kyoto University, 606-8501 Kyoto, Japan
e-mail: lindh@i.kyoto-u.ac.jp

Y. Murakami
Unit of Design, Kyoto University, Kyoto, Japan

R. Hishiyama
Faculty of Science and Engineering, Waseda University, Tokyo, Japan

Y. Nakajima
Department of Information Science, Toho University, Tokyo, Japan

T. Takasaki • Y. Mori
NPO Pangaea, Kyoto, Japan

T. Maeno et al. (eds.), *Serviceology for Designing the Future*,
DOI 10.1007/978-4-431-55861-3_31

1 Introduction

Service has been defined as value co-creation among service providers and customers [1]. However, it is often difficult to identify service providers and customers at the very early stage of service design. To fully reflect the various behaviors and requirements of customers, it is necessary for service designers to identify the individuals or organizations that really are involved [2]. When getting into real fields that are bedeviled by complex issues, service design tends to become a process of trial and error due to the lack of local information.

Let us consider the example of agriculture support projects that are intended to understand agricultural economics, increase productivity, reduce environmental burdens and so on, in developing countries. The process of service design for such projects often adopts a PDCA cycle, with the phases of design, execution, evaluation, and redesign.

In this paper, however, we introduce a vision that goes beyond the usual PDCA cycle: additional services may need to be introduced to satisfy newly recognized requirements, or existing services may need to be redesigned when hidden conflicts among them are revealed. If cross-effects are expanded among related services in both positive and negative ways, how can we grasp the whole picture needed for service design?

We note that the designed service must be sustainable at the final stage of the project [3]. In other words, it is necessary to clarify the cross-effects among new and existing services, reorganize the complex relations among them, and guarantee the sustainability of services as a whole, so as to make it easier to transfer the leadership of the project to local stakeholders.

In the area of service design, some textbooks contain various ideas and methods for customer-centered service design [4]. In this paper, however, we focus on the very early stage of service design. We introduce the concept of *field-oriented service design* based on our field experiences, show how important it is to create a contention-robust design process; we propose a *multiagent approach* for designing sustainable services in real fields.

This paper is also intended to bridge service science (social activities are understood by regarding service as the fundamental basis of exchange [1]), and artificial intelligence (realizing intelligent mechanisms by creating computational models for problem solving). The key idea of connecting the two disciplines is to view service as a social implementation of problem solving. This view can provide a theoretical background for applying problem solving technologies to service design. In particular, technologies of multiagent systems, which model collections of autonomous agents, can be adopted to understand social activities in different fields. In fact, multiagent systems have been used to analyze and simulate different kinds of social activities [5].

The following sections explain the fundamental concepts including field, problem, problem solving, and service design, and then discuss the mapping between problem solving and service design. A multiagent approach including role playing games, participatory simulations, and gaming simulations is explained in detail. Finally, two agricultural support projects in Southeast Asia will be detailed to illustrate how the multiagent approach can contribute to field-oriented service design.

2 Field-Oriented Service Design

This section starts by defining field. We then explain how to understand field-specific problems and to design services that solve those problems.

2.1 What's a Field

You may think field research is a common approach in the service design process, but the term "field" has a wide range of meanings. *Field* here is defined as follows. This definition is given by Osamu Katai [5] and reflects the tradition of cultural anthropology in Kyoto University.

> A spatio-temporal area that is hard to be mastered by any analytical and/or engineering approach due to various individuals and entities being coexistent, which causes the unexpected happening of accidental events thus necessitating our continuing commitment and care.

This definition may confuse researchers or engineers who regard analytical and/or engineering approach as their professions. The definition does not deny the possibility of using analytical and/or engineering approaches to contribute to the solution of problems in the field. However, since problems are layered due to the coexistence and interdependency, problems are spatio-temporally triggered by unexpected events. Therefore, the above definition of field suggests that designers should continue to commit to the problems, since the process of problem solving can never be fully completed. Therefore, fields are troublesome, especially for researchers and engineers, the most functionally divided professionals.

Similar discussions have already appeared in the area of design for a fairly long time. Designers used the term *wicked problem* [6] to describe the difficulties of the design situation that is encumbered with incomplete and changing requirements. Wicked problem means that solving one problem reveals other problems due to the interdependencies among problems.

In the area of artificial intelligence, the term *frame problem* [7] is used to express the difficulties of identifying the effects of actions without representing beliefs that might not be affected. Frame problem is also a philosophical proposition; is it possible to determine the boundary of knowledge and thus fully elucidate the effects of actions.

The definition of field suggests that it is a wicked problem, and the activity of service design in solving it encounters the frame problem. In the following subsections, we first discuss problems and problem solving, and then define field-oriented service design as a process of problem solving and forming problem solving organizations.

2.2 Problem Solving

Field-oriented service design deals with the process of understanding and solving problems in real fields containing complex issues. We define problem, problem solving process and problem solving organization in such situations as follows.

- A *problem* is extracted from complex issues in the field. The expectation is that a problem is always resolvable. That means a problem should be shaped to be solvable.
- A *problem solving process* is literally a process of finding solutions to problems. Since problems are already premised solvable, however, the difficulty of problem solving often lies in the process of shaping problems. In other words, problem solving and problem shaping (sometimes called problem finding) processes are two sides of the same coin.
- A *problem solving organization* is a team of stakeholders tasked with solving problems, or in other words, a team for shaping solvable problems.

As represented in Fig. 1, problems extracted and shaped from the field are often dependent on each other. In such cases, problem solving processes and problem solving organizations turn out to be interdependent. Moreover, the interdependences between problems, problem solving processes and problem solving organizations become clearer but also change as time advances.

In real fields, since problems are networked and layered, it is difficult to fix a description of any problem and to find a permanent solution. Therefore, it is essential to design flexible problem solving processes to deal with continuous problem transfiguration. In other words, problem solving in the field does not aim

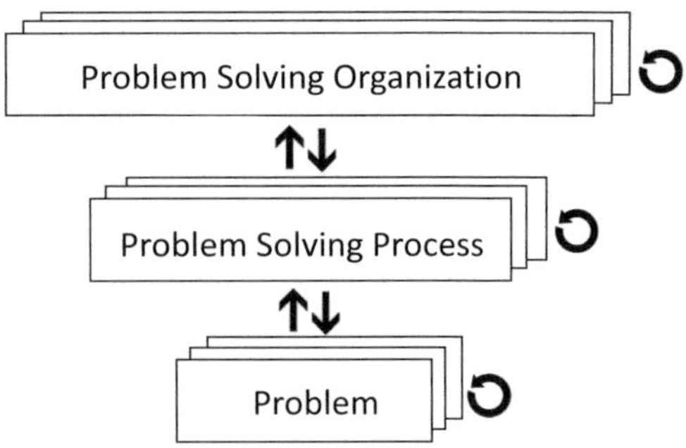

Fig. 1 Interdependency in problem solving

at solving currently known problems but at designing a continuous process for solving problems. Also, it is necessary to maintain a flexible networked organization that can handle the continuous process of problem solving.

2.3 Service Design

In this paper, we regard *service* as an instance of problem solving in society. Since problems are networked and layered, services to solve those problems are also networked and layered. In Fig. 2, layered white circles represent layered services, black circles represent service providers or customers, and directed arrows represent dependencies between services and providers/customers.

Since problem solving is a continuous process, the crucial issue for service design is to analyze the interdependencies among stakeholders, and establish a continuous process for designing services. In this paper, we use the term *stakeholder*, to mean a customer of some services while possibly providing services to others. Furthermore, stakeholders include not only agents who provide and consume services, but also various individuals and organizations related to the service to be designed such as funding agencies, government offices, community representatives and so on, all of who affect service design both positively and negatively.

When we consider service as the co-creation of value-in-context by service providers and customers [1], it is necessary to have an experimental basis for such value co-creation among stakeholders in the field to develop a continuous process for designing services.

We propose a multiagent approach with the intent of providing a basis for stakeholders to collaboratively conduct experiments. Macro modeling might be applied to service design, if stakeholders and their relations are fixed over time. Unfortunately, in the fields, stakeholders and their interdependencies are often altered over time. Given this, it is effective to model individual stakeholders and their interactions with a micro perspective and to adopt a multiagent approach for service design.

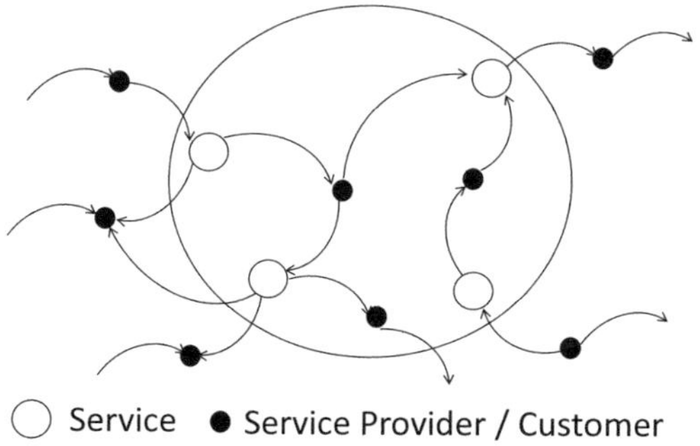

Fig. 2 Networked and layered services

3 Multiagent Approach

This section describes a multiagent approach for field-oriented service design. We first introduce a role playing game, and a participatory simulation that includes humans and software agents. We then propose a gaming simulation, which applies the participatory simulation technology to role playing games.

3.1 Role Playing Game

Role playing games (RPGs) are a well-known multiagent approach, where stakeholders participate in a game and mutually confirm their decision makings on a game board. There are several ways to utilize RPG as follows.

- For consensus building, RPG enables stakeholders to compromise and reach optimal decisions for the community by representing and sharing individual decisions on a game board.
- For system design, team members can learn the requirements and environments of the systems to be designed by simulating the decision processes of various stakeholders.
- For field analysis, a more accurate model of the decision processes of the community can be obtained by observing stakeholder decisions during RPG sessions.

Figure 3 illustrates the modeling process of players' decision making. An initial model is created from relevant literature and surveys. RPG sessions are conducted using a board that represents the players' environment. Decision making process can be understood based on the logs obtained during the game. The reasons behind the decisions made in the RPG are exposed by interviewing the players after the game. Finally, the decision making model is refined by analyzing the RPG log data and the interviews. Running the RPG several times can improve the obtained models.

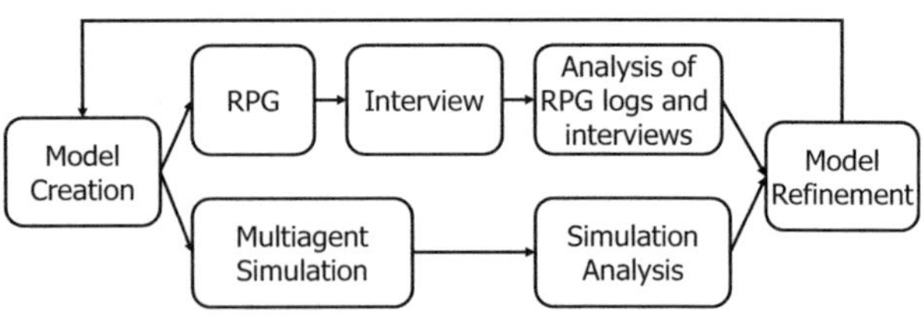

Fig. 3 Modeling farmers' decisions [9]

3.2 Participatory Simulation

Multiagent simulations are getting popular as a method of micro simulation in various research areas. In multiagent simulations, agent behaviors are determined by scenarios, which can either be described by programming languages or scenario description languages with embedded decision-making models. The scenario processor interprets agent scenarios and requests agents in a virtual space to perform sensing and acting functions.

Participatory simulations were invented as an extension of multiagent simulations. We can easily extend multiagent simulations to yield participatory simulations by replacing some of the scenario-guided agents with human-controlled avatars. As in Fig. 4, a participatory simulation consists of (1) *agents* for modeling users, (2) *avatars* to represent human subjects, (3) *scenarios* for modeling interactions, (4) *human subjects* to control avatars, (5) *virtual space* to represent real space, and (6) a *monitor* to visualize simulations underway in the virtual space. In this situation, human subjects and agents can cooperative in performing a simulation. Just as with video games, human subjects can join the simulation by controlling avatars via joy sticks, mice, or other input devices. To analyze simulation results, we monitor the entire process of the simulation by visualizing the virtual space. Recording human behavior is useful for analyzing the simulation results and for improving the agent decision making models.

3.3 Gaming Simulation

Gaming simulation is conducted by fusing participatory simulation with RPG. Although RPG requires all the stakeholders to participate in the game, this is not always practical. Moreover, it might be more efficient for a limited design team to conduct the simulations to understand the nature of problems and services at the early stage of service design.

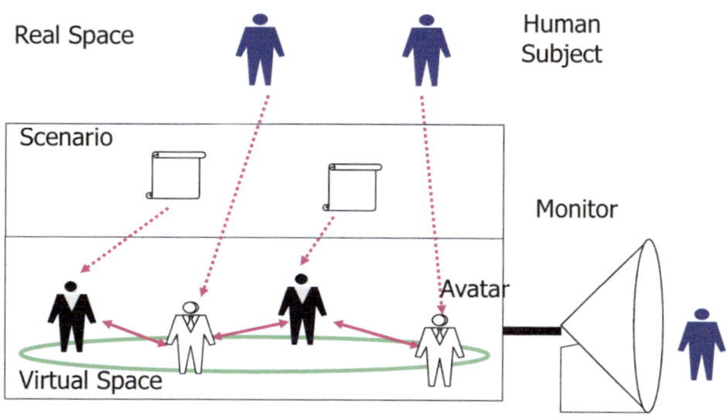

Fig. 4 Participatory simulation [11]

One of the advantages of gaming simulations is that they support seamless design activities: from internal consideration within a service design team to RPG with all stakeholders. Moreover, it is easy to deal with situations where additional services are required or conflicts occur among new and existing services. Those advantages come from the fact that multiagent simulations are a type of micro simulation [8], as is described below in more detail in a comparison with macro simulation.

Macro simulation is intended to reproduce a phenomenon based on a macroscopic viewpoint in that the entire simulation target is represented as a single model and its behavior is defined by governing equations. Macro simulations allow the observation of behaviors or changes in the overall system, but the local properties of individual elements or interactions among elements are not reproduced. For example, system dynamics are often used for macro modeling and macro simulation; they deal with internal feedback loops, stocks and flows that affect the behavior of the entire system.

On the other hand, micro simulations reproduce a complex social phenomenon by accumulating the microscopic behaviors of models of social entities including their interactions. Assuming that human society consists of a lot of decision-making entities, it seems natural to use micro simulations to predict the behavior of society. Micro simulations have manifested their ability to clearly present a variety of individual core behaviors in the reproduction and analysis of complex collective behaviors. The multiagent approach can thus straightforwardly represent the actions of individual people and organizations. It can be applied to social, economic and cultural problems rooted in human decision making and that have been difficult to hold experiments on.

4 Field Experience

The proposed concepts and approaches are based on two agriculture support projects in Southeast Asia that we joined and are active in. In these projects, RPG and gaming simulation were used at the early stage of service design. In this section, we use these two experiences to illustrate that the multiagent approach is promising for field-oriented service design.

4.1 Modeling Agriculture Economics in Thailand

The first case study illustrates how useful it is to introduce the multiagent approach in shaping a problem.

We conducted an RPG to understand farmers' land-use decisions in upper northeast Thailand in a collaboration project with IRRI (International Rice Research Institute). At that time, it was reported that while some farmers used

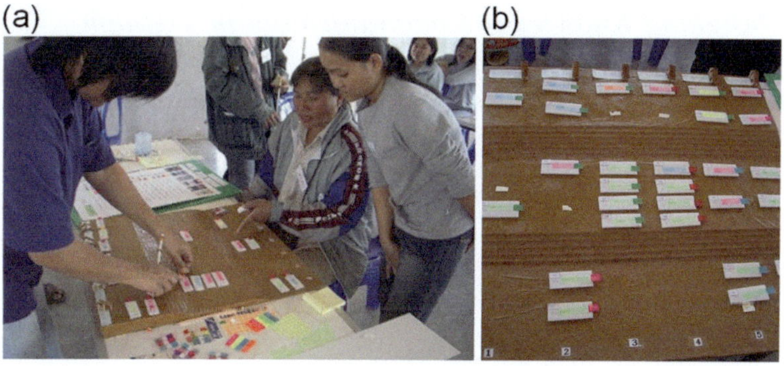

Fig. 5 Role playing game [9] (**a**) playing the game (**b**) RPG board

their own rice seed, more farmers bought seeds and changed seeds frequently. Since the government's production capacity was low, many organizations and companies joined in the rice seed supply system. Unfortunately, no integrated view of a comprehensive seed management system including market, government, and seed producing companies had been reported.

The first step in improving rice seed production was to investigate the farmers' decision making processes. The last three decades have seen an expansion in upland cash crops through application of the rain-fed lowland rice ecosystem. We thought the reason was that rice prices dropped while sugarcane prices remained high. This project aimed to validate such hypotheses, understand the decision making processes of farmers, and identify how land should be used in the future.

Through the RPG, we achieved the participation of stakeholders and obtained a decision making model that closely approached reality. Figure 5a shows the particulars of the RPG that reproduced the environment surrounding the stakeholder on the board. A board representing farm land was created as Fig. 5b. RPG sessions using the board were organized, and the log data of 25 stakeholders were recorded. The results of the RPG were reflected in the decision making model and then a land use simulation was run. Farmers, who had participated in the RPG, evaluated the validity of the model and the simulation.

RPGs were conducted several times in this project. Since stakeholders have not been clarified, though they are expected to include seed sellers, government institutions and companies, they were replaced by researchers in the project. Through the series of RPGs, we obtained a social network of stakeholders and the farmers' decision making processes that interacted with other stakeholders [9].

The RPG approach is flexible enough to be extended later to include delegates of government and industries to better reflect decision making by farmers. Also, more RPGs were conducted by this project in different types of villages for a comparative evaluation. Again, the multiagent approach is flexible enough to broaden the spatial area to include neighboring villages for investigating mutual impact of economic activities.

4.2 Designing Knowledge Communication in Vietnam

The second case study confirms the difficulty of shaping a problem and why we need the multiagent approach.

We have been participating in an agriculture support project in Vietnam with agriculture experts since 2011. The project aims at designing new services for multi-language knowledge communication via the Internet. The goal is to transfer agriculture knowledge from Japanese experts to Vietnamese farmers in rural areas with low literacy rate, to increase rice productivity and to decrease the environmental burdens caused by excess use of agrichemicals. The motivation of Japanese experts, who work for JICA to support developing countries, in using ICT is they cannot physically travel to all rural areas that need their advice.

There exists a huge gap between Japanese experts and Vietnamese farmers, not only in agricultural knowledge but also language and culture. Furthermore, the farmers have difficulties in using computers and indeed in reading/writing messages. Therefore, as shown in Fig. 6, a youth-mediated communication (YMC) model was proposed and applied to bridge the gap of knowledge, language and cultures [10]. In the YMC model, the children received ICT training at a local telecenter, and then acted as mediators between the Vietnamese farmers (parents) and the Japanese experts. The YMC is a breakthrough idea to bridge the gap, but it introduces another difficulty in communication: the computer literacy and background knowledge of children determine the boundaries of agricultural knowledge communication. To cope with all the problems, we formed a project including a variety of stakeholders: agricultural experts and language processing experts in universities, farmers and their children, NGOs targeting the education of children in developing countries, and Vietnamese national and local ministries.

At the beginning, we designed services for knowledge transfer from Japanese experts to Vietnamese children, where the experts are service providers and the children are customers. Multi-language knowledge communication services were developed and managed by the project. These services have been realized based on different motivations from various stakeholders. The NGO staffs working in the field were motivated to educate the children. Soon after the project started, however, we observed that the experts became more and more motivated by obtaining

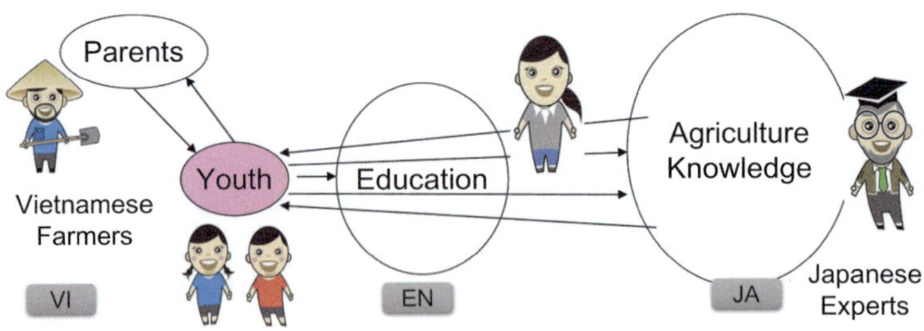

Fig. 6 Youth mediated communication

Fig. 7 Field activities (**a**) collecting field data (**b**) reading expert's advice

field data from the children: a complementary service was added where the children are service providers and the experts are customers. Meanwhile, the local government highly praised the project because they found that it became easier for them to form communities in low literacy areas. Figure 7a shows a child collecting field data, and Fig. 7b shows the local government officers involved in helping the children to read the experts' advice.

The YMC project has been highly recognized and supported by Vietnamese national and regional governments. It has resulted in community formation with local stakeholders and researchers, among which staffs from the local government are in charge of working with the NGOs to train children. During the successful project, however, we recognized the difficulties in understanding the relationships among newly created services and existing services. Here is one episode.

> One of the goals of this project is to reduce the amount of agrichemicals by enhancing knowledge communication among agricultural experts and farmers. Employees of the local government became heavily involved in the project by helping farmers' children to learn ICT. However it was found that local government workers usually work part-time, and they normally have second jobs, such as selling agrochemicals and fertilizers to farmers.

Moreover, it is also difficult to estimate the effects of services. Here is another episode.

> This project aims at increasing the yield of rice. One project member, an agricultural expert, recognized the lack of nitrogen in this particular rice field, and suggested the use of a little more fertilizer. The farmers did not follow this advice, since they believe bugs would gather from neighboring fields, if they increased the amount of fertilizer.

These episodes shows how it is difficult to shape the problem and design services in a single project. Though it is not practical to complete service design in a limited time, the multiagent approach can make it possible to deal with the unexpected interdependency of problems and services by widening the scope of stakeholders and forming a sustainable problem solving organization to continue the process of service design.

5 Conclusion

This paper proposed an approach to service design in the field where complex issues are often faced. At the very early stage of service design, it is necessary to understand the problems in the field, and propose services for solving those problems. However, designing services in a real field is always difficult due to the interdependency between problems and their changes over time. Therefore, developing a continuous problem solving process is more important than solving currently known problems. In other words, the key to service design in the field is to provide a service design process and form a flexible team that can support the process. In this context, we proposed a multiagent approach including role playing games, participatory simulations and gaming simulations as a basis of experiments for designing services with various stakeholders.

The purpose of running gaming simulations is to design sustainable services by understanding the interdependencies of existing services, and to reach agreement on the design of future services. When new services emerge from simulations, stakeholders learn what had been previously overlooked. We have to be careful, however, when interpreting the results of gaming simulations. We should be sensitive to what is important and what is not in the results from simulations. Since simulation results depend on those stakeholders that participate and the results of modeling agents and environments, result validity should be examined both practically and theoretically.

Field-oriented service design does not deal with just the introduction of new services in developing countries. Similar phenomena are observed in general cases, where the service designer cannot completely understand the complex relations and behaviors of stakeholders. Therefore, the lessons learned in our studies can also benefit service design in various situations.

Acknowledgement We thank Francois Bousquet for his collaboration in the Thai project, and Seishi Ninomiya, Takaharu Kameoka and their colleagues and students for joining the YMC Vietnam project as agricultural experts. This research was partially supported by Service Science, Solutions and Foundation Integrated Research Program from JST RISTEX, and a Grant-in-Aid for Scientific Research (S) (24220002) from Japan Society for the Promotion of Science.

References

1. Lusch R, Vargo S (2014) Service-dominant logic: premises, perspectives, possibilities. Cambridge University Press, Cambridge
2. Nicola M (2006) Developing new Product Service Systems (PSS): methodologies and operational tools. J Clean Prod 14:1495–1501
3. Reiko H (2013) Sustainable empowerment models for rural pastoral communities in Kenya. ASEAN Conf Environ Behav Stud 85:432–442
4. Stickdorn M, Schneider J, Andrews K (2011) This is service design thinking: basics, tools, cases. Wiley, Hoboken

5. Ishida T (ed) (2012) Field informatics. Kyoto University Field Informatics Research Group. Springer, Berlin/Heidelberg
6. Rittel H, Webber M (1973) Dilemmas in a general theory of planning. Pol Sci 4(2):155–169
7. McCarthy J, Hayes P (1969) Some philosophical problems from the standpoint of artificial intelligence. Mach Intell 4:463–502
8. Gilbert N, Troitzsch K (1999) Simulation for the social scientist. Open University Press, Maidenhead/New York
9. Torii D, Ishida T, Bousquet F (2006) Modelling agents and interactions in agricultural economics. International joint conference on autonomous agents and multiagent systems, 2006, Hakodate, pp 81–88
10. Yumiko Mori et al (2012) Youth Mediated Communication (YMC) – agricultural technology transfer to illiterate farmers through their children. World conference on computers in agriculture, 2012, Taipei
11. Toru Ishida, Yuu Nakajima, Yohei Murakami, Hideyuki Nakanishi (2007) Augmented experiment: participatory design with multiagent simulation. International joint conference on artificial intelligence, 2007, Hyderabad, pp 1341–1346

Experience Plot: A Template for Co-Creating Customer Journey

Yoshitaka Shibata and Takumi Matsuda

Abstract One of the challenges that large manufacturers are facing is to develop human resources that can bring about innovation by facilitating many employees to take a service design approach. Hitachi Design Division developed a template called "experience plot" for not just designers, but also non-designers who are involved in service design to create a customer experience story in a logical way. Experience plot has been used for customer co-creation projects, as well as for internal education programs for non-designers. The template is proven to be effective in building compelling stories from a user-centred perspective.

Keywords Customer journey • Storytelling • Service design

1 Objectives

1.1 Challenges in Manufacturing

The industrialization, as well as declining communications and logistics costs, of developing countries has given rise to surpluses. This means that even industrial products requiring advanced technology have become rapidly commoditized. In the manufacturing sector, it is important to not just create products with outstanding features and performance, but also to solve customer problems with those products (Masuda [1]). The sector must identify strategies to promote acceptance and long-term use of its products, and create service-industry style packages around them.

Recently, increased attention is being paid to the design-thinking approach, in order to promote innovation that is aimed at driving the change in service mentality in manufacturing. Design thinking can be described as a user-oriented, iterative, exploratory approach used when designers are involved in product development. A unique thing about this user-oriented thought process is how it thinks in terms of stories about the users' lifestyle or behaviour over a period of time. It focuses on the sequence of processes from the user first encountering the product, using it, and

Y. Shibata (✉) • T. Matsuda
Hitachi, Ltd, Akasaka Biz Tower, 3-1, Akasaka 5-chome, Minato-ku, Tokyo 107-6323, Japan
e-mail: yoshitaka.shibata.bd@hitachi.com

T. Maeno et al. (eds.), *Serviceology for Designing the Future*,
DOI 10.1007/978-4-431-55861-3_32

then seeing a change in his or her life, which allows us to consider the necessary strategies between the user and the product, that is, the strategies around the product mentioned above.

1.2 Distilling the Wisdom of Stakeholders

Another feature of the design-thinking approach is to drive the thinking process by involving a large number of stakeholders through means such as workshops. There are two reasons to take this type of approach. First, when taking a story-based approach to consider the necessary strategies between the user and the product, it is more effective to gather many stakeholders than to think about the product in isolation. This allows one to understand the situation, behaviors, and other issues under consideration, including user lifestyle scenarios, supply chain, and support mechanisms. Second, having stakeholders experience the new service themselves allows them to take ownership of the service, which becomes the driving force behind a project that requires considering a new service.

However, the number of people in manufacturing who are accustomed to creative work, such as building stories in groups, is limited. For the manufacturing sector to truly accomplish a transition to a service orientation, many employees must acquire the basic business skills needed for the creative work of building stories. That in turn, we believe, calls for tools for practicing how to develop stories.

1.3 Tools for Developing Stories and the Challenges Associated with Them

A good example of a tool for supporting the creative work of considering services is the business model canvas. This model goes beyond simply organizing the necessary elements when we consider a business model; it provides a shared framework, as its use becomes more widespread, for gathering people with a variety of knowledge and facilitating the integration of ideas through discussions.

The business model canvas is a scheme for sketching out the services needed to bring to reality the values customers demand. However, there is no tool that provides a shared language to many people for sketching out services as user experience stories embodying those values [2].

Service design tools include service blueprints and customer journey maps as means to capture and document stories. However, these are primarily designed to analyze the current reality. In addition, although they share high-level concepts, there is no agreed-upon approach in terms of specific documentation methodologies. Service designers have to start with the documentation methodology each time they create a customer journey map. We have storyboards and other ways to portray

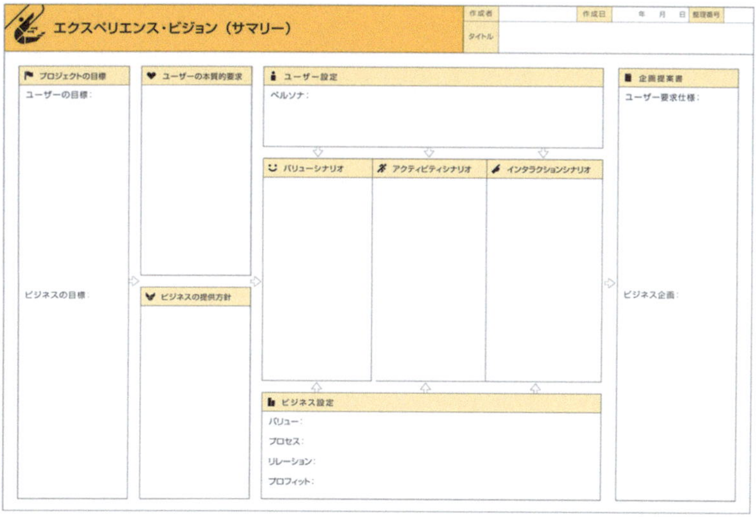

Fig. 1 Experience vision (summary)

how we think things should be, but even here there is no real standard way to represent these things [3, 4].

For story development to become a basic skill among people in various professions, and not just story designers, a tool that defines a standard notation and can serve as a shared language is needed.

Of course, there has already been research into tools for story development. Yamazaki et al's "Experience Vision" is one such frequently used tool [5] (Fig. 1). It consists of an overall picture of the items to be considered and a format for looking at each item. One of its features is the way it divides stories into "value scenarios," "activity scenarios," and "interaction scenarios," and documents each. However, when it comes to writing each scenario, the approach is to write text in big boxes, and in actual use, this freedom in documenting makes development more difficult. The many formats it involves are effective at the idea refinement stage, but it is not well suited to quick discussions about initial concepts among multiple people.

One tool that shows how to write stories is the Customer Journey Canvas attached to Marc Stickdorn et al's "This is Service Design Thinking [4]" (Fig. 2). One feature of this tool is its use of time phases labeled "Pre-Service Period," "Service Period," and "Post-Service Period," all of which provide a perspective on how to introduce the product or service being proposed and how to maintain the situation where it has found a following. Although it may seem like common sense, without this perspective, discussions of experience tend to gravitate toward the Service Period. This tool, however, provides no specific viewpoint on how the "Pre-Service Period," "Service Period," and "Post-Service Period" are to be examined.

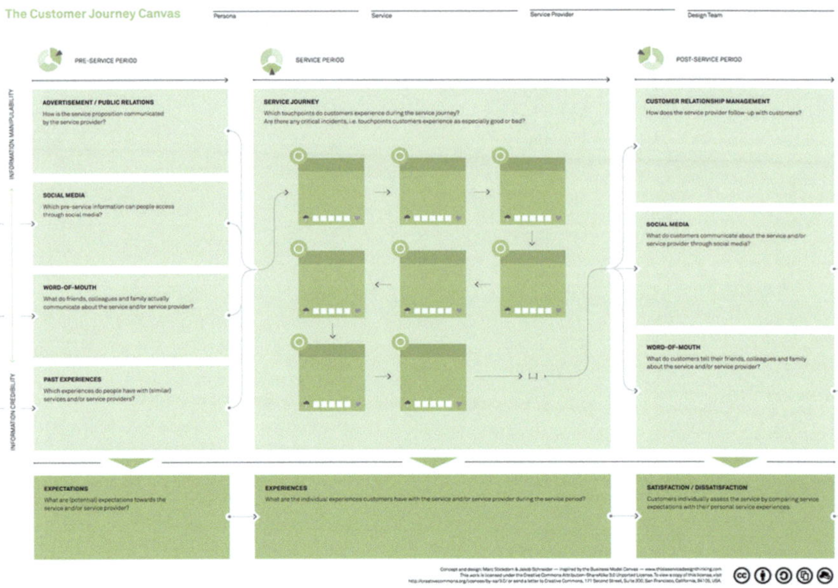

Fig. 2 Customer journey canvas

The tools we just described are good examples of providing overall frameworks for what must be considered when we develop stories. Designers comfortable with creative work will be able to produce good results leveraging the meta-framework and freedom provided by the tools. However, to bring the basic skill of developing stories to people not accustomed to creative work, it is necessary to show them how to put stories together concretely and in a streamlined fashion. This study aims to develop a template that makes it easier for more people to develop creative stories.

2 Topic and Objectives of the Study

The topic of this study is the means for creating service concepts in the initial stages of considering service projects, specifically tools for developing user experience stories.

We consider speed and teamwork requisites for the development of service concepts. Concept development does not call for refining ideas; it is important to perform the development rapidly and pursue a wide range of possibilities. It is also important, when we develop stories as a team, to share points of discussion and throw out and integrate constructive ideas. Unplanned divergence may at times yield wonderful ideas, but is inefficient and risky in a business scenario.

3 Development of the Tool

To address the above objectives and goals, we have developed a template for the ideation of service concepts from user experience stories. We call it the "experience plot." Below we outline its development process.

3.1 *Analysis of Past Projects*

Starting in 2006, Hitachi has, in concert with internal stakeholders and external clients and in the form of workshops, been holding discussions on how services can bring user experience stories to life and implement visualizations to assess such stories [6]. This approach has been applied to diverse fields ranging from finance and insurance to public services and industry, medicine, and transportation. We have analyzed such visualization exercises carried out in the past and identified the following three requirements for developing stories together with large groups of stakeholders.

(1) Standardized approach to documentation

In past visualization exercises, stories were often created using a prepared framework similar to a customer journey map, with time on the horizontal axis. Then, the items discussed in the design workshop would be written on slips of paper and pasted at the appropriate place. At first glance, the slips of paper pasted here and there gave rise to a feeling of real work being done, and exerted a unifying influence on the project members. Documenting detailed user behaviours over time made it possible to expose each and every user issue. Yet as all the things people in different positions noticed or thought of were included, the amount of information exploded. The more the details were captured, the harder it became to grasp what was being said. People who could not attend the workshop often found it difficult to decipher materials that had been a central focus for those who had been there. After the workshops, we actually looked at the approach to documenting the topics discussed, and prepared materials in the optimal format for each issue. We need to improve the clarity of the materials and make the process more efficient by finding a way to organize on the spot the opinions voiced during the workshops.

(2) How to capture experience

In past cases of visualization, as with the tools discussed earlier, there was considerable freedom in how to document. Creating new ways of documentation for each theme is an area where designers can exercise their creativity. However, frameworks that have no particular rules other than a time axis and a way to show relationships among stakeholders end up depending on intuition and experience to gain insights into feelings from users' product- and service-usage processes. There is an undeniable possibility that sometimes these insights can be superficial. It is easy to fall into debates trying to find an optimal isolated solution to easy-to-understand issues, making it harder to grasp the larger meaning from a single overall perspective, and to debate and generate ideas based on that perspective. This makes the job of the workshop facilitator that much harder. What is needed is to discuss the experienced value generated by the service as a whole? Rather than limiting discussions to the happiness created by individual strategies appearing in the service, the facilitator should make it easier to discuss the meaningful aspects of experienced value formation that is realized by the service overall.

(3) Ways to generate ideas about how things should be

Story visualization at Hitachi began with the objective of conceiving and expressing how things should be, with designers participating in the development of the service projects. As they began to be used more in creative collaborations with clients, we found more opportunities to analyze problems in current user experience, and worked to develop more sophisticated methodologies for that purpose. In contrast to identifying issues based on analysis of the current state of affairs and considering how to solve them, conceptualization of how things should be, independent of the current reality, is more difficult for those not accustomed to creative work. The debate over whether the user experience portrayed in a story is desirable can easily end up in a vague discussion based on the subjective views and impressions of those creating it and those reading it. Thus, we need to define approaches for discussing amorphous experienced value.

(1) With regard to organizing documentation, as with the business model canvas, identifying the points of discussion contributes to the solution. We construct templates where the points to be discussed in each box are defined. This makes it clear what the opinion written on each slip of paper is about. Listing up all the items to be considered also allows us to share an overall picture.
(2) For capturing experiences and generating ideas about how things should be, we adopt findings from scenario writing. Scenario writing is a methodology for studying the overall composition of a story, and is used when one creates movie scripts, for example. Creating user experience stories is different from writing catchy plots, but we can glean hints about how to make stories easier to understand, more empathetic, and more compelling. This methodology of considering the story from start to end provides a perspective that can improve the situation described earlier, in which the debate fails to extend to the Pre-Service Period and the Post-Service Period reached after the user has passed through a series of experiences.
(3) With regard to generating ideas about how things should be, the notion of Service Dominant Logic proves helpful. It is difficult to come up with ideas about specific solutions using only insights from scenario writing. We thus adopt Service Dominant Logic, which takes into account user resources and the networks users build. It also considers the role of the server provider, which leverages those resources and networks to create customer value. This is an

important perspective for considering services that will be easy for users to accept in their lives.

3.2 Application of Scenario Writing

Below we summarize concepts from scenario writing that will be of use in developing user experience stories.

(1) Framing the main topic

The main topic can be described as a message that is not made explicit in the story, but conveyed through it.

Momotaro is a Japanese nursery tale where a boy born from a peach (momo) leads a dog, monkey, and pheasant in defeating an ogre. The main topic of this tale can be said to be "rewarding good and punishing evil," but can be equally well portrayed as "the story of a young boy growing and learning the importance of friendship." The latter interpretation would give a different meaning to the interactions between the protagonist Momotaro and the dog, money, and pheasant, and their battle with the ogre. This is the effect of defining the main topic. When developing a user experience story to be realized by a service, failing to establish the main topic will result in a service that is a mere aggregation of local optimizations, solving problems one by one as they arise. The service might nevertheless be viable, but would lack the potential to innovatively create new experiences. We must learn, from the wisdom of scenario writing, what kind of value for the user lies in what we are trying to create through the service, define the main topic, and think through each individual strategy embodying that.

(2) Forms of the story

The customer journey canvas described earlier can be said to be based on the three-act structure basic to scenario writing. The first act is the Pre-Service Period (exposition), the second act is the Service Period (rising action), and the third act is the Post-Service Period (resolution). Although the second act is the most interesting part of the story, the first and third parts play an important role in organizing the story.

Having found common elements in the many myths that he had studied, Joseph Campbell demonstrated in the "Hero's Journey" that most stories can be said to share the same basic structures [7]. Movie stories, writing that is considered extremely sophisticated creative work, still fit into the same well-defined pattern. What we can glean from this is that it is not the framework of the story that is important, but rather the new ideas that are brought into the pre-defined framework. When developing user experience stories that a service is to bring to reality, thinking from scratch without using a framework means that it may take longer to construct the outline of the story, or result in necessary elements being left out.

What we can learn from the insights of scenario writing is that, in order to create a captivating story, we should first have a robust outline with a skeleton prepared in advance. Ideas generated for specific scenes can then be inserted into that skeleton.

(3) Coming up with ideas about difficulties

To generate empathy with a story, story writing introduces elements such as giving the protagonist a handicap or having him face some difficult situation [8]. This creates a richer story and facilitates emotional involvement. It is often desirable when thinking about a new service to adopt this type of thought process. If a user starts using a new service, that person has changed his or her behavior and habits. For the user to be receptive to the new service, there must naturally be some type of difficulty to overcome. When we develop a user experience story to be accomplished by a service, one might proceed on the assumption that the newly conceived idea will be adopted by the user, and the story ends up being centered around the ones proposing it. It is important to go so far as to think about whether the newly conceived idea might lead to some difficulty, and build a story in which that difficulty is overcome, thereby creating a realistic story centered on the user.

3.3 Applying Service Dominant Logic

Management thinkers Michel et al, in studying the fluidity of resources for value creation and the reorganization of integrated networks from a user perspective, propose three useful perspectives for encouraging service innovation, following the Service Dominant Logic framework [9]. In the context of developing stories, when thinking about the overall direction of the solution, we need to take advantage of these notions to generate new user-centric solutions.

(1) Bringing expertise and technology into products

Service Dominant Logic thinks of products as an intermediate means to provide the user with some service. This means that rather than the user using the expertise and technology of the entity providing the service directly, it is built into the product so that he or she can freely take advantage of that expertise and technology to create value for himself anywhere in time and space. With the advances in information technology, it is predicted that the fluidity of the value resources for users accomplishing value creation will only increase. How to provide value resources such as specialized expertise and technology to users equipped with those value resources, and obtaining ideas from the combination of those value resources, provides us with hints for conceptualizing new experiences.

(2) Changing role of resource integration

Between the entities that provide and that use a service, there is a change in the division of labor in resource integration for value creation. According to Christensen et al. (2004), a given act of value creation requires an equivalent resource integration task. Since the overall magnitude between the provider and customer remains the same, if the burden on the enterprise increases, then that on the customer decreases. If the burden on the enterprise becomes lighter, the role shifts to the client side to the same extent. This is the rule of preservation of resource integration [10]. For instance, IKEA is based on the idea that the user assembles the furniture, which massively changed the retail furniture model. In the same way, to create customer value from services, being conscious of what is integrated by whom using resources in what way allows us to dramatically change the traditional meaning of services.

(3) Reorganizing value networks

Michel et al propose the organization of networks of joint value creation connected by new relationships between the customer and an entity providing other diverse resources, going beyond the direct relationship between the service providing entity and the customer. They believe that the building of human networks is a major factor in internal customer motivation. This is an important notion when thinking about the necessary peripheral strategies for turning products into services.

The strategic perspective outlined by Michel et al is also useful for examining what type of entity offers what type of resources; who integrates those resources and to what degree; and how modifying, combining, and allocating those resources can create new value for the stakeholders, notably the customer. It must also be a story development technique for moving ahead with such considerations.

3.4 *Structure of Experience Plot*

Based on the above, we construct the experience plot, a template for developing user experience stories to be brought to life by services (Fig. 3). Below, we describe each constituent element.

(1) Service name

Enter the product or service being proposed. Make it a name explicitly describing what is being offered by the organization. Basically, after filling out the sheets, this will be entered as a word or phrase summarizing the overall concept.

(2) User

Since experience is something felt by a user, we consider here the changes in one user or one group of users. The user fields are constructed as follows.

(a) Basic attributes

Enter demographic attributes such as gender, age, and profession, as well as other attributes intimately connected to the story being portrayed. For example,

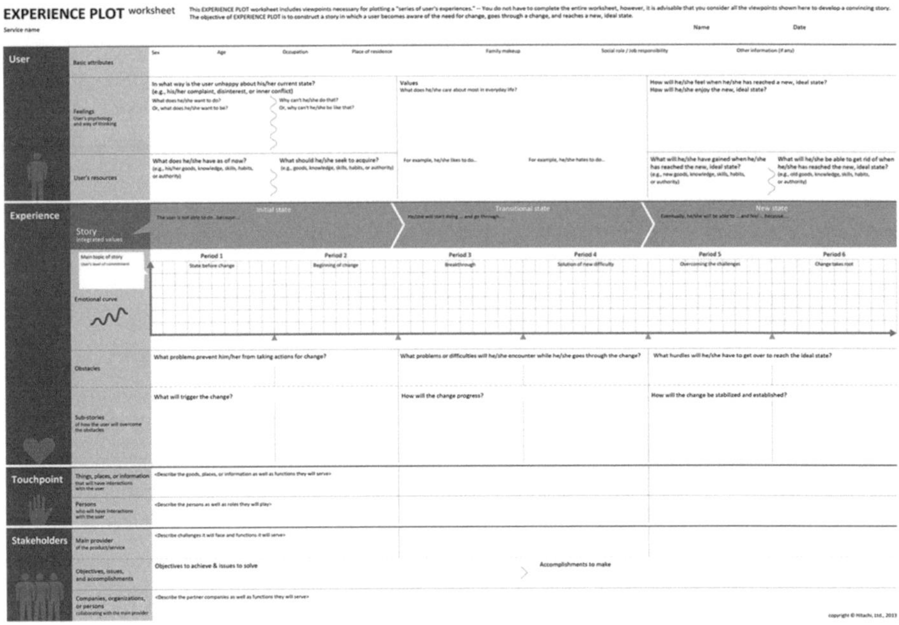

Fig. 3 Experience plot

when developing a diet service targeting a metabolic syndrome, enter a number for waist size.

(b) Feelings

Think about and write down what the user starts off wanting to do and why he cannot do it, or what goal the user will accomplish in terms of outcomes, by using the product or service being proposed. The key point here is not to just indicate the user's goal, but also to raise the question of why he cannot accomplish it now. This is based on the scenario-writing perspective of coming up with difficulties, but can also be considered as a way of facilitating deeper discussion from the user's perspective.

(c) Values

Think about and write down the ways of thinking that the user holds consistently and that form the foundation of his behavior. To encourage thinking about this, as you did with the "why the user cannot" item under "Feelings," we also provide the perspective for thinking about both positive and negative aspects of the user's concrete behavior and preferences, what he is actively undertaking and what he is consciously not doing.

(d) Resources

Using the notions of Service Dominant Logic, focus on the user's resources. Think about and write down the possessions, knowledge, technology, and work authority that the user might be able to exploit with the product or service being proposed, as well as the possessions, knowledge, technology, and work authority that he might like to have but is lacking. Also, think about what the user will ultimately gain through the product or service being proposed, and consider the things that will become unnecessary. This information will provide valuable input when thinking about the specifics of the product or service.

(3) EXPERIENCES

This section indicates the receptiveness of the user to the product or service being proposed, and forms the core of the experience plot. This makes the amorphous, hard-to-grasp notion of experience easier to deal with by placing it in the form of a story. The experience section is broken down as follows.

(a) Story (integrated values)

What you write here can be considered the objective of the experience plot. Following the three-act structure, construct the rough plot of the user becoming aware of the need for change (original state), experiencing change (how the change happens), and reaching the solution and growth (state of accomplishment). You need to document the summary of the proposed product or service within "how the change happens." The kind of experience the user obtains through the product or service being proposed can be understood by just reading this story.

(b) Emotional curve/main topic of story/form of story

Consider the most important user emotion that fulfills the main topic through the proposed product or service, and discuss and illustrate how it progresses over time. Illustrating the emotion makes it easier to discuss to what degree the product or service is required to work strongly on the user's emotions. As a guideline for drawing the emotion curve, we have established the following six periods, dividing each of the three acts further into two parts. These periods form the skeleton for coming up with an "interesting story." Based on these periods, you generate scenes appropriate to the theme in question. (Table 1)

(c) Substories about difficulties and overcoming them

Note what difficulties the user faces in each period, and ideas for overcoming them. This allows you to make clear the points to be considered later when thinking about the details of the product or service, and give the entire story a feeling of reality and empathy.

(4) Touchpoints

Think about items for each act that can actually be proposed for the product or service. Without considering details of the experience plot, write down just a high-level overview of how the experience described in the story can be created,

Table 1 Six periods

Period 1	State before change	Think about the initial state of the most important user emotion that will be affected by the product or service.
Period 2	Beginning of change	Think about the scene where the user takes the first step, becoming aware of his own difficulty and becoming receptive to change (beginning to use the product or service).
Period 3	Breakthrough	Think about what happens when the user changes his lifestyle and begins to use the product or service in earnest, and record the course of emotional change.
Period 4	Solution of new difficulty	Think about difficulties (such as boredom) that might occur when the user becomes accustomed to using the product or service, and record the course of emotional changes until they are overcome.
Period 5	Overcoming the challenges	Think about the problems that will occur regularly during the use of the product, issues that the user will factor in order for the product to take root in his life, such as difficulties in continuing, and situations where experience accumulated to date will be ruined. Write down the course of emotional change until these problems are overcome.
Period 6	Change takes root	Document the emotional state when the product or service has completely taken root in the user's life. There may be cases where the scenario is the product or service completing its role and ending.

describing what roles are required, when they should be played, and by whom. This section contains the following two items.

(a) Things/Places/Information

Think about and write down the form and timing of the points where the user touches the product or service. It may be a "thing," such as a smartphone or a mobile app; a "place," such as a website or the waiting room in a train station where the product or service is touched; or the existence of "information," such as geospatial information or information on train delays.

(b) Persons

Persons around the user, who are visible to the user and who are necessary when thinking about the story, such as a salesclerk in a store.

(5) Stakeholders

Think about and write down the organizational entity providing the product or service, and organizations or people playing a supplementary role. This section consists of the following subsections.

(a) Organizational entity providing the service, its challenges, and objectives

Write down the entity offering the product or service, such as an enterprise of a government office, and the point in time at which it makes its appearance and plays its role. Also, write down the objectives of that organization and the challenges it faces. What framing of issues led it to put out the product or service? What is its goal? This is a simple field to fill out, but without it, the experience story could end up being a fantasy lacking support from a service or product supplier.

(b) Partner organizations and people

Think about and describe organizations that supplement the product or service being provided by the principal organization, such as when record companies offer songs on Apple's iTunes music store. Also, put down here information about reviewers on social networks, such as users who write reviews about Amazon.com products, since they are crucial for forming an important touchpoint.

The structure given above promises to make it easier to elaborate, from an experiential perspective, on the jointly discussed ideas about basic architecture and stakeholder relationships and build an empathetic story.

4 Results

4.1 Application to Projects Carried Out in the Past

We printed out the experience plot on large sheets of paper and conducted simulation-style workshops on two service development projects carried out in the past within the Hitachi Group to see what the effect would be (Fig. 4).

One was a business process re-engineering project for retail staff for a consumer-oriented product. We established the consumer as the protagonist of the story and the building of relationships between the consumer and retail staff as the main topic. The largest difference compared to not using the experience plot was that we were able to look at the story where the service is first introduced, in periods 1 and 2, and the story in periods 5 and 6, where some crisis occurs before the service becomes established and is then overcome. This allowed us to come up with strategies we had not thought of before, and to build a proposal more robustly embodying the customer experience values to be created.

The second was a process re-engineering project for routine operations in the operations and maintenance field, which did not involve any end user. We made a veteran staff member the protagonist of the story, and his growth the main topic. Without using the experience plot, it was difficult to hold a discussion on experiential values in routine operations, but thinking about the items suggested by the experience plot allowed us to present a story, starting with the veteran staff member becoming receptive to a new style of work up until the point where the behavioral changes took root, and allowed us to demonstrate the requisite strategies for creating these experiential values.

Fig. 4 Workshop with the use of experience plot

4.2 Application to Educational Programs

We applied the experience plot to the new business development training within the Hitachi Group. Our intent in doing so was to have technologists whose perspectives tend to have a technical bias, experience a design-based approach to coming up with user experience stories, and thereby gain ideas about creating new businesses. However, it is difficult for trainees not comfortable with such an approach to get their heads around the viewpoints to be debated in coming up with experience stories, and the thinking process thus takes more time.

Applying the experience plot allowed us to complete the program for obtaining ideas for creating new businesses generated from user experience stories in one half day instead of the pervious one day. This reduction in time is not just a matter of efficiency, but also leads directly to increasing the number of chances to perform this kind of thinking in the series of processes for new business development, which is highly significant.

Digging into the discussions more deeply, we found that the groups of trainees came up with stories from the perspective of the experience of the envisioned user and gained service ideas, meeting the original objective.

The study results were also used effectively as a story telling tool during presentations. Starting at the top and explaining the experience plot in order allowed us to explain the overall framework by using the three-act structure, showing what kind of problem is faced by a user, what kind of service he or she uses to overcome the problem, and what kind of ideal state is reached as a result. In tandem with that flow, specific issues and solutions are presented in each scene.

At present, we are starting to go beyond the validation of the experience plot on past projects and in the training program by rolling it out in actual client work.

5 Considerations

In this section, we summarize the benefits of the experience plot.

(1) Planned process of thinking about how things should be

Providing the items to be considered in advance makes it possible to carry forward the discussion on each item without having to worry about what should be discussed. Being able to see the total amount of thinking to be done in advance makes it easier to estimate the required time. Compared to approaches that allow more freedom in terms of how ideas are expressed and how much can be expressed, the experience plot makes it easier to establish goals for the thinking process as participants simply fill in fields in order to tell a story.

(2) Building compelling stories

The guidance provided through the items and questions presented enables us to construct service usage experience stories leading to desirable ideals, founded on insights into the needs and issues of the target user. The three-act structure and six periods allow us to capture the relationship between the user and the service broadly along the time dimension, from the point in time when the user first becomes aware of the need to use the service through to the scene where the service takes firm root in the user's life. We are able to obtain compelling stories rather than mere bullet lists of convenient assumptions, by making sure to generate ideas about the difficulties that are likely to come up in any given scene, and coming up with ideas for overcoming those difficulties.

(3) Ease of discussing experiences

Discussing user experience, which was traditionally difficult, becomes easier when you clearly establish the main topic when conceptualizing stories, based on a designated target user. User experience, which used to be abstract, can be made visible and shareable by imagining it as an emotion curve, drawn against a vertical axis representing progress toward an ideal state for the experience value, which is the main topic, and a horizontal access representing time.

6 Conclusion

The experience plot has been proven to yield good results, but we have also identified the following issues.

The first is how to speed up the process of getting participants in workshops to understand the templates. Since it will be the first time participants in workshops will see the experience plot templates, it is important to devote time at the beginning of the workshops to explain the experience plot itself. There is a trade-off between having more items on the template to encourage the ideation process on part of the participants and provide more specific perspectives, and the time required to explain the items to newcomers.

The second issue is related to techniques for making the discussions more visual. In order to make group discussions easier, we are currently basing the process on making participants write on slips of paper. However, visual elements can make it easier to see what is going on. Just encouraging people to draw pictures during sessions is somewhat effective, but it would be desirable to have some systematic technique for making the process more visual.

References

1. Masuda T (2011) Susumu Seizougyou no Sabisu-ka “Ima Nani ga Okotte Irunoka” (Servitization of manufacturing “What is happening now?”). TBR Sangyou Keizai no Ronten (Toray corporate busi-ness research issues of industrial economy) 11–13:2–3
2. Osterwalder A, Pigneur Y (2012) Business model generation – a handbook for visonaries, game changers, and challengers. Shoeisha Co.,Ltd, Tokyo
3. Martin B, Hanington B (2013) Research & design method index: 100 ways to research comples problems, develop innovatice ideas, and design effective solutions. Bug News Network, Tokyo, pp 196–197
4. Stickdorn M, Schneider J This is service design thinking. Basics – tools – cases. Bug News Network, Tokyo, pp 158–161, 186–189, 204–207 and 373
5. Yamazaki K, Ueda Y, Go K, Takahashi K, Hayakawa S, Yanagida H (2012) EXPERIENCE VISION – User wo Mitsumete Ureshii Taiken wo Kikaku suru Vision Teian-gata Design Shuhou- (EXPERIENCE VISION – vision proposal design method, plannning a joyful experience for users -). MARUZEN Publishing Co., Ltd, Tokyo
6. Kitagawa H, Takada M, Mibe R, Ishikawa T (2013) Development of experience-oriented approach to information and telecommunication systems business. Hitachi Rev 62(6):44–51
7. Voglar C (2002) Shinwa no Housoku (The writers journey: mythic structure for writers). Story Arts and Sciences Laboratory Inc, Tokyo
8. Nakamura A (2014) Ayaemo research institute. http://ayaemo.skr.jp/story/plotbook_formula03_3act.html. Accessed Apr 28 2014
9. Michel S, Brown SW, Gallen AS (2008) An expanded and strategic view of discontinuous innovations: deploying a service dominant logic. J Acad Mark Sci 36:54–66
10. Christensen CM, Anthony SD, Roth EA (2004) Seeing what’s next, using the theories of innovation to predict industry change. Harvard Business School Press, Boston

Kizashi Method – Grasping the Change of Future User's Values

Takuya Akashi and Yukinobu Maruyama

Abstract Hitachi Design Division has developed Kizashi method for grasping the change of future user's values to design products and service from a user-centered perspective. Through the process, Hitachi identified 25 Future Signs for 2025. Combining the discovery of Kizashi with the creation of ideal future scenarios attracted many external partners, and encouraged vision-driven collaboration, such as in the fields of healthcare and mobility. For the future, not just the national government, but also private companies like Hitachi are expected to design visions for an ideal society, and realize the visions with the local governments and business partners.

Keywords Future • User's values • Vision

1 Significance of Foresight in the Future Social System

Complex social issues, such as low birth rate, aging population, and environmental sustainability, necessitate a futuristic view of social infrastructure.

There is growing demand for updating and reforming social systems in both developed and developing countries. However, realizing a new social system requires wide-ranging technology, finance, and technical expertise. Public-private cooperation, such as the PPP (Public-Private Partnership), has previously been explored for large-scale public works projects that were initiated by government agencies. In future, it will become increasingly vital to have a social framework, in which public entities and private companies that provide the services for an ideal life style contribute to the creation of a humanistic social infrastructure through the design of "visions."

Studies and research on issues, such as the impact social changes will have on people in future or the type of change in values and the issues that people will face in future, are critical when considering the creation of such social infrastructure.

T. Akashi (✉) • Y. Maruyama
Hitachi, Ltd, Akasaka Biz Tower, 3-1, Akasaka 5-chome, Minato-ku, Tokyo 107-6323, Japan
e-mail: takuya.akashi.ts@hitachi.com

T. Maeno et al. (eds.), *Serviceology for Designing the Future*,
DOI 10.1007/978-4-431-55861-3_33

The future is very hard to predict, and trends do not move smoothly among the past, present, and future. Rather, they are a reaction to changes caused by unforeseen factors; thus, proactive efforts are required to direct change towards what is considered to be an ideal future form.

In our research, we created a model for a common human value system for the future, and applied it to develop the Kizashi method, which can be used by many businesses for the creation of visions. This paper summarizes the process of the Kizashi method, applicable examples, results, and implications for the future.

2 Kizashi Method

2.1 Comparison with Previous Forecasting Methods

This paper describes the features of the Kizashi method in comparison with three forecasting methods: the "Delphi method," "scanning," and "scenario planning."

The Delphi method poses questions to experts and uses their opinions to elicit intuitive conclusions. The opinions are collected through interviews and compiled statistically to predict technological innovation and social change on targeted themes. Previous research [1] has focused on the debate regarding the considerable dependence on the cultural knowledge of the experts when using the Delphi method. Moreover, the research has noted the high levels of specialization and the dilemma of like-mindedness in their overarching perspective of various issues. Furthermore, the Delphi method tends to result in predictions that current trends will continue since future predictions are based on accumulated knowledge.

"Scanning" is a method of creating an overarching perspective to derive multiple discontinuous future scenarios [2]. According to previous research [1], "scanning" forcibly generates ideas that clash with certain themes by gathering factors of social change, which exist outside of these themes. The focus is on seeking the possibilities caused by impact dynamics.

"Scenario planning" is the same up to the point of compilation of the factors that cause social change. Subsequently, it selects the factors that are important, are highly likely to cause change, or are highly unpredictable. It then looks at combinations of different possible future scenarios to derive multiple scenarios afterwards. Therefore, by clarifying the extent of the future change, the focus is on having a comprehensive way to deal with unpredictability [3].

The Kizashi method is not a forecasting method that assumes continuity among the past, present and future, as typically represented by the Delphi method. It is more of a forecasting method similar to scanning because by detecting unforeseen factors, it seeks to discover what will be an ideal future form. The Kizashi method is the same as "scanning" and "scenario planning" in terms of compiling the factors that cause social change, and it is the same as "scanning" in terms of the final result of multiple discontinuous future scenarios. However, a major feature of the Kizashi method is its intermediate output of a description of the future change in users' values. This method does not entail only one future scenario. It is rather a tool to be applied flexibly and dynamically multiple times for multiple themes for creating multiple future scenarios.

2.2 Characteristics of Kizashi

A feature of the Kizashi method is its insight into changes in users' values and the use of this insight as an output. The results obtained by this method are referred to as Kizashi (future signs). The essential characteristics of Kizashi are as follows:

- Kizashi must indicate the constantly changing underlying trends in various areas of society to be both flexible and dynamic in relation to a particular field.
- Since Kizashi is a tool for creating multiple future scenarios, it needs to be abstract and far reaching.
- Although Kizashi encompasses a variety of aspects, attention should also be given to the finer details of description and uniformity of expression in order to describe users' values.

3 Steps in the Kizashi Method

3.1 Steps in the Kizashi Method

The Kizashi method can be divided into two major processes. The first process considers the impact on people from the factors causing social change that are compiled using P-E-S-T analysis. P-E-S-T is a method for understanding the external environment from the political (P), economic (E), social (S), and technological (T) perspectives. The second process involves an extensive research of the impact on people by investigating personal viewpoints.

A further analysis of the two processes is described in the five steps.

3.1.1 Step 1: Defining the Research Themes

The research themes are defined at the initial stage of the Kizashi method. They are defined before the P-E-S-T research to identify the target field and social issues that need to be covered in the research. It is counterproductive to narrow the field when we are looking for perspectives. Therefore, some ingenuity is required to ensure there is awareness that these research themes are merely guidelines for conducting a P-E-S-T research. When there is not enough prior knowledge about the target field for the Kizashi method or when the field is too broad, the research can be improved by sharing macro perspectives that need to be taken into consideration.

The following two points should be addressed when it comes to proposing the research themes.

(1) The implementation of corporate strategy in relation to the research theme

From the standpoint of the company's business strategy, candidates for the target fields and social issues should be determined.

(2) Reference to existing knowledge concerning the research theme

The scope of the targeted theme can be defined by citing documents and interviews with experts. For example, when the targeted field for the Kizashi method is "finance," there needs to be an understanding of the extent of what is meant by "finance" and what are the current issues involved in "finance." Thus, the scope can be determined by citing the table of contents of multiple documents. At this point, it is critical to list documents written from as many different perspectives as possible.

3.1.2 Step 2: P-E-S-T Workshop

P-E-S-T analysis is often used with SWOT (strengths, weaknesses, opportunities, and threats) analysis. P-E-S-T analysis method is reported to be very effective when taking an objective and comprehensive approach for considering the factors that have a major influence on global trends when proposing the opportunities and threats in a SWOT analysis. P-E-S-T analysis method classifies external conditions into four areas: Political (P), Economic (E), Social (S), and Technological (T). Factors are listed with relevance to these four areas, which can increase the comprehensive nature of the analysis.

The aforementioned qualities indicate that a P-E-S-T analysis itself does not provide any clear solutions. Under the P-E-S-T method, news articles, white papers, statistical data, and other sources are studied to detect factors that could cause social change and are likely to have relevance to the established research themes. Each factor is recorded on a separate data card, which is referred to as a P-E-S-T card.

A debate is conducted at a P-E-S-T workshop, in which the P-E-S-T cards that have been created by many research participants are brought together. The most basic material used at the workshop is the enormous time line to which the P-E-S-T cards are affixed (Fig. 1). The four categories, i.e., P, E, S, and T, are arranged vertically with the P-E-S-T cards arranged horizontally on the time line. This arrangement is created to avoid creating categories at the initial stage by, for example, putting all the energy-related cards together because the nuances of the change factors and their impact on society vary along the time line.

When creating P-E-S-T cards, the content must not be biased and lack objectivity. This study uses a format as shown in Fig. 2. The format includes the title, P-E-S-T category, period of occurrence, details, information source, and name of the

Fig. 1 P-E-S-T cards

person who completed the card. It is desirable that participants in the P-E-S-T workshop come from a variety of backgrounds because this will help diversify differences in participants' views and opinions even when the research is focusing on only one item. During the workshop, only facts are recorded as data, and the research participant's own view is incorporated in the card's title. Attaching a title that indicates future direction, such as "what will increase or decrease," rather than data labels, means it is extremely important for research participants to express their views.

3.1.3 Step 3: The Kizashi Concept

The initial concept for Kizashi is created by using the card arrangement. In this procedure, we assign key words to the characteristic cards and groups of cards, but there are two ways to proceed.

(1) Detect particular factors of change

Select the most inspiring cards and share the reason for selection with participants by discussing the factors that are characteristic of change.

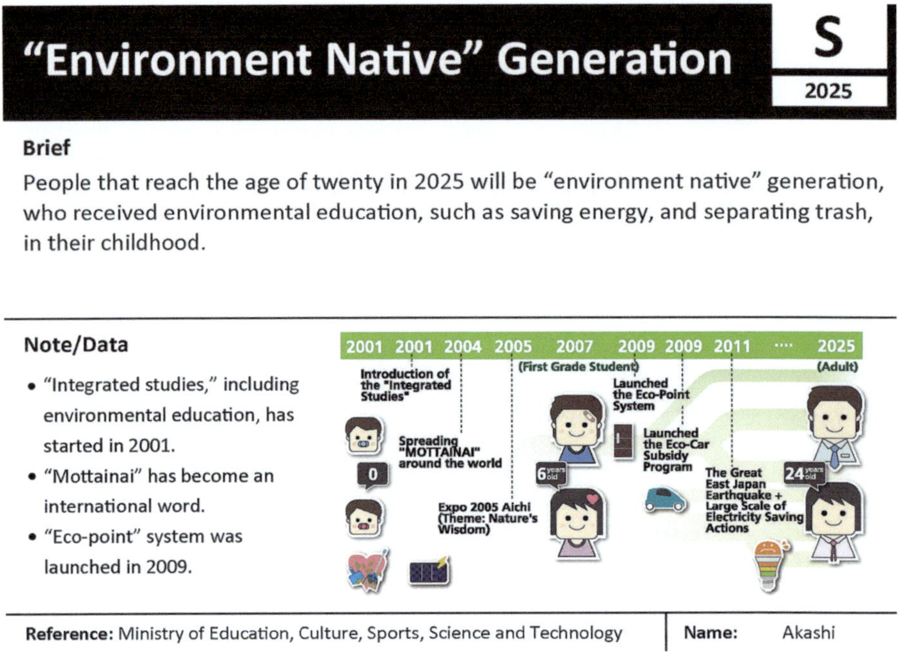

Fig. 2 Example of P-E-S-T card

(2) Detect relationships between factors of change (cards)

Look for which cards result in a similar impact and note their relationship. This refers to two types of relationships. The first is the realization that the relationships straddle P-E-S-T fields. For example, in the case of “a decision being made on XX system 3 years hence (P) with the emergence of various organizations such as the YY non-governmental organization 5 years hence (S),” there is also a relationship along the time line. The second is the equality of the benefit level for ordinary citizens. For example, the card for rubbish collection vehicle is actually related to the healthcare of individuals, whereas growth in the online home-delivery market relates to an increase in aged householders living alone. Note that these are just a few of the many possible examples. The task of detecting relationships is best performed by members with widely varying backgrounds to ensure that discoveries can be made by a number of people looking through different eyes and coming up with different interpretations (Fig. 3).

The initial concept for Kizashi is then finalized. Using this format creates an awareness of the causal relationships behind future social change and their impact on users’ values (Fig. 4).

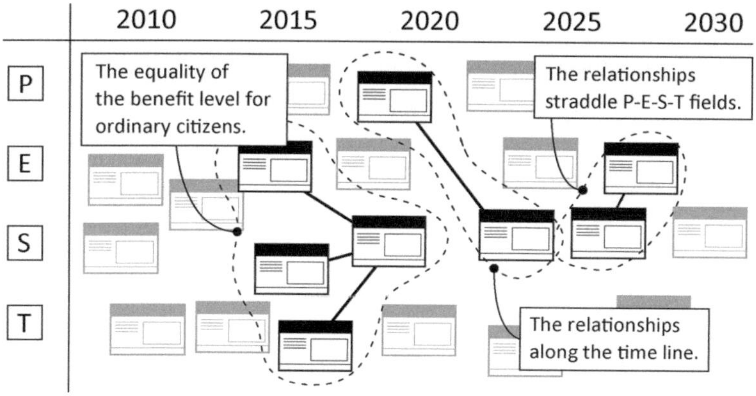

Fig. 3 Identify relationships between factors of change

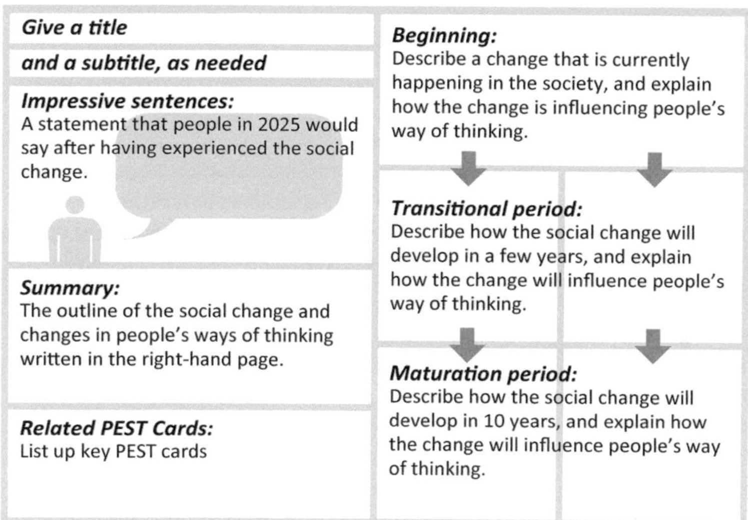

Fig. 4 Format for describing Kizashi

3.1.4 Step 4: Field Research and Expert Interviews

Confirm the initial concept and identify areas for improvement. This step confirms the appropriateness of the initial concept by filling any knowledge deficiencies to improve accuracy. This step consists of the following three parts.

(1) Field research

Confirm the accuracy of Kizashi through data gathering, desktop researches, and telephone interviews at the location, where the concept is actually occurring.

(2) Expert interviews

Listen to specialists with similar opinions to confirm Kizashi's accuracy.

(3) Compile P-E-S-T cards for areas that are deficient

It is advisable to add P-E-S-T cards for any further findings such that the P-E-S-T cards can be re-used after the project and are not limited only for that particular project.

3.1.5 Step 5: Kizashi Development

Brush up on the initial concept. The excesses and deficiencies of the Kizashi concept are discussed based on the knowledge obtained through research in the previous stage, and revised to a final number. Rather than elaborating on the final wording at this point, the focus of discussion is on a key word level. The next phase is to add forecasts using the Kizashi concept.

(1) Envisioning the change in users values

Gain insight into the forcible generation of ideas that match similar concepts and the impact that such concepts will have on future life styles. The forcible generation of ideas is a method for detecting new types of ideas by forcibly adding to and changing the thinking of particular products or ideas. There are various methods for this, a famous one being Alex Osborn's checklist.

In this study, we use the forcible generation of ideas to organize the Kizashi identified in the previous section. The Kizashi that were identified are presented on the vertical and horizontal axes to express new scenarios when combining two Kizashi using a round-robin type method (hereinafter referred to as the "hypothetical scenario"). This concept of the hypothetical scenario is the original idea behind using the forcible generation of ideas method; however, in our study, the forcible generation of ideas is not directly linked to this hypothetical scenario or to the business opportunity scenario. Rather, it is used to examine the details (such as genre, level of feasibility, persuasive power) of the hypothetical scenario to sort out Kizashi. In other words, the forcible generation of ideas provides feedback to Kizashi.

Specifically, we used the forcible generation of ideas to enumerate all the possible combinations of Kizashi at the initial stage (Fig. 5).

(2) Envisioning the realm of opportunities

Envision the realm of opportunities for solutions when this concept is used. As in (1) above, the results are not used for ideas and solutions, but rather to identify the elements that should be incorporated into Kizashi, which are added at the key word level.

Finally, the aforementioned details were summarized in written and visual form in a booklet, which is outlined below in Sect. 3.2.

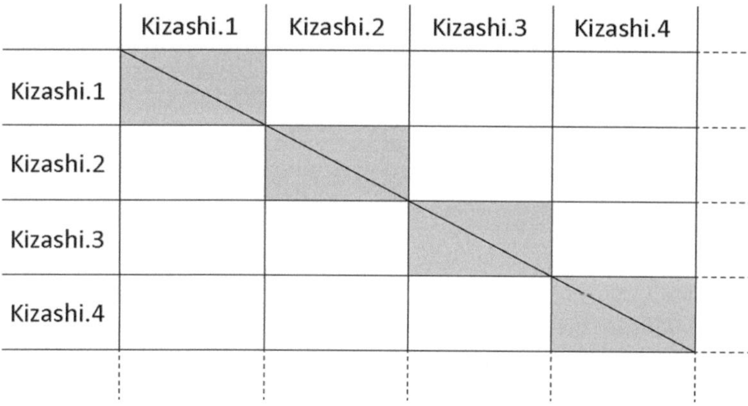

Fig. 5 Matrix for ideation

3.2 25 Kizashi for Japan in 2025

Twenty-five Kizashi were compiled relating to the future trends of Japan in 2025 with booklets created in Japanese, English, and Chinese. In addition, the corporate perspective of the next society has been conveyed. In conjunction with local governments and industry partners, we have created a picture of what the future should look like. This information has been distributed and made widely available in both Japanese and English on the Internet [4].

As noted earlier, the Kizashi method is characterized by the compilation and output of the expected change in users' values. Kizashi comprises the titles, subtitles, phrases that indicate users' values; charts that present key statistical data; and explanations. However, the phrases that indicate users' values are the most distinguishing feature. Let us consider the example of "Beyond Green."

Because of governmental measures to reduce the cost of solar power generation and promote next-generation automobiles, next-generation energy sources will permeate people's everyday lives through field tests and institutional reforms. Furthermore, consumer behavior is changing with people starting to choose life styles with a lower environmental impact according to various values such as convenience and personal principles. These values will become increasingly diverse as a new generation reaches adulthood. People who are approaching adulthood in 2025 will be called the "Environment Native" generation, who have experienced large-scale electricity conservation during their formative years and received environmental education such as the separation of rubbish; thus, they will be well informed about reducing their environmental impact. This generation will have the capacity to comprehend environmental policies and demand justice and transparency in environmental policies at the same time as the diversification of values in relation to the environment increases.

The image of the value in this period can be expressed in a phrase, as shown in the black rectangle in Fig. 6: "We are already aware that there is no justice in

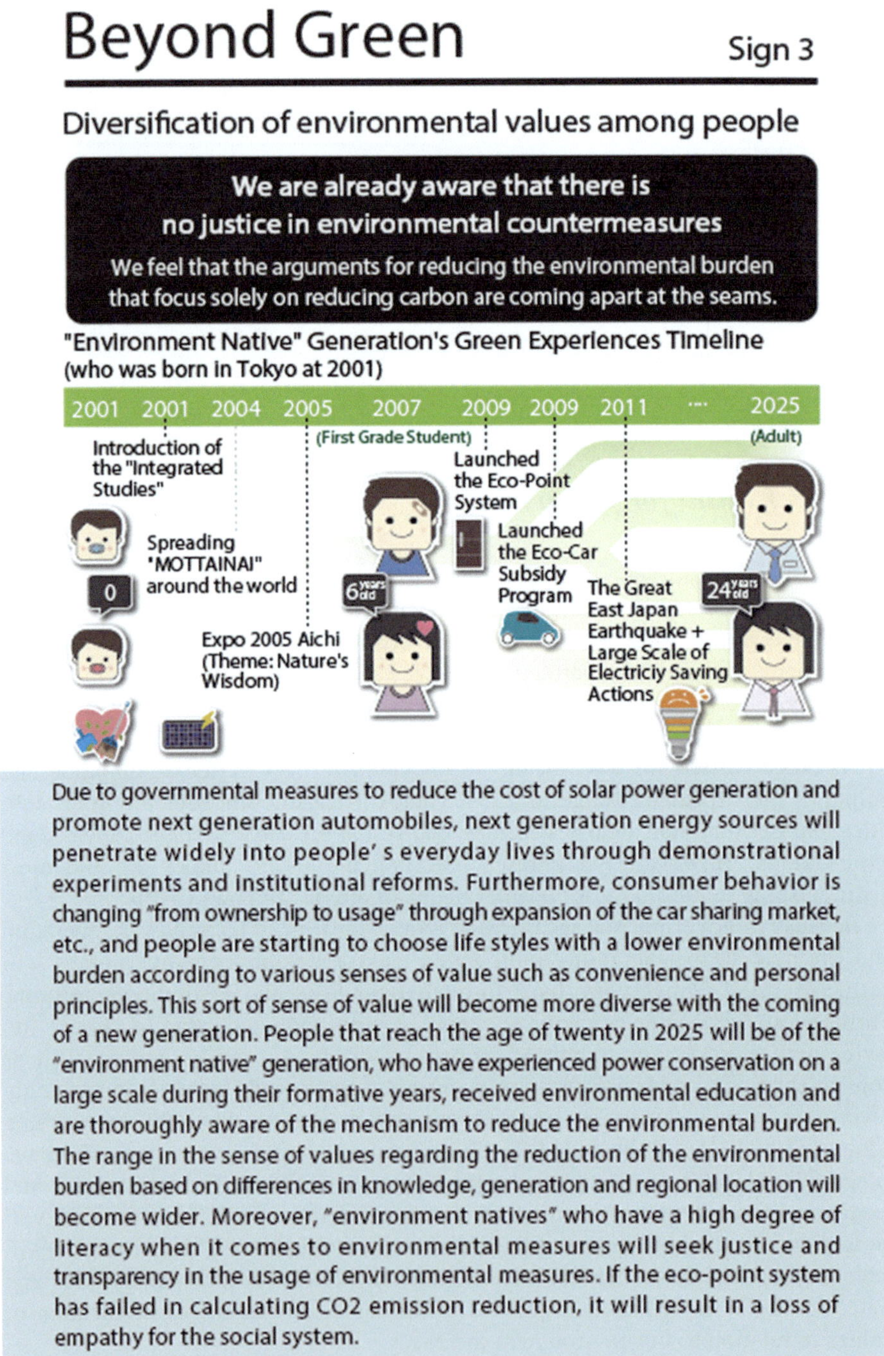

Fig. 6 Sign 3 "Beyond Green"

Sign.1	Sign.2	Sign.3	Sign.4	Sign.5
Civic Mining	Mottainai Space	Beyond Green	Micro Eco	The Slower, The Smarter
Rediscovery and utilization technology for urban resources	Surplus space in the city	Diversification of environmental values among people	Improvement of eco-decomposition capability	The needs for both city and nature
Sign.6	Sign.7	Sign.8	Sign.9	Sign.10
Ownership	Civic Price	Emerging Community	DIY Society	Reshaped Infrastructure
From ownership to usership	Commoditization of cities	New relationships with developing countries	DIY oriented society	Reviewing the aging infrastructure
Sign.11	Sign.12	Sign.13	Sign.14	Sign.15
Sense of Safety	Borderless Medical	The Right to Mobility	Power of a Name	Civic Literacy
Hyperawareness of perceived public security/observed public security	Eliminating the borders for medical services	Secure and convenient new mobility	Change in value by naming	The need for social literacy technology
Sign.16	Sign.17	Sign.18	Sign.19	Sign.20
Community Education	Post Price	Hospitality Crisis	Overdiligent	NEET to go
Promotion of education to improve regional power	Diversification of remunerative value	Limitation of services	Necessity for comprehensive worker care	Transition of NEET population
Sign.21	Sign.22	Sign.23	Sign.24	Sign.25
Global Empowerment	Singleship	Wishloops	Model of Family	Missing Reality
Necessity for international human resources	Single person safety net	Diversification of last wishes	Diversification of the concept of marriage/concept of family	Ambiguity in boundaries of awareness in communication

Fig. 7 25 Future Signs for 2025

environmental countermeasures." A list of the 25 signs that include other Kizashi is shown in Fig. 7.

In this way, the details of the Kizashi have stopped with the forecast of change in users' values. In particular, discussing the 25 items enables a shared understanding of the problems in the arguments and the application of various themes such as mobility and healthcare. Note that this not only a future vision but also is a tool that can be applied flexibly and dynamically multiple times in relation to multiple themes to create multiple future scenarios.

4 Application of the Kizashi Method

4.1 Using Kizashi to Review Research and Business Strategies

Discovering the business opportunities that use the "Kizashi" gained from a qualitative insight into the future change in users' values is a feature that can be achieved within a range of interpretation from limited examples and expressed in a few sentences. Business opportunities can be investigated through a hypothetical discussion of subject matter (not directly indicated by "Kizashi") by applying the details of "Kizashi" to the technologies and areas of interest of Hitachi's business divisions and its customers. Here, we consider the application of "Kizashi" in cases where research institutions and business divisions have already determined their own research and business strategies. Research strategies and business strategies are usually aimed at a long-term sustainability of existing businesses with strategies managed in accordance with a road map. On the other hand, when designing a strategy, the focus is on competitive analysis and analysis of the internal environment (a company's own resources). The market expects a company's narrative to be one of continuous successes; thus, there tends to be an optimistic view of changes in external conditions. In contrast, "Kizashi" identifies the issues that future people, as generators of change, are likely to experience in regard to long-term technologies and business plans anticipated under the road map, which makes it possible to enhance existing research and business strategies.

4.2 *Creating a New Vision Using Kizashi*

The significance of using "Kizashi" to create a new vision is that "Kizashi" acts as a focal point for considering particular medium- to long-term issues in regional societies and in the business arena when determining the vision. Those involved in development can more easily understand problems that are more closely aligned with their own life styles if they review the issues from the user's perspective. This accelerates the period from the concept to agreement on proposals for systematic solutions that resolve specific problems and service ideas that provide new benefits.

Two examples are presented below. The first example is "Tsukuba City Kizashi," which is an example of a future vision for regional society. The local government created this from input received during the basic planning process, and it was jointly created by members of the local government and Hitachi Ltd. The second example is the "healthcare vision," which is an example of a future vision for a field of business. This vision was independently created by Hitachi to consider the means of healthcare for a very aged society. It involved plans presented to medical administrators in the UK, in which a database of public information on diagnostic data already exists. This led to the execution of a contract for a joint empirical project. This shows how long-term relationships can be built with development partners to resolve problems by sharing visions based on future changes in users' values.

4.3 Sales Offices Can Promote the Value of Proposed Products Using Kizashi

Locations where sales are conducted usually promote the benefits of purchasing a particular product to address the needs of the customer; however, the benefits of such products anticipated at the time of purchase have not been realized in recent years because of the extreme changes in society, which have led to unintended inconvenience for customers. In this regard, proposals with little medium-term risk can be developed by analyzing the trends in customer sales based on the business specialty. However, it is very hard to investigate such a phenomenon based on an understanding of the reasons why a product sells. Enumerating the future social issues for the field in which product is sold, in accordance with "Kizashi," fosters an opportunity for a company to proactively create a story that asserts the long-term benefits of purchase to the client.

5 Conclusion and Further Discussion

In this paper we have described the features of the Kizashi method as an intermediate output for compiling analysis of users' values, and compared it to previous forecasting methods. We have also examined the five-step process on the method and its applications. Kizashi method can be applied flexibly and dynamically multiple times in relation to multiple themes to create multiple future scenarios.

5.1 Benefits of Analyzing the Change of User Values

Kizashi method collects a number of factors from various aspects of society, and integrates them into abstract concepts from a perspective of users. This method allows widely varying interpretations among readers, therefore encourages a lot of unique hypothetical scenarios to be developed.

Moreover, a feature of the Kizashi method is its insight into changes in users' values and the use of this insight as an output. Kizashi shows both original insight and its output, so readers can feel connected to either the output (i.e. visions that we designed) or the insight (i.e. user values on which the vision is based). By sharing the process of analysis on user value changes, we have successfully encouraged customers to build strong and long-lasting relationships with us.

5.2 *Improving the Handling of Kizashi Method*

On the other hand, some people may feel it difficult to understand the causal structure of Kizashi, because it is described in a narrative way. For discovering new Kizashi or applying Kizashi that fits a particular situation more efficiently, the handling of Kizashi should be improved so that workshop facilitators and participants use the Kizashi method and contents more effectively. Hitachi Design Division intends to improve the handling of Kizashi to increase more Kizashi user experts, as well as to develop a method for describing the structure of Kizashi contents in a logical manner.

References

1. Washida Y, Mitsuishi S, Horii H (2009) A future scenario generation experiment in socio-technological problems using scanning method. Sociotechnica 6:1–15 (in Japanese)
2. Strategic Business Insights (SBI) http://www.strategicbusinessinsights.com/scan/
3. Schwartz P (1991) The art of the long view: planning for the future in an uncertain world. Currency-Doubleday, New York
4. 25 Future Signs for 2025. http://www.hitachi.com/rd/design/25future/index.html. [Beyond Green] of Sign 3 is a registered trademark of UGL Services Pty Ltd

Service Practices as Organizational Phenomena

Nozomi Ikeya

Abstract Understanding services as organizational phenomena is explored from the actors' point of view, following ethnomethodology's program. Services are conceptualized as organizational constructs and examined how these are experienced by members engaged in practical actions from within services. Using the case of business service in Japan's public libraries, the study's findings demonstrate how a service can be presented as an organized solution by an organization. The study also shows how actors involved in service practices can experience these as constituting different kinds of environments.

Keywords Service innovation and design • Public service • Service practices • Public library • Ethnomethodology

1 Introduction

1.1 Frameworks for Understanding Services

Creating a new service involves not just coming up with great ideas but also interweaving ideas with existing values and services. A new service can be seen as a solution provided by an organization to achieve its objectives at a point in time. It may be the result of a series of developments over time in the organization, and related developments may be on-going [1].

It has been some time since the need for a different framework for understanding services was first examined. For example, Vargo et al. proposed a shift from product-dominant to service-dominant logic [2]. Another framework presented by Normann and Ramirez was a shift from "value chain" to "value constellation" [3].

> The offering is the physical and 'in-person(s)' embodiment of assets made up of knowledge and experience, in themselves the result of myriad activities performed by many people dispersed in time and space [3:49–50].

N. Ikeya (✉)
Department of Humanities and Social Sciences, Keio University, 2-15-45 Mita, Minatoku, Tokyo 108-8345, Japan
e-mail: nozomi.ikeya@keio.jp

T. Maeno et al. (eds.), *Serviceology for Designing the Future*,
DOI 10.1007/978-4-431-55861-3_34

The offering, as Normann and Ramirez explain, is the result of activities carried out by different stakeholders and will produce value when it is put into action. As their explanation implies, what is difficult about understanding a service is that it is usually not designed in a vacuum but rather in an organizational context. This means that, first, the design is not completed at once. Instead, service ideas may be developed over time by different people, interacting with each other in a variety of circumstances. Thus, the offering is "the result of myriad activities performed by many people dispersed in time and space." Given this plurality of people involved in creating value—interacting with each other in different spaces and times—these authors propose the notion of "value constellation" to replace "value chain" because, in the latter, value is added in a linear way.

Concerning the plurality of actors involved in a service's development, researchers who propose to study strategy from a practice-based approach also point out a similar issue [4]. Johnson et al. argue that strategy development is not so much dependent on individuals or even small groups but on people at different levels in the organization, as well as people from outside the organization (e.g., consultants and bankers). Since service design can be regarded as a kind of strategy development, their approach can be applied to research on services.

What Johnson et al. also point out, similarly to Normann and Ramirez's findings, is that strategy practices include a wide range of activities, including strategy planning, development, and enactment [4]. Based on this framework, they argue that research foci need to be varied.

In addition, these authors clearly recognize that strategy practices are grounded in organizations. For example, they argue that "strategy practices such as strategic planning, strategy workshops or consultancy practices need to be understood as institutional phenomena that influence what organizational actors do and in turn how strategies develop in organizations" [4:13]. How individuals behave in organizations can be, therefore, another area on which to focus. These researchers also share their interest in "understanding what people do both within and as an influence on institutional and organizational contexts" [4:13].

1.2 Services as Practical Constructs

The stance taken in this paper is similar to the above point made by Johnson et al., which recognizes that strategy practices are grounded in organizational contexts. This paper takes this point extremely seriously: People involved in a service are always acting in an organizational context throughout planning, development, and enactment, with "no time out," to use Garfinkel's phrase [5].

Consequently, the focus of this paper will be on how constructs of services may appear and are then handled as part of organizational members' practical actions. In other words, understanding services as practical constructs will be the focus of this paper, instead of services as theoretical constructs, following Garfinkel's ethnomethodological studies of work programs [6]. This is a reasonable choice as services are actually applied as practical constructs before being theorized.

As an attempt to understand services as practical constructs, this paper will seek to take the actors' points of view seriously. It also means that the resulting understanding will be defined in relation to activities. This is significant in the sense that the understanding will not be decontextualized from the actual activities, and services can be taken both as results of and as parts of various activities involving various actors.

The business support service—a new service in Japan's public libraries—will be examined. Through analysis, the practical reasoning will be shown as it operates through the various activities related to this service, alongside some organizing principles. In addition, this analytical examination will reveal what is new and innovative about the service in the actual environment where it has been developed and implemented.

2 Approach and Methods

2.1 Ethnomethodological Studies of Work

The ethnomethodological study of work is an approach originated in sociology, that examines exactly how people manage to accomplish tasks. Using this method, what participants know—and how they know it—is studied, alongside their practical reasoning while they work. The word "member" is used to mean any participant with competency in and knowledge about accomplishing tasks [5].

In his study of projects, Sharrock takes Garfinkel's advice on researching organizational practice, treating organizational constructs as entirely practical [7]. He then examines projects as constructs within organizations, in relation to different junctures of service tasks. Instead of using theory to define the concept of service and global criteria for its assessment, services as practical constructs are treated as a part of members' carrying out service tasks, which range from planning and developing services to providing them.

Services as organizational constructs can appear when members explain how the services are designed to meet organizational objectives. How do services as organizational constructs appear and how are they handled when members of two organizations seek to collaborate as part of the services' development? How do services as organizational constructs appear to members who seek to use the services and how do these members handle these constructs? In other words, how organizational constructs appear and how they are treated differently by different actors, depending on circumstances, is this study's research question.

2.2 Business Support Service in Japan's Public Libraries

The business support service is library services with the additional function of supporting businesses, including start-up businesses. Support is provided by using accumulated library resources and digital information available on the Internet and databases, while training librarians to manage this information [8].

Takeuchi specifies that this service is provided by librarians who are trained to search and manage library information resources available to the public. The objective of this service, defined as "supporting business," implies a radical transformation in Japan's public libraries for two reasons. First, it adopts a view of organizing services in which public libraries support citizens' problem-solving activities. This perspective often contrasts with the idea that book circulation services should be the core of public library services—a debate with which this paper is not concerned. Second, the business support service implies openness in the ways the service is designed.

It is important to note that "information service" is not included in the literal translation of the service name, i.e., 'business support service'. The terminology and the stated goal do not define the library's role or means of providing the service. It does not say, for example, that the library's role is to provide information. The intent of the service's name when it was created and introduced to stakeholders is something that needs further investigation. However, different independent organizations were given major roles in creating the service with the library, and the diversity of the service has been noteworthy. In this sense, it is fair to say that business support is an example of 'open innovation' in Japan's public libraries.

It is interesting to note that the service movement itself was initiated by a mix of people who were not necessarily direct stakeholders in the public library: journalists, academics, and government officials. This is quite unique, as library services have typically been viewed as, and have actually been, designed solely by librarians. However, this does not imply that librarians were forced to provide the service by external parties. In actuality, a number of librarians shared the view that services needed to be reorganized so that the public library was not just a place to find good books to read but also a place to search for information to solve life's problems. Some librarians were aware that the quantity of male users in their fifties to seventies had increased, and some reference questions clearly had to do with solving work-related problems [9, 10].

In addition, public library budgets have been decreasing in most parts of Japan due to local governments' economic difficulties. Increasingly, librarians recognized the need for libraries to contribute directly to solutions for problems that people in the local community face in their everyday lives [11]. They decided one way to reorganize library services to be more useful would be to create services supporting local area businesses, thus enabling libraries to contribute to the local economy [8].

Consequently, the business support service is specifically designed to support problem solving in business or work-related contexts by utilizing library information resources and librarians' research and information management skills, and by taking advantage of the library as a place open to all citizens. Organizing a library service based on this problem-solving model is clearly innovative for Japan's public libraries, where a strong emphasis has long been placed on book circulation services. Advocates of this service also generally agree that the mission of public libraries is to deliver this kind of support by facilitating information seeking and usage [12]. The issue of how to design and provide actual services still remains, but this is up to each library to decide.

2.3 Methods

Materials and data used for the analysis presented in this paper comes from research that goes back to 2005. After having conducted a survey of libraries in nationwide, and visited libraries that had started the business support service, the research team eventually focused on four libraries that were actively providing business support services, a new emerging service in Japan's public libraries. Interviews were conducted with librarians and users at the four libraries throughout 2006–2007 and again, in 2008–2009, at two of the four libraries but with members of other organizations in addition to librarians and library users [13, 14]. Various related documents were collected from the libraries, as well as from organizations that were involved in developing the service. Interviews were audio recorded, photographs were taken at the service settings, and observations of users and services were conducted. Observations were made whenever possible by attending seminars held by the library, by attending service consultation sessions whenever permitted, and by walking around in the libraries attending to how people use collections and other information resources. Interviewers asked the following questions:

Interviews with users

1. What was the last use you made of the public library business support service?
2. What was the goal of your use?
3. What process led to the generation of your goal?
4. What use was made of other resources to meet your goal?
5. Can you give a detailed description of your use of the business support service for your goal?
6. How would you evaluate the service?

Interviews with librarians

1. What kind of business support service is currently provided in your library?
2. What kind of users do you expect and encounter in reality?
3. What kind of uses or values do you think users get from the service?

Interviews with members from specialized institutions

1. How did your organization come to work with the library?
2. What did the collaboration involve?
3. How did your organization and its members deal with the collaboration?

4. What do you think about the collaboration?
5. How do you normally use the library, if you do: as part of your work or private life?

Tape-recorded interviews were transcribed and analyzed to identify how the new set of services was designed and implemented in collaboration with different organizations. Some results have been reported elsewhere [e.g., 13, 14].

In this paper, ethnomethodology's approach was adopted in the analysis [11]; and more specifically, the focus was on identifying the practical reasoning or understanding commonly held by different types of stakeholders when engaging in the new service. In so doing, the author examined ways in which services as organizational constructs are handled, specifically as a part of activities.

3 Services as Solutions

3.1 Services as Solutions Provided by Organizations

The business support service can be defined as a solution provided by public libraries. When treated as a *solution*, it is as a new way of meeting the main objective of public libraries: "linking knowledge and people." However, as the service is based on existent public library services, members involved in the program do not treat it as an *alternative* to existing services.

As a solution, it consists of three methods of meeting the goal of "linking knowledge and people." While these methods have long been available as a traditional part of libraries' methods, the ways in which they are actually implemented may be different from the past as it can be seen in the below. The business support service is seen as comprised of the three adjusted methods combined together:

(1) Locating, collecting, and managing knowledge
(2) Curating knowledge
(3) Helping users navigate through the library system

3.2 Locating, Collecting, and Managing Knowledge

3.2.1 Creating a Special Section for Business

In creating the business support service, the libraries typically create a section dedicated to this subject area, where books, journals, pamphlets, magazines, and often databases are brought together. These can otherwise be scattered in different parts of the library. This can be understood as 'curating knowledge,' through the selection and presentation of reference material.

At the time this research started in 2005, the business support service was still relatively new and unusual in Japan's public libraries, in the sense that the service had narrowed its subject area to "business." The act of providing a service under a particular topic was unusual at that time, while later, other services on specific topics—such as law and health—followed suit.

3.2.2 Collecting Knowledge

The majority of the special section is mainly books and journals. Librarians collect them as a source of knowledge people might find useful in solving their problems. Librarians also collect various pamphlets and brochures on relevant information, including business fairs, seminars on starting businesses, financial advice, and other related events and resources. The libraries obtained these information pamphlets mainly through the network the libraries created with business-related organizations.

3.2.3 Locating Knowledge

Locating knowledge is another thing librarians engaged with this service often pursue. Creating a web link collection for a specific subject area is one way of locating knowledge. The collection can include specialist organizations, as well as pathfinders available at the National Diet Library.

As a part of developing this service, librarians contact business-related organizations. Some librarians find it critical to connect with the business network. For some, getting to know people in specialist organizations is another way of locating knowledge in the local area. They can take advantage of these connections to help users solve their problems.

3.2.4 Managing Knowledge

As part of creating and maintaining the special business section, some libraries ask for specialists' help. The specialists decide whether key books for one area are included in collections. As some information needs to be quickly kept up-to-date, books need to be reviewed and updated regularly. In addition to making sure that updated and reliable information is available in the library collections, specialists can also contribute to selecting books. Librarians may ask for specialists' opinions when purchasing expensive books (i.e., which one to choose from amongst candidate books).

While librarians do most of the management of library collections and other information resources, they sometimes ask specialists to help, to make sure that their collections are appropriate. Since specialists may also refer their visitors to the libraries' collections, and they themselves may use these resources, it can be critical that librarians manage collections in collaboration with specialists.

Maintaining a network with specialists in the local area may be another aspect of knowledge management if the librarians consider specialists as additional important resources for their libraries. Traditionally, networking was not usually considered part of their work, but it has become a key part for some of the librarians who are engaged in the business support service.

3.3 *Curating Knowledge*

3.3.1 Organizing Seminars and Lectures

In addition to setting up a special section, the librarians typically organize seminars and lectures on business-related topics. The topics can range from how to start a new business to intellectual property, information resources, and methods of market analysis. Lecturers are those with specialized knowledge in the local business community. The events are organized by the libraries or by other groups or organizations, with the libraries hosting the events as a co-sponsor.

Holding seminars and lectures is a way of 'curating knowledge,' in the sense that the libraries select people with specific knowledge, often from the local business community. Seminars and lectures may attract people interested in immediate problem solving or exposure to something of potential interest now or in the future. In this sense, these events can attract people who have different reasons for obtaining specific knowledge, who may or may not be 'library users.'

3.3.2 Organizing Displays

As part of the business support service, the libraries organize different displays. Book displays on current topics are typically organized in or near the related section. In addition to book displays, libraries organize displays of industrial products related to the local area, such as a display of furniture made from local cedar trees or products from local companies that have won a national prize. They create these displays in collaboration with local industry promotion centers. Thus, creating a display is another way of 'curating knowledge' about a business subject in books and the local business community.

3.4 *Helping Users Navigate Through the Library System*

3.4.1 Navigation Through Mechanisms

Libraries have traditionally provided users with navigational mechanisms for searching collections and other information resources so that they can access knowledge on their own. Arranging bookshelves with signs is one way of helping users navigate. However, inevitably, books and magazines on one subject may not always be together, or books classified under "commerce" (380 in the Dewy Decimal Classification) are shelved some distance from books under "business enterprise" (338.7). Creating a special section for business is one new way of helping users who are specifically interested in searching the collection and other information resources within this subject area.

3.4.2 Personal Navigation

While librarians expect people to find relevant information using the libraries' standard navigation tools, they traditionally also provide opportunities for personal navigation through, for example, reference services. In addition, libraries also provide business consultation services. They usually collaborate with government organizations as well as nongovernmental groups—such as the Chamber of Commerce and Industry or other specialist groups to set up this consultation service. These specialist groups or organizations generally regard libraries as having access to people who may feel reluctant to visit their specialist organizations unless they have a specific purpose or developed plan.

4 Services as Organized Environments

4.1 *The Service as an Environment Organized Under an Organizational Scheme*

As part of developing business support services in libraries, librarians strive to make their capabilities and their library collection more visible to the public by referring to the business support service as an organized environment developed under an organizational scheme.

By creating a special section dedicated to this subject area, which contains books, journals, pamphlets, magazines, and often databases that can be otherwise scattered in different parts of the library, the collection is certainly made more visible. Librarians also promote their capabilities by placing signs next to the business collection, with a message that users can ask for help at the counter. Further, by holding seminars and lectures on business-related topics in the libraries, they try to let people know about the service and attract those who are not necessarily customary users of the libraries. What a librarian has also been doing at each event is explaining their service with reference to relevant books they bring to the event. They do this in the events held outside the library as well.

For example, a service user wanted to start his own business by developing some products based on his ideas. He had been an employee of a medical equipment company that was not related to the area of his new business and started looking for further opportunities to develop his ideas, attending business seminars, looking for information in libraries, and so forth, for 2 years. He was doing well in taking advantage of learning opportunities by attending seminars, but he was obviously struggling to develop and back up his ideas.

He then recognized his opportunity in the service a librarian was describing at the end of a seminar he attended at the Chamber of Commerce and Industry: "If someone came to the library and asked us to lend them money, we would tell the person which places to go to in order to borrow money. Please ask any question". "Lend me money" is certainly not the kind of request one would normally expect to make to a library, but by describing how the librarian portrayed the available services to help a person trying to start a business, the user revealed how he came to learn that the library could help him develop his ideas. The librarian was able to show that the library could provide information other than what was in books, at least in the business area. He recognized the opportunities through this personal navigation service, so he went to the library immediately after the seminar. In such cases, librarians treat the service as an organized environment developed under the organizational scheme for business support, making their capabilities and the collection visible to citizens.

4.2 The Service as an Environment for Collaboration

In addition, as part of developing the service, librarians try to identify potential stakeholders and to connect with them. By informing users of the service the libraries are about to start or when it has started, the librarians can introduce themselves and explain the service to organizations that may not be familiar with the new program. The service as an organizational construct provides an environment where opportunities for collaboration become available but are still open to exploration. The stakeholders include such organizations as the relevant section of the local authorities, some specialist organizations (e.g., the Chamber of Commerce and Industry), and the Local Industry Promotion Center.

The relationships may start with stakeholders asking for their pamphlets to be placed in the library but may eventually develop into "partnerships," when members of these specialist organizations understand the value of the librarians' capabilities and the libraries' collection. Further, members of these organizations may recognize the librarians' intention to support their interest in developing and accelerating the economy and business activities in the local area. As public libraries have traditionally been known as a place for reading for pleasure, these specialist organizations' members may not easily recognize this on their own. Besides, the hurdles may be high for librarians as well, since traditionally they have mostly targeted their services for users who come to read books for pleasure.

Some libraries have already overcome these hurdles. In one case, after having contacted specialist organizations, the librarians were eventually invited to these organizations' informal gatherings, where they and others—including staff from the local bank—regularly discuss issues in developing local business and maintaining networks. The library eventually formed a committee consisting of some members from these specialist organizations, to generate ideas for designing the new service. The library regularly demonstrates what it can do by showing how librarians are helping citizens solve work-related problems, using the business collection.

Members of the specialist organizations eventually realized, first, that citizens who come to their organizations may not necessarily overlap with those who come to the library and, second, that they can benefit from visiting the other organizations. The specialist organizations are well aware that people often feel hesitant to contact them unless they have developed relevant plans and have specific problems. That they are closed during weekends is, they know, also a problem. In addition, they are aware that some of their visitors could benefit from going to libraries and conducting research on topics.

Thus, members of these organizations eventually came to realize that they can reach out to citizens more effectively if they collaborate with libraries, especially when they recognize that far more people go to the libraries than to their own organization. They then join the organization of events, such as seminars, lectures, and library displays.

4.3 *The Service as an Environment for Accessing Capabilities*

For some, the business support service is an environment where they can look around for books and magazines in the special business section—including pamphlets on seeking jobs—and as a useful part of the libraries' larger collection. For these users, the special section is organized in such a way that they can keep up-to-date with the current state of business in general or the job market. Only when necessary, they may choose to seek personal assistance from reference librarians.

For others who visit the libraries, because they can ask a librarian for assistance to find specific information for developing their business, the service is an environment where personal assistance is available in accessing knowledge.

> Small- and medium-sized companies need to create innovation to survive, but, unlike large companies, we cannot afford to hire clever people who can do this kind of research. . .It is as if I have two such clever people.

In addition, for those who come to the library to attend a seminar on a business-related topic or to see a display of local industrial products, it may be their first visit to the local public library. Alternatively, they might have come to the library with their children but never thought of doing research on work-related topics.

Depending on their circumstances, users encounter the service, and the business support service is viewed and experienced differently. In all cases, it is experienced as an environment where librarians' capabilities are available (e.g., the business section and the personal navigation), especially to access knowledge on business-related topics, whether users need to solve problems immediately or to gain knowledge for the future.

For the librarians involved in developing or providing the business support service, this is an environment where they provide assistance in accessing knowledge. The service itself does not dictate to libraries exactly what assistance to offer, and it is up to each library what service menu they provide. Since typical choices have been shared through seminars for librarians, held by the Business Support Library Association, there is some standard assistance most libraries provide, as described above.

At the same time, their service menus vary. One library offers a series of lectures on starting a business. People who attend find that through attending the lectures, they not only learn knowledge they need to start their businesses, but also that various local resources are available, such as specialist organizations, seminars on more specific topics, and people who have similarly started their own businesses. They also recognize that, through attending the lectures, they have gotten to know people with similar goals, and they have been supporting each other ever since.

In another library, librarians attempt to provide access to knowledge in various resources as seamlessly as possible, even beyond their collection and other reference materials. When they find that the person needs knowledge from other resources outside the library, they normally try to arrange a way to reach a specific source. This includes opening up a path to reach a specific specialist outside the library, to make sure that the person can access the needed knowledge without much difficulty.

5 Conclusions

Practices within the business support service recently developed by Japan's public libraries were examined. The motivation for this study has been to arrive at a clearer understanding of services as they are practiced in organizational contexts. This paper has provided descriptions of services as organizational constructs, as understood from actors' points of view. The findings have shown that, as organizational constructs, services are available in two ways that are part of their practical applications. First, services can be understood as a set of constructs with which organizations attempt to achieve their objectives in a particular way. For example, a librarian who introduced the business support service in a seminar on business startups presented it as a 'solution' the library provided for assisting people's problem solving, via various methods of "linking knowledge and people" that is an overall objective of the library. Thus, a service is treated as a 'solution' their organization has provided at a point in time.

Second, services can be understood as environments of different kinds where practical actions take place. For example, librarians who are planning to start the business support service in their library treat the service as an environment in the process of being realized, and members try to make this environment understandable to others, based on how it will be if they introduce the service. In addition, librarians explain to people who have never used the service what kind of environment they will enter, if they decide to access it. A second example is that librarians may consider that starting a new service means they are now in an environment where they can seek to collaborate with the relevant partners available to their organizations. Traditionally, and probably like other public services, libraries have tended to provide services in a self-sufficient manner. However, starting a new business service itself provides the justification for approaching other organizations. A third example appears when people actually approach the service. They look into the business section in the library and search for relevant books, or they attend a lecture on business startups. On the other hand, a librarian, when approached by people asking for specific information, tries to provide relevant information. The service, in this case, is experienced as an environment that allows access to capabilities arranged by the library.

Thus, looking at how a service as a solution is encountered by members—as well as how it provides environments that allow members to conduct various activities—provides a clearer understanding of how it is practiced and experienced as an organizational phenomena. As any service innovation needs to be located in the organizational context to be recognized as actual innovation, examination of services demonstrated in this paper may be of benefit.

References

1. Dougherty D (2004) Organizing practices in services: capturing practice-based knowledge for innovation. Strateg Organ 2(1):35–64
2. Vargo SL et al (2008) On value and value co-creation: a service systems and service logic perspective. Eur Manag J 26:145–152
3. Normann R, Ramirez R (1994) Designing interactive strategy: from value chain to value constellation. Wiley, Chichester
4. Johnson G et al (2007) Strategy as practice: research directions and resources. Cambridge University Press, Cambridge
5. Garfinkel H (1967) Studies in ethnomethodology. Polity Press, London
6. Garfinkel H (1986) Ethnomethodological studies of work. Routledge, London/New York
7. Sharrock W (2011) The project as an organizational environment for the division of labour. In: Rouncefield M, Tolmie P (eds) Ethnomethodology at work. Ashgate, Farnham, pp 19–35

8. Takeuchi T (2005) Sogyo oyobi chusho kigyo no bijinesu wo shien suru kokyo toshokan [Public libraries supporting business start-ups and small businesses]. Kokumin Seikatsu Kinyu Koko Chosa Geppo 531:38–41
9. Toshokan O (2009) 'Sai charenji' wo oensuru toshokan [Libraries supporting 'trying again']. Shakai Kyoiu 2:20–23
10. Tokoyoda R (2003) Kokyo toshokan ni okeru bijinesu shien sabisu no genjo [The current status of business information service in public libraries]. In: Tokoyoda R (ed) Urayasu Toshokan ni Dekiru Koto [What Urayasu public library can do]. Keisoshobo, Tokyo, pp 160–172
11. Kobayashi T (2008) Shigoto ya seikatsu ni yakudatsuto ninchi sareru toshokan ni narutameni [Towards libraries being recognised as useful at work and in daily life]. In: Ogushi N (ed) Mondai Kaiketsugata Sabisu no Sozo to Tenkai [Problem solving oriented service: its creation and implementation]. Seikyusha, Tokyo, pp 21–52
12. Tamura S (2008) Bijinesu shien sabisu [Business information service]. In: Tamura S, Ogawa T (eds) Kokyo Toshokan no Ronten Seiri [Issue on public libraries]. Keisoshobo, Tokyo, pp 35–58
13. Ikeya N et al (2011) In search of facilitating citizens' problem solving: public libraries' collaborative development of services with related organizations. Inf Res 16(1). http://www.informationr.net/ir/16-1/paper468.html. Accessed 15 Apr 2014
14. Ikeya N, Tamura S (2010) Phenomena of value co-creation in service. In: Proceedings of the 2nd International Service Innovation Design Conference (ISIDC 2010), held at Future Hakodate University, 18–20 September, pp 113–120

The Findings from the First Service Design Projects

Yuriko Sawatani, Naoto Kobayashi, and Yoshinori Itoh

Abstract The design focus is expanded from physical products to service systems due to the service economy. The research designing a service system becomes more important. The scope of design becomes wider, more complex and including more interactions among various stakeholders. Because of these shifts, the new research on service system design is emerging. The paper shares the evaluation results and findings of the initial service design projects at Waseda University. The service design projects are evaluated based on the service design management framework. The findings include the importance of the joint ownership of the objectives and the deep argument for innovative idea generation. In addition, the design process of service design projects (meta-design process) such as the team configuration and the theme setting is the key to increase the success of service design projects. Lastly, the prototyping of a service/social system is not enough studied and needs to be explored further.

Keywords Service innovation • Service design • Prototyping • Management • Engineering and Design (SSMED)

1 Introduction

The society has been fast advancing toward a service-based economy. This phenomenon, common to both developed and developing countries, results from the growth of the service sector's share of the economy, spurred by rapid growth in service industries consequent to increased social sophistication and diversification. The growth of services encouraged the researchers to study the fundamental business changes from product-based systems to service economies [1, 2]. Besides the follow-up sales of product-based services, the business strategies of companies had to shift their focus to services as a key differentiator in their new business models [3, 4]. Such a shift does affect the design focus.

Y. Sawatani (✉) • N. Kobayashi • Y. Itoh
Waseda University, Waseda-Tsurumakicho 513, Shinjuku-Ku, Tokyo 162-0041, Japan
e-mail: yurikosw@gmail.com

T. Maeno et al. (eds.), *Serviceology for Designing the Future*,
DOI 10.1007/978-4-431-55861-3_35

The design focus is expanded from physical products to service systems due to the service economy. The research designing a service system becomes more important. One of difficulties of designing a new service system is how to share values among many involved stakeholders. The scope of design becomes wider, more complex and including more interactions among various stakeholders. Because of these shifts, the research on service system design is emerging.

The paper shares the evaluation results and findings of the initial service design projects. The service design projects are evaluated based on the service design management framework. The findings include the importance of the joint ownership of the objectives and the deep argument for innovative idea generation. In addition, the design process of service design projects (meta-design process) such as the team configuration and the theme setting is the key to increase the success of service design projects. Lastly, the prototyping of a service system is not enough studied and needs to be explored further.

2 Service Design Projects

The service design projects aim to create a new service system in the future, which is positioned as the intersection of the meso scope and the open system (Fig. 1). The target of attendees of the service design projects includes students, teachers, and professionals from companies. The six project teams are formed and develop ideas and the prototyping from Dec. 2013 to March 2014.

2.1 Design Process

At the project level, the following steps are executed.

1. Problem identification and definition
2. Idea generation

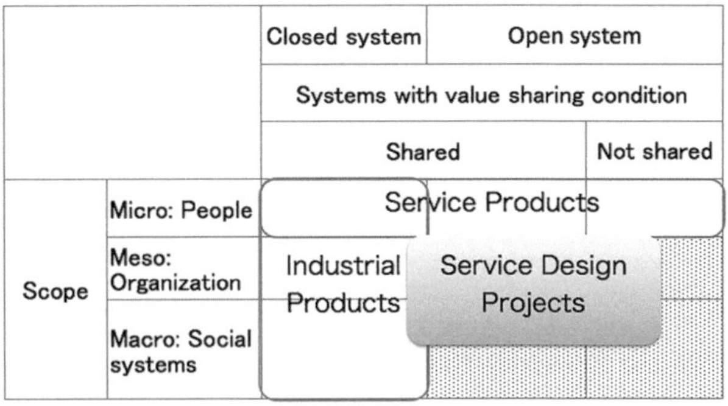

Fig. 1 The positioning of service design projects [5]

3. Hypothesis inspection based prototyping
4. Prototype testing and feedback

Step1: problem indication and definition is to define the clear goals from the broader task objectives. The task to given to the projects is to create a future living place, which is comfort and attractive. Then five companies, which are related with a living place, provide the future problems and directions. Each team defines the problem statement, which is to solve what for whom by 2020.

Idea generation, step 2, is executed by the innovation workshops using various ideation methods. There are various existing methods and guide books for idea generation, so after these workshops, each team execute idea generation itself based on the given method books.

This program puts stress on concept, product and experience testing by prototyping. So that step 3 and 4 are important, which include plans for the prototyping, the execution and the feedback to the original prototype. The mid-team team presentation is done to show each team's main idea and the prototyping plan. The teams present the overall project progress and the prototype and the feedback from the prototyping execution at the final team presentation.

2.2 Meta-Design Process

Activities to design, manage and evaluate projects are design processes of the projects, called as meta-design process. At the meta-design process level, we use the following service design management framework, which is extended the input-process-output model [6] from service system viewpoints (Fig. 2).

The inputs of the service design projects are team group composition, such as heterogeneity, size, group tenure, KSA(knowledge, skills, and abilities) and task complexity. The progress of projects is majored by surveys at each event, such as the kick off, idea generation workshops, the mid-term team presentation, and the final team presentation. The outputs are tested by the judges from social, business and technical point of views.

2.3 Methods

In this project, we focus on the following research questions:

1. What kinds of processes are going on to create innovative solutions by a team?
2. How can a management system help?

To investigate these research questions, a semi-structured interviews and surveys are done as the following:

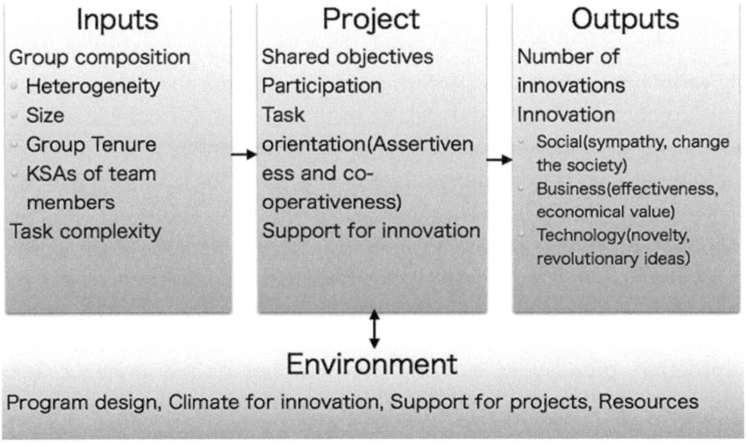

Fig. 2 The service design management framework

- Semi-structured interviews

  Target: six project leaders
  Hours: one hour
  Time: Just after the project

- Surveys

  Target: All attendees of a project
  Time: 5 times (Each at workshop, total three times, After the mid-term presentation, After the final presentation)

3 Result

Table 1 shows the evaluation results of the service design projects. The team size is from 2 to 6. The team F is a special case, which is crated in the middle of projects. The task given to the teams are highly complex. KSAs of teams are from middle to high, however, heterogeneity of teams are various from low to high. The team configuration is done by the attendee's choice without controls from the meta-design process level.

3.1 Outputs

From the outputs, the following points are considered. The total assessment was high when there was an outstanding score at the evaluation. In the service design projects, innovativeness is evaluated from social, business and technology point of

Table 1 Evaluation results of service design projects

	Inputs											Outputs			
	Group composition					Project						Innovation			
	Heterogeneity	Size	Group tenure	KSAs of team members	Task complexity	Shared objectives (mid)	Shared objectives (final)	Participation	Task orientation (assertion)	Task orientation (co-operative)	Innovation orientation	Total	Social	Business	Technology
A	H	5	3 months	M	H	L	M	H	L	M	M	4	M	M	L
B	H	5	3 months	H	H	H	M	M	H	H	H	3	M	M	M
C	M	5	3 months	H	H	H	M	M	M	H	M	2	H	M	M
D	L	4	3 months	M	H	M	H	H	H	M	H	1	M	M	H
E	L	6	3 months	M	H	M	H	H	H	H	M	5	L	M	M
F	L	2	2 months	M	H	M	M	M	M	H	M	6	M	M	L

Note: *H* high (higher than average), *M* middle (around average), *L* low (lower than average) the number at total shows the order of total evaluation points (1: high to 6: low)

Fig. 3 The task orientation

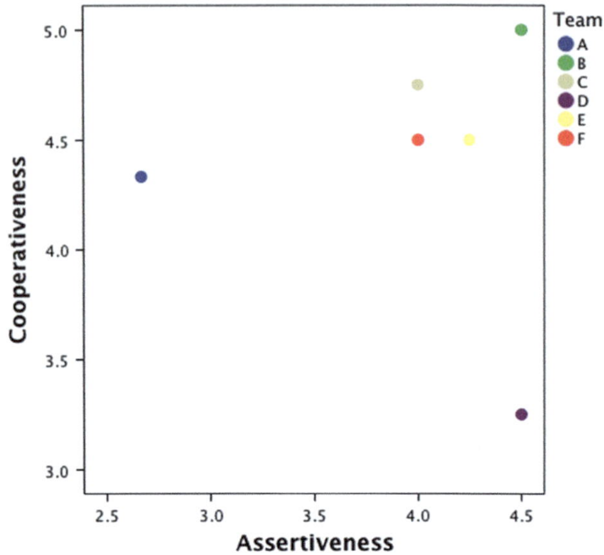

views, however each judge has the different perspectives of each view. It needs more detail understanding how to evaluate innovativeness.

3.2 Task Orientation

Figure 3 shows the task orientation. The most of teams is the right top corner, which is high assertiveness and cooperativeness.

The team D has the stronger assertion task orientation than co-operative task orientation. The team A is opposite. Team D was evaluated as the top at the final team presentation that has a strong technology score.

The deep argument needs to be considered as an important element for innovative idea generation [7]. The excessive sense of cooperation may be rather associated with a sharp idea in an argument reversely.

3.3 Shared Objectives

One of difficulties designing a new service system is problem statement development. Without clear definition of goals, team members are not able to share project objectives enough. However, the goal is determined too precisely, then idea generation does not work effectively. Moreover, how to share project objectives and

Fig. 4 The shared objectives

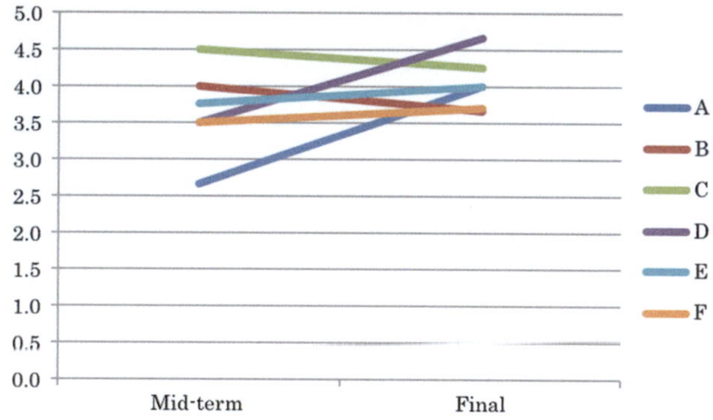

values delivered from a service system is important since various stakeholders are involved in the system (Fig. 4).

Team D and team C shared project objectives with team highly, which is ranked the first and the second at the final team presentation. The joint ownership of the objectives might be important to start the innovative idea generation. When a purpose is not shared enough, arguments may not be carried out effectively. As a result, an idea may become commonplace.

Team D and team A increase the point of shared objectives from the mid-team to the final team presentation. However, team C and team B decrease the point. In team B and C case, various comments given to the proposed idea at the mid-team team presentation might affect the project activates after that.

On the other hand, team D was relatively a low point at the mid-team team presentation. For team D case, even though a purpose is not shared enough, the team is more homogeneous and keeps the same quality (all members of team D come from the same discipline, similar school year), then deep arguments might be possible to generate a sharp idea.

3.4 Prototyping

Project teams are encouraged to test idea through prototyping. The followings are definition of prototyping via Wikipedia.

"A prototype is an early sample, model or release of a product built to test a concept or process or to act as a thing to be replicated or learned from."

"A prototype is designed to test and trial a new design to enhance precision by system analysts and users."

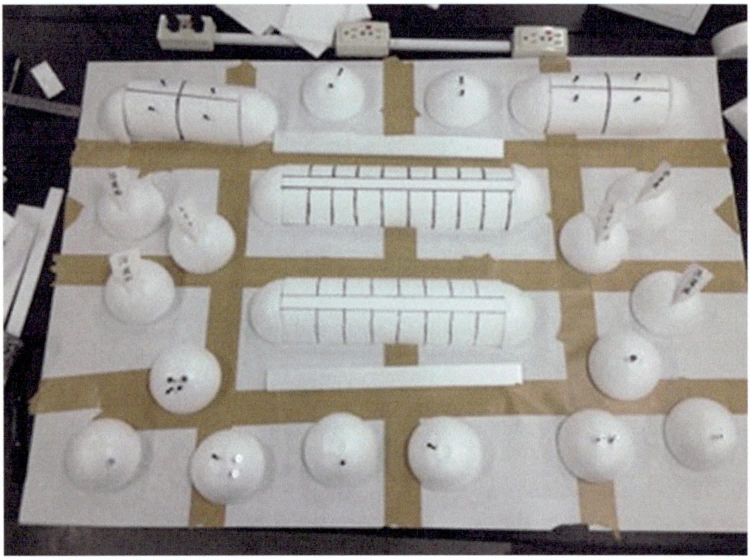

Fig. 5 The concepts prototype

It covers various products (physical or service) and systems from various points of views (user or business point of views). However, people are used to limit the scope of prototyping to the function realization testing. So that the program tries to extend the scope of prototyping, not only to product prototyping for the function testing, but also concept and experience prototyping for service or social systems.

The following pictures show the various prototypes developed by teams (Figs. 5, 6, and 7).

Each team tried to innovate the way to prototype its idea using various methods. The prototypes are used to communicate the idea with various stakeholders and to learn from the feedbacks. However, only a limited knowledge of service and social system prototyping methods and tools is available [3, 4, 8–14]. The prototyping of a service/social system is not enough studied and needs to be explored further (Table 2).

Fig. 6 The product prototype

Fig. 7 The experience prototype

Table 2 Prototyping from objectives and the target

<table>
<tr><td colspan="3" rowspan="3"></td><td colspan="5">Objectives of prototyping</td></tr>
<tr><td colspan="3">User viewpoints</td><td colspan="2">Business viewpoints</td></tr>
<tr><td>Value</td><td>Process</td><td>Interaction</td><td>Realization</td><td>Vision, strategy, market, society</td></tr>
<tr><td rowspan="3">The target of prototyping</td><td rowspan="2">Product</td><td>Goods</td><td rowspan="3">Concept, storyboard, scenario, sketch, video, skit, paper prototyping, business model</td><td></td><td>Interface, use case, scenario</td><td>Paper prototyping, foamed styrol, clay, 3D printer, CAD, software</td><td rowspan="3">Business model canvas</td></tr>
<tr><td>Service</td><td colspan="3">Experience prototyping, storyboard, scenario, sketch, video, skit, blueprinting</td></tr>
<tr><td>System</td><td>Service/social system</td><td colspan="3">System prototyping, storyboard, scenario, sketch, view, skit, simulation</td></tr>
</table>

4 Conclusions and Future Research

From the early research on the service design projects, the findings include the importance of the joint ownership of the objectives and the deep argument for innovative idea generation. In addition, the design process of service design projects (meta-design process) such as the team configuration and the theme setting is the key to increase the success of service design projects.

From the interviews of attendees at the service design projects, prototyping of the service system was challenging. The most of the service systems, which teams created, were not a products, but service systems. The areas of prototyping are expanding from products to concepts and experience. The operationalization from the idea and the concept model to the prototyping is effortful.

Even though the prototyping of a service/social system is important, but it is not enough studied yet. The methods and tools supporting the service and social system prototype need to be researched more in the future.

Acknowledgement This work was supported by "CI3: Center for Idea Interacted Innovation" of the Ministry of Education, Culture, Sports, Science and Technology (MEXT), Japan.

References

1. Chesbrough H, Spohrer J (2006) A research manifesto for services science. Commun ACM 49 (7):35–40
2. IfM, IBM (2007) Succeeding through service innovation: a service perspective for education, research, business and government. University of Cambridge Institute for Manufacturing, Cambridge
3. Sawatani Y, Nakamura F, Sakakibara A, Hoshi M, Masuda S (2007) Innovation patterns. In: Proceeding of IEEE international conference on services computing, pp 427–434
4. van der Aa W, Elfring T (2002) Realizing innovation in services. Scand J Manag 18:155–171
5. Sawatani Y, Arai T, Murakami T (2013) Creating knowledge structure for service science. In: Technology Management in the IT-Driven Services (PICMET), 2013 Proceedings of PICMET '13, pp 2440–2445
6. West MA, Aderson NR (1996) Innovation in top management teams. J Appl Psychol 81 (6):680–693
7. Badke-Schaub P, Goldschmidt G, Meijer M (2010) How does cognitive conflict in design teams support the development of creative ideas? Cogn Conflict Des Teams 19(2):119–133
8. Bitner MJ (1992) Servicecapes: the impact of physical surroundings on customers and employees. J Mark 56(2):56–71
9. Holmlid S (2007) Interaction design and service design: expanding a comparison of design disciplines, Nordic design research conference, NorDes, Stockholm, Sweden
10. Marger B (2004) Service design: a review. KISD, Cologne
11. Blomkvist J, Holmlid S (2010) Service prototyping according to service design practitioners. In: Proceedings of the ServDes 2010, Linköping, Sweden
12. Houde S, Hill C (1997) What do prototypes prototype? In: Helander M, Landauer T, Prabhu P (eds) Handbook of human-computer interaction, 2nd edn. Elsevier Science B. V, Amsterdam

13. Carleton T, Cockayne W (2009) The power of prototypes in foresight engineering. In: Proceedings of the 17th International Conference on Engineering Design (ICED'09), Stanford, Calif
14. Shostack L (1984) Designing services that deliver. Harv Bus Rev 62(1):133–139

An Evaluation Method of a Service Business Model Using Wants Chain Analysis

Kazuto Imazeki, Toshiyuki Yasui, and Takashi Maeno

Abstract We proposed a novel evaluation method for a service business model by incorporating customer's quantified wants to the method. In this method, values which customers feel to provide service are transformed into "WANTS" that they have. The Wants Chain Analysis (WCA), a design and structuring method of business based on system thinking, is applied for visualizing WANTs. And the Analytic Hierarchy Process (AHP), a quantifying method for subjective selections of stakeholders, is also applied for quantifying stakeholders' WANTs.

We applied this proposed method to an example of actual service, and effectiveness of proposed method was confirmed.

Keywords Wants chain analysis • Analytic hierarchy process • Quantitative analysis • Satisfaction of wants

1 Introduction

In recent years, consumer's values have changed and diversified in the most of industrialized counties. These phenomena put services providers in the awkward position when they develop new services for fitting just to the values which customers' feel. One of key problems for services providers here is how to quantify values that customers feel for provided services. If services providers can evaluate rightly their services provisions models with a quantified method, they can construct services models suitable to their customers' needs. A questionnaire investigation is a commonly-used method to analyze and quantify customers' need. However it is not suited to use in a development process of service because it requires a much time and budget. In the Value Engineering for product development, value is defined as the

K. Imazeki (✉) • T. Yasui
Graduate School of System Design and Management, Keio University, 4-1-1 Hiyoshi Kohoku-ku, Yokohama-shi, Kanagawa-Prefecture 223-8526, Japan
e-mail: kazuto_imazeki@a3.keio.jp

T. Maeno
Graduate School of System Design and Management, Keio University, Yokohama 223-0061, Japan

T. Maeno et al. (eds.), *Serviceology for Designing the Future*,
DOI 10.1007/978-4-431-55861-3_36

function of product divided by the cost of the product [1]. However, a service, not a product, has a various functions and a quantification of these function are difficult. A value of service is Customers' values can be translated to their WANTs, if Maslow's hierarchy of needs theory [2] is applied where WANTs are "needs or wants". This study focuses on WANTs and proposes a concept design method for a business model [3]. Therefore, it is useful to use WANTs instead of value. There is a clear requirement for a suitable evaluation method to quantify the customer's WANTs for a developing new service.

The proposed model is conceptually rooted on two methods; the Wants Chain Analysis (WCA) [4, 5] and Analytic Hierarchy Process (AHP) [6]. WCA, a design and structuring method of business based on system thinking, is for visualizing and structuring relations among all stakeholders and their WANTs in a service model. AHP, a quantifying method for subjective selections of stakeholders, is for quantifying stakeholders' WANTs. WCA and AHP should be applied complementarily for this evaluation model. Because WCA visualizes stakeholders' WANTs and their classifications, but it does not quantify the size or volume of WANTs. A quantification of WANTs is difficult because the WANTs are subjective itself. Therefore, the AHP, a subjective decision method, is applied to the method for quantification of the WANTs which the WCA visualizes and structures in a certain service design.

As mentioned above, in this research, WANTs are quantified by applying the concept of AHP to WCA. And we propose a quantitative evaluating method of business model by using quantified WCA.

2 Existing Knowledge

2.1 Want Chain Analysis (WCA)

WCA is a method that analyzes and visualizes relationships among stakeholders [4]. WCA is based on CVCA (Customer Value Chain Analysis) [7]. In WCA, we consider not only things that are considered by CVCA but also stakeholder's WANTs. Figure 1 shows an example of WCA. In WCA, WANTs are divided into 4 classes. Classification of WANTs is shown in Table 1.

The procedure of WCA is shown below.

A: Conduct of CVCA

A-1: The stakeholders relevant to the product or service are enumerated and written in graphic chart.

A-2: Substances that are exchanged between stakeholders such as information and money are written in graphic chart using arrow symbol.

B: filling up of WANTs

About each stakeholder's exchange, WANTs that caused actions are written in the start point of arrow symbol.

C: Evaluation of Satisfaction

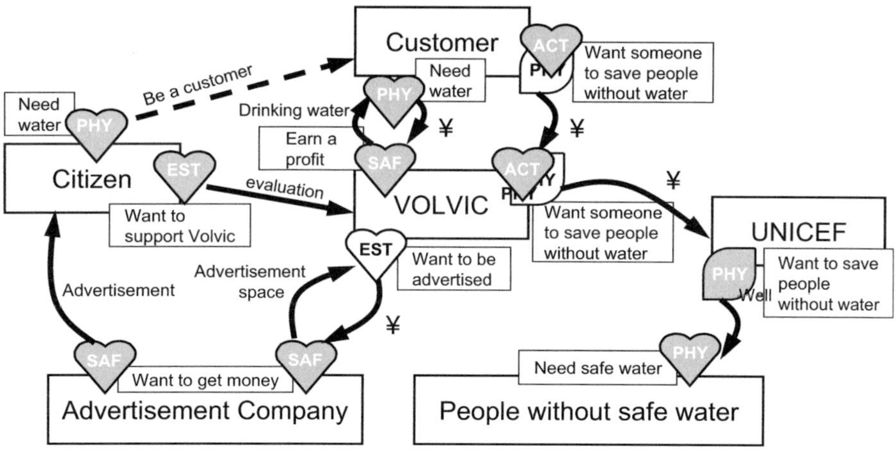

Fig. 1 Example of WCA

Table 1 Classification of WANTs

		Object of needs	
		Self	Others
Subject of needs	Self	α	γ
		I want to feed myself	I want to feed another person
	Others	β	δ
		I want someone to feed me	I want someone to feed me

A satisfaction state of each WANTs is checked. The condition of satisfaction is different with classification of each WANTs. Details of condition are omitted from this paper.

Through this evaluation, if all WANTs are satisfied, the product or service that is intended to analysis is effective. If there is some unsatisfied WANTs, the product or service is insufficiency.

2.2 *Knowledge of WANTs*

In Maslow's classification, the WANTs are classified into 5 (+2) Levels [2]. Figure 2 shows the classification of WANTs.

This Maslow's classification is not proof strictly, but used widely because of its simplicity. In a part to use in engineering, this classification is useful and effective. In Maslow's classification, each WANTs are making the layered structure. However, "desires to know and understand" and "aesthetic needs" are independent of others. These 2 WANTs exist always regardless of other WANTs. In the layered structure, "Physiological needs" is a lowest level and "self-actualization needs" is a

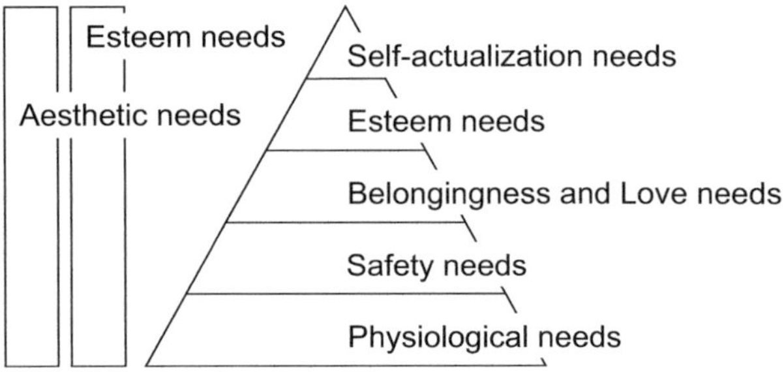

Fig. 2 Maslow's classification of WANTs

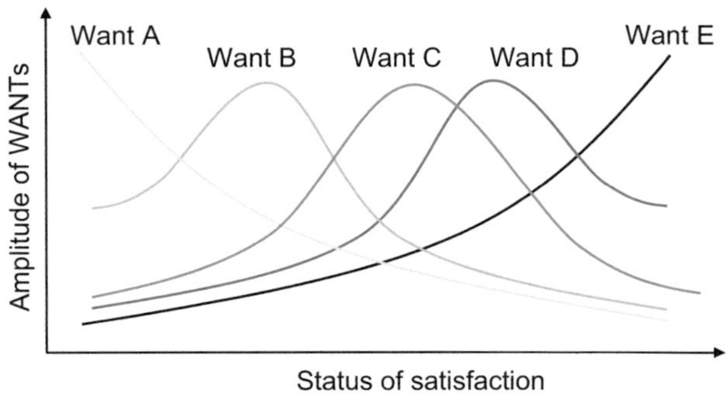

Fig. 3 Transition behaviour of WANTs

highest level. If low-level WANTs are satisfied, high-level WANTs are increasing. As just described, WANTs have amplitude and the amplitude is changed by situation.

These relationships between low-level WANTs and high-level WANTs are not strict. Satisfaction of low-level WANTs and increase of high-level WANTs advance gradually, and do not show a digital behaviour. Therefore, low-level WANTs and high-level WANTs may exist in same time. The transition behaviour of WANTs is shown in Fig. 3.

These WANTs are satisfied when the state corresponding to WANTs is acquired by some action. In this paper, the rate of the WANTs that is satisfied with some action is called "Satisfaction of WANT". The relationship between "Satisfaction of WANT" and action is changed by "amplitude of WANT". For example, there are two persons, the one of them wants to get a 100 on a test, the other one wants to get a 50 on a test. If they get a 50 on a test, the former does not satisfy the "WANT" but, the latter satisfy the "WANT". As just described, same "Satisfaction of WANT" are not obtained by same action.

When stakeholder has many WANTs, the satisfaction of stakeholder is decided in consideration of each degree of WANTs satisfaction synthetically [8]. In this paper, to distinguish from "satisfaction of each WANTs", "stakeholder's satisfaction" is called "Overall Satisfaction".

The influences of "satisfaction of each WANTs" to "Overall Satisfaction" are not same. For example, when we buy water, "Want to drink water" is satisfied and

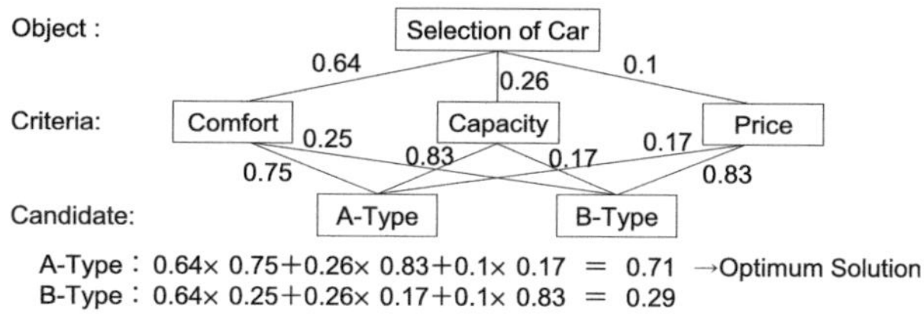

Fig. 4 Example of AHP – purchase of the car

"Want to save money" is not satisfied. However, the influence of "want to drink water" satisfaction is larger than the influence of "want to save money", Therefore, the "overall satisfaction" is increase. In this paper, the influence of "satisfaction of each WANTs" to "Overall Satisfaction" in called "Importance degree of WANTs".

2.3 Analytic Hierarchy Process

AHP is a tool which can be logically made decisions by combining a subjective judgment and systems approach [6]. An example that applied AHP to the purchase of the car is shown in Fig. 4.

At first, problems to be solved are dissolved into 3 hierarchies, Object, Criteria and Candidates. Object is the problem which it is going to solve. In the example of "purchase of car", Object is "Selection of Car". Criteria are things that should be considered for decision such as "Comfort", "Capacity", "Price" and so on. Candidates are alternatives of decision such as types of vehicle.

Next, the relationship of each hierarchy is decided by pair comparison method. The relationship is expressed by "numerical weight". By using pair comparison method, even if quantification is a difficult, relationship can be clarified. In Fig. 4, there are some numerical examples of weight. About the object of "selection of car", the weight of criteria is, comfort: 0.64, capacity: 0.26, price: 0.1. A summation of all weights should be 1.

Finally, priority of each candidate is calculated using relationship of each hierarchy. The candidate that has highest priority is selected for optimum solution. In the example of Fig. 4, the vehicle type-A is selected by AHP as optimum solution.

Through this process, making decisions logically is possible using AHP.

3 Introduction of Proposed Method

3.1 Quantification of WANTs Using AHP

In this section, we introduce the method of AHP to WCA.

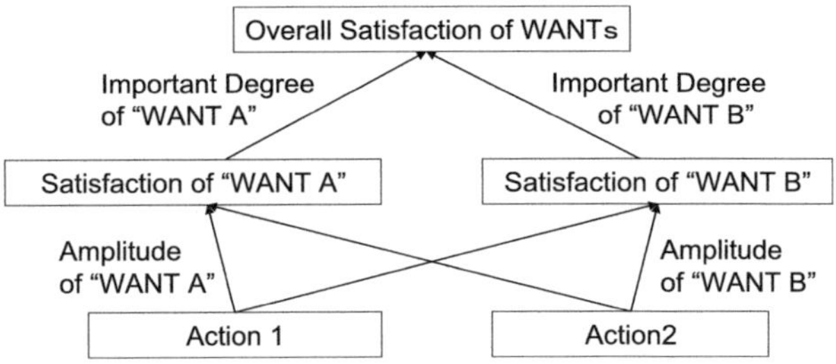

Fig. 5 Model of "Overall Satisfaction of WANTs"

Objective quantification is difficult for the element of WANTs described in Sect. 2.2. This is because that WANTs itself are subjective. On the other hand, AHP described in Sect. 2.3 can carry out appropriate decision-making with a logical procedure and subjective judgment. In this paper, Quantification of WANTs is performed by AHP.

Figure 5 shows the model of "Overall Satisfaction of WANTs".

"Overall satisfaction of WANTs" is calculated from "Satisfaction of WANTs" and "Important degree of WANTs". And "Satisfaction of WANT" is calculated from "Amplitude of WANT" and "Action".

"Important degree of WANT" and "Amplitude of WANT" is determined by a pair comparison method.

The procedure of calculation of "Overall Satisfaction of WANTs" is shown below.

A: Enumeration of WANTs

The WANTs of intended stakeholder is enumerated.

B: Derivation of "Amplitude of WANT"

The "Amplitude of WANT" is derived by a pair comparison method as shown in Fig. 6.

C: Calculation of "Satisfaction of WANT"

The "Satisfaction of WANT" is calculated from "Amplitude of WANT" and "Action". The "Satisfaction of WANT" is denoted by the following formula.

$$S_w = B_w / A_w \tag{1}$$

Where S_w is "Satisfaction of WANTs", B_w is a income and outgo of intended WANTs and A_w is "Amplitude of WANT". For example, in "Desire for food", the "Satisfaction of WANT" is calculated by following formula.

(Satisfaction of desire for food)

= (Amount of eaten food)/(Amplitude of desire for food)

D: Derivation of "Important degree of WANT"

"Amplitude of WANT"

Amplitude of "WANT B" is same as "WANT C"

	Want A	Want B	Want C	Average	Weight
Want A	1	1/5	1/5	0.34	0.090
Want B	5	1	1	1.71	0.455
Want C	5	1	1	1.71	0.455
sum	-	-	-	3.87	(1)

Amplitude of "WANT C" is larger than "WANT A"

Fig. 6 Example of derivation of "Amplitude of WANT"

The "Important degree of WANT" is derived by a pair comparison method. This procedure is almost same as procedure B.

E: Calculation of "Overall Satisfaction of WANTs"

The "Overall Satisfaction of WANTs" is denoted by the following formula.

$$Q = \sum_i w_i s_i \tag{2}$$

Where Q is "Overall Satisfaction of WANTs", w_i is "Important degree of WANT" of want "i",S_i is "Satisfaction of WANT" of want "i".

3.2 *Procedure of Proposed Method*

In this section, the procedure of proposed method is described. By using proposed method, it becomes possible to evaluate a service quantitatively. This proposed method is based on quantitative WCA described in Sect. 3.1.

The procedure of proposed method is shown below.

A: Analysis of intended service using conventional WCA

Conventional WCA of intended service is performed. The result of this process called "Proposed Model (PM)"

B: Derivation of a present service model

The new part of intended service is eliminated and the present service is derived from PM. The result of this process called "Existing Model (EM)"

C: Calculation of "Overall Satisfaction of WANTs"

Using method described in Sect. 3.1, the "Overall Satisfactions of WANTs" are calculated for all stakeholders. This process should be performed to PM and EM

D: Calculation of increment of "Overall Satisfaction of WANTs"

The increment of "Overall Satisfaction of WANTs" is calculated from taking a difference of PM and EM. This process should be performed to every stakeholder.

E: Evaluation of the intended service

If all increments of "Overall Satisfaction of WANTs" of all stakeholders are positive, the intended service is effective.

Furthermore, when there is a competitive service, we can judge which service is more effective to perform the additional procedures.

The additional procedure is shown below.

A′: Analysis of competitive service using WCA

Conventional WCA of competitive service is performed. The result of this process called "Competitive Model (CM)"

B′: Check of a present service model

The new part of competitive service is eliminated and the present service is derived from CM. The result of this process should be same as EM.

C′: Calculation of "Overall Satisfaction of WANTs"

"Overall Satisfactions of WANTs" are calculated for all stakeholders in CM.

D′: Calculation of increment of "Overall Satisfaction of WANTs"

The increment of "Overall Satisfaction of WANTs" is calculated from taking a difference of CM and EM. This process should be performed to every stakeholder.

E′: Evaluation of the competitive service

If all increments of "Overall Satisfaction of WANTs" of all stakeholders are positive, the competitive service is effective.

F: Comparison between intended service and competitive service

Both of increment of "Overall Satisfaction of WANTs" calculated by procedure D and D′ are increment from EM. Therefore, it is possible to compare the intended service with competitive service by comparing increment of each service. The increments of main customers' "Overall Satisfaction of WANTs" are compared between intended service and competitive service. If the "Overall Satisfaction of WANTs" of intended service is larger than competitive service, the intended service is more effective than the competitive service.

4 Validation of Proposed Method

4.1 *Procedure of Validation*

In this chapter, we apply proposed method to existing example of service.

Example of service used for verification is "Table for Two" program [9].

"Table for Two" program (TFT) is a restaurant service model that intermediates restaurant guests to donate a part of their usual meal fees for starving children in Africa. This model is based on restaurant customers' altruistic wants to help children in needy. The customer of advanced nations has a problem resulting from food satiation, such as overweight. On the other hand, there are many starving children in Africa. In this program, the customer in advanced nations eats healthier TFT meals, and US$0.25 per meal is donated to TFT. A child in need receives a healthy school lunch from TFT.

As a competitive service, a conventional donation program of food is considered.

The example settings of each service are shown in Table *. In the TFT program, funds for donation are generated by decreasing quantity of meal. In the conventional donation program, funds for donation are generated by increasing price of meal. The validation of proposed method will be performed based on Table 2.

4.2 *Application of Proposed Method*

In this section, we perform the proposed method according to a procedure described in Sect. 3.2

A: Analysis of intended service using conventional WCA

The result of this procedure is shown in Fig. 7.

B: Derivation of a present service model

The result of this procedure is shown in Fig. 8.

C: Calculation of "Overall Satisfaction of WANTs"

Table 2 Settings of each service

	Table for two program (PM)	Conventional donation program (CM)	Usual meal (EM)
Price of meal	$5.00	$5.25	$5.00
Donation per meal	$0.25	$0.25	–
Quantity of meal	Smaller	Usual	Usual

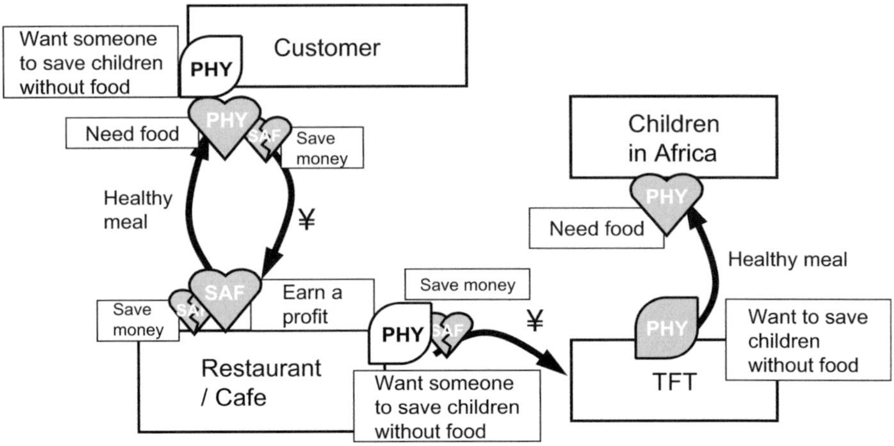

Fig. 7 Result of WCA about "Table for Two" (PM)

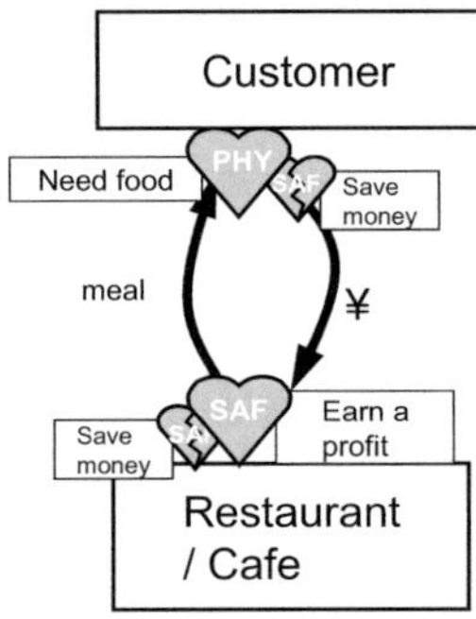

Fig. 8 Result of WCA about present service model (EM)

"Children in Africa" and "TFT" do not exist in EM. It is obvious that increment of "Overall Satisfaction of WANTs" is positive. Therefore, calculation about "Overall satisfaction of WANTs" is performed about "Customer" and "Restaurant / Cafe".

The result of calculation about Customer of PM and EM is shown in Figs. 9 and 10.

The result of calculation about Restaurant / Cafe of PM and EM is shown in Figs. 11 and 12.

D: Calculation of increment of "Overall Satisfaction of WANTs"

The increment of each "Overall Satisfaction of WANTs" is calculated as follows.

Customer: $0.64 - 0 = 0.64$
Restaurant/Cafe: $1.07 - 0 = 1.07$

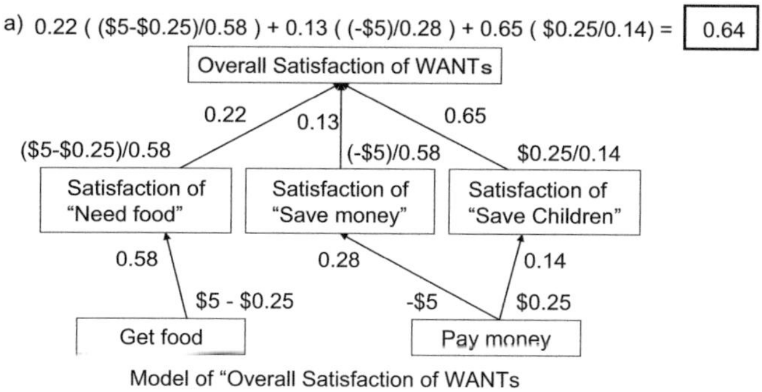

b)

	Need food	Save money	Save Children	Average	Weight
Need food	1	3	3	2.08	*0.58*
Save money	1/3	1	3	1	*0.28*
Save children	1/3	1/3	1	0.48	*0.14*
sum	-	-	-	3.56	(1)

Derivation of "Amplitude of WANT"

b)

	Need food	Save money	Save Children	Average	Weight
Need food	1	3	1/5	0.84	*0.22*
Save money	1/3	1	1/3	0.48	*0.13*
Save children	5	3	1	2.47	*0.65*
sum	-	-	-	3.87	(1)

Derivation of "Importance degree of WANT"

Fig. 9 Calculation about customer of PM

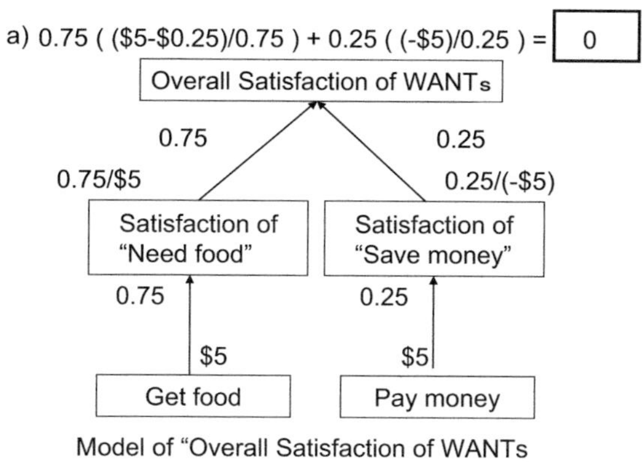

b)

	Need food	Save money	Average	Weight
Need food	1	3	2.08	*0.75*
Save money	1/3	1	0.48	*0.25*
sum	-	-	3.56	(1)

Derivation of "Amplitude of WANT"

b)

	Need food	Save money	Average	Weight
Need food	1	3	2.08	*0.75*
Save money	1/3	1	0.48	*0.25*
sum	-	-	3.56	(1)

Derivation of "Importance degree of WANT"

Fig. 10 Calculation about customer of EM

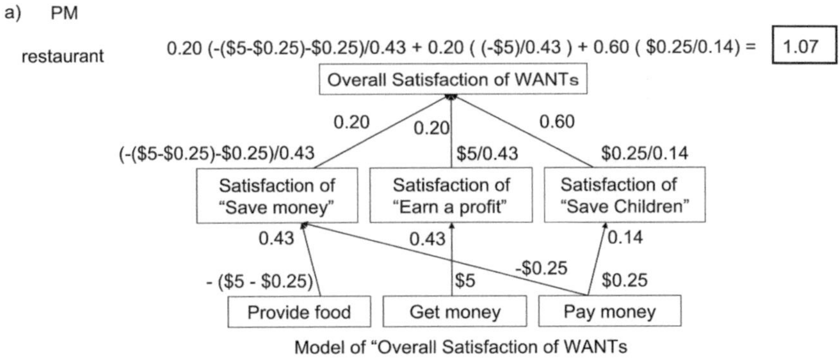

b)

	Save money	Earn a profit	Save Children	Average	Weight
Save money	1	1	3	1.44	*0.43*
Get money	1	1	3	1.44	*0.43*
Save children	1/3	1/3	1	0.48	*0.14*
sum	-	-	-	3.36	(1)

Derivation of "Amplitude of WANT"

b)

	Save money	Earn a profit	Save Children	Average	Weight
Save money	1	1	1/3	0.69	*0.20*
Get money	1	1	1/3	0.69	*0.20*
Save children	3	3	1	2.08	*0.60*
sum	-	-	-	3.46	(1)

Derivation of "Importance degree of WANT"

Fig. 11 Calculation about Restaurant/Café of PM

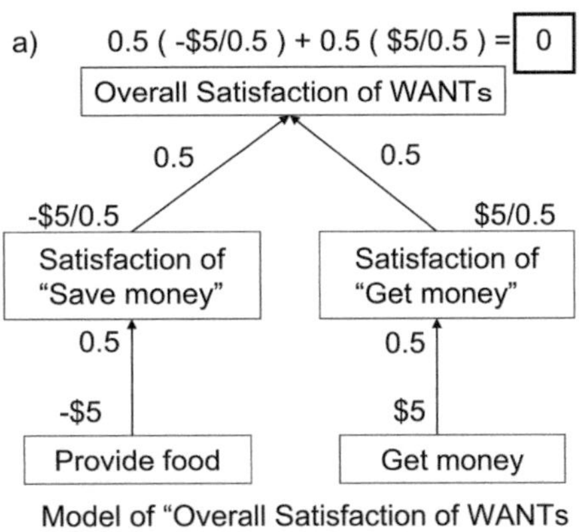

b)

	Save money	Get money	Average	Weight
Save money	1	1	1	*0.5*
Get money	1	1	1	*0.5*
sum	-	-	2	(1)

Derivation of "Amplitude of WANT"

b)

	Save money	Get money	Average	Weight
Save money	1	1	1	*0.5*
Get money	1	1	1	*0.5*
sum	-	-	3.56	(1)

Derivation of "Importance degree of WANT"

Fig. 12 Calculation about Restaurant/Café of EM

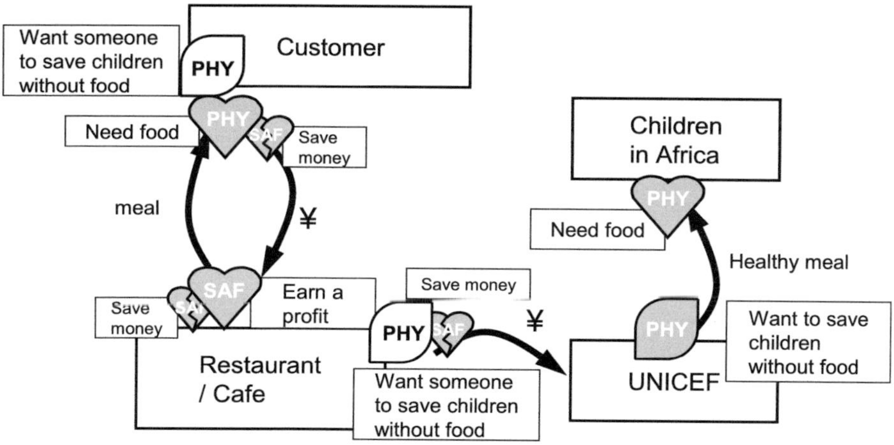

Fig. 13 Result of WCA about conventional donation (CM)

E: Evaluation of the intended service

As shown the result of procedure D, All increments of "Overall Satisfaction of WANTs" are positive. Therefore, the effectiveness of the intended service was conducted.

A′: Analysis of competitive service using WCA

The result of this procedure is shown in Fig. 13.

The result of this procedure, relationship of conventional donation program between stakeholders is almost same as TFT.

B′: Check of a present service model

We eliminated the new part of competitive service from the competitive service, and derived the present service. The present service derived in this process is same as EM. The detail of the result is omitted in this paper because of the result is same as EM.

C′: Calculation of "Overall Satisfaction of WANTs"

The result of calculation about customer of CM is shown in Fig. 14.

The result of calculation about Restaurant / Café of CM is shown in Fig. 15.

D′: Calculation of increment of "Overall Satisfaction of WANTs"

The increment of each "Overall Satisfaction of WANTs" is calculated as follows.

Customer: $0.62 - 0 = 0.62$
Restaurant/Cafe: $1.07 - 0 = 1.07$

E′: Evaluation of the competitive service

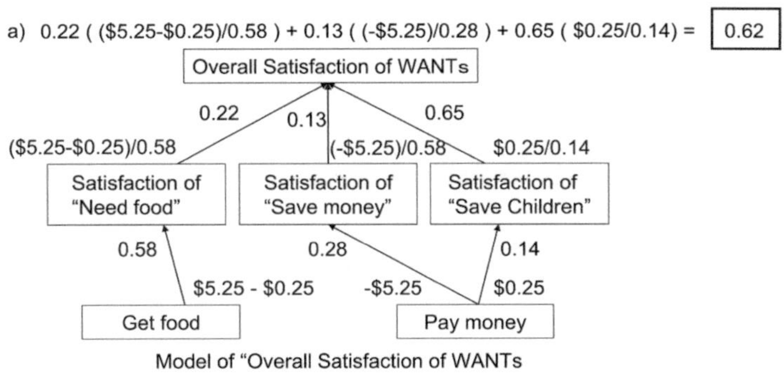

b)

	Need food	Save money	Save Children	Average	Weight
Need food	1	3	3	2.08	*0.58*
Save money	1/3	1	3	1	*0.28*
Save children	1/3	1/3	1	0.48	*0.14*
sum	-	-	-	3.56	(1)

Derivation of "Amplitude of WANT"

b)

	Need food	Save money	Save Children	Average	Weight
Need food	1	3	1/5	0.84	*0.22*
Save money	1/3	1	1/3	0.48	*0.13*
Save children	5	3	1	2.47	*0.65*
sum	-	-	-	3.87	(1)

Derivation of "Importance degree of WANT"

Fig. 14 Calculation about Customer of CM

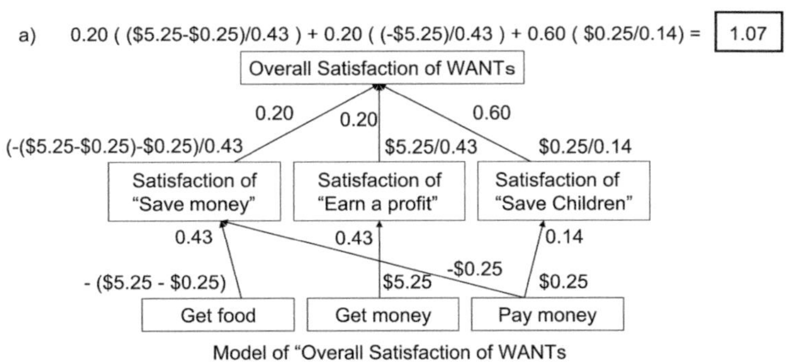

b)

	Save money	Earn a profit	Save Children	Average	Weight
Save money	1	1	3	1.44	*0.43*
Get money	1	1	3	1.44	*0.43*
Save children	1/3	1/3	1	0.48	*0.14*
sum	-	-	-	3.36	(1)

Derivation of "Amplitude of WANT"

b)

	Save money	Earn a profit	Save Children	Average	Weight
Save money	1	1	1/3	0.69	*0.20*
Get money	1	1	1/3	0.69	*0.20*
Save children	3	3	1	2.08	*0.60*
sum	-	-	-	3.46	(1)

Derivation of "Importance degree of WANT"

Fig. 15 Calculation about Restaurant/Café of CM

Table 3 Comparison between PM and CM

Service	Increment	Notes
Intended service (PM)	0.64	Calculated in procedure D
Competitive service (CM)	0.62	Calculated in procedure D′

As shown the result of procedure D′, All increments of "Overall Satisfaction of WANTs" are positive. Therefore, the effectiveness of the competitive service was conducted.

F: Comparison between intended service and competitive service

The increment of "Overall Satisfaction of WANTs" about the customer is shown in Table 3. As shown in Table 3, the increment of PM is larger than CM. Therefore, the intended service is more effective than the competitive service.

4.3 Discussion

As performed through the Sect. 4.2, quantitative analysis and evaluation of actual service are possible using the proposed method. When competitive service is assumed, the difference between intended service and competitive service was able to be quantified clearly.

The TFT, exampled service in this section steadily grow and became quite popular among restaurants users in Japan [9]. This remarkable growth of this service in itself proves validation of the model. In the result of evaluation by proposed method, the validity of TFT is shown in Sect. 4.2. Therefore, the validity of proposed method was confirmed by validation through Sect. 4.2.

4.4 Future Research

In this paper, the validation of proposed method is performed by only one application, demonstrated in Sect. 4.2. Future research could futher validate the proposed method using more examples.

Ideally, the validity could be tested further through the design of actural service.

5 Conclusions

In this paper, we proposed a novel evaluation method of service based on quantification of WCA. The proposed method was validated by case study of application of the proposed method.

References

1. Miles LD (1961) Techniques of value analysis and engineering. McGraw-Hill, New York
2. Maslow AH (1987) Motivation and personality, harpercollins college div, 3rd edn. Harper and Row, New York
3. Fujita T, Sugiyama N, Kanke M, Yasui T, Shirasaka S, Maeno T (2013) Concept design of cause related marketing using wants chain analysis and co-creation workshops. In: Proceedings of the international conference on serviceology, October 2013, Tokyo, pp 36–41
4. Maeno T, Makino Y, Shirasaka S, Makino Y, Kim SK (2011) Wants chain analysis: human-centered method for analyzing and designing social systems. In: Proceedings of the international conference on engineering design, Copenhagen, August, pp 302–313
5. Maeno T, Yasui T, Shirasaka S, Bosch O (2012) Social system design using models of social and human mind network -CVCA, WCA and Bayesian network modeling. In: Proceedings, 4th international conference on Applied Human Factors and Ergonomics (AHFE2012), San Francisco, USA, 21–25 July
6. Saaty TL (1990) How to make a decision: the analytic hierarchy process. Eur J Oper Res 48:9–26
7. Donaldson KM, Ishii K, Sheppard SD (2006) Customer value chain analysis. Res Eng Des 16:174–183
8. Engel JF, Blackwell RD (1982) Consumer behavior, 4th edn. The Dryden Press, Chicago
9. http://www.tablefor2.org. Table for two program, TABLE FOR TWO international, last access date 30 April 2014

Design Method of Target Customer's WANTs for a Service Based on Classification of Services Using WANTs

Kazuto Imazeki, Toshiyuki Yasui, and Takashi Maeno

Abstract We proposed a target customer's WANTs design method for service according to a service classification. Using this proposed method, we can design a target WANTs of a customer when developing a service. First, we classified services using the concept of value engineering and Maslow's hierarchy of needs. Next, a hypothesis about relationships between the classification of service and successful service are proposed. Furthermore, a designing method for "target WANTs" of service is introduced.

We validated the hypothesis by analyzing successful services. We also applied the proposed method to an example of a service in order to validate the proposed method.

Keywords Design method • Wants • Customer satisfaction • Classification

1 Introduction

In recent years, consumer's values have changed and diversified, increasing the difficulty in designing products, services and business models to meet the complexity of today's marketplace.

To cope, product developers conduct a requirements analysis using systems engineering [1] or value analysis [2]. Moreover, customer requirements analysis is a popular area of research, especially relating to service development. Analyzing the relationship between stakeholders can clarify customer requirements, and the Customer Value Chain Analysis (CVCA) is a visualization and analysis tool to map these relationships [3]. While this tool is useful for analyzing a business model, it is not suited to new business model design.

K. Imazeki (✉) • T. Yasui
Graduate School of System Design and Management, Keio University, 4-1-1 Hiyoshi Kohoku-ku, Yokohama-shi, Kanagawa-Prefecture 223-8526, Japan
e-mail: kazuto_imazeki@a3.keio.jp

T. Maeno
Graduate School of System Design and Management, Keio University, Yokohama 223-0061, Japan

T. Maeno et al. (eds.), *Serviceology for Designing the Future*,
DOI 10.1007/978-4-431-55861-3_37

This diversification of values can be put into other words as a diversification of WANTs, where WANTs are "needs or wants". Therefore, when designing a business model, it is more helpful to clarify the target WANTs within the business model. This study focuses on WANTs and proposes a concept design method for a business model [4]. This method effectively designs a business model concept when stakeholder's WANTs are very clear, however, when stakeholder's WANTs are not clear, there is no starting point of analysis. Due to the diversification of WANTs, there are few cases where customers' desires are clear. Therefore, effectiveness of this method is limited.

There is a clear requirement for a suitable design model to clarify the target's WANTs for a new service.

This paper is organized as follows. First, services were classified based on analysis of WANTs. Second, a description of the proposed business model design process is provided. Third, the model is tested and validated. Finally, the paper offers some concluding remarks, including study limitations and opportunities for future research.

2 Classification of Services

2.1 Wants Chain Analysis

In this paper, the Wants Chain Analysis (WCA) is used to analyze the stakeholder's WANTs, a method that analyzes and visualizes relationships between stakeholders [5]. The WCA is based on CVCA (Customer Value Chain Analysis), though the CVCA considers the exchange of money and materials to analyze the relationship between stakeholders, whereas the WCA also includes the stakeholder's WANTs. Figure 1 shows an example of WCA diagram.

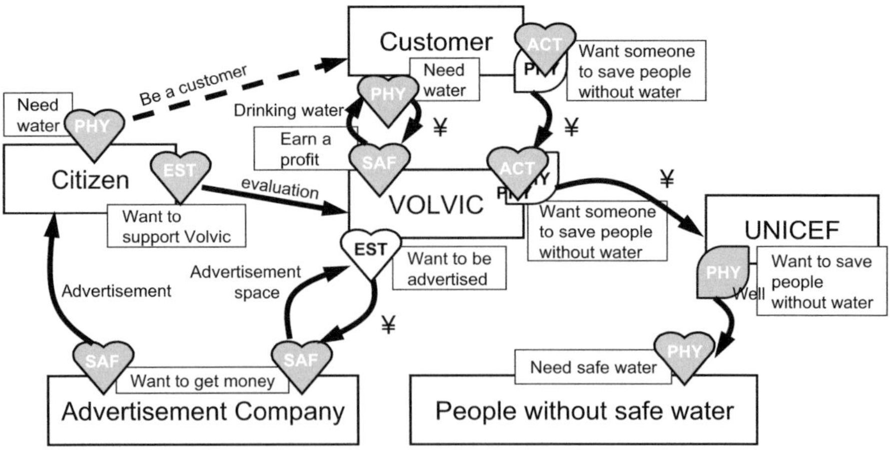

Fig. 1 Example of WCA

The WCA analysis involves three steps. First, the CVCA of the intended product or service is conducted by enumerating the stakeholders and the relationships between them. Second, the WANTs for each stakeholder are listed, clarifying the WANTs that lead to action. Finally, the state of satisfaction for each WANTs is verified against the proposed product or service. If the product or service satisfies all WANTs of all stakeholders, the proposed or service should be effective. If there are unsatisfied WANTs, the product or service is insufficient.

Using WCA, it comes to be able to analyze relationships between stakeholders. Therefore, WCA is useful tool to analyzing a business model.

2.2 *Maslow's Classification of Needs*

The WCA classifies WANTs into 5 (+2) levels based on Maslow's research [6], shown in Fig. 2. While not strictly proven, Maslow's classification is widely used because it is intuitive and easily understood, and is therefore adopted in this study. Maslow's classification begins with basic "physiological needs" advancing toword "self-actualization nees". As each of the lower-level needs are satisfied, an individual will seek the satisfaction of a higher-level need. These relationships are not strict. The satisfaction of low-level WANTs and an increase in high-level WANTs advance gradually shown in Fig. 2b, and do not show a digital behaviour. Therefore, low-level WANTs and high-level WANTs may exist simultaneously. "Desires to know and understand" and "aesthetic needs" are independent of other "needs" existing regardless of the satisfaction status of the other needs.

Using Maslow's classification, it comes to be able to analyze the customers' behaviour based on various WANTs.

2.3 *Relationship Between Value and Wants*

Value Engineering [1, 2] discusses the enhancement of value in product development. However, it is also effective in service development by defining the value of the service and looking for ways to enhance the value.

The mathematical definition of value in value engineering, the function of the product divided by the cost of the product is shown in Fig. 3.

To increase the value of product, either the function must increase or the cost must decrease. The function of the product can be increased through either improvements to current functionality, or the addition of new product functions. In the case of service development, "product" in the sentence should be replaced with "service".

Figure 4 shows a value engineering example of increasing the value in product and service.

A detailed business model analysis can be conducted by considering stakeholders' WANTs, and analyze customer satisfaction in terms of the customers' WANTs as mentioned in Sect. 2.1. Value engineering analyzes a product's ability

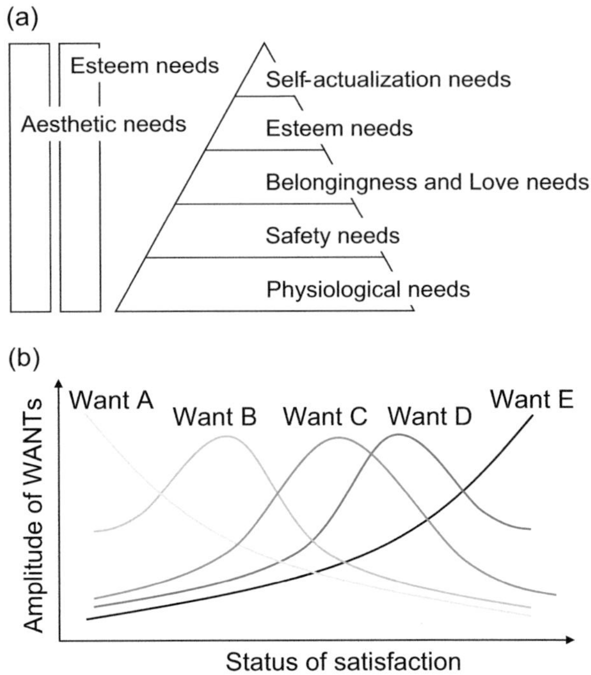

Fig. 2 Classification of WANTs. (**a**) Layered structure of WANTs. (**b**) Transition of WANTs

$$\text{Value} = \frac{\text{Function}}{\text{Cost}}$$

Fig. 3 Definition of "Value"

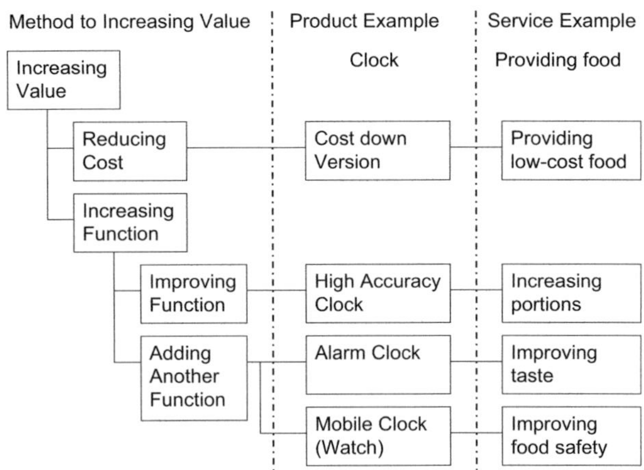

Fig. 4 Example of increasing the "Value"

to satisfy the customer's needs, so this paper replaces "increasing value" with "increasing the satisfaction of the customer's WANTs". To illustrate this position,

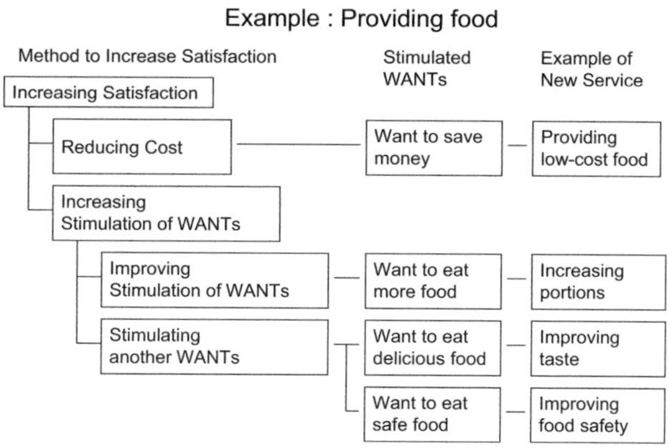

Fig. 5 Example of increasing satisfaction

the service example provided in Fig. 4 is adjusted in Fig. 5 to demonstrate the stimulation of WANTs.

2.4 *Classification of Services Based on WANTs*

Based on the result of Sect. 2.3, classification of services is conducted in this section.

The simplest service is a service that constructed to satisfy the customer's clear WANTs. This paper classifies this type of service as a "WANTs response type service." For example, a service providing food for hungry customers belongs to this type. However, the actual services satisfy various customer WANTs, regardless of whether these are clear or potential.

A service that anticipates potential WANTs is classified as a "WANTs stimulus type service." For example, a service of providing a food can satisfy not only a desire for food but also other WANTs such as to eat delicious food or to eat safe food. Therefore, the actual service has the character of both "WANTs response" and "WANTs stimulus" type services. Present value is captured in the WANTs response type service and additional value in WANTs stimulus type services. In conventional analysis, "WANTs response type service" applied in many cases because the analysis was provided for an existing service. However, to design a new business model, it is important to consider both WANTs stimulus and WANTs response type services.

With the increasing diversification of customer WANTs and sense of value, it is important to design services that stimulate a complex WANTs and decide which WANTs to stimulate. The WANTs stimulus type service can be divided into 2 classes, simple or complex with complex WANTs stimulating services further divided into 3 classes based on the Maslow's classification [6]. In the first, the WANT stimulated by the new service is of a lower-level than that stimulated by the conventional service. In the second, the new and conventional services stimulate

Fig. 6 Classification of service

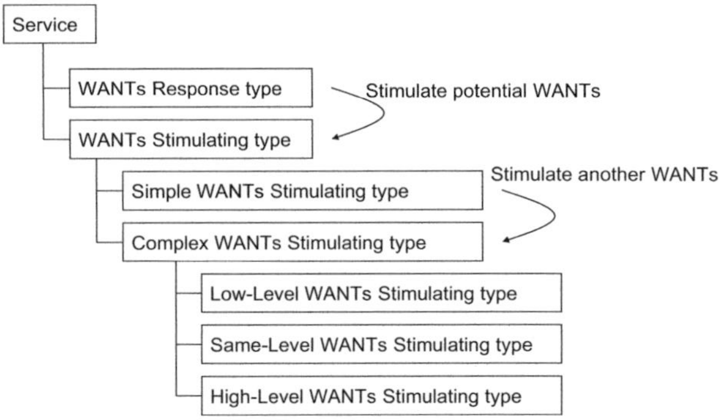

same-level WANTs. In the last, the new service stimulates higher-level WANTs than the conventional service. These service definitions are illustrated in Fig. 6.

3 Business Model Design Method

3.1 Hypothesis

As low-level WANTs are satisfied, a desire for high-level WANTs increases. This concept applies to service design. In other words, when a customer's WANTs are satisfied by the conventional service, the customer's high-level WANTs increase.

Therefore:

Hypothesis: The service becomes effective when the service is designed to satisfy higher-level WANTs than those satisfied by the conventional service.

3.2 Proposed Design Method

As mentioned in Sect. 2.1, WCA is a useful method to analyze a current service and to clarify the WANTs. In the proposed method, WCA is used to both clarify the WANTs satisfied by the conventional service and to evaluate the new service post-design.

Figure 7 illustrates the procedure of the proposed method. First, a WCA of the conventional service is conducted; clarifying the customer's satisfied WANTs. Second, a list of potential customer WANTs are generated during an idea-generating session, such as brainstorming. Next, a target WANT of the new service is selected from listed WANTs of the customer that fulfils the hypothesis proposed in Sect. 3.1. And new service is designed using selected target WANT. Finally, the new service is evaluated with a WCA.

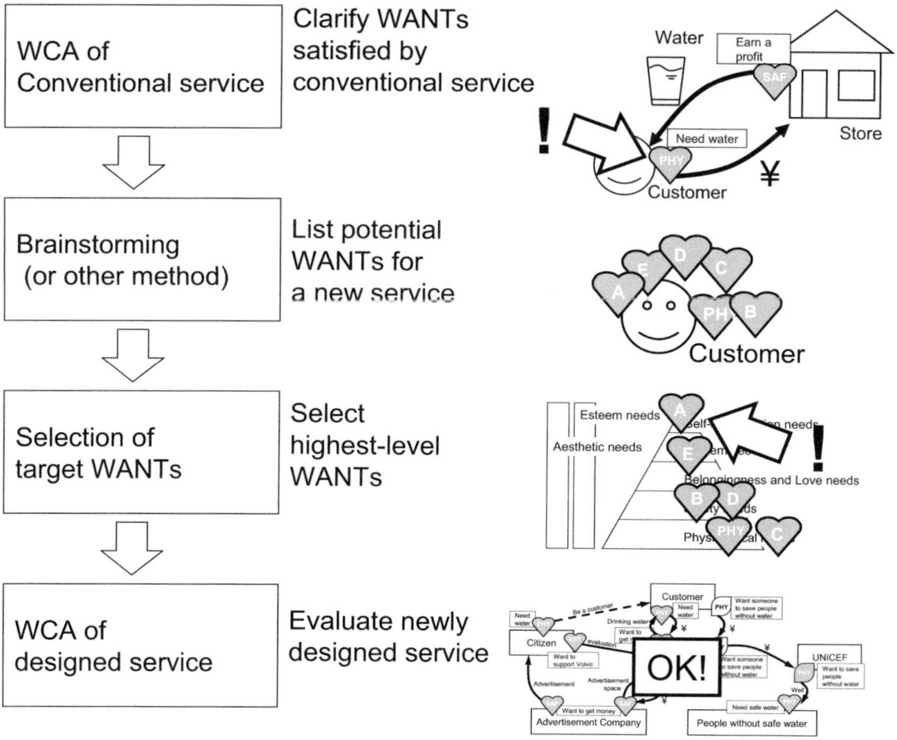

Fig. 7 Procedure of proposed method

4 Validation

4.1 Validation of Hypothesis

To validate the hypothesis, an analysis of successful social business models selected by Ministry of Economy, Trade and Industry of Japan [7] was conducted. This report includes various examples of business model, and each business model has novelty and runs successfully. Therefore, this report is suitable to use validation.

The report includes both non-profit (NPO) and for-profit (non-NPO) type enterprises. The extract is summarized in Table 1.

In this paper, Non-NPO business models are used to verify the hypothesis, because the proposed method is intended to design new services.

For the 26 Non-NPO type enterprises, the WANTs were analyzed to determine the corresponding level in Maslow's hierarchy [6] stimulated by the service. Figure 8 shows an example analysis of "Future-ing Network" enterprise. The added stimulus WANTs "Want to belong to a local community" belong to "Belongingness and Love needs" and conventional WANTs "Want to get information"

Table 1 Extraction of business models

Type	Number of cases
Non-NPO business model	26
NPO business model	29
Total	55

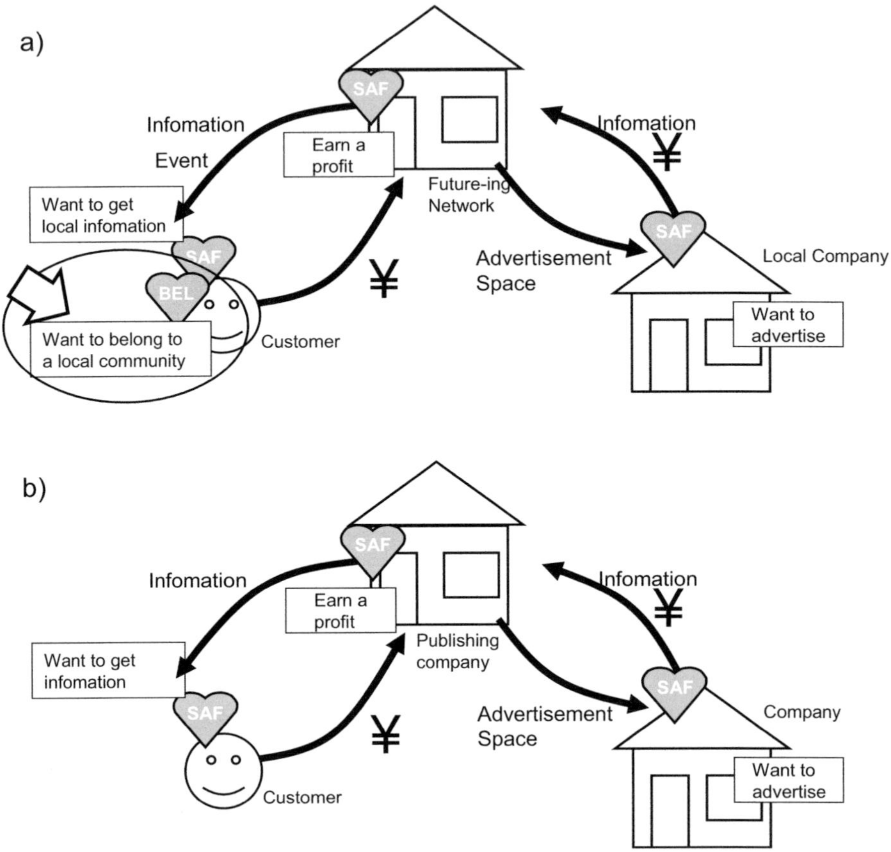

Fig. 8 Example of analysis "Future-ing Network". (**a**) Future-ing Network. (**b**) Conventional service [8]

belong to "Safety needs". Therefore, in this example, the added WANTs belong to a higher-level than conventional WANTs. The analysis of the 26 extracted business models are conducted same as example.

A stimulus situation, that is, a difference between the current service and the conventional service is extracted for each service. The stimulus situations were divided into 5 classes based on the classification of services described in Sect. 2.3. The classification results are shown below.

A: The business model stimulates the WANTs of new customers.
B: The business model stimulates the WANTs of current customers.

Table 2 Result of classification

Class		Number of cases
A	Stimulates new customer	3
B-1	Stimulates lower-level WANTs	0
B-2	Stimulates same level WANTs	3
B-3	Stimulates higher-level WANTs	19
C	Same as conventional model	1
total		26

B-1: The business model stimulates lower-level WANTs than the conventional business model.
B-2: The business model stimulates the same level WANTs as the conventional business model.
B-3: The business model stimulates higher-level WANTs than the conventional business model.
C: The business model is same as the conventional business model.

Table 2 shows numbers of cases for each class of business model.

In the analysis, class B-3 returned the most results, matching and validating the hypothesis.

4.2 Validation of the Proposed Design Method

A case study further validates the proposed method. The "Drink 1, Give 10" is a campaign from VOLVIC [9], where the company contributes some earnings toward charitable work. Conventionally, the customer is interested in water, and a company sells water at a profit to satisfy this WANT. This campaign was chosen because it has few stakeholders, so it is easy to understand, and the campaign is successful.

First, a WCA of the conventional service was performed, showing that the service satisfied "Want to drink water" a physiological needs. The WCA diagram is presented in Fig. 9.

Second, other candidate, or potential, customer WANTs were listed shown in Table 3.

Next, target WANTs from the new service was selected from the potential list of WANTs in Table 3. Number 5 in this list stimulated the same-level WANT as the conventional service, and thus, unsuitable for a new service. The remaining candidates were suitable for new service, stimulating a higher-level WANT than the original service. The most suitable candidate, number 1 in the list, stimulated the highest-level WANT, and is therefore the most suitable for a new service. The example service "1 litter for 10 litters" is the best service among those proposed in the last column of Table 3.

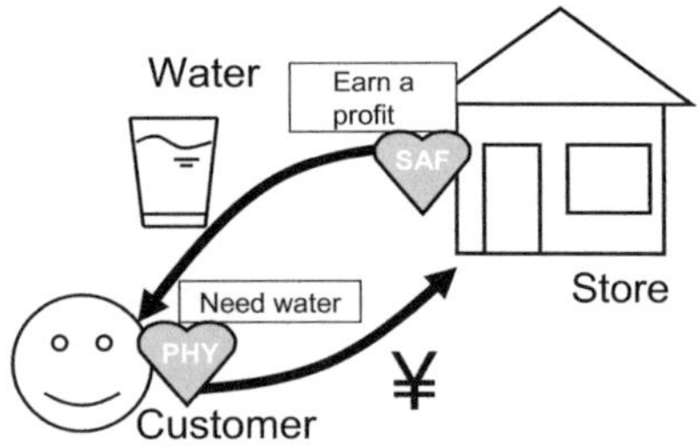

Fig. 9 WCA of conventional service

Table 3 Candidate of WANTs and services

No	Class	WANTs	Example
1	Self actualization	Want to help someone	Drink 1, Give 10 campaign
2	Esteem	Want to esteem someone	–
3	Belongingness & Love	Want to belong local community	Using locally produced water
4	Safety	Want to drink safe water	Advertisement of Water quality
5	Physiological	Want to drink more water	Increase quantity of bottles sold

Finally, a WCA evaluation of the new "Drink 1, Give 10" campaign was conducted. The WCA result, that the new service satisfies the WANTs of all stakeholders, is shown in Fig. 10.

The result show that the proposed method can design a new service by targeting selected higher-level WANTs, further validating the proposed method.

4.3 Discussion

In the previous sections, the hypothesis was validated using an analytical approach and the proposed method was validated using a case study. In the case study, the potential services were selected based on a consideration of WANTs, with the suitable service selected from many candidates. The selected service " Drink 1, Give 10" is a actual campaign by VOLVIC since 2007, and campaigns in 2014. So, the validity of selected service is proved. Therefore, the validity of proposed method was confirmed by validation through Sect. 4.2.

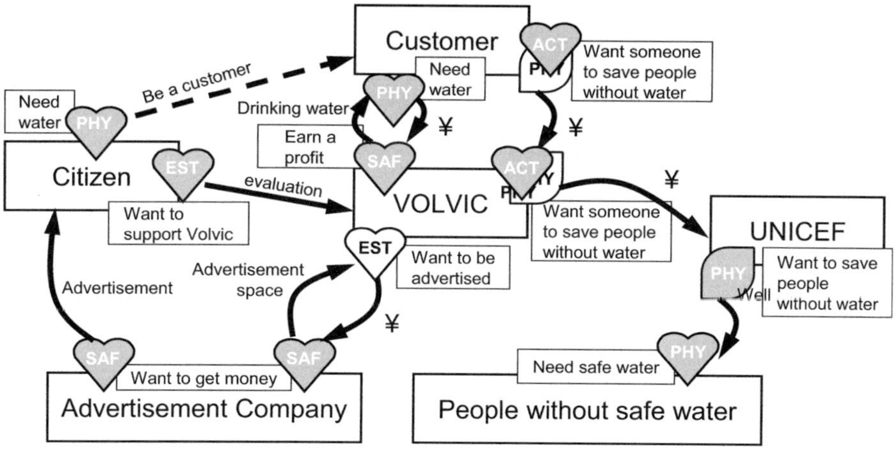

Fig. 10 WCA of "Drink 1, Give 10" campaign

4.4 Future Research

In this paper, the validation of proposed method was validated with only one application, demonstrated in Sect. 4.2. Future research could further validate the proposed method using more examples.

Ideally, the validity could be tested further through the design of actual service.

5 Conclusion

In this paper, we introduced a hypothesis about the relationship between the classification of services and the success of the service, proposing a design method based on customers' WANTs. The hypothesis was validated by analyzing examples of successful social business. And proposed method was validated by a case study of application of the proposed method.

References

1. SE Handbook Working Group (2010) Systems engineering handbook. Incose, San Diego
2. Miles LD (1961) Techniques of value analysis and engineering. McGraw-Hill, New York
3. Donaldson K, Ishii K, Sheppard S (2006) Customer value chain analysis. Res Eng Des 16:174–183
4. Fujita T, Sugiyama N, Kanke M, Yasui T, Shirasaka S, Maeno T (2013) Concept design of cause related marketing using wants chain analysis and co-creation workshops. In: Proceedings of the international conference on serviceology, October 2013, Tokyo, pp 36–41

5. Maeno T, Makino Y, Shirasaka S, Makino Y, Kim SK (2011) Wants chain analysis: human-centered method for analyzing and designing social systems. In: Proceedings of the international conference on engineering design, Copenhagen, August, pp 302–313
6. Maslow AH (1987) Motivation and personality, Harpercollins College Div, 3rd edn. Harper & Row, New York
7. http://www.meti.go.jp/policy/local_economy/sbcb/sb55sen.html (2009) 55 example of social business, Ministry of Economy, Trade and Industry of Japan (in Japanese). Last access date, 28 April, 2014
8. http://www.meti.go.jp/policy/local_economy/sbcb/data55sen/p016_017.pdf (2009) Example of "Future-ing Network", Ministry of Economy, Trade and Industry of Japan (in Japanese). Last access date, 28 April, 2014
9. http://www.drink1give10.com, Volvic Drink 1, Give 10 Campaign, Volvic. Last access date, 28 April, 2014

Business Model Generation Canvas as a Method to Develop Customer-Oriented Service Innovation

Akane Matsumae and Karl Burrow

Abstract A business loses its customers when it no longer meets customer demands. However, even if a manager has noticed the necessity to innovate their timeworn services, s/he often has difficulty knowing how to recapture that customer demand. It is highly productive to involve community members in these decision-making processes, especially if the business aims to take a certain social role in its community.

To address the problem of customer loss, while incorporating community voices, we had groups of a business's potential customers of diverse background co-create its service innovation by using Business Model Generation Canvas. We utilized Business Model Generation Canvas not only to co-design business models but also to take most advantage of its secondary effects of developing customer-oriented service innovation; visualization of thoughts, joy derived from co-creation, emotional attachment to or sense of deep involvement in the creation.

Business Model Generation Canvas enabled potential customers to design business models directly based on their own needs as well as on customer demands. In the process business managers were also helped to develop customer-oriented service innovation.

Positive side-outcomes were that potential customers were spontaneously turned into actual ones, or into supporters of a business due to their emotional attachment or sense of deep involvement in the business co-designed by themselves.

Keywords Customer-oriented • Co-create • Collective knowledge • Business Model Canvas • Service innovation

---

A. Matsumae (✉)
Saga University, Saga 840-8502, Japan
e-mail: akane@entr.cc.saga-u.ac.jp

K. Burrow
Karllestone Capital, Tokyo, Japan

T. Maeno et al. (eds.), *Serviceology for Designing the Future*,
DOI 10.1007/978-4-431-55861-3_38

1 Introduction

Some social problems might be solved in a dependent way, such as a grant, a donation, or charity. However, in most cases, such dependent solutions end up as just a symptomatic therapy and can seldom provide fundamental solutions. Without professional services neither those social problems nor the sustainable and effective impact towards these social problems could be solved or achieved. In contrast, changing our business from profit-oriented business to people-oriented business is essential and effective way to solve the social problems caused from the overwhelming traditional business [1].

How can an appropriate business be generated in its society? How can a business be sustainable and deeply involved with its customers or society members? Especially if the business aims to take a certain social role in its society, it is highly productive to involve society members in these decision-making processes.

In this study, the worldwide method, Business Model Canvas [2], was applied with groups of its potential customers to answer those questions. Some of the workshops from February 2012 to October 2014 are introduced with following discussions in terms of co-creating customer-oriented service innovation.

2 Method

2.1 Business Model Canvas

A business model describes the rationale of how an organization creates, delivers, and captures value, and a business model shows the logic of how a company intends to make money. It does not matter if it is a non-profit or a for-profit enterprise. Social enterprises may be mission-driven, focused on delivering social impact versus a financial return on investment, but they still need a sustainable model to scale over time. Within the age of social interaction, business models must have the capability to adjust more quick and elastically.

The Business Model Canvas (Fig. 1) helps mapping, discussing, designing and inventing new business model as well as questioning existing models among various people. It is a tool used as a conversation piece for establishing how you interact with your customers. It is also a framework that helps us understand how different entities of a business come together to create value for customers.

The complexity of business models makes it difficult to argue about them. A business has many interrelated moving pieces, thereby, it is easy for a person and team to miss something when creating a business model. This complexity and possibilities easily lead to misunderstanding each other when people try to invent new business models.

The Business Model Canvas helps in generating new ideas by asking a few key questions and can be described through nine basic building blocks centred on the

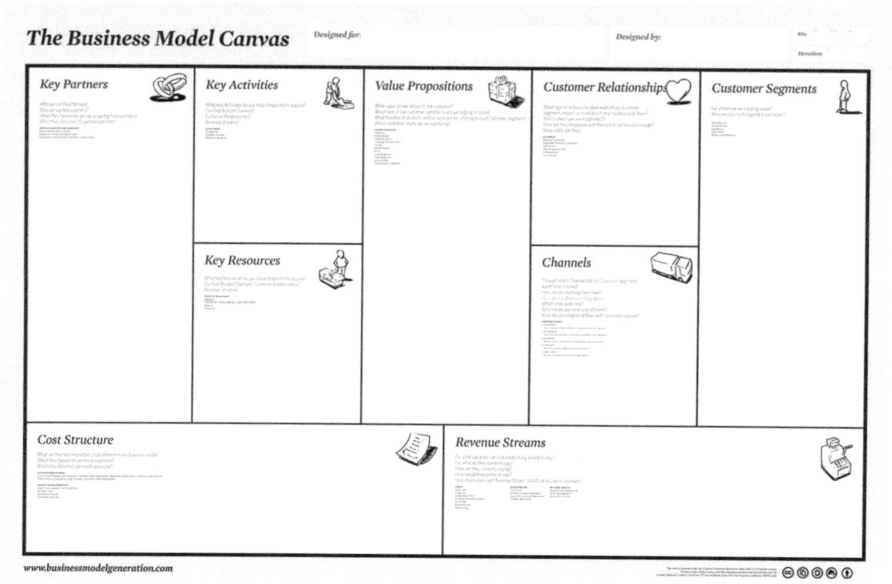

Fig. 1 Business Model Canvas

main block called the Value Proposition. It is a flexible template for capturing, analyzing, and creating a shared language between the nine building blocks of a business model.

2.2 *Basic Workflow*

2.2.1 Preparation

The participants are informed of the business they are going to work on, including its mission and problems, and the concept of the Business Model Canvas a week before the workshop. Short guidance on the target business and the methodology is offered at the beginning of the workshop [3, 4].

2.2.2 Empathy Map

The Empathy Map is a useful tool to understand customer's environment, behaviours, concerns, and aspiration [2]. To co-design a customer-oriented business model, participants start from understanding its customers. After discovering as many customer segments as possible in order to focus on the business model from the aspect of the customer segments as well as the customer perspective, each group

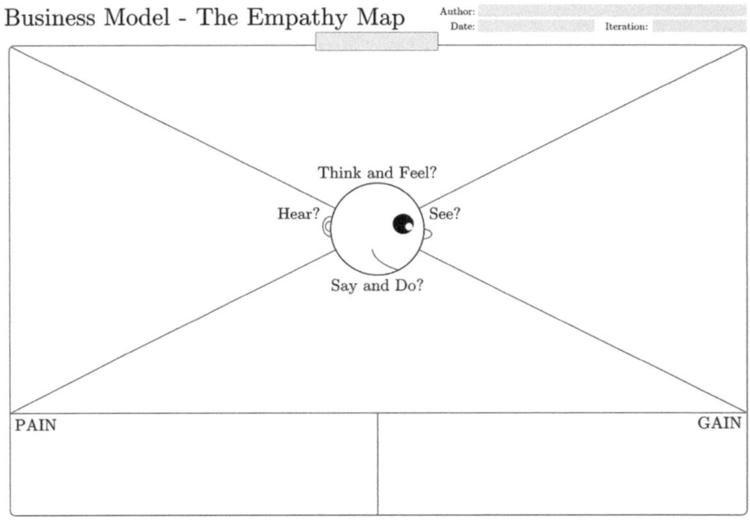

Fig. 2 Empathy Map

chooses the most promising customer segment to be explored by using the Empathy Map (Fig. 2) [3, 4].

2.2.3 Business Model Canvas

Each group is encouraged to find gain creators and pain relievers for the customer segment they choose on the basis of their Empathy Map as a part of the Value Proposition, the core of the Business Model Canvas. Participants are encouraged to produce "What if ?" questions for each element of the Canvas to pivot their basic idea and generate innovative business model (Fig. 3) [3, 4].

2.2.4 Step for Implementation

At the end of the workshop, the director of the target business, who participated in the workshop with his position concealed, states his opinions about the business models generated there and how he would absorb them into his real business [3, 4].

2.3 Participants

In terms of the freshness of the business model ideas, participants should be from various business units, of different ages, with different areas of expertise, of

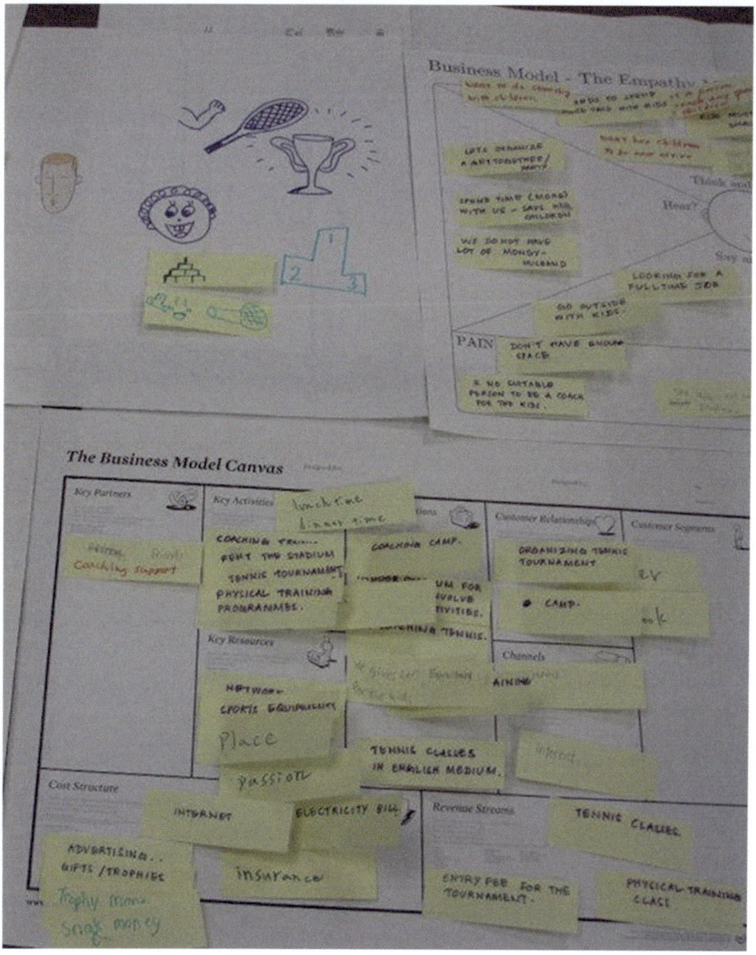

Fig. 3 Empathy Map and Business Model Canvas worked with pictures to share image of new service in a team

differing levels of seniority, with a mixture of experiences, and from different cultural backgrounds [2].

In the workshops reported in this paper, participants are culled from various fields, professions, seniorities and ages. The balance among group members are carefully paid attention when they are grouped.

The workshops reported in this paper were held in different languages, in Japanese and in English, so that our workshops can involve multicultural participants as society members.

2.4 Environments

A series of workshops were carried out with best efforts to offer actual business environments which enabled participants to realize and understand the target business they were working on. Workshops were held on actual business sites, used streamed video images from actual business sites, or in the presence of a key person around whom the business has been based.

3 Case Studies

3.1 Case 1

3.1.1 Target Business in Workshop

Participants worked on a local tennis school. Its director has qualified for the all-Japan championship five times and the national athletic meet five times, and also has the experience of a national team coach of a top junior tennis tournament. He runs his tennis school to convey the joy of tennis widely in the local community whether someone wants to be a professional player or not.

The director has difficulty gathering students and wants to know what his potential customers really expect him to do in the local community. This problem is one of the key questions for his business success in terms of both profits and social impacts.

3.1.2 Workshop Participants

Participants were diverse in age, nationality and profession: ranging in age from 19 to 64, three nationalities (Japanese, Sri Lankan, Chinese), diversity in profession (university students, a medical doctor, a house wives/husbands, a social care manager, a nursery school director, a confectioner, a retired, a local government employee). They were divided into three small groups ensuring the diversity in profession and age (Fig. 4).

3.2 Case 2

3.2.1 Target Business in Workshop

The same business was targeted as in Case 1.

Fig. 4 Workshop Scene: Case 1 with streamed video image from actual business site

3.2.2 Workshop Participants

Participants were multi-cultural: in age ranging from 8 to 48, of five nationalities (Japanese, Sri Lankan, Italian, Chinese, and American), in profession (university students, house wife/husbands, a NPO member, a job seeker, a teacher, and a local government employee, and an elementary school student). They were divided into two small groups ensuring diversity in profession and age (Fig. 5).

3.3 Case 3

3.3.1 Target Business in Workshop

Participants worked for a newly opened café. The café's mission is to serve to sustain for the health of people, community, and global environment through running its business. The cafe owners were experienced and the staff also willingly shared the its mission. A collaboration with local farmers had just started.

The café owner was unsure what kind of services his customers really expected in the local community. This problem should be one of the key questions for their business success in terms of both profits and social impacts.

Fig. 5 Workshop Scene: Case 2 with diverse participants in nationality, age, gender and profession

3.3.2 Workshop Participants

Participants were diverse: in age ranging from early 20s to late 50s, in profession (university students, a professional singer, a CEO of traditional products shop, a web-contents designer, a confectioner, an entrepreneur, and a local government director). They were divided into five small groups ensuring diversity in profession and age (Fig. 6).

3.4 Case 4

3.4.1 Target Business in Workshop

Participants worked for an entertainment business project with a comedian sent from the biggest entertainment company in Japan. The goal of their project is to help promoting local brand with its comedian.

The comedian has been in this region for a couple of years but still has no idea what kind services the local community expects to him to offer (Fig. 7).

Fig. 6 Workshop Scene: Case 3 in actual business site

Fig. 7 Workshop Scene: Case 4 with a comedian sent to help promoting local brand with his service

3.4.2 Workshop Participants

All of participants were university students, however, still diverse in their major (engineering, agriculture, economics, education, and medical department). They were divided into multi-disciplinary and gender -mixed teams of three.

4 Results

Here we obtained five findings:

(1) All the participants were able to get involved and excited while co-designing business models with Business Model Canvas, although it is often said Japanese people are not used to free discussion among a diverse group.
(2) A potential customer group for the target business was formed, who deeply understand the target business, including its mission, problems and possibilities.
(3) The business director or manager understood what their potential customers expect from them and found clues to the customer-oriented business model to be implemented.
(4) Most of the new services co-created and proposed in those workshops were willingly implemented by directors.
(5) 92 % of participants who had never known about the target business before the workshop kept their interest in it at least a week after the workshop.

5 Discussions

5.1 Understanding Customers

It is widely heard among interviewed business managers that they have difficulty understanding customers' needs. It does not necessarily mean that they do not have any information on their customers. Actually, they usually have had plenty of explicit knowledge on them already. To say more accurately, what they have had little opportunity is to capture the context of their customer's behavior, tacit knowledge.

Since Business Model Canvas enables visualizing a business model which has been difficult to pull out of one's mind and transfer to another, it becomes much easier for business managers to understand the context of their customers' behavior by joining these workshops.

5.2 Understanding Business

On the other hand, it is also common for customers to have less opportunity to understand the context of the services offered to them. First, it is simply because business enterprises are seldom provided with the opportunity to introduce their mission and how it reflects their current services offered to customers. Second, customers usually pay less attention to how the services are delivered than what the services are.

The process of our workshops offers opportunities for business enterprises to explain their mission and how it is reflected in their services. In addition, customers can understand the whole structure of the business by using Business Model Canvas, how their services are delivered to them.

5.3 *Interaction Between Services and Customers*

As productive and sane interpersonal relations are based on both deep understanding of the other side and interactive communication between both sides; the relations between services and customers will be productive and sane when they have both understanding for the other side and the interaction between them.

As mentioned in Sect. 2.1, Business Model Canvas can not only visualize a business model but also help mapping, discussing, designing and inventing new business models as well as questioning existing models among various people. Therefore, Business Model Canvas could be a communication tool to interact between services and customers and enable to lead a customer-oriented service innovation.

Understanding and involvement with the business brings attachment to it.

5.4 *Process Management: Grouping*

It is obvious that grouping is the key factor to enrich group work to co-create innovative services. It is important to keep horizontal power balance among group members as well as to involve members with various back ground.

5.5 *Process Management: Ideation*

5.5.1 Value Proposition Canvas

Value Proposition Canvas (Fig. 8) was published in 2013 after workshop Case 3 had finished. Value Proposition Canvas can take a similar role to an Empathy Map in this workshop and it is likely to be helpful to explore and deepen value proposition of business. However, it seemed to be too detailed for usual citizens to use it as just a communication tool between services and customers.

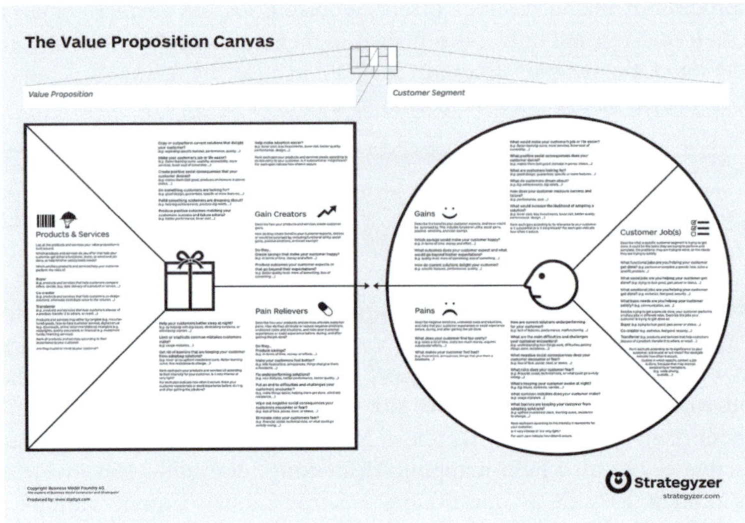

Fig. 8 Value Proposition Canvas [6]

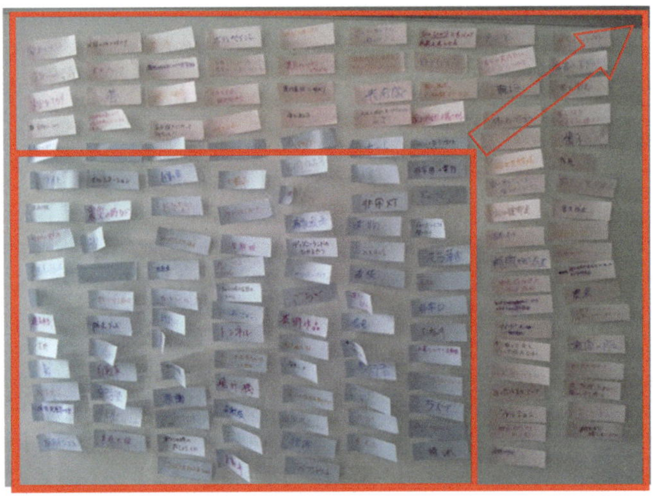

Fig. 9 Increase of ideation in quantity and quality

5.5.2 Customer Characterization

Pink sticky notes are the output from idea generation work after characterizing customer's details (Fig. 9). Customer characterization allows participants to imagine actual business situation and helps them to generate business ideas in terms of quantity and quality.

5.5.3 Environments

Most of co-created services in these workshops were willingly implemented by business managers. This result could be arisen from the environments prepared in workshop. Real environments helped participants understand the actual image of target businesses and led to reasonable Innovative ideas.

5.6 Process Management: Implementation

To settle innovative ideas into society, it is important to gather participants who are expected to be involved in the target business: An implemental body, a creator, an expert, students and various citizens in each group will accelerate the innovation [5].

6 Conclusion

We utilized Business Model Generation Canvas not only to co-design business models but also to take most advantage of its secondary effects of developing customer-oriented service innovation; visualization of thoughts, joy derived from co-creation, emotional attachment to or sense of deep involvement in the creation.

Positive side-outcomes were that potential customers were spontaneously turned into actual ones, or into supporters of a business due to their emotional attachment or sense of deep involvement in the business co-designed by themselves. In the process business managers were also helped to develop customer-oriented service innovation.

6.1 Summary

Groups of potential customers could easily transform the timeworn services into customer-oriented services which meet the needs of society with Business Model Canvas; Business Model Canvas can lead innovative social values by realizing interactive conversation between business and customers.

Grouping, environment, and the implementation system of the workshop are the key factors to lead actual innovation.

6.2 Limitations

The conclusion above should be well-applicable and effective when a target business aims to solve social problem through its business with society members.

6.3 Suggestions for Future Research

This method has also successfully confirmed effectiveness as a mediator between university-industry collaboration and entrepreneurship education (Fig. 10).

As the next step, we plan to apply this method to co-create more professional and specialized business using university's innovation seeds and external funding system. Each team will be grouped by roles and by types of business as Table 1.

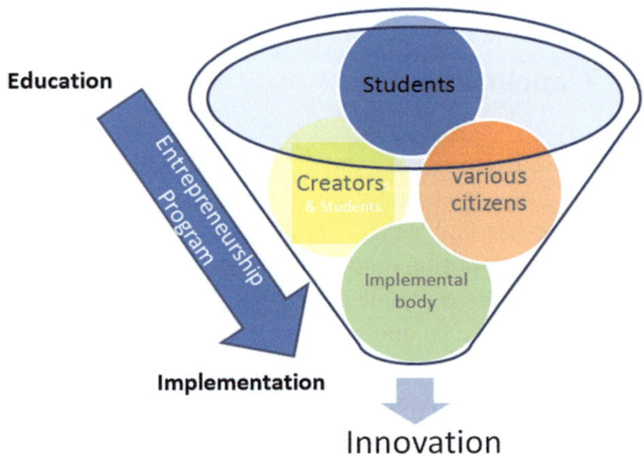

Fig. 10 Fusion of entrepreneurship education and university-industry collaboration

Table 1 Workshop grouping

	Technology Business	Health Business	Agricultural Business	Social Business	Free
Implemental body					
Professional creators					
Potential customers					
Researchers					
Students					

Acknowledgements We would like to take this opportunity to express our gratitude to Prof. Ratnayake of the Department of Economic Systems of Saga University for his cooperation in Cases 1 and 2. Case 3 was hosted by Nabeshima Renaissance Project Office of Saga University sponsored by Japan's Ministry of Education, Culture, Sports, Science and Technology, which is gratefully appreciated.

References

1. Ratnayake P (2012) Japanese grass-roots technical assistance towards development of social business and alleviation of rural poverty: concepts, performance, opportunities and challenges. Ninth international symposium, pp 18–28
2. Osterwalder A, Pigneur Y (2010) Business Model Generation. Wiley, Hoboken
3. Matsumae A, Burrow K (2012) Application of the Business Model Canvas to a Japanese local community to co-design the social Business Model. In: Proceedings of the international conference on convergence content, Saga, Japan, 14–16 December
4. Matsumae A et al (2012) Application of the Business Model Canvas to a multi-cultural community to co-create a business model innovation. In: Proceedings of the international conference on convergence content, Saga, Japan, 14–16 December
5. Matsumae A (2014) BMGen in Japan as a mediator between university-industry collaboration and entrepreneurship education. In: Proceedings of the University industry interaction network international conference 2014, Barcelona, Spain, 23–25 April
6. Business Model Foundry (2014) The value proposition Canvas [online] available from http://www.businessmodelgeneration.com/downloads/value_proposition_canvas.pdf, 30 April 2014

Aligning Product-Service Offerings with Customer Expectations

Paolo Gaiardelli, Giuditta Pezzotta, Barbara Resta, and Lucrezia Songini

Abstract The rapid servitization of the market calls heavy-truck companies for the development of a proper Product-Service (PS) portfolio able to capture customer requirements and needs. In such a context heavy-truck companies are required to monitor strictly customer attitude towards services. The control of the importance given to service quality determinants can be critical to dynamically verify and optimize the completeness and consistency of existing service offerings. An empirically-based typology, built upon an extensive survey involving about 1.000 Italian heavy-truck customers, shows the existence of different types of customers having a different awareness towards service quality determinants that, in turn, can influence their orientation towards PS offering choices.

Keywords Servitization • Service portfolio configuration • Typology • Survey • Cluster analysis • Customers

1 Introduction

Market globalisation, consumers' behavioural change, introduction of new rules and regulatory barriers, have certainly had a great impact on business configuration and organisation in many manufacturing industries. Companies have realised that complementing sales with the provision of value added services can be an important lever to prosper in markets affected by weak demand, hard competition and decreasing margins [1].

An attentive management of all the information on customers' needs and expectations, coupled with a strong control of how customers perceive and evaluate services, are then requested to develop a proper Product-Service (PS) portfolio for fulfilling specific and heterogeneous market requests [2–5].

P. Gaiardelli (✉) • G. Pezzotta • B. Resta
CELS – Research Group on Industrial Engineering, Logistics and Service Operations, University of Bergamo, 24044 Dalmine (BG), Italy
e-mail: paolo.gaiardelli@unibg.it

L. Songini
Department of Economics and Business, Eastern Piedmont University, Novara, Italy

T. Maeno et al. (eds.), *Serviceology for Designing the Future*,
DOI 10.1007/978-4-431-55861-3_39

Focal companies and all the other actors responsible for delivering service components [6], such as their commercial branches, service locations and independent workshops and partners, are thus requested to monitor strictly the market awareness towards service quality determinants. This can allow them to verify dynamically whether the existing PS offering is aligned with customer needs, and to identify the list of potential key services that should be introduced and/or improved in the PS portfolio.

In such a context, in order to understand how service quality determinants can influence customers' service choices, an empirically-based typology developed through a cluster analysis is proposed.

The results are used to shed light on the following research questions:

- Is it possible to identify customer types on the base of the importance given to service quality determinants?
- Are there specific descriptive variables able to represent each individuated customer type?
- To what extent can a major importance, given to a specific set of service quality determinants, influence customers' orientation towards a particular PS offering?

The study is drawn upon an extensive survey carried out in the Italian heavy-truck industry, where companies have been embarking a servitization journey to respond to the recent evolution of the market dynamics and the transformation of customer requirements.

In the reminder of this paper, Sect. 2 introduces a literature review aiming at creating a better understanding about service quality determinants. A brief summary about the existing classification methods of PS portfolios are also presented. The description of the research design, the adopted methodology and the selected industry are reported in Sect. 3, while the survey data analysis and the discussion are introduced in Sect. 4. Finally, Sect. 5 reports the conclusions, the limitations and the further developments of this study.

2 Literature Review

Since this research focuses on understanding the relationship between the importance given to service quality determinants by customers and their PS orientation, a literature review on service quality determinants and PS portfolio configuration is reported in the following.

Service Quality Determinants

Service quality plays a key role in a competitive environment [7–9], since it is a positive element in the customer value creation process [10–12] and is in a direct relationship with the customer satisfaction [13]. An outstanding customer satisfaction drives customers and solution providers to endure their relationship over time, which results in high degree in customer loyalty and retention [13, 14]. These factors, from a company's point of view, lead to a long-term profitability and growth [15, 16].

There is no consensus in literature whether service quality influences customer satisfaction [17–21] or vice versa [22, 23]. Rust and Oliver [20] state that this relation is rather bi-directional, a "virtuous experience-based performance circle". However, all literature contributions agree that service quality and customer satisfaction are multidimensional and dynamic concepts influenced by personal needs, word-of-mouth communication, and past experiences [24, 25] and are based on the following trade-offs: (i) real vs. perceived performances; (ii) customer expectation vs. (dis)confirmation.

Many authors have tried to explicit the different dimensions making up the service quality. Lehtinen and Lehtinen [26] identified three quality dimensions: (i) physical, which refers to all tangible elements of service; (ii) corporate, relevant to company image and vision; and (iii) interactive, related to the relationship between customer and service provider. Following the study of Parasuraman et al. [25], other contributors investigated upon the main determinants of service quality [27]. In particular, Johnston and Silvestro [28] identified 17 determinants which, using the classification proposed by Grönroos [15], can be further declined into:

- Technical determinants (such as flexibility, availability, timeliness and responsiveness, accessibility and comfort, reliability, technical competence cleanliness and tidiness), refer to the results or outcomes of a service, i.e. what the customer is actually receiving;
- Functional determinants (such as customer care, attentiveness/helpfulness, friendliness and courtesy, information, communication, integrity, security and commitment), relate to the manner or the process by which a service is provided, i.e. how the service is delivered.

Product-Service Portfolio

Servitization is now widely recognised as an increasingly relevant strategy for western manufacturers to improve their competitive advantage in the market. Current servitization literature provides different studies that explore the varying degrees of services offered by manufacturing companies along with different dimensions [29–33]. Among others, Gaiardelli et al. [34] propose an unified and schematic descriptive classification model that captures and critically combines three relevant dimensions found in literature, with a complementary and integrated perspective, as summarised in Fig. 1.

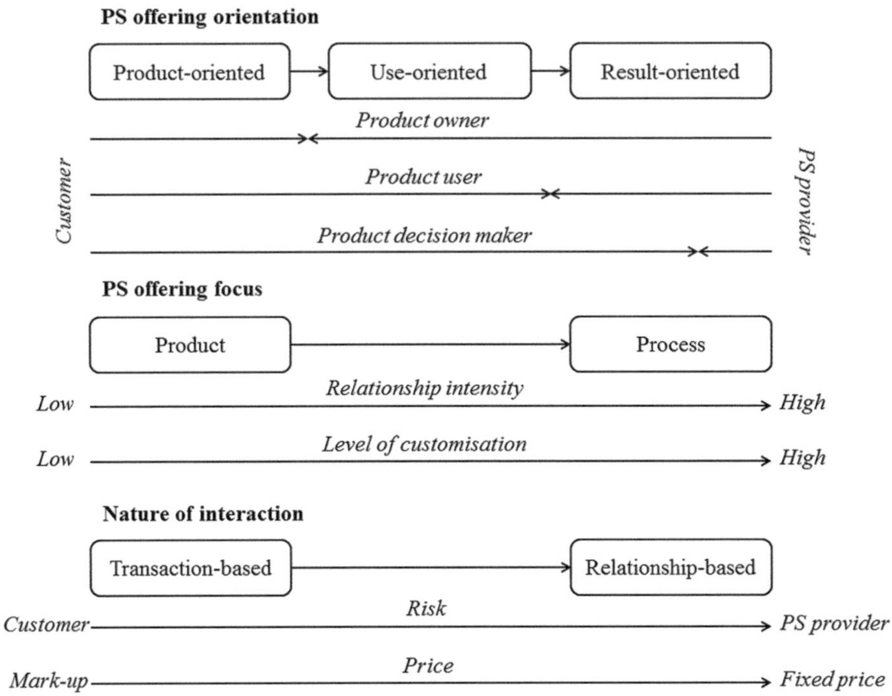

Fig. 1 Main characteristics of PS offering dimensions

3 Research Design

An extensive survey in the heavy-truck industry was carried out. The industry is characterised by a huge service market with high potentials in terms of margins, profitability and competitive advantages [35]. Moreover, the growing dynamism, the enlargement of markets, the changes in the strategic position of suppliers, as well as the introduction of new rules and regulatory barriers, have had a great impact on recent firm configuration and organisation [36]. In particular, heavy-truck companies have progressively moved from transactional to relational business models, established through the provision of value around new service solutions and addressing larger transportation challenges [37].

3.1 Selection of Service Quality Determinants and Identification of PS Portfolio

First of all, the different service quality determinants were identified through a literature review. Then, the preliminary list was presented to the 5 out of the 7 most

Table 1 The most relevant service quality determinants in the heavy-truck business

Service quality determinants	Type
Accessibility	Technical
Attentiveness and courtesy	Functional
Flexibility	Technical
Information and communication	Functional
Integrity and frankness	Functional
Promptness	Technical
Reliability (service)	Technical
Reliability (spares)	Technical
Service Availability (service mix)	Technical
Timeliness	Technical

important players operating in the Italian market and 2 representatives of transportation companies. The received feedback was used to build a final list describing the most important service quality determinants in the heavy-truck business. The list is reported in Table 1.

Then the existing PS offerings were mapped. As reported in Table 2, the same list of 23 services classified by Gaiardelli et al. [38], in accordance with the dimensions described in Fig. 1, was adopted.

3.2 *Survey Method*

In order to answer the proposed research questions, an extended survey was carried out. The survey aimed at developing a typology of heavy truck customers based on the importance given to service quality determinants.

A kick-off meeting with top senior service managers of 4 heavy-truck companies, 2 representatives of franchised and independent service assistance workshops and 2 representatives of transportation companies, was held at the beginning of the research project to design the sample and define questionnaire structure, wording and scaling [39]. In particular:

- The sample of the analysis was drawn from the national lists of freight carriers located in Italy and operating in the European market;
- The questionnaire was divided into four main sections: company general information, importance given to service quality determinants, service customer orientation (current and future). Then the questionnaire was tested to examine its measurement properties and the viability of survey administration. It was submitted to colleagues, industry experts and a sample of key customers selected by the experts involved in the kick-off meeting. Validity and reliability of measure were evaluated and the questionnaire was modified accordingly, to get over the observed measurement errors.

Table 2 Existing PS offerings in the heavy-truck business

	PS offering orientation	PS offering focus	PS nature of interaction
Brand experience services	PO	PR	RL
Driver repatriation	PO	PR	TR
Express maintenance / repair (fast-fit)	PO	PD	TR
Extended warranties	PO	PD	RL
Fidelity card	PO	PR	RL
Financing scheme for product repair services	PO	PD	TR
Help desk	PO	PR	TR
Legal support	PO	PR	TR
Online documentation	PO	PD	TR
Online monitoring of repair activities	PO	PD	RL
Remote monitoring and diagnostics	PO	PD	RL
Repair and maintenance contracts	PO	PD	RL
Road assistance	PO	PD	RL
Satisfied or refunded guarantee	PO	PD	TR
Service for alarm systems	PO	PD	TR
Software to support maintenance activities and spare parts management	PO	PR	RL
Spare parts supply and maintenance (all inclusive)	PO	PD	TR
Take back service	PO	PD	TR
Training for customers (with their own fleet workshop)	PO	PR	RL
Truck rental	UO	PR	RL
Visibility on workshop activities	PO	PD	TR
Web community service	PO	PR	RL
24/7 repair and maintenance services (not for emergency)	PO	PD	TR

PO Product-Oriented, *UO* Used-Oriented, *RO* Result-Oriented, *PD* Product-Based, *PR* Process-Based, *TR* Transaction-based, *RL* Relationship-based

Measures were developed in accordance with standard scale development processes [40] as follows:

- A 5-point Likert scale anchored by not significant (1) and vital (5) was adopted to measure the importance of each service quality determinant, given by the customer;
- A binomial variable was used to understand whether a service is known (0 = not known; 1 = known);
- A 5-point Likert scale anchored by useless for business (1) and vital for business (5) was adopted to measure customers' current and future orientation towards services.

The final questionnaire was sent by email to the sample and data collected between May and July 2013.

4 Data Analysis

4.1 *Sample Description*

Out of about 9.000 submitted questionnaires, 960 were collected (11 %). Of these, 12 were eliminated because of incomplete responses. Therefore, the final usable sample was composed of 948 responses. The main characteristics and features describing the analysed sample are reported in Table 3.

4.2 *Service Quality Determinants*

The general analysis of data shows that heavy-truck customers are really focused on service quality determinants (total sample average = 4.15/5.00). On average,

Table 3 Main characteristics and features of the analysed sample

Sample characteristics		Feature
Customer activity	48 %	Truck owner
	52 %	Truck owner and driver
Type of vehicle(s) (tonnage)	5 %	<3.5 tons (commercial)
	3 %	$3.5 \div 16$ tons
	61 %	>16 tons (heavy)
	31 %	Multi-tonnage fleet
Fleet dimension	52 %	$1 \div 5$ vehicles
	28 %	$6 \div 15$ vehicles
	10 %	$16 \div 30$ vehicles
	4 %	$31 \div 50$ vehicles
	6 %	>50 vehicles
Type of route	7 %	Local
	15 %	Regional
	51 %	National
	14 %	International
Number of owned brands	42 %	Mono-brand
	30 %	Two brands
	14 %	Three brands
	8 %	Four brands
	5 %	Five or more brands
Main choice factor (Product purchase)	20 %	Product price
	63 %	Product quality
	17 %	Service quality

Table 4 Service quality determinants: importance evaluation

Service quality determinants	Evaluation	Ranking
Accessibility	4.03	8°
Attentiveness and courtesy	3.98	9°
Flexibility	4.09	7°
Information and communication	4.22	6°
Integrity and frankness	4.23	5°
Promptness	4.38	3°
Reliability (service)	4.52	1°
Reliability (spares)	4.28	4°
Service Availability (service mix)	3.93	10°
Timeliness	4.41	2°
Total	4.15	–

customers are more interested in technical determinants rather than in functional ones. Indeed, as reported in Table 4, the first four important service quality determinants are all technical. This consideration is valid for each type of selected customer, independently of its main characteristics.

4.3 Cluster Analysis

In order to answer the research questions, a grouping technique was applied. The techniques available for grouping individuals into segments/types on the base of multivariate survey information, are wide and various. In this research, the cluster analysis was selected because it is the most popular and applied method [41].

The cluster analysis was performed with the aim at developing a typology of heavy-truck customers based on the importance given to service quality determinants.

In order to identify the optimum solution, a two-stage sequence of analysis was adopted:

1. A hierarchical cluster analysis using Ward's method and applying squared Euclidean Distance was chosen to determine the optimum number of clusters. Ward's method uses an analysis of variance approach to evaluate the distances between clusters. The Ward's method was selected because the analysed dataset did not include any outliers [42];
2. The hierarchical cluster analysis was re-run with the selected number of clusters. This second step allowed the allocation of each case to a particular cluster.

Out of the 948 cases, only 864 were considered valid in the clustering method. Using an agglomeration schedule and a dendrogram, four clusters were identified: 78 respondents were included in cluster #1, 288 in cluster #2, 394 in cluster #3 and 104 in cluster #4.

Table 5 One-way ANOVA and Tukey's post hoc tests: results ("Accessibility" and "Attentiveness and courtesy")

Dependent variable		Sum of squares	Sig.
Accessibility	Between groups	57.170	0.000
	Within groups	261.642	
	Total	318.813	
Attentiveness and courtesy	Between groups	77.174	0.000
	Within groups	277.435	
	Total	354.609	

Dependent variable	(I) Ward method	(J) Ward method	(I-J) Mean difference	Sig.
Accessibility	1.00	2.00	−0.55235*	0.001
		3.00	−1.01393*	0.000
		4.00	−1.62179*	0.000
	2.00	1.00	0.55235*	0.001
		3.00	−0.46158*	0.000
		4.00	−1.06944*	0.000
	3.00	1.00	1.01393*	0.000
		2.00	0.46158*	0.000
		4.00	−0.60787*	0.000
	4.00	1.00	1.62179*	0.000
		2.00	1.06944*	0.000
		3.00	0.60787*	0.000
Attentiveness and courtesy	1.00	2.00	−0.55235*	0.001
		3.00	−0.78419*	0.000
		4.00	−1.11597*	0.000
	2.00	1.00	−1.44231*	0.000
		3.00	0.78419*	0.000
		4.00	−0.33178*	0.001
	3.00	1.00	−0.65812*	0.000
		2.00	1.11597*	0.000
		4.00	0.33178*	0.001
	4.00	1.00	−0.32634*	0.038
		2.00	1.44231*	0.000
		3.00	0.65812*	0.000

*The mean difference is significant at the 0.05 level

A one-way ANOVA test was then conducted to verify whether the identified groups were significantly different in terms of importance given to service quality determinants. With a significant ANOVA and four clusters, a Tukey's post-hoc test was performed to determine where the differences were lying [43]. Table 5 reports an extract of the one-way ANOVA and the Tukey's post-hoc tests carried out for the "Accessibility" and "Attentiveness and courtesy" service quality determinants. The analysis shows that the identified four clusters are significantly different (at the level of 0.05) in terms of importance given to service quality determinants.

Therefore, on the base of the achieved results it is possible to affirm that heavy-truck customers can be clustered upon the importance given to service quality determinants and the customer typology is composed of four statistically relevant types, which are briefly described in the following:

1. *Unenthusiastic*: this cluster (10 % of the sample) refers to customers who are totally unconcerned about service quality determinants. They do not present a clear and organized approach in performing their decisions.
2. *Meditative:* customers belonging to this cluster (33 % of the sample) consider quite relevant only technical service quality determinants. Indeed, their decisions are mainly based on service reliability and timeliness.
3. *Sensible:* this type of customers (45 % of the sample) considers relevant both technical and functional service quality determinants. Their decisions are based on the analysis of the reliability and timeliness of the intervention and on the integrity and frankness of the provided service activities.
4. *Passionate:* all customers of this cluster (12 % of the sample) consider all service quality determinants fundamental for their decision process.

The developed typology was then used to shed light on the second research question. In particular, each variable belonging to the general information area was tested to verify if it was able to describe a specific type of customer. To perform such analysis a Discriminant Analysis (DA) was conducted and applied to the results of the clustering phase. This combination of approaches was adopted for descriptive purposes, assisting with the cluster interpretation. Indeed, DA provides a way to summarize group differences and similarity with respect to the general information.

In particular, a stepwise DA was carried out, because it is the best approach to identify predictor variables from a large number of alternatives in an exploratory situation, as in this case. According to the stepwise DA procedure, the most correlated independent variable was entered first, then the second, and so on until an additional dependent variable added no significant amount to the canonical R squared. The criteria for adding or removing were the setting of a critical significance level for 'F to remove'.

The analysis suggested that no general characteristic (such as the customer activity, the truck tonnage, the fleet dimension, the type of route, the number of owned brands or the choices factor influencing the truck purchase) was statistically able to discriminate, and therefore, to describe each identified type of customer.

Finally, a one-way ANOVA test and a Tukey's post hoc test were performed to understand to what extent a major attitude on specific service quality determinants influences customers' orientation towards PS offerings. The analysis revealed that,

Table 6 List of services statistically different among the four identified customer types (current and future)

PS offering	Current	Future
Brand experience services	YES	YES
Driver repatriation	NO	NO
Express maintenance / repair (fast-fit)	YES	YES
Extended warranties	YES	YES
Fidelity card	YES	YES
Financing scheme for product repair services	NO	NO
Help desk	NO	YES
Legal support	NO	NO
Online documentation	YES	YES
Online monitoring of repair activities	NO	NO
Remote monitoring and diagnostics	YES	YES
Repair and maintenance contracts	YES	YES
Road assistance	YES	YES
Satisfied or refunded guarantee	YES	YES
Service for alarm systems	NO	NO
Software to support maintenance activities and spare parts management	YES	YES
Spare parts supply and maintenance (all inclusive)	YES	YES
Take back service	NO	NO
Training for customers (with their own fleet workshop)	NO	YES
Truck rental	YES	YES
Visibility on workshop activities	YES	YES
Web community service	NO	YES
24/7 repair and maintenance services (not for emergency)	YES	YES

YES: the PS is statistically different among the four customer types
NO: the PS is not statistically different among the four customer types

for the majority of the tested variables, the differences between the means of each group were significantly relevant at the 0.95 level of confidence (p value < 0.05). In particular, as shown in Table 6, if the current situation is observed, it emerges that 14 out of 23 services are significantly different among the four identified customer types. When the future orientation is analysed, the number of differentiating services moves from 14 to 17.

Therefore, it can be stated that there is a significant relationship between the identified customer types and their orientation towards PS offering. In particular, moving from an Unenthusiastic to a Passionate customer, the attention on services and their role in actual and future business increases, as summarized in Table 7 and reported in Tables 8 and 9, that include the average customers' current and future orientation towards differentiating services.

Table 7 Current and future service orientation of each customer type

	Service orientation (Current)	Service orientation (Future)
Unenthusiastic	2.69/5.00	3.04/5.00
Meditative	2.88/5.00	3.04/5.00
Sensible	3.10/5.00	3.26/5.00
Passionate	3.38/5.00	3.54/5.00

Table 8 Current service orientation (differentiating services)

PS offering	Unenthusiastic	Meditative	Sensible	Passionate
Brand experience services	2.64	2.63	2.77	3.43
Express maintenance / repair (fast-fit)	3.31	4.00	4.36	4.75
Extended warranties	2.93	3.54	3.61	3.95
Fidelity card	2.35	2.73	2.72	3.28
Online documentation	2.66	2.70	3.19	3.43
Remote monitoring and diagnostics	2.57	2.48	2.82	3.34
Repair and maintenance contracts	2.74	3.06	3.09	3.51
Road assistance	2.77	3.20	3.51	3.65
Satisfied or refunded guarantee	3.03	3.46	3.51	3.94
Software to support maintenance activities and spare parts management	2.63	2.68	3.08	3.32
Spare parts supply and maintenance (all inclusive)	2.73	3.13	3.39	3.90
Truck rental	2.47	2.62	3.11	3.11
Visibility on workshop activities	2.42	2.47	2.69	3.04
24/7 repair and maintenance services (not for emergency)	2.86	3.54	3.95	4.10

Moreover, as shown in Table 10, that reports for each PS offering category the service orientation provided by each customer type, two additional considerations can be outlined:

- the service orientation increases, independently of PS categories, but the growth is higher moving from an Unenthusiastic to a Passionate customer;
- moving from an Unenthusiastic to Passionate customer, although a PS orientation towards the relational process-based services emerges, a transactional product-based perspective is still prevalent among all the customer types.

Table 9 Future service orientation (differentiating services)

PS offering	Unenthusiastic	Meditative	Sensible	Passionate
Brand experience services	2.73	2.82	3.09	3.66
Express maintenance / repair (fast-fit)	3.42	4.04	4.42	4.70
Extended warranties	2.84	3.58	3.79	4.02
Fidelity card	2.62	2.93	2.85	3.55
Help desk	2.85	2.88	3.23	3.35
Online documentation	2.76	3.02	3.27	3.41
Remote monitoring and diagnostics	2.91	2.73	3.01	3.30
Repair and maintenance contracts	3.03	3.33	3.50	3.87
Road assistance	2.88	3.13	3.48	3.65
Satisfied or refunded guarantee	3.00	3.48	3.76	4.13
Software to support maintenance activities and spare parts management	2.73	2.97	3.10	3.57
Spare parts supply and maintenance (all inclusive)	2.97	3.33	3.57	3.95
Training for customers (with their own fleet workshop)	2.70	2.57	2.98	3.00
Truck rental	2.79	3.11	3.27	3.51
Visibility on workshop activities	2.76	2.58	2.82	3.15
Web community service	2.88	2.59	2.74	3.13
24/7 repair and maintenance services (not for emergency)	3.06	3.82	4.13	4.07

Table 10 Current and future service orientation of each customer type (PS offering categories)

PS offering categories	*Unenthusiastic*	Meditative	Sensible	s
PO	2.74	3.05	3.28	3.66
UO	2.47	2.62	3.11	3.11
PD	2.80	3.16	3.41	3.76
PR	2.52	2.67	2.92	3.29
TR	2.83	3.22	3.52	3.86
RL	2.64	2.87	3.09	3.45

PO Product-Oriented, *UO* Used-Oriented, *RO* Result-Oriented, *PD* Product-Based, *PR* Process-Based, *TR* Transaction, *RL* Relationship

5 Discussion

Although the heavy-truck market is characterised by a rapid servitization, customers seem to have a different feeling and awareness about services. Indeed, this research has demonstrated the existence of four types of customers, characterised by a different importance given to service quality determinants: 'Unenthusiastic', 'Meditative', 'Sensible' and 'Passionate'. Moreover, the statistical analysis has shown that the importance given to a specific set of service quality determinants influences customers' service orientation and, therefore, the selection of PS offerings.

In particular, Unenthusiastic are uninterested in services, and traditionally benefit only from maintenance activities, that are strictly necessary to fulfil the laws in force. Meditative tend to adopt product-based services like maintenance, repair and spare parts, essentially to maintain their vehicles in a good shape. Nevertheless, this type of customer is not particularly interested in building long term relationship with the service provider, through multi-year contracts, because it considers services essentially a cost that should be avoided or postponed at the latest. Therefore, Unenthusiastic and Meditative customers appear not particularly strategic for heavy-truck companies. Indeed, the services they need are characterised by a very low profitability and the competitiveness around them is strong. Moreover, according to a recent research [38], all needs of these two types of customers could be satisfied by any heavy-truck workshop operating in the market. On the contrary, assistance centres able to fulfil Sensible and Passionate PS needs, not only are decisively fewer (about 22 % of the total market only), but are also operating in a more profitable business, since their served customers are willing to pay premium prices to benefit from high quality solutions. However, 'all that glitters is not gold'. The provision of a more comprehensive PS portfolio, involves structured organizations and calls for the development of competent and managerial service networks that require large investments in time, resources and money.

6 Conclusions, Limitations and Further Developments

In a market affected by weak demand and hard competition, the provision of a coherent set of services is critical to enable any company to servitize in a coherent fashion. A strict monitoring of customers attitude towards service is fundamental to verify dynamically and optimize the completeness and consistency of the existing PS offerings. In this paper, an empirically-based typology of the heavy-truck customer, based on the importance given to service quality determinants, has been proposed. Four types of customers have emerged ('Unenthusiastic', 'Meditative', 'Sensible' and 'Passionate'), each one characterised by a different awareness towards technical and functional service quality determinants, and by a diverse current and future service orientation. The study has underlined that, moving from an 'Unenthusiastic' to a 'Passionate' perspective, a progressive enlargement and renovation of the PS portfolio, with a higher adoption of process-based, emerges. This observed customers' behavioural change is in line with the results carried out from a twin analysis recently conducted on the heavy-truck assistance network. In particular, contrasting the analyses of these two researches, it is possible to state that assistance workshops, providing a more comprehensive set of services, could catch a major quantity of potential customers and then have a major probability of success, though this requires the development of a more complex organisation. However, the results presented in this paper cannot be considered exhaustive. An in-depth study should be carried out to understand (i) why are more sophisticated organisations required to satisfy customers characterised by a higher awareness with service quality determinants?, (ii) what is the optimal type of customer, in order to achieve the best trade-off between revenues and costs?

References

1. Juehling E, Torney M, Herrmann C, Droeder K (2010) Integration of automotive service and technology strategies. Ind Prod Serv Syst – IPS2 3(2):98–106
2. Baines TS, Lightfoot HW, Whitney D, Kay JM (2009) Servitized manufacture: practical challenges of delivering integrated products and services. Proc Inst Mech Eng B J Eng Manuf 223:1–9
3. Brax S (2005) A manufacturer becoming service provider – challenges and a paradox. Manag Serv Qual 15(2):142–155
4. Martinez V, Bastl M, Kingston J, Evans S (2010) Challenges in transforming manufacturing organisations into product-service providers. J Manuf Technol Manag 21(4):449–469
5. Neely A (2009) Exploring the financial consequences of the servitization of manufacturing. Oper Manag Res 1(2):103–118
6. Schweitzer E, Mannweiler C, Aurich JC (2009) Continuous improvement of industrial product-service systems. In: Proceedings of the 1st CIRP Industrial Product-Service Systems (IPS2)
7. Dawkins P, Reichheld F (1990) Customer retention as a competitive weapon. Dir Board 14:42–47
8. Zeithaml VA, Berry LL, Parasuraman A (1990) Delivering quality service: balancing customer perceptions and expectations. Free Press, New York
9. Reichheld FF, Sasser WE Jr (1990) Zero defections: quality comes to services. Harv Bus Rev 68:105–111
10. Raimondi M (2005) Marketing del prodotto-servizio. Hoepli, Milano (in Italian)
11. Berry LL (1995) On great service: a framework for action. The Free Press, New York
12. Grönroos C (1995) Relationship marketing: the strategy continuum. J Acad Mark Sci 23:252–254
13. Storbacka K (1994) The nature of customer relationship profitability: analysis of relationships and customer bases in retail banking. Research report no. 55. Swedish School of Economics and Business Administration, Helsinki
14. Parasuraman A, Grewal D (2000) Serving customers and consumers effectively in the twenty-first century: a conceptual framework and overview. J Acad Mark Sci 28(1):9–16
15. Grönroos C (1990) Service management and marketing. Lexington Books, Toronto
16. Storbacka K, Strandvik T, Grönroos C (1994) Managing customer relationships for profit: the dynamics of relationship quality. Int J Serv Ind Manag 5(5):21–38
17. Cronin JJ Jr, Taylor SA (1992) Measuring service quality: a reexamination and extension. J Mark 56:55–68
18. Oliver RL (1993) A conceptual model of service quality and service satisfaction: compatible goals, different concepts. In: Swartz TA, Bowen DE, Brown SW (eds) Advances in services marketing and management: research and practice., vol 3. JAI Press, Greenwich, pp 65–85
19. Parasuraman A, Zeithaml VA, Berry LL (1994) Alternative scales for measuring service quality: a comparative assessment based on psychometric and diagnostic criteria. J Retail 70 (3):201–230
20. Rust RT, Oliver RL (1994) Service quality: insights and managerial implications from the frontier. In: Rust Roland T, Oliver RL (eds) Service quality: new directions in theory and practice. Sage, Thousand Oaks, pp 1–19

21. Strandvik T, Liljander V (1994) Relationship strength in bank services. Unpublished paper presented at 1994 research conference on relationship marketing, 11–13 June, Atlanta, GA
22. Bolton RN, Drew JH (1991) A multistage model of customers' assessment of service quality and value. J Consum Res 17:375–384
23. Bitner MJ, Hubbert AR (1994) Encounter satisfaction versus overall satisfaction versus service quality: the consumer's voice. In: Rust RT, Oliver RL (eds) Service quality: new directions in theory and practice. Sage, Thousand Oaks
24. Parasuraman A, Zeithaml V, Berry A, Leonard L (1985) A conceptual model of service quality and its implications for future research. J Mark 49(4):41–50
25. Parasuraman A, Zeithaml VA, Berry LL (1988) SERVQUAL: a multiple-item scale for measuring consumer perceptions of service quality. J Retail 64:12–40
26. Lehtinen JR, Lehtinen U (1982) Service quality: a study of quality dimensions. Service Management Institute, Helsinki
27. Finn DW, Lamb CW (1991) An evaluation of the SERVQUAL scales in a retailing setting. Adv Consum Res 18:483–490
28. Johnston R, Silvestro R (1990) The determinants of service quality – a customer-based approach. In: Proceedings of the decision science institute conference, San Diego, CA, November
29. Oliva R, Kallenberg R (2003) Managing the transition from products to services. Int J Serv Ind Manag 14(2):160–172
30. Davies A (2004) Moving base into high-value integrated solutions: a value stream approach. Ind Corp Chang 13(5):727–756
31. Gebauer H, Fleisch E, Friedli T (2005) Overcoming the service paradox in manufacturing companies. Eur Manag J 23(1):14–26
32. Davies A, Brady T, Hobday M (2006) Charting a path towards integrated solutions. MIT Sloan Manag Rev 47(3):39–48
33. Mathieu V (2001) Product services: from a service supporting the product to a service supporting the client. J Bus Ind Mark 16(1):39–58
34. Gaiardelli P, Resta B, Martinez V, Pinto R, Albores P (2013) A classification model for product-service offerings. J Clean Prod 66:507–519
35. Bundschuh RG, Dezvane TM (2003) How to make after-sales services pay off. McKinsey Quarterly 4, pp 116–127
36. Porter ME (2008) The five competitive forces that shape strategy. Harv Bus Rev 86(1):78–93
37. Rishi S, Gyimesi K, Burek C, Monday M (2009) Truck 2020. Transcending turbulence. Research report, IBM Institute for Business Value, October 2009
38. Gaiardelli P, Pezzotta G, Resta B, Songini L (2014) Product-service portfolio configuration vs. economic and financial results: an empirical analysis in the Italian truck industry. In: Masaaki M, Ueda K, Takenaka T (eds) Serviceology for services – selected papers of the 1st international conference of serviceology. Springer, Tokyo, pp 125–132
39. Forza C (2002) Survey research in operations management: a process-based perspective. Int J Oper Prod Manag 22(2):152–194
40. Churchill GA (1979) A paradigm for developing better measures of marketing constructs. J Mark Res 16:64–73
41. Donicar S (2003) Using cluster analysis for market segmentation-typical misconceptions, established methodological weaknesses and some recommendations for improvement. Aust J Mark Res 11(2):5–12
42. Ward JH Jr (1963) Hierarchical grouping to optimize an objective function. J Am Stat Assoc 58(301):236–244
43. Keppel G (1973) Design and analysis: a researcher's handbook. Prentice-Hall, Englewood Cliffs

User-De-centeredness in Service Design

Yutaka Yamauchi

Abstract User-centeredness is a fundamental principle of design in general and of service design in particular. The current paper offers an alternative view of this concept. Here, the "user" is seen not as a self-evident and static subject that is firmly centered but as part of a performative interaction through which the subject is transformed—i.e., the subject is de-centered. As service involves users as people and not as objects, the agency of persons involved needs to be fully acknowledged. Based on previously reported empirical studies of service encounters, this study proposes the thesis that service should be seen as a "struggle" rather than harmonious totality. The subject "user" is an outcome of this struggle, not its a priori condition. Therefore, a dialectical process by which the subject develops must be designed. This perspective allows for design that is different from, or even opposite to, user-centered design. This paper discusses the theoretical framework and key design principles of user-decentered service design.

Keywords Service design • User-de-centered • Service as struggle • Dialectics

1 Introduction

User-centeredness or human-centeredness has been an important topic in academic discourse for many years, and user-centeredness is firmly incorporated into current educational curricula focused on designing products for consumer use. The goal of this paper is to re-examine this concept. Specifically, I will discuss that in service design, the notion of "user" is in fact rather precarious. Accordingly, I propose reconsidering service design and emphasizing a user-de-centered approach as service design cannot be confined in the framework of user-centeredness.

In this paper, theoretical concepts and frameworks are discussed in terms of user-de-centeredness. First, problems of user-centeredness are discussed. Second, the basic thesis of this paper, "service is a struggle", is outlined. Then, the concept

---

Y. Yamauchi (✉)
Graduate School of Management, Kyoto University, Yoshida-Honmachi Sakyo-ku, Kyoto 606-8501, Japan
e-mail: yamauchi@gsm.kyoto-u.ac.jp

T. Maeno et al. (eds.), *Serviceology for Designing the Future*,
DOI 10.1007/978-4-431-55861-3_40

of user-de-centeredness is discussed in detail followed by proposals for design methodology from this perspective. Some remarks on possibilities and impossibilities of service design are incorporated into the discussion. Finally the conclusion looks at possible future directions.

2 User-Centered Design

The discussion of user-centeredness began as a response to a common situation: many designs were conceived without much consideration of users. Designers tended to ignore actual user needs and capabilities and impose designs that they considered right for the users. Subsequently, some designers and scholars proposed putting users at the center of design [18]. To be sure, there are debates as to how to put humans at the center. It is not difficult to see that soliciting design ideas from users is inherently problematic. Users can only give ideas from within their current work context and their design requests tend to be minor improvements. In addition, there is some debate as to defining the actual users because there are a variety of stakeholders. Those who hire designers are often not users, and these non-users determine design.

A related concept is user participation [6, 14, 21]. One method of user-centeredness is to have users participate in the design process. Yet, even if users participate in these discussions, they may be discouraged from expressing opinions. This could happen in a direct manner, as designers may intentionally exclude the user from certain discussions, or could be done indirectly through tactics such as the use of technical jargon.

A recent service design textbook by Stickdorn and Schneider [23] lists five principles for service design; the first of these is "user centered," which is followed by "co-creative," "sequencing," "evidencing," and "holistic." Currently the basic concepts in service design aim to move beyond designing products and consider the whole process of user experience. To this end, design focuses on customer touchpoints, and therefore it is not surprising that user-centeredness is the first principle of service design.

In this paper, I seek to reexamine the assumption of this theoretical concept. Specifically, the tendency to substantiate and hypostatize the user into a simple and self-evident human being needs to be re-evaluated. Methodological individualism is an appealing approach. As academics, we often seek a foundation from which to discuss modern society, and in society the individual appears evident and clear; that is, individuals exist as self-evident entities. That a certain person is here in front of us is factual, and based on this factuality, we tend to accept individuals as a fully integrated totality and discuss society as a collection of these entities. This idea is difficult to resist.

However, is the notion of an individual so self-evident? This idea has been contested by Georg Simmel [22], who claimed that individuals are a complex combination of various factors that interact with each other. This occurs in the same manner as we see in a society, which most of us understand as consisting of interrelated factors. Individuals are not necessarily firmly grounded at a center. Marx had previously written in this vein, "the human essence is no abstraction inherent in each single individual. In its reality it is the ensemble of the social relations" [12]. This relationalist view of the individual is currently widely accepted in the social sciences [8, 10]. A more radical perspective also denies subjectivity in individuals; post-structuralist scholars have offered the most provocative take on this view. For instance, Jacque Lacan explains that a person's desire is created by society, which induces a lack in that person and the person then accepts others' desires as his or her own [9]. Subjectivity is not given a priori but created performatively [2]. To be more precise, subjectivity does not create a certain entity, there is only the performativity.

Hypostatizing users is problematic as the users are thereby deified: the user becomes an absolute existence that transcends us. That is to say, design is seen to fulfill user "needs," which are the ultimate and unquestionable goal. Of course, it is widely known that users often cannot express their needs well. Nonetheless, to say that designers must "discover" the users' "hidden needs" and derive designs from these needs would not be controversial. However, this leads to the heart of the problem; we need to reject substantiation of users as self-evident subjects and reconsider discourse that revers users as absolute entities and seeks their needs that are veiled in mystery.

While advocating human-centered design, Donald Norman discussed an opposite design approach called "emotional design" [17]. Norman wrote, "to the uninitiated, walking into the Diesel jeans store on Union Square West feels a lot like stumbling into a rave... There are no helpful signs pointing to men's or women's departments, no obvious staff members in sight" (p. 92). Accordingly to Norman, Diesel "deliberately confuses and intimidates" (p. 93) customers. He reports that a Diesel's director of retail operation said, "We didn't design our stores to be user-friendly because we want you to interact with our people. You can't understand Diesel without talking to someone" (p. 92). Then, Norman writes, "To the practitioners of human-centered design, serving customers means relieving them of frustration, of confusion, of a sense of helplessness. Make them feel in control and empowered. To the clever salesperson, just the reverse is true" (p. 92). It is interesting to observe that for Norman, "reflective designs" that places emphasis on "self-image" are a "reverse" approach from human-centered design.

3 Service Is a Struggle

Before moving on to the discussion of user-de-centeredness, we need to review the concept of service itself. Based on previous research by my colleague and myself, I propose a thesis stating; "service is a struggle." In everyday discourse on service, we hear such terms as hospitality, and empathy. These terms imply an all-positive, harmonious totality. We need to be critical of such mysterious concepts. The thesis of service as a struggle is an attempt to turn these concepts on their head.

To begin, we should acknowledge the basic nature of service: it is an interaction that involves humans. When humans are involved in service production, this is always an increase in uncertainty [7, 13]. When consulting with a doctor at a clinic, studying at school, or detailing wedding plans, specific customers and providers are involved. The same service is never repeated. On the other hand, we observe extreme standardization and routinization in service. McDonald's and other fast food franchises represent a typical Weberian rationalization and bureaucratization of service [20]. Here humans are treated as objects. Nonetheless, it is interesting to observe that at many of these restaurants, human employees smile as they interact with customers; despite all the automation, this part of service has not been replaced by machine [11].

Then, how should we treat customers? One strategy has been obvious to most people looking at service. This is the strategy in which customers can be venerated as absolute beings and then service providers seek to fulfill their needs. However, the flip side of this is that the providers end up not viewing a specific customer in front of them as human. The goal of this type of service is to increase customer satisfaction, and the customer is treated as a box that displays its level of satisfaction. In this service, the human element is excised from customers as well as providers. Service is designed to be mechanical. Only when a problem occurs, which is treated as an exception, are human agents engaged in the process. Therefore, while it is the fundamental nature of service to involve humans, many service systems are designed to exclude them.

What does it mean to involve humans qua humans? Jacques Derrida's [4] discussion on hospitality is helpful here. He suggested that hospitality is impossibility. The Latin etymology of hospitality is traced back to "hospes," which comes from two different sources. It is a combination of "hostis" and "pets." "Hostis" originally referred to unknown others and it became "hostilis," somebody hostile. "Pets" referred to owning power, related to such words as potentia. That is to say, it is assumed that a guest is an adversary over whom power is exercised. This is not just play of words; we are all familiar with this tension when we invite guests to our house. Nobody accepts guests wholeheartedly. When one says, "Please make yourself at home," one does not mean that the guest can really do so.

On the other hand, words that carry the beauty of harmony, such as hospitality, are in fact necessary precisely because we understand these concepts represent impossibility. We need to believe that they are possible; otherwise it is difficult to make the effort that good service requires. Service providers believe in this myth and can see in this belief that they are doing something special. Customers also believe in this myth, which allows them to justify paying for such service. That said, even if we need to deceive ourselves to some degree while engaging in service activities in our everyday lives, when theorizing about the nature of service, we must look at the practice directly.

In order to fully grasp the concept of service as a struggle, we should turn to Hegel's dialectics. The above description bears a resemblance to Hegel's [8] lord and bondsman dialectic. In his argument self-consciousness has desire and the object of desire is other self-consciousness, i.e., other people. Self-consciousness requires recognition as its condition. Then, metaphorically, by demanding others' recognition self-consciousness seeks others' death. If this self-consciousness wins the life-or-death battle, it becomes a lord. If it loses, it becomes a bondsman, who exists only for the lord. The lord dominates the bondsman and acquires its power through bondsman's labor. The bondsman on the other hand is tied to objects that he produces through labor. However, the lord eventually learns that he is completely dependent on the bondsman's labor and the bondsman learns that his self-consciousness is objectified through labor and his fear of the lord and anxiety about existence are cleared. It is labor's essence to give form and here the bondsman finds himself. The relationship between the lord and bondsman is reversed.

When we treat customers as gods or kings/queens, we need to understand this dialectic. In service, the basic premise that we seek others' recognition underlies the customer-provider relationship. That this relationship involves a battle between these two entities parallels Derrida's argument. Here, we understand what it means to have power; that is, we present our power over others and obtain their recognition.

However, there is more to this dialectic. It shows that we cannot assume individuals as subjects, but we must examine how individuals come to acquire subjectivity through the dialectical process. The driver of this process is the contradictory nature of the interactions; we should assume there is not harmony but contradiction, negation, and lack. Of course, we should not accept *aufhebung* or sublation of the opposing forces too easily, as deconstructionists argue and later materialist dialectics showed. When somebody acquires subjectivity, it really means that he or she never achieves full subjectivity, he or she is always full of the potential of these contradictions thus always on the move. This is what service actually constitutes.

This acts to clarify the thesis of service as a struggle. My colleague and I conducted empirical studies at traditional sushi restaurants in Tokyo, and as part of this research we set up several video camcorders and analyzed interactions between chefs and customers [24]. Sushi is a peculiar service. First of all, customers feel anxious when opening the door and entering the restaurant. Corson [3] wrote,

> Many Americans walk into a sushi restaurant and opt to sit at a table because they find the sushi bar intimidating. Sitting at a table feels familiar, as does ordering from a menu...Turning one's back on familiarity and choosing to sit at the sushi bar requires courage, but the experience is more interesting...*Not* knowing what to expect, either with the ingredients or the order in which they are served, is part of the fun...Americans can take solace in the knowledge that they are not alone. Many Japanese people also find the sushi bar intimidating. (p. 317, emphasis in original).

While Corson wrote about typical sushi restaurants in the U.S., traditional sushi establishments in Tokyo are much more "intimidating." We do not witness a flattering smile on the face of the chef. As soon as we are seated, we are asked, "What would you like?" The chef has not provided any explanation of the sushi bar. Nor has he or she provided a written menu. No price is shown for each item and we are anxious until we receive the bill at the end of the meal. There are a number of customs and rules that customers need to know, for instance, one should choose leaner and lighter fish first before moving to stronger and fatter fish, and diners should eat the sushi by hand. In his work, following the passage above, Corson gave a detailed guide for how to eat sushi.

In short, Sushi services are contentious. Chefs do not make the service 'easy' for the customers, and they assess the type of customers that have entered their establishments. One of the chefs we worked with said that the interaction with the customer is a "duel." They either overwhelm the customers or are overwhelmed by the customers. When I suggest that service is a struggle, I have this in mind. Customers who are seem to understand this dynamic will try to step up their performance during the meal. This is a unique experience that exists at these establishments. As Corson wrote above, ginning up your courage and eating at the bar is "more interesting."

This kind of struggle is not necessarily irrelevant to other kinds of services. For example, in Japan there is some tension in interactions at Starbuck's coffee shops. Customers feel that they need to behave appropriately there, that going to Starbucks might call for cool or stylish clothes. Customers also try to be experienced in the system of service. Some people learn how to order from the somewhat complicated Starbuck's menu, which includes items such as "Iced triple tall café Americano" and "Dark moca chip Frappuccino chocolate sauce." At Italian restaurants in Japan, the menu includes many dishes with names cryptic to most Japanese, for example "Pizza Melanzane" and "Salsiccia piccante." The restaurants use these names knowing that most customers will not understand them, and only a few, special customers are aware of the meaning. Other customers try to read the menu and feel a need to present more sophisticated versions of their selves. To be sure, the struggle is only at the surface level, in terms of the sign. Nonetheless, we can sense that the struggle is a relevant aspect of most service interactions.

4 User-De-centeredness

Based on this understanding, we can now see that users are not what we can assume a priori but are themselves the outcomes of the services. Of course, the user as an outcome is not a fully realized subject, but one that contains contradictions. When the object of design has been products or graphics, designers might have been able to assume relatively static users. In contrast, when services become the object of design, and when we recognize the reality of service as a struggle, we need to examine what kind of users and providers individuals become through service. Designing the processes of struggle is critical for service design. In this sense, I propose "user-de-centeredness" as opposed to "user-centeredness."

In order to de-center users, we need to treat the phenomena of users not as self-evident and complete entities but as performance. That it to say, the subject of user is shaped through performativity. One aspect of this performativity is how users present themselves. As Goffman [5] showed, when we are in the presence of others, we present our definition of the situation and thereby present our "self." In this sense, the action becomes performance. Others, e.g., service providers, then accept the definition or challenge it. Here, the notion of identity feels empty. Instead of assuming identity, we must pay attention to the process by which the subject is constructed performatively through actions.

Therefore, we do not assume users as absolute entities that have a distinct center. The subject of the user is mutable, possessing incompleteness and contradictions. As a result, user-de-centered design may not lead to user-friendly designs. As high-end sushi services are challenging to ordinary customers, a duel with a customer through which mutual recognition can be arrived at is an important goal of the design. By asking a first-time customer who has just been seated, "What would you like?" without presenting any information as to what is available, the chef is defining the situation: his or her customers can respond to this question without any difficulty. Experienced customers can succinctly say, "Hot sake, please." Other customers may say, "U::m, we::ll, then beer::?" By producing the latter utterance while observing how the chef responds and expecting positive feedback, these customers "give off" the impression that they do not feel at home [5]. Such customers are not treated as those who have needs to fulfill; the gap between who they are and what they should be is produced, and this gap then urges the customers to strive for better performance.

Customer satisfaction is often cited as the ultimate goal of service. Yet, if customers are in fact completely satisfied, they do not recognize value-added in the service. Rather, customers who are satisfied but still encounter a number of aspects of service they do not completely understand or those who are overwhelmed by the service would see value-added. Of course, the argument is not to reduce satisfaction. In the case of high-end sushi, most customers enjoy the service but at the same time feel overwhelmed during the "duel", or perhaps the satisfaction may stem from lasting through the duel itself. Sushi chefs are certainly oriented to entertaining and satisfying customers. The problem lies in the assumption that the goal of design is to achieve the harmony of fulfilled needs and customer satisfaction. Without maintaining a nuanced and multi-faceted meaning of satisfaction, designers may make the mistake of jeopardizing the value-added qualities that they have set to achieve. Alternatively, it is also risky to offer overly high quality service that is well beyond customers' expectations. Overreaching customer expectations is not sustainable, as each subsequent visit would require, even a higher level of service.

Furthermore, there has also been a historical production of the assumption that users have latent needs. Here, designers are assumed to be transcendental beings that fulfill users' needs, whereas in actuality the crucial point of service is not customer needs but the customers themselves. These customers are worth engaging, and this engagement is often a struggle; they should not be unilaterally flattered. The goal of services in which customers matter as humans is to engage in a duel, and then achieve mutual recognition, although this recognition is not completely harmonious and remains to some degree antagonistic. Instead of a duel in which there is an assumed outcome, designers need to open themselves to various possibilities that cannot be pre-defined.

In viewing service as a struggle, we could expect the following dynamic in the long-term, which needs to be examined in future work. Customers may become oriented to acquiring more experience to meet to a chef's expectations. The future behavior of such customers would then theoretically put pressure on the chef to improve his or her skills further. A well known chef in Tokyo wrote in his book [15]:

> If the number of customers who seriously delight in sushi decreases, the chef will loose the sense of tension and the sushi bar's quality of food will decline... Only when customers enjoy superlative quality of food and atmosphere, can they appreciate the real value of the sushi bar. This is a quality that nourishes customers' sense of taste, disciplines the chef who earnestly faces customers, and elevates the quality of the restaurant. (translation by the present author)

This two-way tension appears to be instrumental in maintaining and improving the level of service. Chefs set a high standard for their customers in order to first elevate customer performance and then, reciprocally, their own actions.

5 How to Design Services

How can we design such services? In the first place, the service is what customers read and decipher. In restaurant services, customers try to discern the food and drinks. For example, they may engage in specifying the region and vintage of wine, quality of fish, and the history and tradition of a particular cuisine. At the same time, qualities such as those listed above should not be too easily comprehensible; the service should be designed so that it cannot be deciphered completely. What is offered in service requires complexity. Foods that are simply delicious are not an object to discern. Instead, complex flavor is needed. To appreciate such complex flavor, customers are required to accumulate experience. This is often called acquired taste. Whisky is a good example, as many people do not like whisky when they try it for the first time. Often people start to appreciate the subtlety of the taste only after acquiring more experience with the drink.

Mr. Kenichi Hashimoto, the owner/chef of Ryozanpaku, a traditional restaurant in Kyoto, says that a combination of ingredients of similar tastes does not necessarily result in overall taste as these ingredients can negate each other. In contrast, ingredients with distinct flavors can be combined to form interesting tastes. In traditional Kyoto cuisine, there are certain dishes called "*Deaimon*" or matching taste. For instance, bamboo shoots from the mountain and seaweed from the sea are matched. The combination of dried cods and shrimp-shaped potatoes is another example. Mr. Seiichi Koshimizu, the chief whisky blender of Suntory Distillery, writes that blending so-called good whiskies creates a blend that is not completely satisfactory. By adding a small amount of so-called poor whisky, the blend achieves a great taste. He says, "100 plus one becomes 200."

These juxtapositions constitute a harmony of disharmony. Paul Ricoeur [19] suggested that narratives contain disharmony and this is integrated into a harmonious configuration. Nietzsche [e.g., 16] contrasted Dionysian and Apollonian values, proposing dialectics between them. Dionysian refers to ecstasy, intoxication, and chaos, and is often characterized as dissonance, while Apollonian refers to brilliance and harmonious beauty. Although Apollonian beauty is necessary, real beauty stems from the Dionysian dimensions. Dissonant and conflicting ingredients are juxtaposed in a configuration engendering a profound taste.

This kind of value is not immediately comprehendible, but is co-created through customers' deep reading. This experience of reading constitutes the value. For instance, when Mr. Koshimizu created the blended 12-year old Whisky Hibiki with the goal of marketing it globally, he added whisky from plum wine barrels. This addition did not drastically improve the taste; the aim was to guide customers to appreciate the taste by searching for the plum wine flavor when drinking the whisky. This reading process creates a special drinking experience. Mr. Koshimizu designed the whisky with this in mind – it is a story that bartenders can tell their customers.

In service design, it is important to induce some element of surprise on the part of users. The aim of this surprise is not only to delight a customer with an unexpected action but also to urge the customer to confront the service. Mr. Hideaki Matsuo, the owner/chef of Kashiwaya restaurant in Osaka, altered the traditional disk of *Wakatake*, a combination of bamboo shoots (*Takenoko*) and seaweed (*Wakame*). In the traditional dish, the unique texture of bamboo shoots is a highlight. He, however, processes the bamboo into a paste. According to Mr. Matsuo, bamboo shoots have much more flavor than they do texture; however many customers do not even realize there are such flavors in this ingredient. This type of unexpected design leads customers to search for the meaning of the dish. This surprise therefore only signals the beginning of the duel. Mr. Hashimoto of Ryozanpaku also relies on surprise when making a subtle vegetable soup. He adds various colorful vegetables to make it beautiful. To this, he adds *Kudzukiri*, or starch noodles, which have little taste but a unique texture. Customers begin to eat without special attention to the ingredients because it is a common dish. However, once they bite into a *Kudzukiri* noodle, their curiosity is raised and they then start searching for other special ingredients in this dish. This kind of design is rarely derived from a user-centered approach.

User-centered design seeks to render the service easy to understand and transparent by delivering clear and sufficient information and providing affordances for the users. In contrast, a user-de-centered approach can aim at designs in which the service is made less transparent; e.g., the written menu is not provided and prices are not indicated. In this design, users are challenged and drawn into a struggle, and as a result are encouraged to treat their own subjectivity as problematic and subsequently to make efforts to transform it. In other words, not explaining the service—not even implicitly with affordance—is a key aspect of design. Mr. Matsuo said, "Each effort we make is very thin but we can build thickness by piling them up. Customers do not clearly understand it but feel it without specific words." This way, an overwhelming "thickness" is created and only a small part of it is conveyed to customers. Customers go home without explicitly knowing the full extent of all elements involved in the service encounter. This dialectic of surprise and silence turns user-centered service design on its head.

6 Is Service Design Possible?

There is, however, a counter argument: User-de-centeredness still centers users who are de-centered. There is no doubt that it is important to take users seriously. Yet, user-centeredness implies the action of centering users. This action is taken from a transcendental standpoint where a particular person, or persons, determines where the center is. In other words, designers put themselves outside this world and outside the struggle. Design is required to be endogenous to the object of design. When we talk about endogenous design in the context of participatory design and inclusive design, we must recognize this problem. The discourse that designers should involve users and obtain their participation is based on the assumption that designers have transcendental power. For endogenous design, designers need to engage themselves in the struggle with users. That participatory design and inclusive design require struggle with users is not paradoxical at all; it is necessity.

Therefore, we come to the conclusion: Design for users is an impossibility. When Derrida [4] pointed out the impossibility of hospitality, he also suggested that this impossibility is the condition for its possibility. If we design genuinely for users, there should not be an action of "designing" at all. This is because the action of designing for users is transcendental and thereby undermines the goal it has set out to achieve in the first place. If we expose ourselves to the struggle, we cannot design. We must expose ourselves to open possibilities; this alone is inconsistent with the notion of design, which is to give form to what ought to be. However, user-de-centered design is possible precisely because of its impossibility. That is to say, by moving past this impossibility, bringing oneself to the struggle, and thereby failing to design the service, we can attain the true value of service.

If we face services in the most fundamental sense, this conclusion appears inevitable. We could seek to design a complete machine where humans are generally excluded and deny customers' involvement in service production. Here the human element is only brought in only in cases of exception; the exception – or problem – requires flattering the customer, and the fact that the act of flattery is the exception serves to maintain the semblance of a complete machine. Having employees talk in a friendly manner with smiles on their faces has no place in the true meaning of service. Certainly, this kind of service design is needed and important. Services of this design have value as certain form of fantasy. Insofar as people are seeking such fantasy, such services may become successful businesses. However, it is also the case that such an approach has reached its limitations. Because such services are so pervasive, people appear to be looking to alternatives. If we see this user-centered service design as the only design approach available for us, and if we characterize traditional services such as sushi and *ryotei* (traditional high-end Japanese restaurants) only as outdated and dying phenomena, then we shut ourselves off from the most interesting and most promising domains of service design.

7 Conclusion

User-de-centered design refers to a service design methodology that assumes users not as self-evident and static subjects but as an outcome of the service and involves specification of the process by which users transform their subjectivity. In proposing this approach, this study outlined a central thesis that "service is a struggle." This conception of service counters the commonsense views that services are harmonious, caring, and empathic. Such characterizations gloss the surface appearances of service, a phenomenon that has a true value in the struggle through which participants are transformed. User-de-centeredness is a unique service design approach that builds on this nuanced understanding of services.

This paper only engages in preliminary discussions of the concept of service as struggle and user de-centeredness. The design methodology needs to be developed further to offer concrete design principles. It remains unclear what specifically service designers can do in the actual design context. For this future direction, we need to specify the theoretical framework in more detail.

The discussion of user-de-centered has been inspired by the analysis of traditional and high-end services such as sushi and ryotei. Yet, this design approach offers more value when it is applied to more common services. We have pointed out that some semblance of a struggle is included in the design of many common services, and much more research can be done in this regard. Furthermore, an interesting question has not been touched on: How can this kind of service be implemented in a different culture? In future work, we can explore these directions.

References

1. Baudrillard J (1998) The consumer society. Sage, Thousand Oaks
2. Butler J (2011) Gender trouble. Routledge, New York
3. Corson T (2009) The story of Sushi. HarperCollins, New York
4. Derrida J (1997) Deconstruction in a nutshell. Fordham University Press, New York
5. Goffman E (1959) The presentation of self in everyday life. Anchor, New York
6. Greenbaum JM, Kyng M (1991) Design at work. CRC, New York
7. Gutek B et al (2002) Achieving service success through relationships and enhanced encounters. Acad Manage Exec (1993) 16(4):132–144
8. Hegel GWF (1977) Phenomenology of spirit. Oxford University Press, New York
9. Lacan J (2006) Écrits. W. W. Norton & Company, New York
10. Latour B (2007) Reassembling the social: an introduction to actor-network-theory. Oxford University Press, New York

11. Leidner R (1993) Fast food, fast talk. University of California Press, Berkeley
12. Marx K, Engels F (1976) The German ideology. Prometheus Books, New York
13. Mills PK, Moberg DJ (1982) Perspectives on the technology of service operations. Acad Manag Rev 7(3):467–478
14. Mumford E (1983) Participative systems design: practice and theory. J Occup Behav 4 (1):47–57
15. Nakazawa K (2007) Human capacity of Sushi Bars (Sushi-ya No Ningen-ryoku). Bungei Shunju, Tokyo
16. Nietzsche F (1999) The birth of tragedy. Cambridge University Press, Cambridge
17. Norman DA (2007) Emotional design. Basic Books, New York
18. Norman DA, Draper SW (1986) User centered system design. Lawrence Erlbaum Associates, Hillsdale
19. Ricoeur P (2012) Time and narrative. University of Chicago Press, Chicago
20. Ritzer G (2000) The McDonaldization of society. Sage, London
21. Schuler D, Namioka A (1993) Participatory design. Erlbaum, Hillsdale
22. Simmel G (1917) Grundfragen der Soziologie. Göschen, Berlin
23. Stickdorn M, Schneider J (2011) This is service design thinking: basics – tools – cases. BIS Publishers, Amsterdam
24. Yamauchi Y, Hiramoto T (2014) Negotiation of selves in initial service encounters: conversation analysis of sushi. In: Mochimaru M et al (eds) Serviceology for services. Springer, Tokyo

A Probe-Based Approach for Designing Inspirational Services at Museums

Kumiyo Nakakoji and Yasuhiro Yamamoto

Abstract Our MESS (Museum Experiences and Service Science) project views a local museum as a place not only for communicating the fact and knowledge about exhibited objects with the visitors, but also for inspiring them. We have designed and set up design probes in a museum exhibition as a way to investigate how visitors got inspired at a museum. The applied design probes include LED-lit candles and a tea ceremony house for viewing old Japanese paintings, an improvisational dance workshop for appreciating an abstract modern-art sculpture, and an improvisational drama workshop for reading old family correspondence. The study has led us to identifying a set of features for inspiration, and moreover, revealed that curators and museum administrators in turn got inspired by the representations produced by the visitors through their engagement in museum experience.

Keywords Service experience design • Probe-based design approach • Inspirational service • Museum service • Collective creativity

1 Introduction

Our MESS (Museum Experiences and Service Science) project focuses on the design of museum experiences as an inspirational service.

Museums are places for public service, which "acquire, conserve, research, communicate and exhibit the tangible and intangible heritage of humanity and its environment for the purposes of education, study and enjoyment" [1]. We are particularly interested in the role of a museum as a cabinet of curiosities, which is said to be the origin of a museum. A cabinet of curiosities is "an encyclopedic collection in Renaissance Europe of types of objects whose categorial boundaries were yet to be defined" [2]. Although a museum is generally believed to be a place where objects of known value are carefully exhibited by the curators, it could also

K. Nakakoji (✉) • Y. Yamamoto
Unit of Design, C-Pier, Kyoto University, Shimogyo, Kyoto 606-8306, Japan
e-mail: kumiyo@acm.org

T. Maeno et al. (eds.), *Serviceology for Designing the Future*,
DOI 10.1007/978-4-431-55861-3_41

be more than a place for the public to learn the fact about the objects; it could be a place for the public to encounter wonders and get curious by being inspired. Our premise is that the physical space of a museum with exhibited objects, the curators of the museum, the visitors who look at and interact with the exhibition, and possible gadgets and computational tools installed in the museum, collectively form the environment where inspirational service experience takes place.

In order to investigate how to design and support such inspirational service experience, we have taken the probe-based design approach. Probes used in HCI (Human-Computer Interaction) design are everyday things and simple gadgets that are installed in an in-situ environment in an evocative manner to elicit inspirational responses from people. We have set up five probes in museum exhibitions, and conducted workshops to observe how people react and interact with them, based on which we have identified five features for inspiration. We have then designed a museum instrument based on one of the features.

This paper starts with anecdotes to illustrate the type of service experience we focus on; namely, how inspirational service experience takes place at a museum. We then describe our probe-based service experience design approach. We introduce five probe studies we have conducted at museum exhibitions, explain the five features for inspiration, and present the designed instrument for inspirational service experience.

2 Motivational Anecdotes

The following two anecdotes illustrate the type of inspirational service experience we are interested in supporting at a museum. The anecdotes are based on one of the authors' personal experience.

Anecdote 1

(1) When Jane visited a modern art museum, there was one sculpture, which had got the grand award from the local government. The object has a shape of a beach bed, with very smooth surface with lots of swirling colors. Jane thought that the bed is interestingly colored, but she did not find it particularly interesting, and she had no idea why the particular bed got the best award among other art works (there were other objects with less significant award shown in the exhibition).
(2) When Jane was about to leave the site, she noticed that the bed has a very rugged surface on the back.
(3) So she carefully looked into the back of the bed, where she found a number of PVC anime figurines (i.e., so called "action figures") hanging on the back of the bed.
(4) In fact, when she looked at the bed more carefully, she has found that the bed is made out of melted PVC figurines.

(5) The bed all the sudden looked very grotesque and shocking to Jane.
(6) When Jane went home, she checked with their exhibition Web site and found that the bed was in fact made out of 17,500 melted figurines.
(7) Once she understood how the bed was made, she imagined a situation in which the artist was melting the tens of thousands of figurines in a large pot on a stove. She then fully understood why the bed got the grand award.

Anecdote 2

(1) Bob was in a small local museum, where there was an exhibition of home use casual dish plates from 50 to 100 years ago. All of the plates were the same size with very similar look. Dozens of similar looking dish plates were categorized into a few groups according to the subtle differences of brushing patterns drawn on each plate. The description of the exhibition said that they were made through the very early industry mass production and the museum seemed to own hundreds of them. Bob found the exhibition okay and about to move to a different room.
(2) Then the curator Mary, who designed the exhibition, came to Bob and started telling him how the curator came to the idea of this exhibition.
(3) Mary kept telling Bob how she prepared the exhibition and how hard it was. She opened dozens of boxes of dishes after the museum was closed in the evening (and therefore it was very cold), she took them all out from the boxes and arranged the plates all over the museum floor, she then became aware of small differences among the mass-produced dish plates, she crawled on the cold museum floor for hours and hours to find matching patterns, and so on.
(4) After Bob heard Mary's story, the exhibition looked completely different and Bob started looking at the exhibition with great excitement.

These are the type of inspirational service experience we think is important, but often missed due to the lack of appropriate support. Curators typically have put lots of thought into their exhibition, with quite a lot of enthusiasm. However, such aspect of the exhibition is not always communicated well with the visitors. Exhibited objects would have significantly different meaning to visitors once the visitors get to know the excitement the curators have.

3 The Probe-Based Experience Design Approach

The MESS project has taken an approach to use *probes* toward designing inspirational service experience. What we call a *probe* in this project is a simple gadget, mechanism, or physical setting, which researchers introduced and settle down within the context under study in order to observe and explore how people use, interact with, react to it, or change behavior or attitude within the context. Such probes are typically small-scaled, casually obtainable artifact.

The use of a probe is introduced in the domain of human-computer-interaction (HCI) design to investigate how people react and interact with a simple object or a mechanism setup in an everyday context. In this section, we first briefly describe the role of probes in HCI design. We then present our probe-based instrumentation approach.

3.1 Probes in HCI Design

Gaver, Dunne and Pacenti [3] first introduced the notion of *cultural probes* in one of their projects to design technology for homes particularly for elderly people. Designing technology for entertainment, and for enriching people's lives calls for different approaches from designing technology for utility. Cultural probes are "collections of evocative tasks meant to elicit inspirational responses from people." An example of their probes was a disposable camera, which was repackaged and labeled with extremely open-ended or even absurd requests for particular pictures, such as "something red" or "something you would like to get rid of" [4]. Since then, a number of design projects have used *probes* in HCI design [5].

Probes are, by definition, to obtain "fragmentary clues about their lives and thoughts" and not to obtain "comprehensive information about them" [4]. A probe study in a design project typically consists of a cycle of expression and interpretation [4]. A designer expresses a probe for a set of study participants, who are then interprets the probe and demonstrates certain expressions as the result of their interpretations. The designer then interprets the demonstrated results by the participants.

Hutchinson et al. have introduced the notion of *technology probes*, "a particular type of probe that combine the social science goal of collecting information about the use and the users of the technology in a real-world setting" and "involve installing a technology into a real use context, watching how it is used over a period of time, and then reflecting on this use to gather information about the users and inspire ideas for new technologies" [6].

3.2 The Probe-Based Instrumentation Approach

The MESS project has been applying the use of cultural probes in museum service experience design. Figure 1 illustrates our probe-based instrumentation approach for designing inspirational service experience for a museum.

Starting from the analysis of field studies at museums, we have identified key elements of inspirational experience at museums. We have then designed probes, each of which is based on a few of the identified key elements of inspirational experience. We have then installed each probe in workshop held at a museum. By observing how people reacted and interacted with the probes, we have extracted

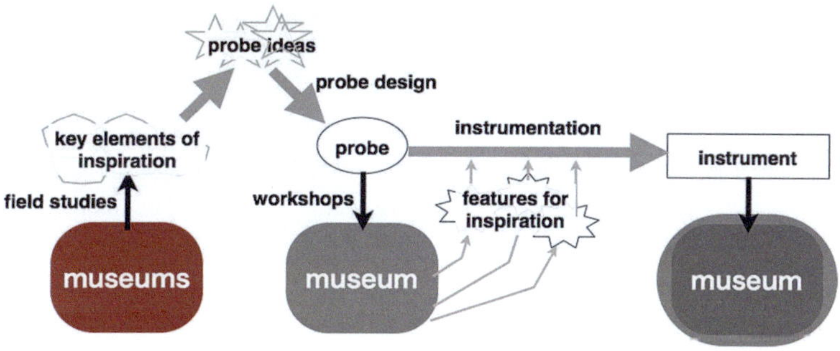

Fig. 1 The probe-based instrumentation approach

features for inspiration. We then instrument each feature so that the designed instrument can be introduced in a museum's practical context.

4 Probe Studies

4.1 Museum Studies

The design of our probes used in the MESS project is based on the key elements for inspiration we have identified through our field studies at worldly renowned 17 museums for high quality museum experience, including Natural History Museum at Bern, Switzerland, and the Exploratorium at San Francisco, USA. Such elements include visual aesthetics of exhibition, dynamically contextualized exhibition, contextual fusion of exhibition, extremely dense exhibition, creative uses of mirrors, multi-layered visitor experience, and visitor interactions as exhibition. More details of our museum studies are reported in [7] and [8].

By using those elements as motivational factors, we have constructed design ideas, some of which have been further developed to be design probes.

4.2 Designed Probes

The MESS project has designed the following five probes and setup in museum workshops (Fig. 2) for investigating inspirational service experience at museums.

(a) photo taking workshop

It is often not allowed for visitors to take pictures in many of Japanese museums. We have brought in a camera with us into the Hakodate City Museum to freely take pictures to observe what pictures are taken through what motives and intentions.

Fig. 2 Designed probes (extract)

(b) improvisational dance workshop

We have conducted a workshop where the participants of the workshop engaged in two 3-h improvisational dance workshop in the same room where Marcel Du Champ's large modern abstract sculpture, "The Bride Stripped Bare by Her Bachelors, Even", so-called "Big Glass" was exhibited at the Komaba Museum.

(c) improvisational drama workshop

We have conducted a half-day workshop where the participants improvisationally played a drama inspired by the Hakodate City Museum's special exhibition on the personal letters exchanged among the one family of father, mother, two small children and the father's mistress about 100 years ago.

(d) candle light workshop

When old Japanese ink paintings were drawn, there was no electric light but only candles. We hanged a Japanese ink painting in a dark room at Hakodate City Museum, and asked the study participants to appreciate the Japanese ink painting by Hakyo Kakizaki, a famous Japanese painter, with hand-held candle light.

(e) teahouse workshop

Some old Japanese ink paintings were painted particularly for being hanged in a teahouse, to be appreciated during a tea ceremony. We hanged a Japanese ink painting in a tea ceremony room, which is also a museum exhibit located in a basement of the Hakodate City Museum, and served an actual tea ceremony for the study participants at the site.

4.3 Features for Inspiration at Museums

We have collected video data, interview data, and other activity data during the above workshops. This subsection summarizes five features for inspiration we have extracted from the results of the above probe studies. Some of the results of detailed analyses of each probe study are published elsewhere [9, 10] and some are under preparation for publication.

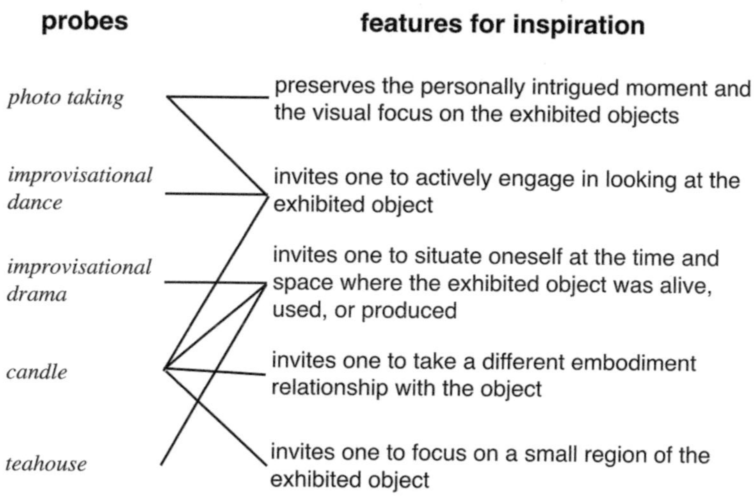

Fig. 3 Probes and extracted features for inspiration at museums

Figure 3 summarizes which probe study has led us to identify which feature for inspiration at museums.

4.4 Instrumentation

While *probes* are means for researchers to investigate inspirational service experience, *instruments* are means for museum practitioners to engage in inspirational service experience production. For museum curators and staff members to independently exercise some of the features for inspiration we have identified above, such features need to be contextualized within a practical setting. Some probes may be directly transformed into a fully functional instrument, but other probes may need to be completely re-designed so that it could be operational in terms of cost and effort needed for museum practitioners. We call the process *instrumentation*.

We are currently instrumenting the first feature for inspiration described in Fig. 3 – to preserve the personally intrigued moment and the visual focus of the exhibition. The result is the semi-automated poster-making instrument [11]. The instrument consists of the 24 high-resolution 360-degree photograph (15 degrees each) of an object exhibited at a museum, and a poster framework, which is composed of ten small image areas and one large image area. A visitor can select ten images (duplicates are allowed) for the ten small areas among the 24 images by browsing them, and select one image and enlarge it to fit into the one large image area.

By using this instrument, a visitor selects personally favorite views of an exhibited object, and produces a poster in a relatively short period of time. In one of our preliminary studies, some students used the instrument and the resulted posters evoked conversations among them to ask each other why a particular shot is chosen and share excitement about their discoveries mediated by the posters [11].

5 Discussion

This paper reports our ongoing effort in the MESS project to design inspirational service experience in museums. We described our probe-based approach, and introduced the currently designed probes and extracted features for inspiration.

Our preliminary finding based on the initial deployment of the designed instrument is that it is not only visitors who get inspired by a museum, but other museum stakeholders, such as curators and designers are also inspired. One of the curators was excited about the posters produced by the students using the instrument, and another curator got inspired by the inspirational drama produced by probe study participants.

We envision that the physical space of a museum with exhibited objects, curators who design an exhibition at the museum, visitors who look at and interact with the exhibition, and possible gadgets and computational tools set up in the museum, collectively form an inspiration-support collectively creative environment [12].

One of the goals of the MESS project is to develop a scheme to evaluate how successful a museum exhibition is in terms of how much the visitors get inspired. A museum exhibition is currently often evaluated only in terms of how many visitors come to see the exhibition, which we think tends to be misleading by missing the important aspect of the role of a museum. Our research challenge will be to assess one aspect of a museum exhibition by evaluating how it serves for inspirational service experience.

Acknowledgments This work is supported by S3FIRE, RISTEX, JST. We are very much grateful to the other members of the MESS project, particularly for Takeshi Okada, Toshio Kawashima, Ken-ichi Kimura, for the discussions, design, data-collections, analyses and reflections on the Probe workshops.

References

1. ICOM, International Council of Museums. http://icom.museum/. Visited on April 2014
2. Wikipedia. http://en.wikipedia.org/wiki/Cabinet_of_curiosities. Visited on March 2014
3. Gaver WW, Dunne T, Pacenti E (1999) Design: cultural probes. Interactions 6(1). ACM Press, New York, NY, pp 21–29
4. Gaver WW, Boucher A, Pennington S, Walker B (2004) Cultural probes and the value of uncertainty, Interactions 11(5). ACM, New York, NY, pp 53–56
5. Boehner K, Vertesi J, Sengers P, Dourish P (2007) How HCI interprets the probes. In Proceedings of the SIGCHI conference on human factors in computing systems (CHI '07). ACM, New York, NY, USA, pp 1077–1086

6. Hutchinson H, Mackay W, Westerlund B, Bederson BB, Druin A, Plaisant C, Beaudouin-Lafon M, Conversy S, Evans H, Hansen H, Roussel N, Eiderbäck B (2003) Technology probes: inspiring design for and with families. In: Proceedings of the SIGCHI conference on human factors in computing systems (CHI '03). ACM, New York, NY, USA, pp 17–24
7. Kumiyo Nakakoji, Yasuhiro Yamamoto (2014) Modes of museum services. In: Proceedings of serviceology 2014, pp 195–200, Hakodate, Hokkaido, Japan, April, 2014 (in Japanese)
8. Kumiyo Nakakoji, Yasuhiro Yamamoto, Toshio Kawashima, Ken-ichi Kimura, Takeshi Okada, Hironobu Shindo, Kyo Kageura (2014) Toward inspirational experience and experiential inspiration at museums. In: Proceedings of the 27th annual conference of the Japanse society for artificial intelligence 2014, 2B3-05, pp 1–4, Matsuyama, Ehime, Japan, May, 2014 (in Japanese)
9. Satoshi Kamegamori, Toshio Kawashima, Ken-ichi Kimura, Kumiyo Nakakoji, Yasuhiro Yamamoto (2014) Tools for promoting visitor's attention to the exhibition in museum. In: Proceedings of the 27th annual conference of the Japanse society for artificial intelligence 2014, 2E1-1, pp 1–2, Matsuyama, Ehime, Japan, May, 2014 (in Japanese)
10. Kumiyo Nakakoji, Yasuhiro Yamamoto, Toshio Kawashima, Ken-ichi Kimura, Yasuharu Koike (2014) An anatomy of Shikakes, AI & society, journal of knowledge, culture and communication, Springer, 2014b (to appear)
11. Kanako Ebina, Ken-ichi Kimura, Toshio Kawashima, Kumiyo Nakakoji, Yasuhiro Yamamoto (2014) Manigraphy: inspirational Communicatino probe from the creative workshop. In: Proceedings of the 27th annual conference of the Japanse society for artificial intelligence 2014, 2E1-3, Matsuyama, Ehime, Japan, May, 2014, pp 1–3 (in Japanese)
12. Nakakoji K, Ohira M, Yamamoto Y (2000) Computational support for collective creativity, knowledge-based systems journal. Elsevier Sci 13(7–8):451–458

MIX
Papier aus verantwortungsvollen Quellen
Paper from responsible sources
FSC® C105338

If you have any concerns about our products,
you can contact us on
ProductSafety@springernature.com

In case Publisher is established outside the EU,
the EU authorized representative is:
Springer Nature Customer Service Center GmbH
Europaplatz 3, 69115 Heidelberg, Germany

Printed by Libri Plureos GmbH
in Hamburg, Germany